BIOLOGY OF BRAIN DYSFUNCTION
Volume 2

A Continuation Order Plan is available for this series. A continuation order will bring delivery of each new volume immediately upon publication. Volumes are billed only upon actual shipment. For further information please contact the publisher.

BIOLOGY OF
BRAIN DYSFUNCTION
Volume 2

Edited by
Gerald E. Gaull

New York State Institute for Basic Research in Mental Retardation
Staten Island, New York
and
Mount Sinai School of Medicine of the City University of New York
New York, New York

PLENUM PRESS • NEW YORK–LONDON

Library of Congress Catalog Card Number 72-80796
ISBN 0-306-35062-9

© 1973 Plenum Press, New York
A Division of Plenum Publishing Corporation
227 West 17th Street, New York, N.Y. 10011

United Kingdom edition published by Plenum Press, London
A Division of Plenum Publishing Company, Ltd.
Davis House (4th Floor), 8 Scrubs Lane, Harlesden, London,
NW10 6SE, England

CONTRIBUTORS TO THIS VOLUME

Andre Barbeau
> Department of Neurobiology, Clinical Research Institute of Montreal, Montreal, Quebec, Canada

John P. Blass
> Departments of Biological Chemistry and Psychiatry and the Mental Retardation Center, Los Angeles, California

Doris H. Clouet
> New York State Narcotic Addiction Control Commission, Testing and Research Laboratories, Brooklyn, New York

Marilyn Louise Cowger
> Albany Medical College of Union University and State University of New York at Albany, New York

Norbert N. Herschkowitz
> Department of Pediatrics, University of Berne, Berne, Switzerland

Charles Kennedy
> Georgetown University Hospital, Washington, D.C.

Marian W. Kies
> Section on Myelin Chemistry, Laboratory of Cerebral Metabolism, National Institute of Mental Health, Bethesda, Maryland

Dorothy T. Krieger
> Neuroendocrinology Laboratory, Division of Endocrinology, Department of Medicine, Mount Sinai School of Medicine of the City University of New York, New York, New York

Louis Sokoloff
> National Institute of Mental Health, Bethesda, Maryland

v

Daniel Steinberg

> *Division of Metabolic Diseases, Department of Medicine, University of California, San Diego, La Jolla, California*

Kinuko Suzuki

> *The Saul R. Korey Department of Neurology and Department of Pathology, Rose F. Kennedy Center for Research in Mental Retardation and Human Development, Albert Einstein College of Medicine, Bronx, New York*

Kunihiko Suzuki

> *The Saul R. Korey Department of Neurology and Department of Pathology, Rose F. Kennedy Center for Research in Mental Retardation and Human Development, Albert Einstein College of Medicine, Bronx, New York*

A. G. Waltz

> *Department of Neurology, University of Minnesota Medical School, Minneapolis, Minnesota*

PREFACE

The growth of neurochemistry, molecular biology, and biochemical genetics has led to a burgeoning of new information relevant to the pathogenesis of brain dysfunction. This explosion of exciting new information is crying out for collation and meaningful synthesis. In its totality, it defies *systematic* summation, and, of course, no one author can cope. Thus invitations for contributions were given to various experts in areas which are under active investigation, of current neurological interest, and pregnant. Although this project is relatively comprehensive, by dint of size, other topics might have been included; the selection was solely my responsibility.

I believe systematic summation a virtual impossibility—indeed, hardly worth the effort. The attempt to assemble all of the sections involved in a large treatise with multiple authors inevitably results in untoward delays due to the difference in the rate at which various authors work. Therefore, the following strategy has been adopted: multiple small volumes and a relatively flexible format, with publication in order of receipt and as soon as enough chapters are assembled to make publication practical and economical. In this way, the time lag between the ideas and their emergence in print is the shortest.

This book is aimed at research workers, students, and physicians interested in the pathogenesis of brain dysfunction. Clinical data have been included only when relevant to such an understanding, but the reader is provided with suitable source material for fuller clinical description and discussion. The emphasis in this book is on *biological* aspects of brain dysfunction. Although much current work is biochemical, there has been a conscious attempt to integrate structural, physiological, nutritional, and immunological approaches.

GERALD E. GAULL, M.D.

CONTENTS

Chapter 3

Effects of Narcotic Analgesics on Brain Function
     Doris H. Clouet

Chapter 4

Genetic Disorders of Brain Development: Animal Models
    Norbert N. Herschkowitz

Chapter 7

Bilirubin Encephalopathy

Marilyn Louise Cowger

Chapter 8

**The Action of Thyroid Hormones and Their Influence on Brain
Development and Function**

Louis Sokoloff and Charles Kennedy

Chapter 9

Biology of the Striatum
André Barbeau

Chapter 10

Pathophysiology of Central Nervous System Regulation of Anterior Pituitary Function
Dorothy T. Krieger

Chapter 1

DISORDERS OF SPHINGOLIPID METABOLISM[1]

Kunihiko Suzuki and Kinuko Suzuki
The Saul R. Korey Department of Neurology and Department of Pathology
Rose F. Kennedy Center for Research in Mental Retardation and Human Development
Albert Einstein College of Medicine
Bronx, New York

I. INTRODUCTION

The last 10 years have witnessed dramatic and accelerating progress in in-vestigations of the group of genetic disorders commonly categorized as lipid storage diseases. Except for a few diseases, including Refsum's disease and Wolman's disease, most of the lipid storage diseases involve inborn errors of metabolism of a particular type of complex lipids that are characterized by acylsphingosine as the common building block, hence the name *sphingolipidoses*. Clinical delineation of individual sphingolipidoses dates back to 1881, when Warren Tay, a British ophthalmologist, first described the characteristic retinal finding of what was to become known later as Tay–Sachs disease.[1] Sachs was the first to recognize that the ophthalmological finding reported by Tay was a part of the manifestations of a more generalized degenerative disorder of the central nervous system.[2] By the end of the first quarter of this century, all of the classical forms of sphingolipid storage disorders had been recognized clinically and histopathologically. They include Gaucher's disease,[3] Fabry's

[1]The work from the authors' laboratories was supported by research grants NS-08420 and NS-09093 from the United States Public Health Service and by research grants 670-A-1 (Inex J. Warriner Memorial Grant for Research on Multiple Sclerosis) and 670-B-2 from the National Multiple Sclerosis Society.

1

disease,[4] Niemann–Pick disease,[5,6] Krabbe's disease,[7] and metachromatic leukodystrophy.[8] There was then a 35-year hiatus before another major sphingolipidosis, G_{M1}-gangliosidosis, was recognized.[9,10] The recognition of G_{M1}-gangliosidosis was greatly aided by progress in analytical techniques for determination of lipids, by which respective sphingolipids involved in these disorders could be identified with certainty. The identification of pathologically accumulated materials in sphingolipidoses began in the mid-1930s with the identification of glucocerebroside in Gaucher's disease[11] and sphingomyelin in Niemann–Pick disease[12] and was essentially concluded by the end of the last decade. On the other hand, application of electron microscopy to investigations of sphingolipidoses was initiated in 1957 by DeMarsh and Kautz[13] on Gaucher's disease and gained full momentum with the discovery of the membranous cytoplasmic bodies in the cortical neurons of a patient with Tay–Sachs disease.[14] The ever-increasing refinements of the ultrastructural and biochemical investigations complemented each other, greatly accelerating the progress of the study of sphingolipidoses. It was not until the advent of the enzymatic investigations, however, that some of the classical disease entities were found to consist of more than one enzymatically distinct disorder. The most notable examples are the four different G_{M2}-gangliosidoses, which were categorized as Tay–Sachs disease until 4 years ago, as discussed later. The history of the sphingolipidosis investigations has been recently reviewed in some detail.[15] All of these intensive and apparently basic efforts have culminated in recent developments of practical value, such as genetic counseling based on enzymatic detection of heterozygous carriers and *in utero* diagnosis of affected fetuses.

II. CHEMISTRY, METABOLISM, AND DISTRIBUTION OF SPHINGOLIPIDS

It is not the intent of this chapter to discuss details of the chemistry and biochemistry of sphingolipids. The following provides the background information essential for understanding of the pathogenesis and pathophysiology of sphingolipid storage disorders. For details of the chemistry and metabolism of sphingolipids in the nervous system, there are many recent review articles available for reference.[16–23]

A. Chemistry

1. Sphingosine and Ceramide

The term *sphingolipids* is derived from the fact that these lipids contain sphingosine within the molecules as the common moiety. Sphingosine is an unsaturated long-chain amino diol, the structure of which is shown in Fig. 1.

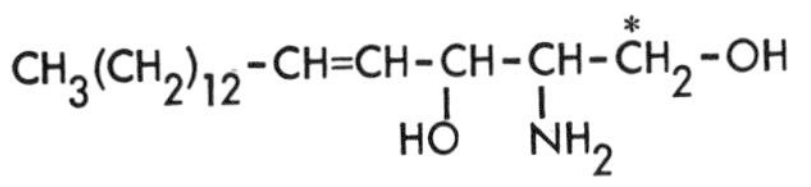

$$CH_3(CH_2)_{12}-CH=CH-\underset{\underset{OH}{|}}{CH}-\underset{\underset{NH_2}{|}}{CH}-\overset{*}{CH_2}-OH$$

Fig. 1. Structure of sphingosine. The carbon atom with the asterisk is C-1. Substitutions occur at this point to form various sphingolipids.

The major sphingosine found in nature is C_{18}-sphingosine. C_{20}-sphingosine occurs in smaller amounts. Small portions of these sphingosines are also present in saturated (dihydro) forms. Different sphingolipids, or even the same sphingolipid from different sources or in different developmental stages, have characteristic sphingosine compositions, but these differences do not appear to be important in relation to metabolism of these lipids in sphingolipidoses.

The amino group of sphingosine is almost always acylated with a long-chain fatty acid, ranging from C_{14} to C_{26}. N-acylsphingosine is generically called *ceramide*, the basic common building block of almost all sphingolipids. Each sphingolipid has its own characteristic fatty acid composition, and some of these will be touched upon later.

2. Individual Sphingolipids

The hydroxy group of C-1 of sphingosine can be substituted by varieties of moieties, thus producing the multitude of sphingolipids. They can be divided into three series of compounds: those in which (a) the substitution is by a phosphorus-containing moiety, (b) the substitution is by galactose, and (c) the substitution is by glucose. Sphingolipids which are relevant in sphingolipid storage disorders are listed in Table 1, showing the substituting moieties and the commonly used trivial names.

a. Sphingomyelin (Ceramide Phosphorylcholine). Sphingomyelin is the only phosphorus-containing sphingolipids present in any significant amounts in mammals. It is therefore a phospholipid as well as a sphingolipid.

b. Galactosylceramide Series. The common structure of the second group of sphingolipids is galactosylceramide (galactocerebroside), which has a β-D-galactose moiety glycosidically linked to the C-1 of sphingosine. Sulfatide is a sulfuric acid ester of galactocerebroside with the sulfate group on C-3 of the galactose moiety. Very small amounts of digalactosylceramide are normally present in the kidney. The second galactose moiety probably has α-anomeric configuration. Galactocerebroside and sulfatide are unusual in that substantial portions of these compounds contain α-hydroxy fatty acid as the acyl moiety.

c. Glucosylceramide Series. The simplest of this series is glucosylceramide (glucocerebroside). The structure is analogous to that of galactocerebroside; a glucose moiety replaces the galactose moiety. Glucosylceramide is the starting compound for numerous sphingoglycolipids that are important in sphingolipidoses. An additional β-galactose moiety makes lactosylceramide. Lactosylceramide is the branching point for the two subseries of glucosylceramide series

TABLE 1. MAJOR SPHINGOLIPIDS PERTINENT TO SPHINGOLIPIDOSES

Major group	Commonly used name	Substituting group[a]
Phosphosphingolipid	Sphingomyelin	–phosphorylcholine
Galactosylceramide series	Galactocerebroside	–gal
	Sulfatide	–gal–sulfate
	Digalactosylceramide	–gal–(α)gal
Glucosylceramide series	Glucocerebroside	–glc
	Lactosylceramide	–glc–gal
Globoside series	Trihexosylceramide	–glc–gal–(α)gal
	Globoside	–glc–gal–(α)gal–galNAc
Ganglioside series	Hematoside	–glc–gal–NANA
	Asialo-G_{M2};	–glc–gal–galNAc
	G_{M2}; (Tay–Sachs ganglioside)	–glc–gal–galNAc 　　　| 　　NANA
	Asialo-G_{M1}	–glc–gal–galNAc–gal
	G_{M1}	–glc–gal–galNAc–gal 　　　　　| 　　　　NANA
	G_{D1a}	–glc–gal–galNAc–gal–NANA 　　　　　| 　　　　NANA
	G_{D1b}	–glc–gal–galNAc–gal 　　　　　　| 　　　　NANA–NANA
	G_{T1}	–glc–gal–galNAc–gal–NANA 　　　　　　| 　　　　NANA–NANA

[a] These are the substituting groups at the C-1 of the ceramide portion of the molecule: gal, galactose; glc, glucose; galNAc, N-acetylgalactosamine; NANA, N-acetylneuraminic acid (sialic acid).

sphingolipids—the globoside series and the ganglioside series. The globoside series is characterized by the third hexose moiety, which is α-galactose, while the ganglioside series has N-acetyl-β-D-galactosamine as the third sugar moiety (Table 1).

Gangliosides are, by definition, sialic acid–containing sphingoglycolipids. Sialic acid is the generic name for N-substituted neuraminic acid. Although N-glycolylneuraminic acid is known to occur in gangliosides of systemic organs or in lower mammalian brains, N-acetylneuraminic acid (NANA) is the only sialic acid found in human brain gangliosides. Monosialogangliosides which have the NANA moiety on the second galactose moiety are known to occur, each corresponding to the respective oligohexosylceramides of the ganglioside series. Additional one, two, or more NANA moieties form several polysialogangliosides, of which the three compounds shown in Table 1 are the major components of normal mammalian brain gangliosides. It is clear by now that there are many more ganglioside species than those in Table 1, and the structures of many have been established. However, all of them are minor components in mammalian brains, and their significance in sphingolipidoses has not been established.

B. Metabolism

1. Biosynthesis of Sphingolipids

a. Sphingosine and Ceramide. Sphingosine is synthesized from serine and palmitoyl-CoA by a system of synthetic enzymes present in the microsomal fraction of young mammalian brains.[24] There appear to be two alternate biosynthetic pathways for the formation of ceramide. One of them was described originally by Sribney[25] and involves sphingosine and acyl-CoA. The other is an apparent reverse reaction of the degradative enzyme, ceramidase, characterized by Gatt.[26] The latter reaction involves sphingosine and free fatty acid.

b. Sphingomyelin. There appear to be two pathways available for the synthesis of sphingomyelin in the brain. The first one, demonstrated by Sribney and Kennedy,[27] is the formation of sphingomyelin from ceramide and CDP-choline. Brady *et al.*[28] described the alternative pathway in which the first step is the formation of sphingosylphosphorylcholine from sphingosine and CDP-choline, followed by its acylation, At present, it is not clear which of the two routes is the major pathway in the brain.

c. Galactocerebroside and Sulfatide. The biosynthesis of galactocerebroside starts from sphingosine. Since there are two moieties to be added to sphingosine —fatty acid and galactose—two alternative pathways are logically possible. One is the route through galactosylsphingosine (psychosine), in which sphingosine is first galactosylated from UDP-galactose, followed by acylation by acyl-CoA. The other pathway requires sphingosine to be acylated first to ceramide, which in turn acts as the acceptor for the galactosyl residue from UDP-galactose. For the first possibility, the formation of psychosine from sphingosine and UDP-galactose was first demonstrated by Cleland and Kennedy[29] and has been repeatedly confirmed. The second step of acylation of psychosine was reported by Brady.[30,31] More recently, Morell and Radin demonstrated that ceramide, both normal and α-hydroxy fatty acid–containing, can be galactosylated by UDP-galactose.[32,33] These authors were unable to confirm the reaction from psychosine to galactocerebroside. Most recently, however, Hammarström[34,35] demonstrated convincingly that such reaction also occurs in brain tissue.* The biosynthesis of galactocerebroside in the brain is therefore analogous to that of sphingomyelin regarding the existence of two alternative pathways. Which of the two is the major pathway in normal brain *in vivo* remains to be determined.

Biosynthesis of sulfatide occurs through galactocerebroside with the "active sulfate," 3'-phosphoadenosine-5'-phosphosulfate (PAPS), as the sulfate donor.

---

*Note added in proof: The formation of galactocerebroside from psychosine as reported by Hammarström appears to have been the result of a chemical, rather than enzymatic, reaction.[329]

The formation of digalactosylceramide by rat kidney homogenate was demonstrated by Hay and Gray as galactosylation of galactosylceramide by UDP-galactose.[36,37]

   d. Glucosylceramide and Higher Oligohexosylceramides. The formation of glucocerebroside is the initial step for the biosynthesis of most of the complex sphingoglycolipids important in sphingolipid storage diseases. Glucocerebroside is formed from ceramide by the transfer of the glucose moiety from UDP-glucose, catalyzed by a glucosyltransferase localized in particulate fractions.[38] The next higher hexosylceramide, lactosylceramide, is formed from glucocerebroside by the addition of galactose, as demonstrated by Hauser *et al.*[39,40] and by Basu *et al.*[38] Enzymatic synthesis of galactosyl-lactosylceramide from lactosylceramide has been demonstrated in spleen[40] and kidney.[36] Globoside is likely to be synthesized from galactosyl-lactosylceramide by an addition of *N*-acetyl-β-D-galactosamine, but this step has not been demonstrated.

   e. Gangliosides and Related Oligohexosylceramides. The major biosynthetic pathway of brain gangliosides appears to start by the addition of a sialic acid to lactosylceramide to form hematoside (G_{M3}).[41] Stepwise additions then of *N*-acetyl-β-D-galactosamine, β-D-galactose, and the second and third sialic acids were indicated by Roseman and coworkers.[41] However, there is evidence for a synthetic pathway in which sequential additions of *N*-acetylgalactosamine and galactose to lactosylceramide can occur before the first sialic acid moiety is added to the molecule.[42,43] Which of the proposed pathways plays the major role in ganglioside biosynthesis *in vivo* in mammalian brains has not been established.

2. Degradation of Sphingolipids

   Most of the degradative steps of individual sphingolipids have been clarified in recent years. The degradation of these compounds generally proceeds by stepwise cleavage of the hydrophilic moieties—phosphorylcholine in sphingomyelin, sulfate in sulfatide, and carbohydrates in all other instances—eventually reaching ceramide, which can be further degraded to sphingosine and fatty acid. Many, if not all, of the degradative enzymes of sphingolipids show acidic pH optima, are often activated by freeze-thawing, ultrasonication, osmotic shock, or detergents, and are stable in the tissue for many years when stored frozen. These are characteristics of hydrolases localized in lysosomes. Although not without exceptions, as discussed later, most of these enzymes have been found to be localized in lysosomes by subcellular fractionation studies. It is often postulated that the enzymes involved in sequential degradation of sphingolipids form a geometrically arranged enzyme complex within lysosomes,[44] which would facilitate orderly stepwise degradation.

   a. Ceramide. All sphingolipids are eventually degraded to ceramide.

Ceramide is further degraded to sphingosine and fatty acid by ceramidase.[26,45] The enzyme has a pH optimum of 4.8, requires bile acids for activity, and is further stimulated by Triton X100.

b. Sphingomyelin. The degradation of sphingomyelin in mammalian tissues occurs by cleavage of the phosphodiester bond between ceramide and phosphorylcholine by an acid hydrolase, sphingomyelinase. This enzyme has been demonstrated in rat liver,[46,47] rat brain,[48] and human spleen.[49] It does not exhibit activity toward lecithin. While a major portion of total sphingomyelinase is found in lysosomes, a smaller fraction may also be localized in mitochondria.[50]

c. Sphingolipids of Galactosylceramide Series. The initial degradative step of sulfatide is cleavage of the sulfate group to yield galactosylceramide and sulfate. The enzyme has been isolated and purified from pig kidney.[51] This enzyme was active toward sulfatide only when it was combined with a heat-stable factor which was eliminated during the preparation of the enzyme. The enzyme, however, exhibited arylsulfatase A activity without the addition of the factor. If not identical, sulfatide sulfatase appears to constitute at least a portion of arylsulfatase A activity, as suggested by electrophoretic studies[52] and by developmental and regional changes[53] of these enzymes.

A series of investigations by Radin and coworkers has greatly clarified the first step of galactosylceramide degradation. The step involves a specific β-galactosidase, galactosylceramide galactosyl hydrolase, which degrades galactocerebroside to ceramide and galactose.[54–57] The enzyme has a pH optimum of 4.5, requires bile salts for activity, and is localized in lysosomes.

d. Gangliosides and Higher Oligohexosylceramides. The pathways of physiological degradation of gangliosides and higher oligohexosylceramides are essentially sequential removal of terminal carbohydrate moieties, with all compounds converging to lactosylceramide. Each step involves a glycosyl hydrolase, specific for the reaction.

The three major polysialogangliosides of the brain, G_{T1}, G_{D1a}, and G_{D1b}, are all converted to the major monosialoganglioside, G_{M1}, by the action of neuraminidase. Apparently due to steric hindrance by *N*-acetylgalactosamine, the last NANA moiety on the internal galactose of G_{M1} is resistant to this neuraminidase. Most neuraminidase preparations, including those from mammalian tissues, are unable to cleave the sialic acid of G_{M1}- and G_{M2}-gangliosides.[58–60] In human brain, neuraminidase appears to be mostly particulate bound.[60] Unlike other glycosidases, however, a substantial portion of human brain neuraminidase has been found to be concentrated in the synaptosomal fraction.[61,62] This is in contrast to the finding of Mahadevan *et al.*,[63] who reported lysosomal localization of rat liver and kidney neuraminidase.

The next step of brain ganglioside degradation appears to be removal of

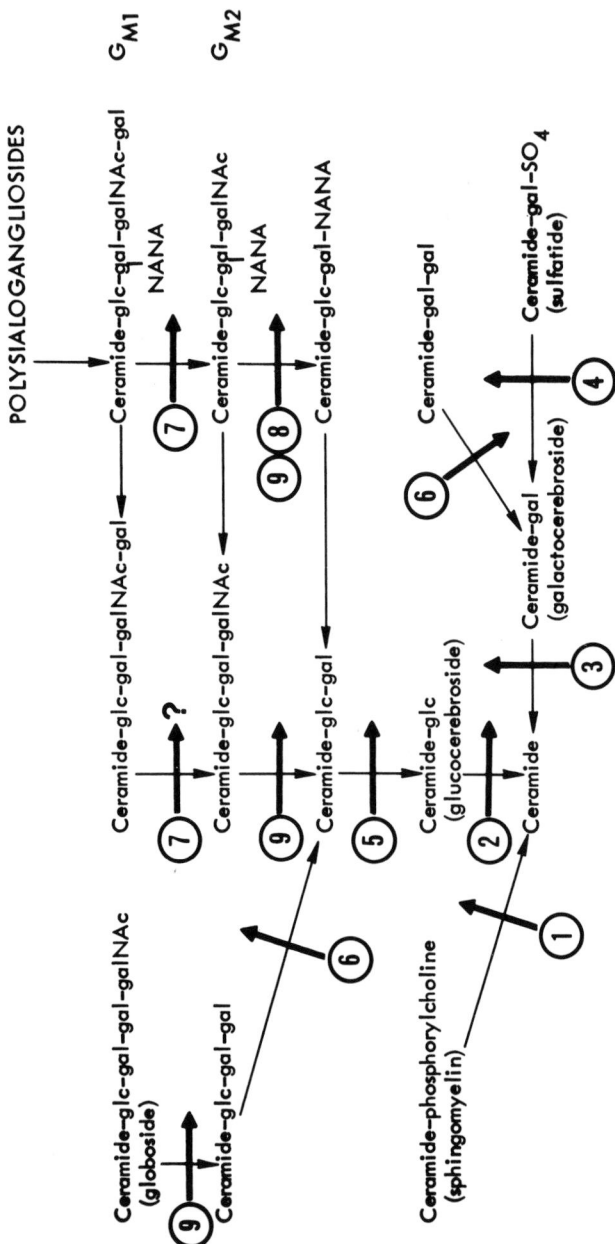

Fig. 2. Interrelationship among sphingolipids and the locations of the metabolic blocks of known sphingolipidoses. The numbers for the sites of the blocks correspond to those assigned to respective diseases in Table 6.

the terminal galactose of G_{M1}- to form G_{M2}-ganglioside,[64,65] which in turn is degraded to hematoside (G_{M2}) by an N-acetyl-β-D-galactosaminidase.[66,67] Once the hematoside stage is reached, the last NANA moiety can be removed by neuraminidase because the steric block by N-acetylgalactosamine no longer exists.[68,69] Through the pathway described above, major brain gangliosides are degraded to lactosylceramide.[69]

The above scheme most probably represents the pathway of brain ganglioside degradation occurring *in vivo*. However, reports of existence of neuraminidase which is capable of cleaving the NANA moiety of G_{M2}-ganglioside suggest possible alternative pathways.[70] Although the level of total activity is much lower than that of the more common neuraminidase, the enzyme was capable of cleaving G_{M2}-ganglioside to its asialo derivative, N-acetylgalactosaminyl-lactosylceramide. Since the asialo derivatives of G_{M1}- and G_{M2}-gangliosides are known to be degraded by sequential removal of galactose and then N-acetylgalactosamine,[65,66,71] the existence of this neuraminidase makes degradation of monosialogangliosides possible through respective asialo derivatives. Globoside(N-acetylgalactosaminylgalactosylgalactosylglucosylceramide) undergoes normal degradation also by sequential removal of terminal sugars, by N-acetylgalactosaminidase, and then by an α-galactosidase to reach lactosylceramide.[66,67,72]

Thus all gangliosides and higher oligohexosylceramides are eventually degraded to the common product, lactosylceramide. Lactosylceramide is further broken down to glucosylceramide by a β-galactosidase which is probably specific for this compound.[64,71,73,74]

Finally, glucosylceramide is degraded by a glucosidase to ceramide and glucose. The enzyme has been purified and characterized from human spleen[75] and ox brain.[76]

Figure 2 depicts the interrelationship of sphingolipids from the viewpoint of their degradative pathways. All of the known genetic disorders involving sphingolipid metabolism have enzymatic blocks at one or more of the degradative steps as illustrated in this figure.

C. Distribution

Many individual sphingolipids show characteristic distribution patterns and developmental changes. While the fundamental underlying causes of sphingolipidoses can be understood as genetic deficiencies of specific degradative enzymes, natural distribution of individual lipids appears to play a decisive role in determining the clinical and histopathological manifestations of these disorders.

Sphingomyelin is ubiquitous in its distribution. It has been found in practi-

cally every mammalian tissue examined, and its concentration is, in most instances, greater than that of sphingoglycolipids.

Unlike sphingomyelin, carbohydrate-containing sphingolipids have preferential distributions within mammalian bodies. Sulfatide and galactosylceramide are highly localized within the nervous system, particularly within the
myelin sheath. Galactocerebroside and sulfatide constitute approximately 20%
of myelin dry weight, and myelin is estimated to make up half of the dry weight
of white matter. In contrast, gray matter contains only a few percent of dry
weight as these galactolipids. Peripheral nerve myelin is similarly rich in galactocerebroside and sulfatide. Compared to the abundance in the nervous system,
these galactosphingolipids are practically absent in most extraneural tissues,
although they have been detected by highly sensitive analytical methods in
several organs, notably in the kidney. This unique and almost exclusive localization of galactocerebroside and sulfatide in the myelin sheath also results in
unusual developmental changes. These two compounds are present in extremely
low concentrations even in the nervous system before the active myelination
period. As newly formed myelin is deposited, the amounts of these lipids
increase rapidly, eventually reaching the adult levels.

Gangliosides are another class of sphingoglycolipids characteristic of the
nervous system. Unlike the galactolipids, they are primarily constituents of
neuronal membranes. A high concentration of gangliosides occurs in synaptic
membranes.[77,78] As a consequence, gray matter is much richer in gangliosides
than white matter or peripheral nerves. In addition to the high total concentration, the molecular distribution of brain gangliosides is also characteristic,
consisting of the four major molecular species, G_{T1}, G_{D1a}, G_{D1b}, and G_{M1}.
Several minor gangliosides occur in mammalian brain, totaling less than 10%
of total brain gangliosides. Some of the review references cited earlier give
detailed accounts of these minor gangliosides. Extraneural tissues generally
contain much smaller concentrations of gangliosides, and, in most instances,
the relatively simple hematoside (G_{M3}, sialyl-lactosylceramide) is the major
molecular species.

Asialo derivatives of G_{M1}- and G_{M2}-gangliosides (galactosyl-N-acetyl-
galactosaminyl-lactosylceramide and N-acetylgalactosaminyl-lactosylceramide)
are present in normal brain only in extremely small amounts, although the levels
are slightly higher in immature brains.[79] The concentrations of these compounds in extraneural tissues are essentially nondetectable. On the other hand,
globoside and galactosylgalactosylglucosylceramide are not normal constituents
of the brain. Globoside was so named because it was first isolated from erythrocytes,[80] but by now it is known that most extraneural organs contain small
but definite amounts of globoside as well as the next degradation product,
digalactosylglucosylceramide. Since lactosylceramide and glucosylceramide are

common degradation products of both globoside and gangliosides, these compounds can be found in the brain as well as in systemic organs. Concentrations of these compounds in the brain are again extremely small. While lactosylceramide is usually detectable in the brain throughout development, glucosylceramide is found only in immature brains. In systemic organs, both lactosylceramide and glucosylceramide are normal constituents. In fact, glucosylceramide is the only cerebroside in most extraneural tissues, in contrast to the nervous system in which virtually all cerebroside is galactosylceramide. Such sharp separation in the distribution of analogous compounds among different organs is the key to the understanding of entirely dissimilar clinical and histological involvement in biochemically analogous disorders involving these two cerebrosides, Gaucher's disease and globoid cell leukodystrophy, as discussed later.

Table 2 summarizes the relative distribution patterns of individual sphingolipids within a mammalian body.

III. INDIVIDUAL SPHINGOLIPIDOSES

The progress in investigations of human disorders of sphingolipid metabolism in the past decade has greatly expanded the number of biochemically well-defined diseases. Some conditions which had been thought to be single diseases are now recognized as consisting of more than one enzymatically distinct disease. Many fundamental questions have been answered. Many practical benefits have been attained, such as definitive diagnosis without surgical biopsy, heterozygous carrier detection, and intrauterine diagnosis. On the other hand, a number of important questions still remain for future investigation. Enzymatic delineation of the clinically and genetically distinct subtypes of some of the sphingolipidoses has not been achieved. The precise mechanism by which brain

TABLE 2. DISTRIBUTION OF SPHINGOLIPIDS[a]

| Sphingolipids | Brain | | Peripheral nerves | Systemic organs |
	Graymatter	White matter		
Sphingomyelin	+ +	+ +	+ +	+ +
Galactocerebroside	±	+ + +	+ + +	− (± in kidney)
Sulfatide	±	+ + +	+ + +	± (+ in kidney)
Digalactosylceramide	−	−	−	+ (in kidney)
Glucocerebroside	−	−	−	+ +
Lactosylceramide	±	−	−	+ +
Trihexosylceramide	−	−	−	+ +
Globoside	−	−	−	+ +
Ganglioside	+ + +	+	+	±

[a]It should be kept in mind that this table gives an extremely simplified overview of the distribution of individual sphingolipids.

dysfunction is brought about in many diseases is still largely obscure. The current state of our knowledge about respective disorders of sphingolipid metabolism is outlined here.

A. Sphingomyelinosis (Niemann–Pick Disease)

The disease category traditionally represented by the eponym *Niemann–Pick disease* is clearly a heterogeneous group, consisting of clinically, genetically, and biochemically distinct disorders. Since enzymatic deficiencies of various forms of the disease have not been completely elucidated, the classification remains tentative.

Incidence, Genetics, and Clinical Manifestations: Age of patients and the presence or absence of central nervous system involvement have been the major criteria for classification of various forms of Niemann–Pick disease. The clinicopathological classification proposed by Crocker[81,82] is widely used. While the following description places more emphasis on the CNS involvement, references will be made to the corresponding types of Crocker when appropriate.

The infantile form with brain involvement is the classical form of Niemann–Pick disease, constituting 85% of all cases, and occurs predominantly in the Jewish population[81] (Crocker's type A). The clinical onset of the disease is anytime between birth and 1 year, and the initial presenting symptom is most often hepatosplenomegaly. There is slow psychomotor development, commonly with feeding difficulty and frequent vomiting, ataxia, tremor, and muscular spasticity or rigidity. Seizures are common. Cherry-red spots at the maculae are observed in some patients. In addition to hepatosplenomegaly, other systemic manifestations include lymph node enlargement, pancytopenia, jaundice, and radiological evidence of pulmonary infiltrations. The clinical course is relatively acute, and most patients die by 3 years of age.

A similar set of clinical features is observed in the juvenile form of Niemann–Pick disease with CNS involvement (Crocker's type C). Systemic manifestations are generally milder than in the infantile neuropathic form, with the clinical onset between 1 and 5 years. Neurological signs and symptoms occur somewhat later, the clinical course is more protracted, and patients die anytime between late infancy and late adolescence. There are a few patients reported in the literature who developed the first symptoms and signs after 5 years and survived for many years. These adult patients usually show only mild systemic signs and may or may not show neurological involvement. When neurological symptoms are present, they are mild, consisting of disturbance of mentation and emotion, ataxia, and other movement and speech difficulties.

In contrast to the above clinical types with neurological manifestations, a fewer number of patients have been described who developed full-fledged systemic manifestations but without showing any signs of nervous system involvement (Crocker's type B). The clinical onset may be anytime from infancy to adulthood. Due to the lack of neurological involvement, patients with this type of Niemann–Pick disease can survive for many years, while the systemic manifestations, such as hepatosplenomegaly, may progress to an extremely severe degree.

Crocker also described another form which was characterized by clinical course and features similar to the juvenile form of Niemann–Pick disease with cerebral involvement (type D). This particular form was genetically distinct in that it occurred in a Catholic population with the family origin in Nova Scotia. For varieties of reasons which will be discussed later, there is serious doubt whether this type should be included in Niemann–Pick disease, defined as sphingomyelinosis.

Morphology: Grossly, affected organs such as liver and spleen are enlarged and yellow or pale gray. The brain is usually atrophic with unusually firm consistency—50–96% of normal weight in 11 patients.[82] The most characteristic feature of the disease is the presence of "Niemann–Pick cells" in systemic organs, particularly in those rich in reticuloendothelial tissues such as bone marrow, liver, spleen, and lymph nodes. The Niemann–Pick cells are large foamy cells with single, and sometimes multiple, nuclei. Histochemical stains indicate large amounts of phospholipid and cholesterol to be present.[82]

In the infantile neuropathic form of the disease, there is extensive and severe neuronal ballooning similar to that in ganglioside storage disorders. Disorganization of the cortical architecture, gliosis, and mild to moderate secondary demyelination are common associated neuropathological changes. There is no histopathological involvement of the nervous system in the nonneuropathic forms of the disease.

Relatively few electron microscopic studies of Niemann–Pick disease are available. Lynn and Terry,[83] in a study of a patient with the adult form, found round or oval abnormal inclusions in Niemann–Pick cells in lymph nodes. The inclusions consisted of fine lamellar structures and closely packed granules of moderate density, similar to those reported by Tanaka et al.[84] In the central nervous system, distended neurons contain numerous electron-lucent vacuoles of 1–2 μ in diameter. They are single membrane bound and contain loosely arranged concentric lamellar structures.[85,86]

Analytical Chemistry: Enormous accumulation of sphingomyelin in Niemann–Pick disease was first demonstrated by Klenk[12,87] in the liver and spleen and has since been confirmed and well established.[88–90] It appears to be the fundamental chemical abnormality of the disease. The degree of accumulation is greatest in organs rich in reticuloendothelial tissues, such as the liver or spleen, which are enlarged and full of Niemann–Pick cells. The concentrations

TABLE 3. SPHINGOMYELIN CONCENTRATIONS OF VARIOUS ORGANS
IN A CASE OF INFANTILE NEUROPATHIC NIEMANN–PICK DISEASE

Organ	Normal control	Niemann–Pick
Liver	1.3	22.4
Spleen	1.1	26.3
Gray matter	2.3	8.2
White matter	3.4	6.5
Myelin	5.5	5.8
Brain MCB[a]	—	42.9

[a]The fraction of the abnormal membranous cytoplasmic bodies isolated from the brain tissue. The values are expressed as percent dry weight.

of sphingomyelin in these organs often reach higher than 20–30 times normal on a dry weight basis. Considering the abnormal size of these organs, the total increase can easily be more than 100 times normal (Table 3). In the infantile neuropathic form of the disease, there is a moderate increase of sphingomyelin in gray matter and a less prominent increase in white matter[81,88,91–93] (Table 3). In one patient with the infantile neuropathic form, isolated myelin was shown to be normal in sphingomyelin content.[93] In older patients with CNS involvement, the concentration of brain sphingomyelin is often normal, despite the presence of ballooned neurons.[81,82,94]

In addition to sphingomyelin accumulation, there are abnormal increases of other lipids in many patients with Niemann–Pick disease. The most prominent of these is an increase of cholesterol, which, in fact, had been recognized even before the discovery of sphingomyelin accumulation.[95,96] Such increase of cholesterol has been observed in both systemic organs and the brain.[82,97–100] In the so-called Nova Scotian variant (Crocker's type D), sphingomyelin accumulation is only mild, but there is a six- to tenfold increase of cholesterol in the liver and spleen.[99] The increase of cholesterol appears to be closely associated with the sphingomyelin increase. The abnormal membranous bodies isolated from the liver and spleen of an infantile neuropathic patient consisted of approximately 60% sphingomyelin and 15–20% cholesterol.[93] More recently, accumulation of another phospholipid, lysobisphosphatidic acid, has been attracting attention. This compound has been shown to increase in sysstemic organs of infantile or juvenile neuropathic patients.[93,101,102] In the brain of patients with the neuropathic forms, there are additional abnormal increases of glucocerebroside and normally minor monosialogangliosides G_{M2} and G_{M3}.[93,103] This phenomenon has been observed even in the brain tissue of a juvenile patient which showed definite histopathological changes but no increase in either sphingomyelin or cholesterol.[94] These glycolipids were highly concentrated in the abnormal membranous bodies isolated from the brain of an infantile neuropathic patient.[93]

Enzymatic Deficiency: An earlier study by Crocker and Mays[104] demonstrated no abnormally increased rate of sphingomyelin synthesis and left a degradative block as the likely cause of Niemann–Pick disease. In 1966, utilizing specifically labeled sphingomyelin as the substrate, Brady *et al.*[105] demonstrated a profound deficiency of sphingomyelinase in the liver of six patients with classical infantile Niemann–Pick disease (Fig. 3). Schneider and Kennedy[106] studied spleen tissues from patients with different types of the disease and found the profound deficiency of the enzyme in Crocker's type A (infantile neuropathic) and type B (visceral) but not in type C (juvenile and adult neuropathic).[106] The Nova Scotian type appeared to show a partial reduction of the enzyme. Since then, the deficient sphingomyelinase activities have been consistently confirmed in types A and B in every tissue examined, including leukocytes[107] and cultured fibroblasts and bone marrow cells.[108] The data on sphingomyelinase activities in types C and D, however, are not consistent. The activity was reported to be partially deficient in leukocytes in type C,[109] and it was within the normal range in cultured fibroblasts in type D.[108] These results appear to be at variance with those of Schneider and Kennedy cited above. Part of this difficulty may be the result of uncertainty in the clinicopathological classification of patients. Most recently, successful prenatal diagnosis of an affected fetus was accomplished on the basis of deficient sphingomyelinase activity in cultured amniotic fluid cells.[110]

Pathophysiological Problems: The dysfunction of the nervous system in the infantile neuropathic form of Niemann–Pick disease might be attributed to the abnormal accumulation of sphingomyelin within neurons. The abnormal membranous bodies isolated from a brain consisted of almost 50% sphingomyelin.[93] However, in many older neuropathic patients, the abnormal accumulation of sphingomyelin cannot be demonstrated in the brain, thereby excluding excess sphingomyelin as the direct cause of brain dysfunction in these patients. Neurons of these patients show abnormal ballooning, and analytical results indicate small but definite increases of normally minor glycolipids, such as glucocerebroside, lactosylceramide, or G_{M3}- and G_{M2}-gangliosides. In an infantile patient, isolated cerebral membranous bodies were also the site of accumulation of these glycolipids.[93] Therefore, at least in older neuropathic

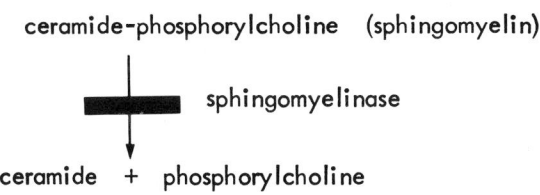

Fig. 3. Metabolic block in Niemann–Pick disease.

forms, these glycolipids might play an important role in producing functional disturbance of the brain. In the absence of positive evidence that excess sphingo-myelin itself causes brain dysfunction in infantile cases, possible roles of these glycolipids must be kept in mind also in the infantile neuropathic form. It is not yet clear how these glycolipids, as well as cholesterol, accumulate in this group of diseases, the underlying cause of which appears to be sphingomyelinase deficiency.

The profound deficiency of sphingomyelinase is found consistently in both types A and B of Crocker. And yet the nervous system is severely affected in one type and not in the other. Mere quantitative differences in the degree of the enzymatic deficiency do not appear to explain this phenomenon. This is essentially the same question being raised for different types of Gaucher's disease (see below). These types are not only clinically and pathologically dif-ferent but genetically distinct. More sophisticated knowledge of the exact genetic alterations of sphingomyelinase in different types would be required to answer this question.

Some of the patients who are classified as having the juvenile or adult neuropathic form (type C) or the Nova Scotian variant (type D) show no sphingomyelinase deficiency. Therefore, at present we cannot include these cases in the category of Niemann–Pick disease, characterized by sphingomyeli-nase deficiency, although future investigation may reveal close biochemical relationship of these disorders to the more classical Niemann–Pick disease. Sphingomyelinase is a specific enzyme within the category of phosphodi-esterases. One promising approach may be a broader survey of phosphodiester-ases in different types of the disease.[111]

B. Glucosylceramide Lipidosis (Gaucher's Disease)

Incidence, Genetics, and Clinical Manifestations: The first description of glucosylceramide lipidosis is credited to Gaucher,[3] who, however, thought that the condition was a neoplasm. Frederickson and Sloan[112] estimate that as many as a thousand patients have been recorded in the literature.

Traditionally, Gaucher's disease has been categorized according to age at clinical onset—infantile, juvenile, and adult types. More recently, Frederickson proposed to classify the disease according to the clinicopathological findings in the central nervous system—chronic nonneuropathic, acute neuropathic, and subacute neuropathic.[112] For practical purposes, the chronic nonneuropathic type roughly corresponds to the so-called adult type, the acute neuropathic type to the infantile type, and the subacute neuropathic type to the juvenile type, but occasionally the acute neuropathic form of Gaucher's disease may occur in adults and the chronic nonneuropathic type in infants.

A majority of patients with Gaucher's disease have the chronic nonneuropathic type, and there is an unusually high incidence of this form among the Jewish population. In all types of Gaucher's disease, the mode of inheritance appears to be autosomal recessive. Only the same form of the disease occurs within one family, indicating that these clinicopathological types are also genotypically distinct.

The major criterion for the chronic nonneuropathic form is the absence of CNS involvement. While most patients are in the older age categories, this form of the disease may begin in early infancy. The most prominent clinical sign is splenomegaly. Other reticuloendothelial organs, such as liver and lymph nodes, are also enlarged. There is depression of bone marrow, resulting in anemia, leukopenia, and thrombocytopenia. These organs contain numerous "Gaucher cells," which are transformed storage cells of reticuloendothelial origin. They are relatively large (20–100 μ) and pale cells with apparent fibrillar content within cytoplasm. The presence of Gaucher cells can make the diagnosis of the disease almost certain. The chronic nonneuropathic Gaucher's disease generally runs a prolonged clinical course, and patients frequently survive for several decades.

On the other hand, the acute neuropathic form takes a much more malignant clinical course. This type occurs most frequently among infants. The disease usually manifests itself within several months after birth with hepatosplenomegaly and additional neurological signs such as retarded psychomotor development or difficulty in swallowing. The systemic findings described above for the chronic nonneuropathic type are also present in this type. Neurological signs commonly observed during the course are spasticity with hyperreflexia, hyperextension of the neck and flexion of the arms, ocular muscle signs, and dysphagia. Seizures are common. In the terminal stage, patients are hypotonic and demented. Most patients die before age 2.

In addition to the above two clinical types, there are significant numbers of patients reported in the literature who showed most of the clinical signs of the acute neuropathic form of Gaucher's disease but who were generally older and had a less acute clinical course. It is not clear, at present, if this constitutes a separate type of Gaucher's disease. In fact, patients who fall in this category appear to be quite heterogeneous.[112]

Morphology: On gross examination, the spleen is always enormously enlarged. Moderate to severe enlargement of the liver and lymph nodes is present. In patients with CNS involvement, the brain weight is usually less than normal with or without obvious cerebral atrophy.

The most characteristic finding on light microscopy is the presence of numerous Gaucher cells in various organs, most prominently in the liver, spleen, and bone marrow. Ultrastructurally, Gaucher cells contain abnormal inclusions, corresponding to the cytoplasmic fibrils seen by light microscopy.

These inclusions are small hollow tubules with a diameter of 200–300 Å and have a right-handed helical structure. Commonly, a group of tubules is found within a membrane-limited space.[13,113–117]

In addition to the visceral lesions, the central nervous system is extensively involved in the acute neuropathic form of the disease. Focal cortical necrosis has been observed.[118] Neuronal degeneration associated with severe gliosis is found in many areas, most consistently in dentate nuclei of the cerebellum.[119] Frequently, neurons are swollen and contain periodic acid–Schiff stain (PAS) positive material. Myelination is generally normal for the age. Typical Gaucher cells are sometimes present in the brain, usually clustered around blood vessels. In a patient with the acute neuropathic form of the disease, Adachi et al.[120] found occasional neurons containing intracytoplasmic tubular inclusions similar to those in Gaucher cells. These authors, as well as Wakutani et al.,[121] also found membranous cytoplasmic bodies composed of parallel membranes in some of the swollen neurons.

Analytical Chemistry: It is now established beyond any doubt that the primary compound accumulating in Gaucher's disease is glucosylceramide. All of the systemic organs which are infiltrated by numerous Gaucher cells contain abnormally increased glucosylceramide. Considering the increase in the size of entire organs, the total amount of glucosylceramide in the spleen or liver could be easily several hundred times normal. This enormous increase of glucocerebroside in reticuloendothelial organs is the most consistent finding among all types of Gaucher's disease, as attested to by numerous analytical studies.[112]

The question of glucocerebroside accumulation within the brain, however, has not been clearly resolved. Svennerholm[122] reported that in a patient with infantile neuropathic Gaucher's disease, 70% of monohexosylceramide in the cerebral cortex was glucocerebroside. Others who reported abnormal increases of glucocerebroside in the brain include Maloney and Cumings[123] and Montreuiel et al.[124] Many other investigators, on the other hand, failed to find a significant increase of glucosylceramide in the brain of patients with CNS involvement.[118,125–127] At present, generalization is not possible regarding the presence or absence of abnormal glucocerebroside accumulation in the brain of patients with Gaucher's disease and regarding any correlation between such accumulation and degree of clinicopathological CNS involvement.

Although by no means consistent, there are frequent reports of moderate increases of other complex sphingoglycolipids in the organs of patients with Gaucher's disease, such as lactosylceramide and G_{M3}-ganglioside (hematoside) in the spleen[128] or in the brain,[122] or fatty acid esters of glucosylceramide in the brain.[129] While accumulation of these compounds is only moderate and not a constant finding, it nevertheless presents an interesting subject of speculation in relation to the pathogenesis of the disease.

Enzymatic Deficiency: An early study of Trams and Brady[130] excluded abnormally active biosynthesis of glucosylceramide as the underlying cause of glucosylceramide accumulation in Gaucher's disease. The alternative possibility of a genetic block of glucocerebroside degradation was then demonstrated in the spleen of 11 patients by Brady *et al.*[131,132] and also in the spleen of four patients by Patrick (Fig. 4).[133] The enzyme is normally present in a variety of tissues, and its activity has been found deficient in Gaucher's disease in every tissue so far examined, including leukocytes[107,134] and cultured fibroblasts.[135,136] Fibroblasts and leukocytes from parents of patients with Gaucher's disease show glucocerebrosidase activities intermediate between those of normal individuals and patients, thus providing a means of heterozygous carrier detection.[134,136] Most recently, intrauterine diagnosis of an affected fetus was accomplished on the basis of deficient activity of glucocerebrosidase in cultured amniotic fluid cells.[137] There appears to be a general impression that the degree of the enzymatic deficiency is more severe in patients with acute clinical courses, but a rigorous systematic survey including a large number of cases, both chronic nonneuropathic and acute neuropathic types, would be required to establish this point unequivocally.

In his original report, Patrick[133] indicated that the deficiency of glucocerebrosidase in Gaucher's disease might be reflected in deficient activities of nonspecific β-glucosidase measured with *p*-nitrophenyl-β-glucoside. Öckerman and Köhlin[138] reported a similar finding of deficient β-glucosidase with the use of 4-methylumbelliferyl-β-glucoside. Beutler and Kühl[134] reported that in leukocytes β-glucosidase activities of patients are greatly deficient when assayed at *p*H 4.0, while they are only moderately deficient at higher *p*H. At present, the specific natural substrate, glucocerebroside, is still preferred for definitive diagnosis of the disease, but chromogenic substrates are also useful, particularly when the specific assay for glucocerebrosidase is not available. Other acidic lysosomal hydrolases, such as acid phosphatase and β-galactosidase, are generally normal or even elevated in Gaucher's disease.

Pathophysiological Problems: The fundamental enzymatic defect underlying Gaucher's disease is analogous to that in globoid cell leukodystrophy (see below). Both diseases involve degradative blocks of monohexosylceramides to

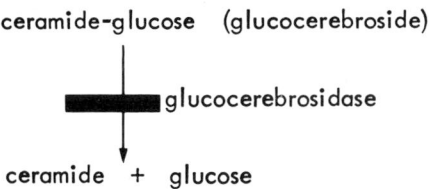

Fig. 4. Metabolic block in Gaucher's disease.

ceramide and respective sugars—glucose in Gaucher's disease and galactose in Krabbe's disease. Yet clinical and morphological characteristics of the diseases are entirely dissimilar. Clinical and morphological manifestations of Gaucher's disease are almost exclusively in systemic organs in the chronic form. Even in the acute neuropathic form, the primary sites of the disease are in the systemic organs, although the brain, particularly gray matter, shows lesions. In contrast, manifestations of globoid cell leukodystrophy are entirely within the nervous system, particularly in white matter. These diseases seem to be excellent examples of how the normal distribution of the involved compound affects the eventual expression of the disease.

Two fundamental and interrelated questions remain to be answered regarding the pathogenesis of Gaucher's disease: the pathogenetic mechanism of the nervous system involvement in the neuropathic form and the fundamental difference in the nature of the enzymatic defects between the neuropathic and nonneuropathic forms.

The defective activity of glucocerebrosidase does not explain directly how such severe neurological manifestations occur. Excess storage of glucosylceramide in the brain does not appear to be consistent even in patients with the neuropathic form. Glucocerebroside is not a normal constituent of the brain except during early development, and consequently even a small amount of storage may be detrimental to normal metabolism of neurons. Then, at least in cases where there is abnormal accumulation of glucocerebroside in the brain, the neuropathic effects may be due to such accumulation. The source of glucocerebroside in the brain is likely to be gangliosides, which are known to turn over actively. The existence of the degradative block itself rather than its consequence, accumulation of glucocerebroside, might adversely affect the regulation of ganglioside metabolism, thus causing functional disturbance of neurons.

The question of the presence or absence of CNS involvement also remains for future investigation. The brain may not require as high a level of glucocerebrosidase activity as systemic organs for maintenance of normal metabolism, and the differences among the various types may possibly be due to quantitative differences in the degree of glucocerebrosidase deficiency. However, information regarding quantitative aspects of the glucocerebrosidase deficiency in different types is at present inadequate to permit a firm conclusion. Furthermore, there are many speculative alternatives other than this simplistic possibility to explain such differences. More detailed studies on the exact nature of genetic abnormalities of the enzyme in different phenotypes would be needed.

C. Leukodystrophies

Galactocerebroside and sulfatide are uniquely concentrated within the myelin sheath. Quantitatively, virtually all of these two compounds in the mammalian body are found in the myelin sheath. It is therefore expected that genetic metabolic disturbances involving these two lipids will manifest themselves as degenerative disorders of white matter and the peripheral nerves, both clinically and morphologically. There are three well-characterized, enzymatically distinct disorders of sphingolipid metabolism which belong to this category—the genetic leukodystrophy. One involves the metabolism of galactocerebroside and the other two, sulfatide.

1. Galactosylceramide Lipidosis (Globoid Cell Leukodystrophy, Krabbe's Disease)

Incidence, Genetics, and Clinical Manifestations: Globoid cell leukodystrophy is rare. Since the first description of the disease by Krabbe,[7] fewer than 100 cases have been recorded, although undoubtedly there have been more patients who were not reported. In the authors' laboratory, approximately a dozen new patients were diagnosed enzymatically during the period from February 1970 to October 1971. This number probably is close to the total number of newly diagnosed patients in the United States during this period. The disease is relatively common in Scandinavian countries, and Hagberg *et al.*[139] reported an incidence of nearly two per 100,000; however, it is worldwide in distribution. The disease is inherited as an autosomal recessive trait and affects both sexes.

Typically, the disease has its clinical onset during the period between 3 and 6 months after birth. It is a steadily and rapidly progressive neurological disorder, and death results usually within 2 years after clinical onset. Signs of white matter involvement are most prominent. There is marked hypertonicity with hyperactive tendon reflexes, extended and crossed legs, and flexed arms. Hypotonicity ensues, and the patient eventually reaches the stage of complete decerebration. Peripheral nerve involvement can always be detected by careful neurological examination and/or by delayed conduction velocity.[140] Spinal fluid protein is almost always increased without an increase in cell count. Clinical manifestations are exclusively neurological, and systemic organs are not involved.

Morphology: As suggested by the clinical features, all significant morphological changes are confined to the nervous system. The brain is usually atrophic. On section, white matter is strikingly reduced in bulk with a whitish-gray appearance and firm, rubbery consistency. This is the result of widespread demyelination with severe astrocytic gliosis. Subcortical arcuate fibers tend to be spared. In contrast, gray matter appears relatively normal.

The most prominent pathological changes seen with the light microscope are in the white matter. The major abnormalities are severe lack of myelin, presence of globoid cells, and astrocytic gliosis. Lack of myelin is often profound, with very few myelin sheaths remaining that are detectable by myelin-staining techniques. More commonly, a thin layer of relatively preserved myelin may be found in subcortical regions. Not only does the myelin sheath disappear, but also very few oligodendroglial cells remain in the terminal stage, and axons also show degenerative changes. White matter contains numerous abnormal cells, globoid cells, the histological hallmark of this disorder. Globoid cells are either round or oval mononuclear cells, or large (up to 50 μ in diameter) irregular multinucleated cells. They are scattered within white matter, often forming perivascular aggregates (Fig. 5). The cytoplasm is moderately PAS positive and exhibits strong acid phosphatase activity.[141,142] The globoid cells are most likely of nonneural, mesodermal origin, being essentially macrophages.[143] The remainder of white matter is filled with dense fibrous astrocytic gliosis.

Globoid cells have numerous pseudopods which are seen on electron microscopy and are characteristic of macrophages. Two types of abnormal inclusions are found in globoid cells. The first, more common type of the inclusion is a moderately electron-dense, straight or curved hollow tubular structure with irregularly crystalloid cross sections.[144] The second type of abnormal inclusion

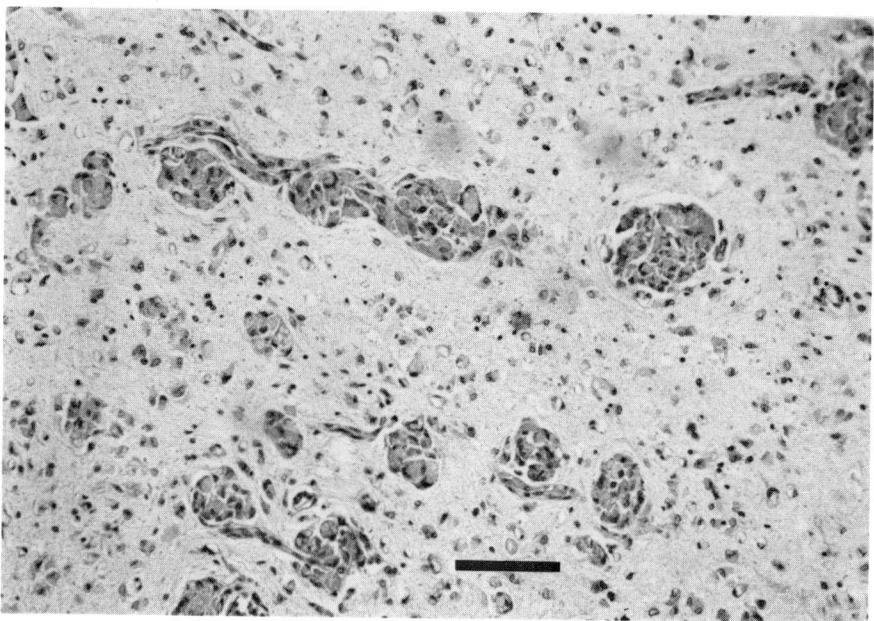

Fig. 5. Light microscopic appearance of white matter in globoid cell leukodystrophy. The numerous abnormal globoid cells stain PAS positive. The remainder of white matter is replaced by severe reactive gliosis, and the myelin sheath is essentially absent. The line indicates the scale of 10 μ.

was first described by Yunis and Lee.[145] It has the structure of right-handed twisted tubules, reminiscent of the abnormal inclusions in Gaucher's disease. These authors pointed out the morphological similarity of this helical structure to negatively stained pure galactocerebroside.[146] The abnormal inclusions are found almost exclusively in globoid cells. Astrocytes and the few remaining oligodendroglia are free of the abnormal inclusions.

Peripheral nerves usually do not show gross abnormalities, but they are always affected histologically. Segmental demyelination is common. There are degenerative changes in axons associated with endoneurial fibrosis and accumulation of foamy histiocytes. Abnormal tubular inclusions, similar to those seen with the electron microscope in globoid cells in the brain, are found within both the histiocytes and the Schwann cells.[144,147,148]

Analytical Chemistry: Abnormalities in chemical composition are also largely confined to white matter. Compared to normal white matter, dry-weight of white matter of patients with globoid cell leukodystrophy is diminished, largely due to loss of lipids. While all types of lipids are decreased, the most prominent loss is of galactocerebroside and sulfatide. These analytical findings reflect the almost total lack of myelin in the tissue. In 1963, Austin[149] and Svennerholm[125] independently reported the important finding that while decreased galactocerebroside and sulfatide are expected in view of the severe myelin loss, the ratio of galactocerebroside to sulfatide was abnormally high in white matter in globoid cell leukodystrophy. When analytical values of white matter from two patients were compared to those in two other similarly devastating white matter disorders, in order to evaluate whether the abnormal ratio was due to an increase of cerebroside or a decrease of sulfatide, there appeared to be a disproportionate preservation of galactocerebroside, rather than lack of sulfatide, in Krabbe's disease.[150]

White matter also contains small but definitely abnormal amounts of glucocerebroside, lactosylceramide, galactosylgalactosylglucosylceramide, and globoside.[151] Globoside had not been detected in the brain previously, and galactosylgalactosylglucosylceramide had been reported in the brain only in patients with Fabry's disease.[152] The presence of these abnormal hexosylceramides, particularly the last two, was attributed by Eto and Suzuki[151] to the numerous globoid cells, consistent with their mesodermal origin.

While myelin is almost absent, the little remaining myelin could be isolated and was shown to have normal galactolipid composition,[153] in contrast to excess sulfatide in the myelin sheath in metachromatic leukodystrophy (see below).

Enzymatic Deficiency: The genetic cause underlying globoid cell leukodystrophy is a deficiency of galactocerebroside β-galactosidase (Fig. 6).[154] The deficiency has been demonstrated in patients' brains, livers, spleens, and

ceramide–galactose (galactocerebroside)

galactocerebrosidase

ceramide + galactose

Fig. 6. Metabolic block in globoid cell leukodystrophy.

kidneys,[154–156] as well as in peripheral leukocytes, serum, cultured fibroblasts, and amniotic fluid cells.[157–159] Not only was the enzymatic activity profoundly deficient compared to controls, but none of the pathological controls, including metachromatic leukodystrophy, Schilder's disease, Tay–Sachs disease, Hurler's syndrome, total hexosaminidase deficiency, Niemann–Pick disease, Gaucher's disease, and G_{M1}-gangliosidosis, showed such deficient activities of galactocerebroside β-galactosidase.

The deficiency of galactocerebroside β-galactosidase is specific in the sense that if enzymatic assays are carried out for β-galactosidase activities using synthetic chromogenic substrates, such as p-nitrophenyl- or 4-methylumbelliferyl β-galactoside, this deficiency cannot be demonstrated. Galactocerebrosidase assays of serum, leukocytes, and fibroblasts not only offer the means of definitive antemortem diagnosis, but also provide the possibility of heterozygous carrier detection, because materials from parents of patients show intermediate activities of galactocerebroside β-galactosidase.[160] A successful intrauterine diagnosis of the disease has been recently accomplished with the use of cultured amniotic fluid cells from a fetus at risk.[159]

In 1967, an apparent deficiency of a synthetic enzyme, cerebroside sulfotransferase, was reported by Bachhawat et al.[161] in the brain and kidney of two patients and was suggested to be the primary genetic defect of globoid cell leukodystrophy. Subsequent studies on additional patients, however, showed that, while the low activity was always found in the brain tissue, it was not found consistently in the kidney.[156] In addition, there are many morphological and biochemical features of the disease that are difficult to explain on the basis of sulfotransferase deficiency. While the precise mechanism is not known, the sulfotransferase deficiency in globoid cell leukodystrophy appears to be a secondary phenomenon.

Pathophysiological Problems: The unique histological feature of Krabbe's disease—the infiltration of globoid cells in white matter—appears to be the result of a normal tissue reaction to excess galactocerebroside. A tissue reaction very similar to human globoid cell leukodystrophy can be experimentally produced by implanting pure galactocerebroside pellets in rat brain.[162] This property is unique for galactocerebroside: other lipids studied thus far do not elicit such histological response. They include sphingosine, ceramide, glucocerebroside, sulfatide, and ganglioside. Experimental globoid cells are indistinguishable from human globoid cells on the ultrastructural level.[163]

The precise sequence of the disappearance of oligodendroglial cells and concentration of excess galactocerebroside within globoid cells remains a major question on the morphological level. Since galactocerebroside is a major constituent of the myelin sheath, its turnover is expected to take place within the cell of its origin, the oligodendroglia. Oligodendroglia, however, not only disappear almost completely, but the few remaining ones do not contain the characteristic inclusion bodies. Another unusual morphological feature is the appearent lack of association of the abnormal inclusions with morphologically or enzymatically identifiable lysosomes. These inclusions are for the most part freely scattered within the cytoplasm of globoid cells, rather than forming single membrane–bound packets.[150]

From a biochemical viewpoint, the most unusual feature of globoid cell leukodystrophy is the lack of accumulation of galactocerebroside beyond the normal level. In all other sphingolipidoses, degradative enzymes for respective lipids are deficient, and the involved lipids accumulate in tissues to an enormous excess. In Krabbe's disease, the total amount of galactocerebroside in white matter is usually in the range of 20% or less of normal, despite the genetic block of its degradation. Cessation of galactocerebroside synthesis appears to be the only logically possible explanation. Since galactocerebroside is an almost exclusive constituent of myelin, it is expected that oligodendroglial cells normally synthesize most of galactocerebroside. Therefore, the cessation of galactocerebroside synthesis may simply be a result of the total disappearance of these cells. This could also be an explanation for the low sulfotransferase activities consistently found in brains of Krabbe's disease patients, since oligodendroglial cells must also be the major site of sulfatide synthesis. The absence of specific accumulation of galactocerebroside in histologically normal kidneys, however, suggests the alternative explanation that the cessation of galactocerebroside synthesis may be due to a specific metabolic regulatory mechanism in the presence of a block in the degradative pathway.[164]

2. Sulfatidosis (Metachromatic Leukodystrophy)

There are two genetically and enzymatically distinct disorders in which a metabolic block of sulfatide degradation results in an abnormal accumulation of the lipid primarily in the brain and also in other systemic organs. They are arylsulfatase A deficiency (classical metachromatic leukodystrophy) and multiple sulfatase deficiency.

   a. Arylsulfatase A Deficiency (Classical Metachromatic Leukodystrophy). Incidence, Genetics, and Clinical Manifestations: Metachromatic leukodystrophy occurs as at least three different clinical types—late infantile, juvenile, and adult. Although no chemical or biochemical criteria have been found to distinguish these forms clearly, family studies indicate that they are genotypically different from each other. Only one form occurs in a particular family. The mode of inheritance is autosomal recessive in all of the three forms, and both sexes are affected. While the disease is rare, there appears to be no ethnic or geographic preponderance.

The late infantile form is the most common, probably comprising two thirds of arylsulfatase A deficiency cases. The clinical onset of the disease is usually from the end of the first year to up to 2 years of age. Slowdown, and then regression of psychomotor development are the most dominant clinical features. As in globoid cell leukodystrophy, signs of white matter involvement are more prominent than cortical signs. Flaccid paraplegia or quadriplegia often develop early, with strikingly reduced or absent tendon reflexes. Regression in mental function results in apathy and poor contact with the environment. Flaccid paresis becomes spastic, and the child becomes blind and decerebrate. This is a steadily progressive disorder, and the entire course rarely exceeds 5 years. Hagberg[165] divided the typical clinical course of late infantile metachromatic leukodystrophy into four stages.

The juvenile form manifests itself at any time between 5 and 15 years but most commonly around 7–10 years. Here, visual and auditory symptoms may precede motor signs, which are characterized by long-tract signs and cerebellar ataxia. Mentation is decreased, and psychotic symptoms may occur. Progression is slow but steady. In the terminal stage, there are marked spasticity and ataxia. The adult form may have its clinical onset at any time after adolescence. The onset is gradual, and behavioral changes such as mental depression or schizophrenia-like signs are often presenting symptoms, making clinical diagnosis difficult during the early stages. Spasticity and other long-tract signs eventually develop. The clinical course of the adult form is much slower than that of the other two earlier forms, and the patient may survive even into his 50s.

The peripheral nerves are involved in all forms, with progressively prolonged conduction velocities. Cerebrospinal fluid protein is commonly elevated.

Morphology: Major morphological changes are observed primarily in the nervous system, particularly in white matter and peripheral nerves. Unlike globoid cell leukodystrophy, however, histological lesions are found in some systemic organs, notably in the kidney, liver, and gall bladder. This is a reflection of less exclusive localization of sulfatide in the myelin sheath compared to galactocerebroside.

Postmortem brains are atrophic at the terminal stage. On section, white matter of cerebral hemispheres is grayish, indicating extensive loss of myelin. Subcortical arcuate fibers tend to be spared. Lesions in the brain stem and spinal cord are often less severe than those in the cerebrum and cerebellum. Cortical thickness is often reduced.

Light microscopic examination reveals most of the significant alterations within white matter. The major findings are extensive loss of myelin, presence of abnormal metachromatic granules, and reactive astrocytic gliosis. The degree of myelin loss varies according to the stage of the disease, but almost total loss of myelin may result at the terminal stage. As in globoid cell leukodystrophy, axis cylinders also degenerate in severely involved areas. Sudanophilic droplets are not present, and inflammatory changes are not a part of regular changes. While the number of oligodendroglia decreases as the disease progresses, the degree of loss is less extensive than that in globoid cell leukodystrophy. A considerable number of oligodendroglia remain even at the terminal stage.

The presence of numerous abnormal granules with an unusual staining characteristic within white matter is the most prominent morphological feature of the disease. These granules are scattered within white matter, often more concentrated in regions where the disease process is more active, such as sub-cortical areas. The name of the disease originated from the staining characteristics of these abnormal granules. When frozen sections are stained with acidic cresyl violet,[166] these granules stain yellow-brown rather than purple, the

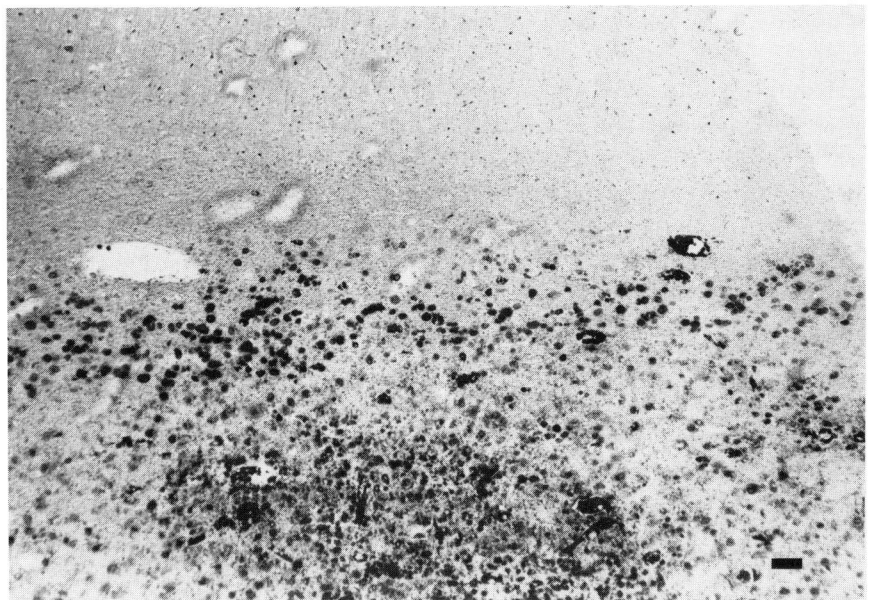

Fig. 7. Light microscopic appearance of the subcortical area in metachromatic leukodystrophy. The upper half of the figure shows gray matter which is relatively normal. In the white matter, there are numerous abnormal granules which stain metachromatically with acidic cresyl violet. Such granules are absent in gray matter. The line indicates the scale of 10 μ.

original color of the dye (Fig. 7). In addition to the characteristic metachromatic staining, the granules are PAS and sudan black positive, indicating that they contain glycolipids. The remainder of white matter is occupied by reactive astrocytic gliosis.

While gray matter is relatively well preserved, neurons in some areas, notably dentate and brain stem nuclei and anterior horn cells, may be moderately distended, containing metachromatic granules similar to those in white matter. Similar metachromatically staining abnormal granules are also commonly found in renal tubules and mucosal cells of the gall bladder. They are less frequently found in the pancreas, testes, liver, and other organs. The metachromatic granules in the renal tubular epithelium give rise to the appearance of such granules in patients' urine and provide a useful clinical screening procedure for diagnosis of the disease.

The peripheral nervous system is always involved in metachromatic leukodystrophy. Segmental demyelination is a common feature. Abnormal metachromatic granules, essentially identical to those in the central nervous system, are found in macrophages, Schwann cells, and, to a lesser degree, in ganglion cells of spinal root ganglia.

Abnormal inclusions are found by electon microscopy primarily within oligodendroglial cells.[167] They consist of lamellar structures with a periodicity of 50–60 Å, either concentrically arranged or in the form of stacked discs (Fig. 8).[168–170] Many are surrounded by a single unit membrane, but some lamellar structures appear to be scattered freely within cytoplasm without limiting membranes. Similarly, in the peripheral nervous system, abnormal inclusions were found in Schwann cells.[171] These were mostly less than 1 μ in diameter and often surrounded by a limiting membrane. They were made up of irregular areas of varying density, and lamellar structures were less frequently seen than those in the central nervous system. Ultrastructural localization of acid phosphatase activities within these abnormal bodies suggests their lysosomal origin.[172,173]

Analytical Chemistry: The distinctive histochemical features of the abnormal metachromatic granules in white matter and peripheral nerves had suggested some essential properties of the abnormally accumulated material in metachromatic leukodystrophy even before the advent of sophisticated analytical techniques. These granules stained with sudan black (lipid), strongly positive with PAS stain (reducing sugar), negative with Bial's reagent (absence of sialic acid), and strongly metachromatic at a low pH (presence of strongly acidic groups). Within the nervous system, sulfatide was the only compound which would fulfill the above criteria.

Positive analytical proof for excess storage of sulfatide came in the late 1950s with the use of thin layer chromatography and infrared spectrosco-

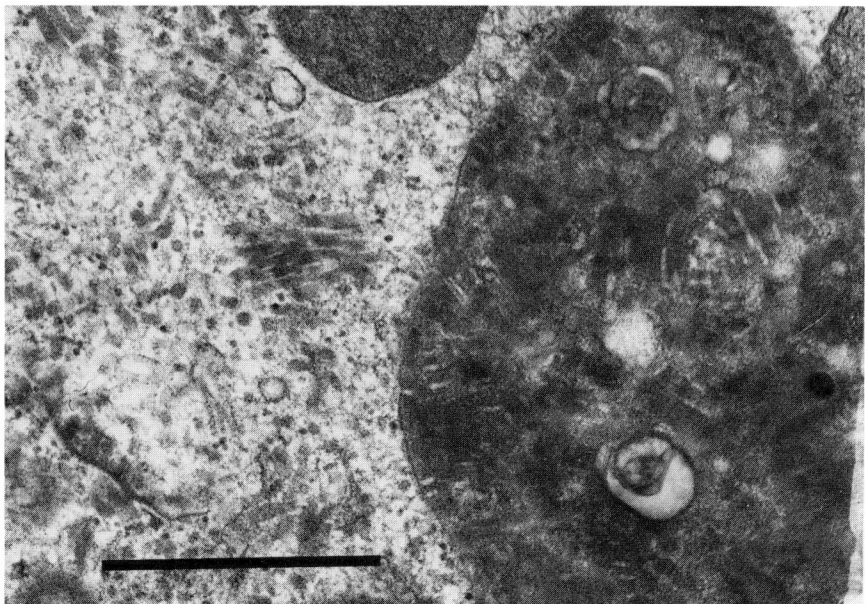

Fig. 8. An electron micrograph of metachromatic granules. The ultrastructure is mostly lamellar in the form of either stacked discs or concentrically arranged bodies. There are two bodies in this picture which are surrounded by single membranes, while the remainder of the cytoplasm is filled with abnormal lamellar structures which appear to lack any limiting membranes. The line indicates the scale of 0.5 μ. [This micrograph was kindly provided by Dr. Robert D. Terry, Albert Einstein College of Medicine, Bronx, New York.]

py.[174,175] Sulfatide abnormally accumulating in metachromatic leukodystrophy was later shown to be identical in chemical structure to sulfatide occurring in normal individuals.[176]

In the brain, white matter shows abnormally increased amounts of sulfatide, with concomitant decreases in cerebroside.[125,177–179] Consequently, the normal cerebroside-to-sulfatide ratio of approximately 4 becomes reversed. Because of the large amounts of these galactolipids normally present in the myelin sheath, the chemical abnormality is most prominent in white matter. Unlike in globoid cell leukodystrophy, the myelin sheath in metachromatic leukodystrophy is chemically abnormal in that isolated myelin fractions contain excess sulfatide and decreased cerebroside.[180,181] Furthermore, the abnormal metachromatic granules from the brain of patients with the disease have been isolated and shown to contain a large amount of sulfatide, providing direct evidence that the characteristic metachromasia probably is due to the presence of sulfatide (Table 4, Fig. 9).[182,183] Because of the severe loss of myelin, white matter shows decreased solids and particularly decreased cholesterol and phospholipids. Gray matter is relatively unaffected.

TABLE 4. LIPIDS IN BRAIN IN METACHROMATIC LEUKODYSTROPHY[a]

	Metachromatic leukodystrophy				Normal controls	
	Granule[b]		White matter	Myelin	White matter	Myelin
	I	II				
Cholesterol	18.4	19.0	16.2	21.2	26.8	25.6
Total galactolipid	47.6	44.5	50.7	37.4	25.0	28.2
Cerebroside	8.0	6.0	5.9	9.0	19.5	24.7
Sulfatide	39.6	38.5	44.2	28.4	5.5	3.5
Phospholipid	34.0	36.5	32.9	36.1	49.4	42.8

[a]The data are expressed as percent of total lipid. The values for white matter and myelin are through the courtesy of Dr. William T. Norton, Department of Neurology, Albert Einstein College of Medicine, Bronx, New York.
[b]The abnormal metachromatic granule fraction isolated from the brain.

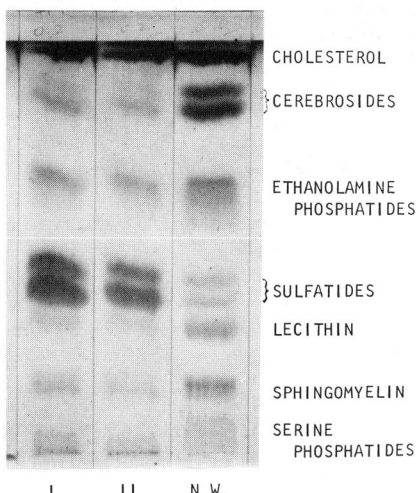

CHOLESTEROL

}CEREBROSIDES

ETHANOLAMINE
PHOSPHATIDES

}SULFATIDES

LECITHIN

SPHINGOMYELIN

SERINE
PHOSPHATIDES

I   II   N.W.

Fig. 9. Thin layer chromatogram of the lipids of the isolated metachromatic granules. I and II are from two separate collections of the granules. N.W., normal white matter. The solvent system was chloroform–methanol–water (70:30:4, v/v/v). Spots were located by 50% sulfuric acid spray and heating. The white matter lipid in metachromatic leukodystrophy would look essentially the same as that of the metachromatic granules. [Reproduced from Suzuki et al.[183] by permission.]

There appears to be no definite correlation between the analytical findings and the onset or duration of the disease in late infantile or juvenile forms. In the adult type, however, Austin et al.[184] found only a moderate increase of sulfatide in white matter (twice normal) and a relatively large increase in gray matter (about four times normal).

Fundamental analytical findings are similar in peripheral nerves.[185] In the kidney, in addition to sulfatide being drastically increased, another sulfated sphingoglycolipid, lactosylceramide sulfate, is also increased.[186] This latter compound is a small but normal constituent of the kidney.

Enzymatic Deficiency: The abnormal accumulation of sulfatide in various tissues prompted a search for a degradative enzyme for sulfatide which might be genetically defective in metachromatic leukodystrophy. Such an enzyme was found to be present in pig kidney.[51] Meanwhile, Austin et al.[187,188] demonstrated a specific deficiency of arylsulfatase A in the brain, liver, and kidney of patients with metachromatic leukodystrophy, using *p*-nitrocatechol sulfate as the substrate. Arylsulfatases B and C were normal in these patients. Then Mehl and Jatzkewitz[189] demonstrated a deficiency of a specific enzyme, cerebroside sulfatase (Fig. 10). The relationship between cerebroside sulfatase and arylsulfatase A was clarified by these workers. An extensive purification of cerebroside sulfatase resulted in a heat-labile fraction active toward *p*-nitrocatechol sulfate but inactive toward sulfatide unless recombined with a heat-stable fraction, eliminated during the purification procedure.[52]

The same enzymatic deficiency has been demonstrated in urine,[190] leukocytes,[191] and cultured fibroblasts.[192,193] Enzymatic assays of these readily obtainable materials provide a useful means not only for diagnosis of affected patients but also for heterozygous carrier detection, because arylsulfatase A activities in heterozygous individuals are generally intermediate between those of normal individuals and affected patients.[194–196]

Pathophysiological Problems: The classical metachromatic leukodystrophy is analogous to globoid cell leukodystrophy in that both of the diseases are caused by genetic defects of adjacent degradative steps of the same family of sphingoglycolipids, both of which are highly concentrated within the myelin sheath. Consequently, they share some common features, such as clinical signs of white matter degeneration and morphological findings of myelin destruction. A careful examination, however, also reveals some intriguing contrasting features which suggest some fundamentally different pathogenetic effects of the respective degradative blocks. Oligodendroglial population is generally better preserved in metachromatic leukodystrophy than in Krabbe's disease. Abnormal storage of the affected lipid, sulfatide, occurs primarily within oligodendroglial cells, often in the form of single membrane–bound granules containing acid phosphatase activities. In Krabbe's disease, abnormal inclusions are present only in globoid cells, and they are, for the most part, free in the cytoplasm without limiting membranes. The myelin sheath in metachromatic leukodystrophy contains excess amounts of sulfatide, while that in globoid cell leukodystrophy

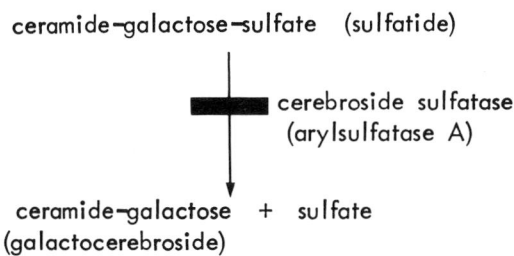

Fig. 10. Metabolic block in metachromatic leukodystrophy.

is normal in galactolipid composition. And finally, greater than normal accumulation of sulfatide occurs in metachromatic leukodystrophy, while such accumulation of galactocerebroside does not occur in globoid cell leukodystrophy. These features of metachromatic leukodystrophy seem to be what is logically expected in a disease in which degradation of one of the major constituents of myelin is impaired genetically and further point to the apparent peculiarity of galactocerebroside in Krabbe's disease.

As stated above, metachromatic leukodystrophy occurs in at least three distinct clinical forms. Biochemical differentiation of these clinical types has not been achieved. Many patients with the adult form of metachromatic leukodystrophy exhibit arylsulfatase A deficiency as severe as in patients with the infantile form. Recent attempts from the laboratory of Austin[197] were directed to more detailed studies of the residual arylsulfatase A activities in urine of patients with different forms of the disease. Older patients tended to have higher residual activities in the urine, and some kinetic differences were observed among different forms. Whether the observed differences will always show one-to-one correspondence to different clinical types and whether the same differences will be found in tissues remain to be answered.

b. Multiple Sulfatase Deficiency. Incidence, Genetics, and Clinical Manifestations: The number of patients having multiple sulfatase deficiency reported in the literature is still exceedingly small—11, including those without definite biochemical characterization.[198-205] Austin reported two patients in one family who might be considered the prototypes of this disorder. Clinical features are complex. The age at incidence ranges from late infantile to juvenile periods: clinical onset is usually around 12 months, and death occurs between 4 and 12 years. The clinical features of classical metachromatic leukodystrophy are always present. Superimposed on such pictures, rapidly progressive dementia, retinal degeneration, abnormal EEG, and other signs of cortical involvement occur. Skin is often thick and dry. Bell-shaped thorax and hepatosplenomegaly are common. X-ray examinations often show minor changes in vertebrae and metacarpals. Characteristically, there is excess urinary excretion of sulfated mucopolysaccharides. Detailed comparison of clinical pictures of previously reported cases was made in a recent article.[205] The genetic patterns of patients are consistent with the autosomal recessive mode of inheritance.

Morphology: The light microscopic findings described above for arylsulfatase A deficiency in white matter and peripheral nerves and the presence of metachromatic granules in the kidney are all present in multiple sulfatase deficiency. Morphological changes in multiple sulfatase deficiency are, therefore, those of arylsulfatase A deficiency plus the following features. The liver and spleen may or may not be enlarged, but Kupffer cells, hepatic parenchymal cells, and the epithelial cells of splenic sinusoids are extensively vacuolated, resembling those in mucopolysaccharidoses. These vacuoles contain, if appropriately stained, water-soluble, strongly metachromatic material.

In addition to the lesions in white matter, gray matter is extensively involved. Cortical thickness is greatly reduced. There is a severe loss of neurons, and remaining neurons are distended with storage material which stains PAS positive but orthochromic. The light microscopic appearance of the cortical neurons is similar to that in gangliosidoses and mucopolysaccharidoses.

Analytical Chemistry: Abnormal accumulation of sulfatide in the brain, particularly white matter and peripheral nerves, and in the kidney is essentially similar to that in classical metachromatic leukodystrophy. In addition, other sulfated compounds are found to be abnormally increased. Bischel *et al.*[200] found an increase of sulfated mucopolysaccharides in the brain and kidney to the same extent as in Hurler–Hunter syndrome. The increased mucopolysaccharide was primarily heparan sulfate, and no dermatan sulfate was detected. More recently, Murphy *et al.*[205] obtained two fractions of hepatic mucopolysaccharides which were greatly increased. Constituent analyses indicated these fractions to be consistent with heparan sulfate and dermatan sulfate. These authors also found abnormally increased cholesterol sulfate in the liver, kidney, and plasma but not in the brain.[205]

In gray matter, there is a moderate increase of total gangliosides, with distinct increases of the two normally very minor monosialogangliosides, G_{M2} and G_{M3},[199,205] essentially the same finding as in Hurler–Hunter syndrome.[206] From the viewpoint of analytical chemistry, therefore, multiple sulfatase deficiency gives the findings of combined metachromatic leukodystrophy and mucopolysaccharidosis.

Enzymatic Deficiency: As early as 1965, Austin *et al.*[188] demonstrated that tissues of their patients were deficient not only in arylsulfatase A but also arylsulfatases B and C. This finding was confirmed by Murphy *et al.*,[205] who also found additional deficiency of dehydroepiandrosterone sulfatase and cholesterol sulfatase activities. The deficiency was present in the brain, liver, and kidney. These authors included liver tissue of one of Austin's original patients and obtained identical enzymatic findings.

There was a moderate deficiency of β-galactosidase in the liver tissues of two patients,[205] the same finding reported for Hurler–Hunter syndrome.[207]

Pathophysiological Problems: Accumulation of a number of sulfated compounds can be understood on the basis of the simultaneous deficiency of multiple sulfatases. The most fundamental question regarding this disorder is how simultaneous deficiency of apparently different enzymes is brought about in a single genetic disorder. Arylsulfatases A and B are both lysosomal enzymes, but they are distinctly different. Arylsulfatase C has an alkaline pH optimum and is localized in microsomes. Whether this apparent inconsistency with the one gene–one enzyme concept is the result of a regulator gene deficiency or due to other mechanisms remains one of the most important questions.

D. Lactosylceramide Lipidosis

Incidence, Genetics, and Clinical Manifestations: Only one patient has been reported recently with an abnormal accumulation of lactosylceramide in tissues, as apparently the primary abnormality.[208–211] The patient was a Negro female who showed slow development and mild hypotonia at age $2\frac{1}{2}$ years followed by accelerated neurological regression. Cerebellar ataxia, spasticity, and Babinski signs developed. Slow development of reddish macular discoloration was noticed. Hepatosplenomegaly was present, and there were abnormal mononuclear storage cells with foamy cytoplasm in the bone marrow. Serum acid phosphatase was elevated. The patient's neurological signs gradually worsened, and she died at age 4.

The parents were not consanguineous, and there was no family history of neurological disorders. There is, therefore, no clue as to the genetic nature of the disease except for the partial reduction of lactosylceramide galactosidase in skin fibroblasts of both parents, as discussed below, which suggests the autosomal recessive mode of inheritance.

Morphology: No morphological description of the patient, either light or electron microscopic, has been published, except for an abstract mainly on the hepatic ultrastructure.[211] Since the patient died very recently, a full morphological report would be expected in the near future.

Analytical Chemistry: Detailed analytical studies showed an abnormal increase of lactosylceramide, on the order of two to eight times normal, in the erythrocytes, plasma, bone marrow, brain, and liver of the patient. Urine sediment also contained excess lactosylceramide. Unexpectedly, there was about a sixfold increase of glucocerebroside in biopsied liver tissue concomitant with an eightfold increase of lactosylceramide. Hematoside (G_{M3}) was also increased moderately (up to twice normal) in some tissues. Cultured fibroblasts showed a twofold elevation of lactosylceramide. The concentrations of galactocerebroside and sulfatide in the brain were reduced, suggesting the presence of demyelination.

Enzymatic Deficiency: From the analytical findings, and by the analogy to other sphingolipid storage disorders, a defect of a degradative enzyme was suspected. The liver of the patient showed the activity of lactosylceramide galactosidase to be 10–15% of normal, while other lysosomal acidic hydrolases were either normal or increased (Fig. 11). Excess lactosylceramide was not inhibitory, and mixing experiments excluded the presence of inhibitors in the pathological tissue. The same deficiency was demonstrated in cultured fibroblasts of the patient. In this instance, a 50% reduction of β-glucosidase was also observed—an interesting finding in relation to the increased glucocerebroside in the liver,

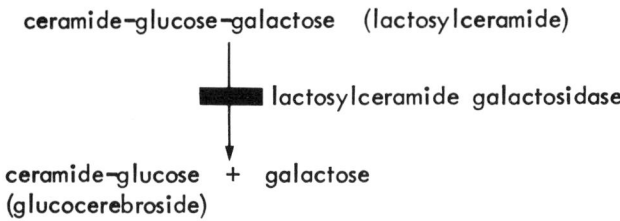

ceramide-glucose-galactose   (lactosylceramide)

lactosylceramide galactosidase

ceramide-glucose + galactose
(glucocerebroside)

Fig. 11. Metabolic block in lactosylceramide lipidosis.

although β-glucosidase was normal in the liver. Cultured fibroblasts from both parents showed activities of lactosylceramide galactosidase intermediate between those of the patient and normal controls.

Pathophysiological Problems: Lactosylceramide occupies the pivotal position at the converging point of degradation of all the sphingoglycolipids of the glucosylceramide series (Fig. 2). The brain is rich in gangliosides, and systemic organs contain primarily globoside and related compounds. Both groups are broken down to the common product, lactosylceramide. Therefore, it is understandable that a disease caused by a metabolic disturbance of lactosylceramide has both neurological and systemic manifestations.

From the single case thus far reported, the disease appears to be relatively mild in neurological manifestations. Visceral accumulation of lactosylceramide is likewise much less severe than that of glucocerebroside in Gaucher's disease, despite the fact that the turnover rates of glucocerebroside and lactosylceramide should be similar since these compounds are next to each other on the degradative pathway of the same series of compounds. The relatively high residual activity of lactosylceramide galactosidase might explain such milder accumulation. The abnormally high glucocerebroside reported for biopsied liver remains to be explained.

E. Trihexosylceramide Lipidosis (Fabry's Disease)

Incidence, Genetics, and Clinical Manifestations: Trihexosylceramide lipidosis was first recognized as a dermatological disorder in 1898 by Fabry[4] and independently by Anderson.[212] Almost all reported cases have occurred in Caucasian populations, but cases are known among the Chinese and Japanese. The genetic pattern of the disease strongly favors an X-linked incomplete recessive mutation, with full clinical and morphological manifestations occurring only in males.[213] Partial penetration in females is observed occasionally, with milder clinical pictures without the typical skin lesions.

The presenting symptom in the majority of cases is the appearance of keratotic skin lesions, 1–5 mm in diameter, most often distributed in the lower trunk region. The clinical onset is usually between 10 and 18 years of age.

Generalized vascular lesions produce complex and widespread clinical pictures. Severe pain and paresthesia in the extremities are frequent. There is a slow progressive impairment of renal and cardiac functions. Neurological symptoms occur in 30% of the patients, being mostly of a vascular nature, such as headache, dizziness, hemiparesis, cranial nerve symptoms, and ataxia.[214-216] Edema of extremities is common. Progression of renal impairment most often determines the prognosis. The clinical course is prolonged, and many patients survive for several decades.

Morphology: Systemic lesions are more prominent than those within the nervous system. Foamy cells filled with PAS-positive lipid are found in the lymph nodes, spleen, bone marrow, and urinary sediment.[217,218] Cardiac muscle fibers and endothelial and smooth muscle cells of blood vessels contain abnormal deposits of lipids.[219-221] Extensive vacuolization of renal epithelial cells is a prominent feature.[222,223] Histologically, the characteristic skin lesions consist of blood-filled lacunae surrounded by hyperkeratotic epidermis.[224] Within the central nervous system, lipid-laden neurons are found most prominently in rostral hypothalamic nuclei, particularly in the supraoptic, paraventricular, and preoptic nuclei.[217,220] Other neurons in hippocampus, brain stem, and spinal cord are also frequently involved. Similar abnormal lipid storage is common in the neurons of the autonomic nervous system. Within the peripheral nervous system, Schwann cells and the perineurial cells occasionally contain abnormal cytoplasmic inclusions,[225,226] and commonly there is a moderate reduction of the number of small myelinated fibers.

The abnormal cytoplasmic inclusions stain PAS and sudan black positive, indicating the glycolipid nature of the storage material. Electron microscopic examination reveals these inclusions to consist of single membrane-bound electron-dense materials. Regular lamellar patterns are frequently observed within the inclusions, with the periodicity of the lamellae varying between 40 and 70 Å, according to different investigators. Lattice-like structures and concentric lamellar structures are also observed occasionally.[227-231]

Analytical Chemistry: Earlier, Fabry's disease was considered to be a phospholipid storage disorder, often based on the inadequately documented personal communication by Kühnau, who studied the patient reported by Scriba[217] and by Ruiter.[232] In 1963, however, Sweeley and Klionsky[233] conclusively established that the disease is a sphingoglycolipid storage disorder. They demonstrated that the major storage compound in the kidney of a 28-year-old patient was a trihexosylceramide with a galactose moiety at the terminal position. The fatty acid composition was similar to that of the corresponding trihexosylceramide in normal kidney. There were no α-hydroxy fatty acids. In addition to the major storage lipid, another minor compound was increased in the kidney. α-Hydroxy fatty acids constituted one third of the total fatty acid.

This compound was characterized also by Sweeley and Klionsky[233,234] as digalactosylceramide. These findings have been confirmed by several investigators.[230,235–237] These two compounds have a common and unusual characteristic—the terminal galactose moiety has an α-anomeric configuration.[238–240]

As discussed earlier, digalactosylglucosylceramide and digalactosylceramide are both primarily constituents of systemic organs. As a consequence, accumulation of these compounds occurs more prominently outside the nervous system, particularly in the kidney, which normally contains relatively large amounts of these compounds. Digalactosylglucosylceramide has been found abnormally increased in almost every tissue of patients thus far examined, including lymph nodes, liver, spleen, arteries, cardiac, striated, and smooth muscles, prostate gland, lung, kidney, pancreas, and brain.[152,230,241] Abnormal accumulation of digalactosylceramide is detected in certain tissues—such as kidney, brain, spleen, and pancreas—but not in others. Within the nervous system, there is a greater accumulation of these compounds in the areas where morphological changes—particularly lipid-laden neurons—are most prominent. Cultured fibroblasts from patients with the disease also accumulate excessive amounts of trihexosylceramide,[242] and there is an excess excretion of the lipid into urine,[243,244] both of diagnostic value.

Enzymatic Deficiency: Brady *et al.*[72] partially purified and characterized an enzyme from rat intestinal tissue which was capable of cleaving the terminal galactose moiety from galactosylgalactosylglucosylceramide. They then demonstrated almost complete lack of activities of this enzyme in the intestinal mucosa of two patients with Fabry's disease (Fig. 12).[245] There was a partial reduction of the activity in the tissues of carrier females. This finding established the specific enzymatic deficiency underlying Fabry's disease. A few years later, an imaginative experiment by Kint[246] unequivocally demonstrated that there was an almost total lack of α-galactosidase activities in leukocytes of patients with Fabry's disease when measured with chromogenic synthetic substrates. The activities were intermediate in female carriers, and other acid hydrolases were normal. This report made it possible to use the readily available synthetic substrate for the diagnosis of the disease and also provided the crucial impetus

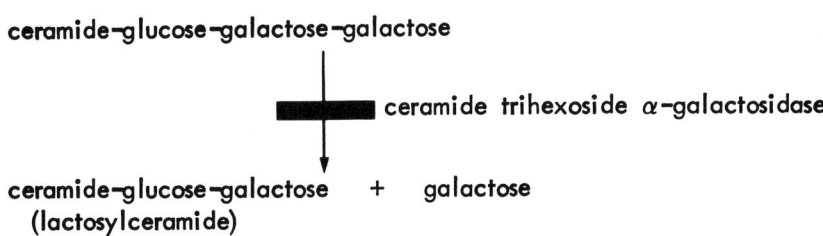

Fig. 12. Metabolic block in Fabry's disease.

for the subsequent chemical and enzymatic investigation of the anomeric configuration of the terminal galactose of digalactosylglucosylceramide.

The same enzymatic deficiency has been demonstrated in cultured fibroblasts[247] as well as in plasma.[248] Prenatal diagnosis of the disease has been accomplished using cultured amniotic fluid cells.[249]

Pathophysiological Problems: The precise pathogenetic mechanism of the nervous system dysfunction in Fabry's disease is not clear. Some of the symptoms seem to be the results of the dysfunction of the cerebrovascular system. However, other symptoms, such as severe paroxysmal pains in the extremities, vegetative disturbance, and diabetes insipidus, may well be caused directly by the functional disturbance of autonomic neurons which contain abnormal amounts of trihexosylceramide. The recent successful trial of phenylhydantoin as the therapeutic agent for the paroxysmal pains also indicates that the phenomenon is not merely vascular.[250]

The most likely source of digalactosylglucosylceramide is globoside. Globoside is a small but normal constituent of most of the systemic tissues and is present at a relatively high concentration in erythrocytes. Its first catabolic product is digalactosylglucosylceramide. The significance of digalactosylceramide accumulation is more obscure. This compound is essentially undetectable in tissues of normal individuals, except in the kidney. Because of this limitation, little is known regarding its biosynthesis and degradation in normal conditions. While its accumulation in Fabry's disease is most likely the result of a genetic lack of an α-galactosidase which normally cleaves the terminal α-galactose moiety, no direct evidence exists as yet to support this conjecture. Furthermore, the question as to whether a single enzyme cleaves the terminal galactose moieties of both compounds or whether two separate specific α-galactosidases exist for respective lipids remains to be answered.

F. Gangliosidoses

Within the past 10 years, one of the most dramatic series of advances in the study of genetic disorders of sphingolipid metabolism occurred in the area of ganglioside storage diseases. Less than 10 years ago, the only known ganglioside storage disorder was the classical Tay–Sachs disease. At present, at least five, and probably six, enzymatically distinct ganglioside storage diseases are known. Ultrastructural studies, basic enzymology of sphingolipids, and analytical and organic chemistry all played indispensable roles in providing the knowledge necessary for identification of the underlying metabolic defects of the respective disorders.

All of the well-characterized gangliosidoses can be divided into two major groups, depending on the molecular species of the gangliosides primarily involved—G_{M1}-gangliosidosis and G_{M2}-gangliosidosis. Within each major category, the state of gangliosidosis nomenclature is at present somewhat disorderly. In this chapter, we shall be uncommitted to any particular nomenclature and present each disease with descriptive names.

1. G_{M1}-Gangliosidosis

Incidence, Genetics, and Clinical Manifestations: Before the advent of thin layer chromatography and detailed structural analysis of gangliosides, there were several clinicopathological reports of patients who had conspicuous visceral and bony abnormalities in addition to classical findings of Tay–Sachs disease.[9,10] Chemical analysis of the ganglioside stored in the brain conclusively established G_{M1}-gangliosidosis as a disease entity fundamentally different from Tay–Sachs disease.[251–253] G_{M1}-gangliosidosis appears to be evenly distributed without geographic or ethnic preponderance. Cases have been known among Negroes and Japanese.

There are two clinically distinct types of G_{M1}-gangliosidosis—type 1 and type 2. The distinction between the two types is, at present, largely clinical. As such, these types are equivalent to various types of Gaucher's disease or metachromatic leukodystrophy. Analogous to these diseases, these two types of G_{M1}-gangliosidosis appear to be also genetically distinct in that only one type occurs in one family. The mode of inheritance appears to be autosomal recessive.

Patients with type 1 G_{M1}-gangliosidosis often show signs of the disease at birth, and, if not, clinical manifestations become apparent by 6 months of age. Both neurological and systemic signs are prominent—the reason for many synonyms, such as *generalized gangliosidosis,*[252] *neurovisceral lipidosis,*[10] *Tay–Sachs disease with visceral involvement,*[9] or *pseudo–Hurler syndrome.*[254] Neurologically, patients are generally slow in psychomotor development, which later regresses. There are weak cries, apathetic appearance, and difficulty in feeding. At an early stage, muscle tones tend to be hypotonic and floppy but become spastic in later stages. Seizures are not infrequent. Cherry-red spots at bilateral maculae, identical to those seen in Tay–Sachs disease, have been recorded in approximately one half of the cases. At the terminal stage, patients become decerebrate, blind, and deaf, totally without contact with the environment. Prominent systemic involvement is the clinical feature which distinguishes type 1 from type 2 G_{M1}-gangliosidosis. Patients have facial abnormalities, often from birth, reminiscent of patients with mucopolysaccharidosis. They include frontal bossing, flat nasal bridge, hirsutism, hypertrophied gums, and marcoglossia. Corneal opacity, however, is not present. Bony abnormalities, also reminiscent of mucopolysaccharidosis, are present, such as kyphoscoliossi, deformed vertebrae and long bones, and cartilaginous hypertrophy, which are apparent both clinically and radiologically. Hepatosplenomegaly is often prominent. Patients die within a few years after steady progression of the neurolog-

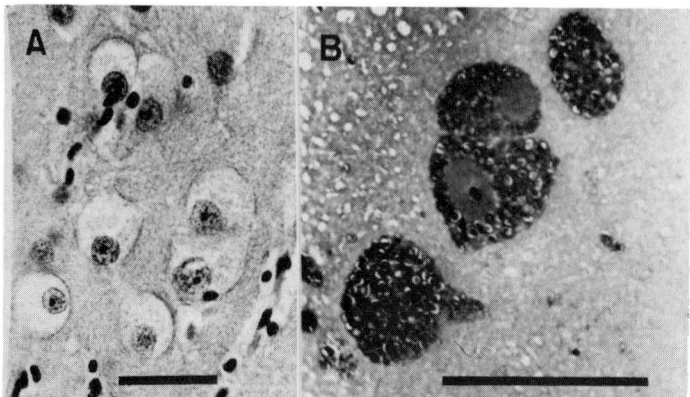

Fig. 13. A, Cortical neurons in G_{M1}-gangliosidosis. The cytoplasm is swollen due to the accumulation of abnormal material, which appears finely granular on hematoxylin–eosin stain. B, The nature of the abnormal cytoplasmic bodies is apparent even by light microscopy. A semithin section was cut from a plastic-embedded specimen prepared for electron microscopy and stained with toluidine blue. The light microscopic appearance of cortical neurons is essentially identical in Tay–Sachs disease. The lines indicate the scale of 5 μ. [Reproduced from Suzuki et al. [255] by permission.]

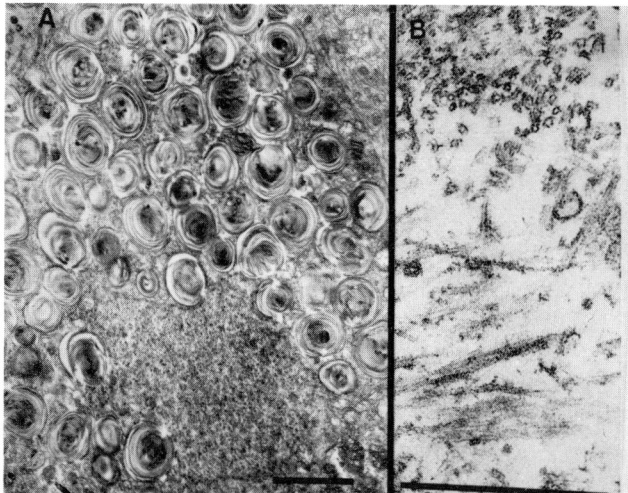

Fig. 14. A, An electron micrograph of a cortical neuron in G_{M1}-gangliosidosis. The cytoplasm is filled with abnormal membranous cytoplasmic bodies. This appearance is essentially indistinguishable from a cortical neuron in Tay–Sachs disease. The line indicates the scale of 5 μ. B, The fine structure of the abnormal material in the foamy cells in the liver in G_{M1}-gangliosidosis shows a hollow tubular structure, completely dissimilar from the membranous cytoplasmic bodies in the neurons. The line indicates the scale of 0.5 μ. [Reproduced from Suzuki et al. [255] by permission.]

ical manifestations. The majority of the patients with G_{M1}-gangliosidosis belong to type 1.

Patients with type 2 G_{M1}-gangliosidosis differ from type 1 patients in the relatively late onset, prolonged course, and most importantly lack of systemic manifestations. Patients often appear to develop normally up to 1 year of age. Presenting neurological signs are motor disturbance, such as ataxia and incoordination, and speech abnormality. There is a steady deterioration of psychomotor functions, eventually reaching the totally decerebrate state. Seizures are common. Facial abnormalities, bony changes, and hepatosplenomegaly are not present. The clinical course is more protracted than in type 1, and patients may survive up to 10 years, although at least one patient had a relatively acute course with clinical onset before 6 months and death at 3 years. This patient should be included in type 2 because, despite the acute course, there were no signs of systemic involvement, even in retrospect.[255]

Morphology: At the terminal stage, the brain may be larger than normal, although the degree of enlargement is generally milder than that seen in Tay–Sachs disease. Normal or even atrophic brains are common. In type 1 patients, there is gross enlargement of the liver and spleen. In one patient with type 2 disease, both the liver and spleen were smaller than normal.[255]

All neurons throughout the nervous system, including those in peripheral autonomic ganglia, are abnormally swollen, filled with finely granular, faintly PAS-positive material with eccentrically located nuclei (Fig. 13).[10,253,255] These granular materials have strong acid phosphatase activity.[255] The light microscopic appearance of neurons is identical to that in Tay-Sachs disease. Various degrees of secondary demyelination are present with sudanophilic materials, as the result of neuronal loss.

Swollen foamy histiocytes are found in the liver, spleen, lymph nodes, bone marrow, intestine, and lungs. They contain strongly PAS-postive granular materials. Many hepatic parenchymal cells are vacuolated. The cytoplasm of renal glomerular ephithelium is swollen, and the subcapsular spaces are widened. It should be noted that these histopathological changes of systemic organs are present even in patients with type 2 G_{M1}-gangliosidosis who show no clinical signs of visceral involvement. In the liver of type 1 patients, there appear to be more prominent hepatic cell vacuolation and less infiltration of foamy histiocytes.

Under the electron microscope, the swollen cortical neurons are seen to be filled with numerous abnormal lamellar bodies ranging from 0.3 to 3 μ in diameter (Fig. 14).[253,255–258] Their ultrastructure is virtually identical to that of the membranous cytoplasmic bodies (MCB) originally described in neurons of patients with Tay–Sachs disease.[14] Normal cytoplasmic organelles, such as mitochondria, endoplasmic reticulum, and ribosomes, are pushed and squeezed between the abnormal bodies. In addition, small glial cells often contain another type of abnormal inclusion which is membrane bound and consists of vesicular myelin-like figures and straight or curved lamellae.[255] Many astrocytes also contain extremely pleomorphic abnormal inclusions. The presence of such abnormal inclusions in glial cells is one ultrastructural characteristic of G_{M1}-gangliosidosis brain, different from classical Tay–Sachs disease. Abnormality of renal tubule epithelium cells is characterized primarily by numerous and mostly empty vacuoles.[259,260] Histiocytes in the liver and spleen are filled with interwoven bundles of fine tubular structures approximately 200 Å in diameter.[255] The vacuoles within hepatic cells are surrounded by a single membrane and contain amorphous material.[261] Thus the ultrastructure of the liver and spleen is quite dissimilar to that of the brain, indicating the different nature of the stored material (Fig. 14).

Analytical Chemistry: The analytical hallmark of G_{M1}-gangliosidosis is the abnormal accumulation of G_{M1}-ganglioside in the brain. The total concentration of brain ganglioside may range from three to five times normal in gray matter and may be as high as ten times the normally low concentration in white matter (Fig. 15).[262] G_{M1}-ganglioside constitutes approximately 80–90% of

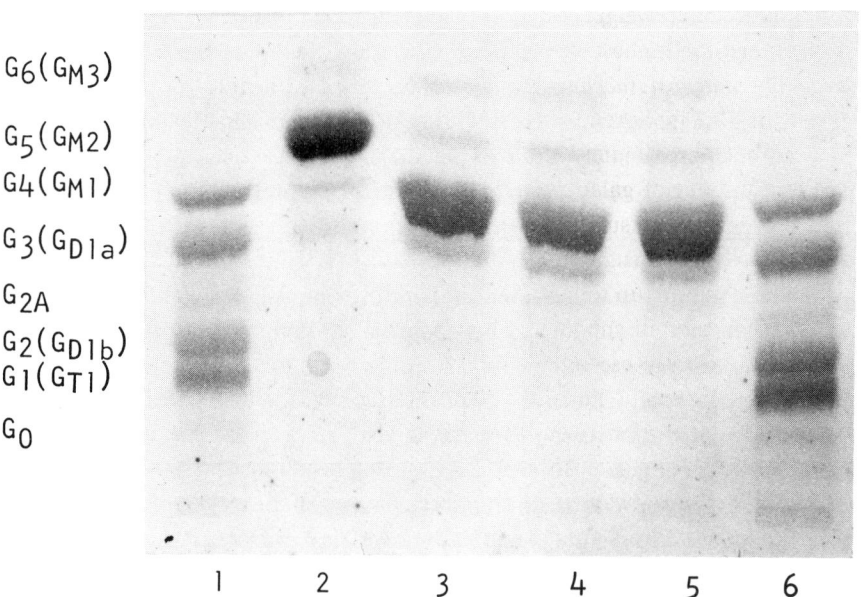

Fig. 15. A thin layer chromatogram of brain gangliosides in gangliosidoses. The solvent system was chloroform–methanol–2.5 N ammonia (60:40:9, v/v/v). The spots were located by resorcinol spray, which visualizes sialic acid. Columns 1 and 6, normal brain gangliosides; column 2, Tay–Sachs disease; columns 3,4, and 5, G_{M1}-gangliosidosis. The nomenclature of gangliosides in this figure is that of Korey and Gonatas [328] with the corresponding Svennerholm nomenclature in parentheses. The latter nomenclature is used as the standard throughout this chapter. [Reproduced from Suzuki *et al.* [262] by permission.]

total ganglioside. Since G_{M1}-ganglioside is about 15% of the total ganglioside in normal gray matter,[206] the increase of G_{M1}-ganglioside in gray matter is actually 15–30 times normal. Identity of this abnormal ganglioside with normal G_{M1}-ganglioside has been established.[263] In addition, there is a moderate increase of the asialo derivative of G_{M1}-ganglioside in gray matter, and lactosylceramide and glucocerebroside are also abnormally high.[262,264] The increases of the last two compounds, however, are not specific for G_{M1}-gangliosidosis and are also seen in Tay–Sachs disease.[264] G_{M1}-ganglioside is also increased in the liver and spleen up to 100 times the normal level. This seemingly enormous increase, however, still represents only a very small amount of G_{M1}-ganglioside in these visceral organs (less than 0.5 mg per gram wet weight), since the normal concentration of G_{M1}-ganglioside in these organs is extremely small. The magnitude of G_{M1}-ganglioside accumulation in the brain is similar in both clinical types of G_{M1}-gangliosidosis, but its accumulation in systemic organs is generally either absent or only slight in type 2 patients, while it is always high in type 1 patients.[265]

In the visceral organs, the main storage material is galactose-rich polysaccharide, which actually appears to consist of a series of related compounds of varying molecular weights. The most insoluble of these compounds, which probably has the highest molecular weight, was first characterized as *keratan sulfate–like mucopolysaccharide*,[266] based on its behavior in preparative procedures, its electrophoretic properties, and its sugar composition. The material consisted of approximately equimolar amounts of galactose and glucosamine, a smaller amount of galactosamine, and traces of mannose, fucose, and sialic acid, and it was undersulfated.[262] There was a more soluble and nondialyzable fraction with similar sugar composition. With both of these combined, the liver and spleen levels are up to 20 times normal in G_{M1}-gangliosidosis. Characterization of these keratan sulfate–like compounds has not been completed.[267,268] In addition, a galactose- and hexosamine-containing fraction, which became dialyzable after proteolytic digestion of defatted livers and spleens, was similarly increased in patients' tissues, suggesting that a number of glycoproteins with shorter carbohydrate chains are also abnormally increased in this disorder.[265] Generally, but not always, the degree of accumulation of these mucopolysaccharide glycoproteins is greater in the systemic organs of type 1 patients.[265]

There is no increase in the keratan sulfate–like material in the brain, but there appears to be a moderate increase of galactose-rich glycoproteins in the brain, as attested to by the increased amounts of the dialyzable sugar fraction after proteolytic digestion.[265]

Enzymatic Deficiency: The prominent accumulation of G_{M1}-ganglioside and its asialo derivative, both with terminal β-galactose, prompted a search for possible β-galactosidase deficiency. Such deficiency was demonstrated in the brain, liver, spleen, and kidney of patients with the use of chromogenic synthetic substrates,[257,265,269-272] as well as with specifically labeled natural substrate, G_{M1}-ganglioside (Fig. 16).[270] Other lysosomal acid hydrolases were either normal or increased when assayed with synthetic substrates, as were more specific sphingolipid galactosidases, such as galactocerebrosidase,[154,273] lactosylceramide β-galactosidase, and trihexosylceramide α-galactosidase.[273]

Liver homogenate of a patient was also shown to be deficient in the ability to release galactose from the keratan sulfate–like mucopolysaccharide prepared from pathological liver, as well as from desialylated fetuin.[274] This finding appears to explain the accumulation of the galactose-rich mucopolysaccharide and glycoproteins in G_{M1}-gangliosidosis. The same enzymatic deficiency has been demonstrated in leukocytes,[267,275,276] urine,[277] and cultured fibroblasts.[267-279] All of these materials can be used for diagnosis of the disease, as well as for detection of heterozygous carriers.

The major question still remaining is biochemical delineation of the two clinical types.[280] A careful analytical study of eight cases indicated generally less accumulation of G_{M1}-ganglioside and the keratan sulfate–like material in the liver and spleen of type 2 patients, but correspondence between the clinical classification and the biochemical criteria was not perfect.[265] β-Galactosidase in normal liver can be electrophoretically separated into three components. O'Brien et al.[281] observed that the liver tissue of their type 2 patient contained well-preserved activity of the fastest-moving A, while tissues from type 1 patients lacked all of the three components, A, B, and C. This preserved component A activity corresponded to a minor peak normally present with a pH optimum of 6.6. A similar study by Suzuki et al[265] showed, however, that the liver tissue of one type 2 patient lacked all β-galactosidase components, and the liver of one clear-cut type 1 patient showed preserved component A activity. There were other patients who showed various degrees of component A preservation.

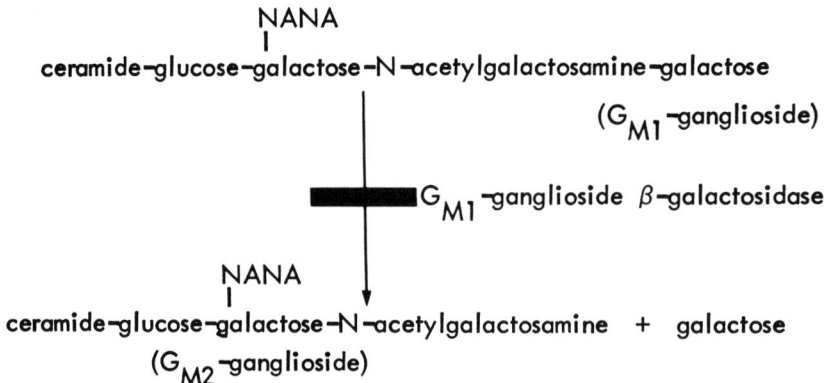

Fig. 16. Metabolic block in G_{M1}-gangliosidosis.

Furthermore, Schafer[282] recently obtained data showing both type 1 and 2 patients with active component A. His type 2 patient actually had pH 6.6 β-galactosidase much higher than normal.[282] Therefore, there appears to be no one-to-one correspondence between the clinical classification and the absence or presence of β-galactosidase component A. In this regard, Pinsky et al.[283] studied the residual β-galactosidase activity in cultured fibroblasts of patients and reported that residual β-galactosidase activity of a type 2 patient had the same pH optimum and thermostability as that of normal fibroblasts but that the enzyme of a type 1 patient had a lower pH optimum and was more resistant to heat inactivation. Whether one-to-one correspondence would hold for these criteria remains to be determined.

2. G_{M2}-Gangliosidosis

At the time of this writing, there are four enzymatically distinct disorders, all of which are characterized by an abnormal accumulation of G_{M2}-ganglioside in the brain as the most prominent biochemical manifestation. The classical Tay–Sachs disease, which is caused by a complete absence of one of the two electrophoretic components of hexosaminidase, is the prototype of G_{M2}-gangliosidosis. All of the other G_{M2}-gangliosidoses were discovered within the past 4 years.

a. Tay–Sachs Disease (Hexosaminidase A Deficiency). Incidence, Genetics, and Clinical Manifestations: For many years, Tay–Sachs disease was the only known ganglioside storage disorder since the original descriptions of Tay[1] and Sachs.[2] The disease occurs predominantly among the Ashkenazi Jewish population with ancestors from a relatively small area of Eastern Europe.[284] One out of 30 individuals in the high-risk population is estimated to be a heterozygous carrier of the disease, while the gene incidence for the rest of the population is one tenth of that in Ashkenazi Jews. While extremely rare, enzymatically well-documented cases are known in many other ethnic groups. The mode of transmittance is almost certainly autosomal recessive.

Patients generally appear normal at birth. Slow growth and motor weakness become apparent by 6 months of age. Hyperacusis and bilateral cherry-red spots at the maculae are very common early signs. Psychomotor retardation and then regression become apparent by 1 year of age. Neurological manifestations progress rapidly, with poor muscle tone, bulbar signs, blindness, and deafness, eventually reaching the totally decerebrate stage without contact with the environment. There is progressive enlargement of the head near the terminal stage. Most patients die by the age of 3 years. Characteristically, there are no systemic manifestations associated with Tay–Sachs disease.

Morphology: Gross appearance of the brain is often atrophic except for the advanced late stage of the disease when the brain is more commonly larger than normal and firm. This is probably due to reactive astrocytosis rather than neuronal swelling. Systemic organs are not affected on gross examination. Practically all neurons are abnormally swollen and filled with finely granular material. Nuclei are eccentrically placed.[285] The cytoplasmic granular materials stain faintly with PAS but show strong acid phosphatase activities.[286] White matter shows mild to moderate demyelination with sudanophilia, most likely secondary to neuronal loss. Throughout the central nervous system, in both gray and white matter, reactive astrocytic gliosis is prominent. Thus neuropathological changes seen by light microscopy are essentially identical to those in G_{M1}-gangliosidosis (see Fig. 13).

The classical work of Terry and Weiss[287] established the most essential aspects of the ultrastructure of Tay–Sachs disease. Swollen neurons are filled with round or oval abnormal inclusion bodies. Nuclei appear ultrastructurally normal except that they are usually pushed aside to the periphery of the cell. Normal cytoplasmic organelles are present, scattered among the abnormal inclusions. The inclusions are commonly called *membranous cytoplasmic bodies* (MCB), because they consist mostly of concentrically arranged lamellar structures, similar to an onion in a three-dimensional representation. They are on the average about 1 μ in diameter, and the periodicity of lamellae is 50–60 Å. In the cerebrum, they almost always lack a surrounding limiting membrane. Some membranous cytoplasmic bodies in the cerebellum, where the disease process is thought to be less advanced, are surrounded by a single membrane and contain histochemically demonstrable acid phosphatase activity.[288] Abnormal lamellar structures are also found occasionally in the liver.[116] It is important to note that the morphological changes in the brain begin long before birth.[289]

Analytical Chemistry: The discovery in the brain, by Klenk,[290,291] of a new family of glycolipids containing neuraminic acid was closely associated with Tay–Sachs disease. He termed the new glycolipid *ganglioside* because it was found predominantly in gray matter of the brain, suggesting a high concentration in neurons. Soon the molecular heterogeneity of brain ganglioside became apparent, and in 1962 Svennerholm[292] identified the specific ganglioside abnormally accumulating in Tay–Sachs brain as G_{M2}-ganglioside, which lacks the terminal galactose of the normal major monosialoganglioside, G_{M1}. The detailed structure of G_{M2}-ganglioside has since been established.[293]

The total concentrations of gray matter ganglioside in brains of patients generally range from three to five times normal, and G_{M2}-ganglioside constitutes 80–90% of the total (Fig. 15). Since G_{M2}-ganglioside is normally a very minor component, the net increase of G_{M2}-ganglioside is a few hundred times normal. White matter may show a relatively more dramatic increase in total ganglioside, with similarly predominant G_{M2}-ganglioside. In addition to G_{M2}-ganglioside, its asialo derivative, N-acetylgalactosaminylgalactosylglucosylceramide, is also moderately increased up to 20 times normal.[264] Much less prominent but significant are abnormal increases of glucocerebroside and lactosylceramide, predominantly in gray matter.[264] The abnormal MCB fraction isolated from gray matter consisted of 30% ganglioside, almost all of which was G_{M2}-ganglioside.[173,262] Asialo-G_{M2}, lactosylceramide, and glucocerebrosides were also enriched in the MCB fraction.[262] These findings provided direct chemical evidence that the characteristic membranous cytoplasmic bodies are the site of the abnormal glycolipid accumulation.

Systemic organs, such as liver, spleen, and muscle, normally contain very little G_{M2}-gangliosides, but G_{M2}-ganglioside accumulation occurs in these organs in Tay–Sachs disease.[262,266,294] Quantitatively, the amounts of G_{M2}-ganglioside in these organs are, on the average, similar to that of hematoside (G_{M3}), normally present in systemic organs.[262] While the relative increase over the normal amount is enormous, the actual amounts of G_{M2}-ganglioside in these tissues are too small to cause clinical or histopathological changes. The degree of G_{M2}-ganglioside increase in systemic organs in Tay–Sachs disease is the same as the G_{M1}-ganglioside increase in these organs in G_{M1}-gangliosidosis.

Enzymatic Deficiency: Because of the accumulation of G_{M2}-ganglioside, and to a lesser degree asialo-G_{M2}, it was long suspected that the genetic cause of Tay–Sachs disease might be a deficiency of an N-acetylgalactosaminidase which would normally cleave the terminal N-acetylgalactosamine moiety from these compounds. However, attempts to assay tissues of Tay–Sachs patients for activities of hexosaminidases with synthetic chromogenic substrates consistently failed to detect any deficiencies. Activities were either normal or, more commonly, higher than normal. This dilemma was finally resolved in 1969 when both Sandhoff[295] and Okada and O'Brien[296] independently reported that one of the two hexosaminidase components separable by starch gel electrophoresis or electrofocusing is essentially absent in various tissues of Tay–Sachs disease patients. This component is the more acidic of the two and is called *hexosaminidase A*, according to Robinson and Stirling.[297] In most normal tissues, both components are present, in roughly equal amounts. Okada and O'Brien[296] demonstrated this hexosaminidase A deficiency in the brain, liver, skin, kidney, leukocytes, serum, and cultured fibroblasts. Hexosaminidase B is often much higher than normal, thus accounting for the frequent increased total hexosaminidase activities in Tay–Sachs disease. The hexosaminidase A deficiency has been consistently found in tissues of patients and has not been observed in normal individuals or in any other neurological disorders. While likely, this finding does not prove that hexosaminidase A normally cleaves the terminal N-acetyl-β-D-galactosamine of G_{M2}-ganglioside. Kolodny *et al.*[298] then showed that normal muscles contain the enzyme to cleave the terminal galactosamine moiety of radioactively labeled G_{M2}-ganglioside and that muscles from Tay–Sachs disease patients are deficient in its activity. Furthermore, the recent substrate specificity study demonstrated that only hexosaminidase A, not B, was active toward the terminal galactosamine of G_{M2}-ganglioside.[67] With all of these findings combined, it is now reasonably well established that the underlying cause of Tay–Sachs disease is the deficiency of hexosaminidase A, which normally cleaves galactosamine from G_{M2}-ganglioside, resulting in its accumulation (Fig. 17).

Parents of patients with Tay–Sachs disease show intermediate hexosaminidase A activities in serum, leukocytes, and cultured fibroblasts.[299,300] This finding has been utilized extensively for detection of heterozygous carriers and subsequent genetic counseling. Finally, with the use of cultured, and in a few instances uncultured, amniotic fluid cells, prenatal diagnosis of the disease has been achieved.[301,302] Hexosaminidase A was found to be almost completely absent in tissues of aborted affected fetuses.[302]

Sandhoff et al.[303] showed early that radioactively labeled asialo derivative of G_{M2}-ganglioside could be degraded normally by tissue homogenates of Tay–Sachs disease patients. More recently, it has been shown that both hexosaminidases A and B are active toward the asialo derivative.[67] Yet moderate accumulation of the asialo derivative does occur in the CNS of Tay–Sachs patients. Perhaps the large amounts of G_{M2}-ganglioside may act as an inhibitor of the enzyme or the asialo derivative may be sequestered within the MCBs and inaccessible to the enzyme.

b. Juvenile G_{M2}-Gangliosidosis. Incidence, Genetics, and Clinical Manifestations: Bernheimer and Seitelberger[304] first reported two sisters who developed neurological manifestations somewhat later than in typical Tay–Sachs disease and died at ages 5 and 9 years, respectively, after slowly progressive deterioration. Biochemical investigations indicated an accumulation of both G_{M2}-ganglioside and its asialo derivative in the brain. Several other patients are known.[305–307] The patient reported by Suzuki et al.[306] is the oldest patient

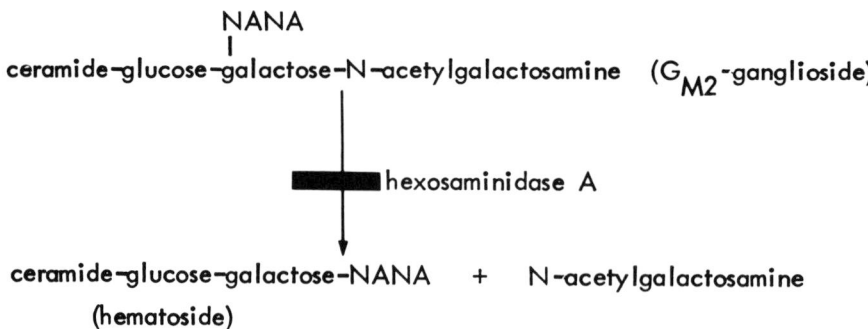

Fig. 17. Metabolic block in Tay–Sachs disease.

known, with clinical onset at 6 years and death at 15 years. None of the patients thus far known are Jewish.

The clinical features of the disease are characterized by late onset, ranging from 2 to 6 years, and a generally protracted course. The onset at a younger age is marked by progressive psychomotor deterioration, but the patient reported by Suzuki *et al.* started with what appeared to be personality changes and slowly progressive dementia. Seizures and cerebellar, extrapyramidal, and then pyramidal signs follow. The disease progresses steadily, eventually reaching decerebrate rigidity and total unresponsiveness to external stimuli. Macular cherry-red spots have not been observed. There are no systemic manifestations.

Morphology: The cerebral hemispheres and cerebellum are atrophic. Many neurons are swollen and appear similar to those in Tay–Sachs disease, although somewhat less severe. Such swollen neurons are more prominent in the thalamus, brain stem nuclei, and spinal cord than in the cerebral cortex. In a younger patient, neuronal swelling was more uniformly distributed. There are severe degenerative changes in the cerebellum. White matter shows moderate secondary demyelination.

Electron microscopic appearance of swollen neurons is complex.[305–307] They contain numerous abnormal bodies of various sizes and structures. Some of them are similar to the membranous cytoplasmic bodies in Tay–Sachs disease. In addition, there are poorly defined lamellar structures, lipofuscin bodies, membranovesicular bodies, and also extremely large—up 15 μ in diameter—complex conglomerate bodies consisting of all of the above abnormal bodies (Fig. 18).

Analytical Chemistry: The chemical abnormalities of juvenile G_{M2}-gangliosidosis are qualitatively identical to those of Tay–Sachs disease, but quantitatively milder.[306–308] The total gray matter ganglioside is usually about twice normal, with G_{M2}-ganglioside constituting approximately 50% of the total. The asialo derivative of G_{M2}-ganglioside also accumulates. Isolated MCB is quite similar in chemical composition to Tay–Sachs MCB.[306] Ganglioside in MCB is predominantly G_{M2}.

Enzymatic Deficiency: A partial deficiency of hexosaminidase A was demonstrated in tissues of three patients[309–311] and in cultured fibroblasts and serum in another.[312] The degree of deficiency in the brain and spleen ranged from 25 to 50% of normal, but it was much more severe in serum and cultured fibroblasts. Serum and fibroblasts of parents showed intermediate activities.[312]

Pathophysiological Problems: The partial deficiency of hexosaminidase A appears to explain the chemical abnormalities of juvenile G_{M2}-gangliosidosis, which are qualitatively identical to those of Tay–Sachs disease, but quantitatively milder. However, the findings on the patient reported by Suzuki and

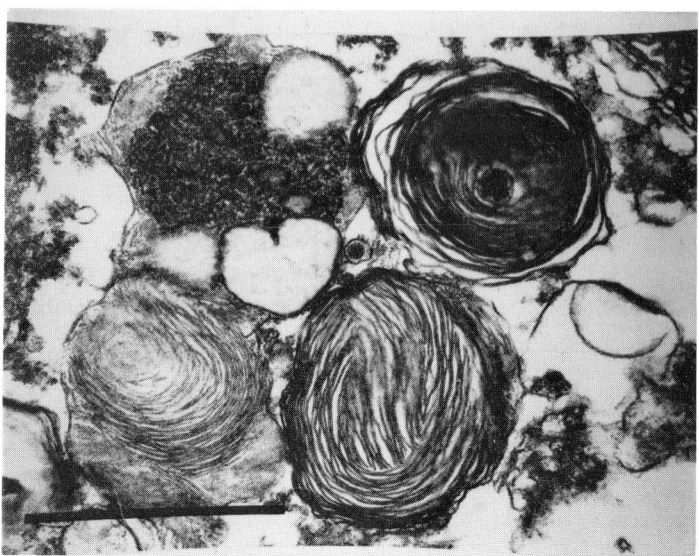

Fig. 18. An electron micrograph of a cortical neuron in juvenile G_{M2}-gangliosidosis. The complex nature of the abnormal inclusion bodies is apparent. Typical granular lipofuscin and Tay–Sachs-like membranous cytoplasmic bodies coexist closely. Sometimes they are surrounded by a common continuous limiting membrane. The line indicates the scale of 1 μ. [Reproduced from Suzuki et al. [306] by permission.]

Suzuki[309] pose a question which requires future investigation. The degree of hexosaminidase A deficiency in the tissues of this patient was in the same range as in heterozygous carriers of Tay–Sachs disease, who are phenotypically normal. This suggests that the partial reduction of hexosaminidase A in juvenile G_{M2}-gangliosidosis may not simply be a reduction of a single enzyme, hexosaminidase A, but it may be more or less complete deficiency of one or more specific enzymes which are constituents of hexosaminidase A. For example, it is quite possible that the degree of deficiency may be much greater if the natural substrate, G_{M2}-ganglioside, is used for enzyme assays. Or if hexosaminidase A represents a single protein, the mutation might have occurred in such a way that it causes preferential loss of activity toward certain natural substrates while retaining reasonable activity toward the synthetic substrate.

 c. *Total Hexosaminidase Deficiency (Sandhoff's Disease). Incidence, Genetics, and Clinical Manifestations:* The existence of total hexosaminidase deficiency was first indicated by Sandhoff *et al.*[303,313] and Pilz *et al.*[314] in 1968. Approximately a dozen patients are known to investigators, although the number of cases already in the literature is still small. The incidence of the disease cannot be estimated, but none of the patients thus far known is Jewish. It is quite possible that a substantial number of non-Jewish patients previously reported as having Tay–Sachs disease may in fact have had this disorder.

To our knowledge, the clinical findings are probably indistinguishable from those in Tay–Saches disease. The clinical onset is usually at about 6 months of age, and death occurs by 3 years, after a clinical course identical to that described for Tay–Sachs disease. Macular cherry-red spots may or may not be present. The liver or spleen may be palpable, but conspicuous hepatosplenomegaly is absent.

Morphology: At the terminal stage, the brain may be larger than normal, as in Tay–Sachs disease. There is no abnormality in the gross appearance of systemic organs. There is significant loss of neurons and severe gliosis in the cerebral cortex. The remaining neurons are extremely ballooned and filled with finely granular materials. In white matter, there was unusually severe myelin loss in both patients for whom details of morphological descriptions are available.[314,315] Whether this extremely severe demyelination is a consistent feature of the disease remains to be determined in future cases. In the cerebellum, there are also severe neuronal loss, complete disappearance of normal cytoarchitecture, and severe gliosis. The swollen neurons contain membranous cytoplasmic bodies quite similar to those in Tay–Sachs disease.

Unlike in other G_{M2}-gangliosidoses, however, there are histological abnormalities in systemic organs such as liver, kidney, and spleen.[314,315] The most conspicuous of these is the presence of strongly PAS-positive material within hepatic cell cytoplasm, Kupffer cells, histiocytes of spleen and lung, and renal tubular epithelium. This PAS positivity largely disappears when sections are stained after ethanol dehydration and paraffin embedding.

Analytical Chemistry: G_{M2}-ganglioside is the major storage substance in the nervous system. The degree of accumulation is similar to that in Tay–Sachs disease. However, the asialo derivative of G_{M2}-ganglioside also accumulates massively, almost to an equal level as G_{M2}-ganglioside itself. While its accumulation in Tay–Sachs brain is ten to 20 times normal, its level in this disease is four to five times that in Tay–Sachs disease.[305,315] G_{M2}-ganglioside also accumulates in the liver and spleen to a similar extent as in Tay–Sachs disease. The most conspicuous analytical finding in Sandhoff's disease is a large accumulation of globoside (*N*-acetylgalactosaminyldigalactosylglucosylceramide) in the systemic organs. While globoside is a normal constituent of these organs, its normal concentration is small. In Sandhoff's disease, the concentrations of globoside may range from ten to 200 times normal, depending on the organs and patients.[303,315] Asialo-G_{M2} is also present in the liver in a high concentration. Its fatty acid composition in one case was dissimilar to either asialo-G_{M2} in the brain or to globoside in the liver.[315]

Enzymatic Deficiency: Sandhoff's disease is characterized by an almost

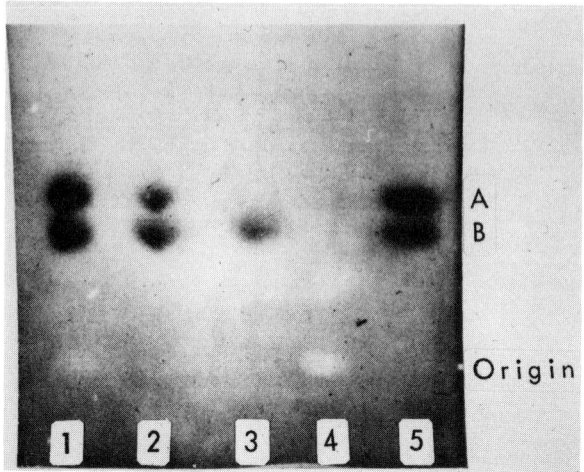

Fig. 19. An electrophoregram of hexosaminidase in liver of patients with G_{M2}-gangliosidoses. Electrophoresis was carried out with a cellulose acetate strip, and the enzyme activities were visualized with 4-methylumbelliferyl-N-acetyl-β-glucosaminide as the substrate. Columns 1 and 5, normal controls; column 2, juvenile G_{M2}-gangliosidosis; column 3, Tay–Sachs disease; and column 4, Sandhoff's disease. Note the total absence of both components A and B in Sandhoff's disease, absence of component A in Tay–Sachs disease, and diminution of component A in juvenile G_{M2}-gangliosidosis. This figure is reproduced as a negative print for better photographic reproduction; black and white are reversed in the actual electrophoregram. [Reproduced from Suzuki and Suzuki[309] by permission.]

complete lack of hexosaminidases.[303,315] This was demonstrated in the brain, liver, spleen, and kidney of the original patient by Sandhoff *et al.*[303] with the use of synthetic substrates as well as with radioactively labeled asialo-G_{M2} and globoside. By electrophoresis or by electrofocusing, it can be shown that both hexosaminidases A and B are almost completely absent (Fig. 19). This deficiency has been demonstrated also in leukocytes, serum, and cultured fibroblasts.[281,315] O'Brien *et al.*[281] found intermediate activities in materials from parents of patients.

By comparing the analytical and enzymatic findings in Tay–Sachs disease and total hexosaminidase deficiency, specificities of hexosaminidases A and B toward natural substrates can be inferred. This aspect has been recently directly answered by substrate specificity studies on purified hexosaminidases.[67] Both globoside and asialo-G_{M2} can be cleaved by either A or B component, but only the A form is active toward G_{M2}-ganglioside. These results explain the analytical findings very well. When only hexosaminidase A is absent, G_{M2}-

ganglioside is the only conspicuous storage substance, while additional massive accumulation of asialo-G_{M2} and globoside results when both components A and B are deficient (Fig. 20). The characteristic hexosaminidase patterns of the three G_{M2}-gangliosidoses are summarized in Table 5.

Pathophysiological Problems: Total hexosaminidase deficiency is another good example of differential distribution of abnormal storage substances according to the normal distribution of these compounds. Gangliosides are primarily localized in neurons, whereas globoside is a normal constituent of many systemic organs but is absent in the brain. The pattern of abnormal accumulation of these compounds in Sandhoff's disease follows this normal pattern of distribution.

The simultaneous absence of two components of hexosaminidases in this disorder evokes the same question raised in multiple sulfatase deficiency in

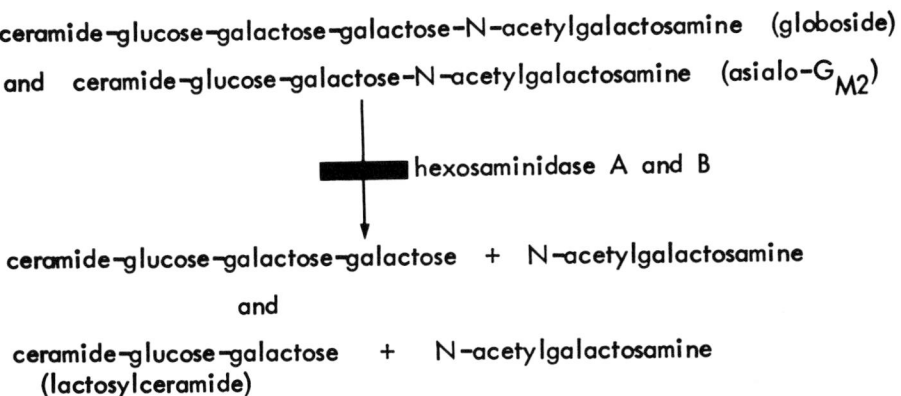

ceramide–glucose–galactose–galactose–N–acetylgalactosamine   (globoside)

and   ceramide–glucose–galactose–N–acetylgalactosamine   (asialo-G_{M2})

hexosaminidase A and B

ceramide–glucose–galactose–galactose   +   N–acetylgalactosamine

and

ceramide–glucose–galactose   +   N–acetylgalactosamine
(lactosylceramide)

Fig. 20. Metabolic block in total hexosaminidase deficiency. This is the additional block to that already depicted for Tay–Sachs disease (Fig. 17).

TABLE 5. *N*-ACETYL-β-D-HEXOSAMINIDASES IN BRAIN
IN G_{M2}-GANGLIOSIDOSES[a]

		N-acetyl-β-glucosaminidase			*N*-acetyl-β-galactosaminidase		
		Total	A	B	Total	A	B
Tay–Sachs disease	1	195	0	195	21.4	0	21.4
	2	57.6	0	57.6	8.8	0	8.8
Juvenile G_{M2}-gangliosidosis		41.7	17.5	24.2	5.5	2.2	3.3
Sandhoff's disease		4.0	Trace	Trace	0.7	Trace	Trace
Normal ($n = 7$)							
Average		56.3	35.5	20.8	7.3	4.5	2.8
Range		29.0–81.1	21.5–49.5	7.5–26.3	4.0–10.1	3.0–6.3	1.0–4.5

[a]The activities are expressed as mμmoles of substrate cleaved/hr/mg wet-weight tissue.

relation to the one gene–one enzyme concept (see above). The answer must await more sophisticated investigation.

   d. Specific G_{M2}-Ganglioside Galactosaminidase Deficiency. There is only one case of specific G_{M2}-ganglioside galactosaminidase deficiency documented in the literature.[295,316] No adequate clinicopathological descriptions are available, but it is said that the clinical course was compatible with classical Tay–Sachs disease. The analytical study also showed chemical abnormalities essentially identical to those in Tay–Sachs disease—a massive increase of G_{M2}-ganglioside and a moderate increase of its asialo derivative, primarily in the brain.

   Enzymatically, however, this patient appeared to be different from all other patients with G_{M2}-gangliosidosis. When synthetic chromogenic substrates were used, hexosaminidases appeared perfectly normal, both in the total activity and in the distribution of A and B components. If anything, hexosaminidase A was slightly higher than normal. It was only when radioactively labeled natural substrate, G_{M2}-ganglioside, was used for enzymatic assay that the activity of the patient's tissue was shown to be only 15% of normal.

   While additional cases would be necessary to establish this entity, this patient raises some of the most intriguing questions regarding the possible complex nature of these electrophoretic components of hexosaminidases.

G. Miscellaneous Disorders

   There are a few disorders which probably should be included as disorders of sphingolipid metabolism. There is also a group of disorders which are often described as if they were sphingolipidoses but which are almost certainly not diseases of sphingolipid metabolism.

1. Lipogranulomatosis of Farber

   In 1952, Farber described three patients with unusual clinicopathological features.[317,318] Less than ten patients are known thus far. The disease is genetically determined and appears to be transmitted as an autosomal recessive trait. Affected infants develop tender swollen joints soon after birth. Generalized multiple subcutaneous nodules appear and become increasingly enlarged. Neurological signs are often present, including flaccid paralysis, hyporeflexia, Babinski signs, and, terminally, severe mental deterioration. Most patients die by 2 years of age, but a few patients are known to have survived for several years, up to 16 years.[319]

   Subcutaneous nodules show histological appearance of granuloma with an accumulation of foamy cells containing PAS-positive cytoplasmic material.

The anterior horn cells and the neurons of the medulla, pons, and cerebellum are more affected than cerebral cortical neurons. Autonomic and visceral ganglion cells are also distended.

Chemical analysis showed generalized increases of ceramide in the lymph-nodes, liver, kidney, lung, and subcutaneous nodules, up to sixtyfold.[319,320] In addition, there were less conspicuous increases of hematoside, G_{M3}, in various tissues.[319,320] This abnormal increase of ceramide has recently been confirmed in an older patient.[321,322] From the analytical viewpoint, therefore, Farber's lipogranulomatosis is characterized by an abnormal accumulation of ceramide and, to a lesser degree, hematoside.

By analogy from other sphingolipidoses, the prime suspect for the genetic cause of the disease seems to be deficient ceramidase. An attempt to demonstrate such a deficiency in one patient was unsuccessful.[319] Therefore, the enzymatic basis of the disease is not yet clear.*

2. Fucosidosis

A disease characterized by genetically determined lack of α-fucosidase was discovered recently. Of the three patients described thus far, two were siblings, and the third was apparently also related to the others.[323,324] Clinically, the disease manifests itself with frequent respiratory infection and delayed psycho-motor development followed by mental deterioration, increasing spasticity, and tremor. The third patient reported by Loeb et al.[324] showed clinical and radiological features akin to the Hurler–Hunter syndrome.

Chemical analysis of the first two cases showed an accumulation of fucose in the form of tetrahexosyl- and pentahexosylceramide with terminal fucose moiety. These compounds exhibited H and Lewis blood group activity. The accumulation was most prominent in the liver (ten times normal) and less in the brain (one-fortieth of the amount in the liver). There were increased concentrations of fucose in the polysaccharide fraction (3 to 36 times normal), particularly in the brain and to a lesser degree in the liver. Lipid analysis of cerebral gray matter from the third patient[324] showed the overall lipid composition to be essentially normal. Total glycolipid was not increased, and cerebroside was the only glycolipid demonstrable on thin layer chromatography. There was no abnormality in brain gangliosides in the third patient, in total amount and molecular distribution. On the other hand, there was an increase in total glycolipids in the liver, but only glycolipids qualitatively similar to those in normal liver were detected on thin layer chromatograms. These results do not exclude the presence of fucose-containing ceramide hexosides such as those detected in

*Note added in proof: A ceramidase deficiency has recently been demonstrated in Farber's disease.[330]

TABLE 6. SUMMARY OF ENZYMATICALLY CHARACTERIZED SPHINGOLIPIDOSES

Disease	Clinical and pathological manifestations	Lipid involved	Defective enzyme	Remarks
1. Niemann–Pick	Four clinical types; mental retardation, hepatosplenomegaly, foamy bone marrow cells, cherry-red spots (30%)	Sphingomyelin	Sphingomyelinase	Some clinical types do not show deficient sphingomyelinase and therefore must be considered different diseases
2. Gaucher's	Generally similar to 1; chronic nonneuropathic and acute neuropathic forms; Gaucher cells in bone marrow, spleen, etc.	Glucocerebroside	Glucocerebroside β-glucosidase	
3. Krabbe's (globoid cell leukodystrophy)	Almost always infantile disease; long-tract signs, spasticity; later flaccid, globoid cell infiltration in white matter	Galactocerebroside	Galactocerebroside β-galactosidase	Accumulation of the lipid focal only in the globoid cells
4. Metachromatic leukodystrophy				
4a. Classical form	Infantile, juvenile and adult forms; mental impairment, long-tract signs, metachromatic granules in CNS, PNS, and urine	Sulfatide	Arylsulfatase A	
4b. Multiple sulfatase deficiency	Additional gray matter involvement; facial and bony abnormalities similar to Hunter–Hurler syndrome; excess urinary mucopolysaccharide	Sulfatide	Arylsulfatases A,B, and C, steroid sulfatase, etc.	Additional storage of sulfated mucopolysaccharide and cholesterol sulfate

5. Lactosylceramide lipidosis	Slowly progressive CNS impairment, hepatosplenomegaly	Lactosylceramide	Lactosylceramide β-galactosidase	Only one patient known
6. Fabry's	Angiokeratoma around buttocks, renal damage; primarily adult and nonneurological disease	Ceramide–glc–gal–gal, ceramide–gal–gal	α-Galactosidase	
7. G_{M1}-gangliosidosis	Neurologically similar to Tay–Sachs; the infantile form shows additional organomegaly and bone abnormality similar to mucopolysaccharidoses; the older form shows little visceral involvement	G_{M1}-ganglioside	β-Galactosidase	Accumulation of keratan sulfate–like mucopolysaccharide in viscera
8. G_{M2}-gangliosidosis 8a. Tay–Sachs	Slow growth, motor weakness, hyperacusis, cherry-red spots, terminally enlarged head; death by 3 years of age	G_{M2}-ganglioside	Hexosaminidase A	Highly prevalent among Ashkenazi Jews
8b. Juvenile	Late onset and slow progression (2–15 years) but otherwise similar to Tay–Sachs	G_{M2}-ganglioside	Hexosaminidase A (partial deficiency)	Not to be confused with so-called juvenile amaurotic idiocy, which is not a sphingolipidosis
9. Sandhoff's	Identical to Tay–Sachs	G_{M2}-ganglioside, globoside (in viscera)	Hexosaminidases A and B	

other cases, because, even though increased to an abnormally high concentration, the actual amounts are expected to be extremely small. There was an increase of glucuronic acid content in the liver, accompanied by a great excess of fucose. This finding suggests that probably both mucopolysaccharides and glycoproteins were increased, with a large excess of terminal fucose.

Van Hoof and Hers[272] studied all three patients for lysosomal hydrolytic enzymes and found no activity of α-fucosidase in the brain, liver, kidney, lung, and pancreas, while many other glycosidic enzymes were hyperactive. In the third patient, α-fucosidase was also undetectable in urine, whereas it was present in normal urine.

Since fucose is primarily the constituent of glycoprotein in normal individuals, the disease should probably be classified within glycoprotein disorders. Quantitatively, accumulation of glycoprotein appears to be more prominent. Yet fucose-containing sphingolipids, including blood group substances, do accumulate in this condition, and the disease can also be regarded as a sphingolipidosis.[325]

3. So-Called Juvenile Lipidosis

There is a group of disorders commonly called *juvenile lipidosis*. Other synonyms include *late infantile, juvenile,* or *adult amaurotic idiocy,* with associated eponyms, such as *Bielschowsky, Spielmeyer–Vogt, Batten,* or *Kuf.* More recently, names such as *lipofuscin storage disease* or *ceroid storage disease* have been used. Because of the clinical and histopathological features which are somewhat similar to those of Tay–Sachs disease, these diseases have been traditionally treated as if they were late-occurring forms of Tay–Sachs disease, particularly in the hands of clinicians and traditional pathologists. This has been often a source of confusion, because despite the clinicopathological resemblance, there is no chemical evidence to indicate that these disorders are gangliosidoses or even sphingolipidoses. It is outside the scope of the chapter to discuss these diseases in any detail, and only a general reference is given here.[326] It is merely emphasized here that these so-called later amaurotic idiocies should not be confused with well-defined later forms of sphingolipidoses, such as juvenile G_{M2}-gangliosidosis or the type 2 of G_{M1}-gangliosidosis.

IV. SUMMARY AND CONCLUDING REMARKS

All of the major sphingolipids are structurally interrelated and can be depicted in a treelike pattern by connecting pairs of compounds which are the same except for the presence or absence of a particular terminal moiety (Fig. 2). The sphingolipid tree thus created then represents the degradative pathways of individual sphingolipids *in vivo*. At each step, degradation of a higher and more complex sphingolipid occurs by removal of the terminal moiety to a one-step-lower sphingolipid, all sphingolipids thus eventually converging to ceramide. All of the degradative steps, possibly with the exception of steps in which removal of *N*-acetylneuraminic acid is involved, are catalyzed by hydrolytic enzymes with acidic pH optima, which are localized within lysosomes. The nature and localization of neuraminidase have not been completely clarified, and at least some neuraminidase in certain organs does not appear to be lysosomal.

All of the well-defined human disorders of sphingolipid metabolism occur as the results of genetically determined enzymatic defects at one or more of these degradative steps of sphingolipids (Table 6). By now, sphingolipidoses are known for every single step of the entire degradative pathway, again with the exception of the steps involving neuraminidases. At the time of this writing, no disorder with a genetic deficiency of neuraminidase is known. Sphingolipidoses belong, as a group, to the "inborn lysosomal diseases," the concept originally proposed by Hers.[327] Thus, despite the seeming complexities in clinical, morphological, and analytical manifestations of these disorders, a fundamental, unifying view is possible for the entire group of sphingolipidoses (Fig. 2). From this viewpoint, with additional knowledge about the unique distribution of some sphingolipids within the body, many clinical, pathological, and analytical complexities of these disorders become much easier to comprehend.

The progress of our knowledge of sphingolipid disorders within the past 10 years has been indeed dramatic. However, we must also keep in mind that there still are numerous unanswered questions about some of the most fundamental aspects of genetic disorders. Some of these questions were discussed above in the sections on individual disorders. Only more sophisticated investigations in the future can provide the answers to these questions.

REFERENCES

1. W. Tay, Symmetrical changes in the region of the yellow spot in each eye of an infant, *Trans. Ophthalmol. Soc. United Kingdom* **1**:55–57 (1881).
2. B. Sachs, On arrested cerebral development, with special reference to its cortical pathology, *J. Nerv. Ment. Dis.* **14**:541–553 (1887).
3. C. P. E. Gaucher, De l'epithelioma primitif de la rate, Thèse de Paris (1882).
4. J. Fabry, Ein Beitrag zur Kenntnis der Purpura haemorrhagica nodularis (purpura papulosa hemorrhagica hebrae), *Arch. Dermatol. Syphilol.* **43**:187–200 (1898).
5. A. Niemann, Ein unbekanntes Krankheitsbild, *Jahrb. Kinderheilk.* **79**:1–10 (1914).
6. L. Pick, Über die lipoidzellige Splenohepatomegalie Typus Niemann–Pick als Stoffwechselerkrankung, *Med. Klin.* **23**:1483–1488 (1927).

7. K. Krabbe, A new familial, infantile form of diffuse brain sclerosis, *Brain* **39**:74–114 (1916).
8. W. Scholtz, Klinische, pathologisch-anatomische und erbbiologische Untersuchungen bei familiarer diffuser Hirnsklerose im Kindesalter, *Z. Ges. Neurol. Psychiat.* **99**:651–717 (1925).
9. R. M. Norman, H. Urich, A. H. Tingey, and R. A. Goodbody, Tay–Sachs disease with visceral involvement and its relationship to Niemann–Pick's disease, *J. Pathol. Bacteriol.* **78**:409–421 (1959).
10. B. H. Landing, F. N. Silverman, J. M. Craig, M. D. Jacoby, M. E. Lahey, and D. L. Chadwick, Familial neurovisceral lipidosis, *Am. J. Dis. Child.* **108**:503–522 (1964).
11. A. Aghion, La maladie de Gaucher dans l'enfance, Thèse de Paris (1934).
12. E. Klenk, Über die Natur der Phosphatide der Milz bei der Niemann-Pickschen Krankheit, *Z. Physiol. Chem.* **229**: 151–156 (1934).
13. Q. B. DeMarsh and J. Kautz, The submicroscopic morphology of Gaucher cells, *Blood* **12**:324–335 (1957).
14. R. D. Terry and S. R. Korey, Membranous cytoplasmic granules in infantile amaurotic idiocy, *Nature* **188**:1000–1002 (1960).
15. R. O. Brady, in "Lipid Storage Diseases: Enzymatic Defects and Clinical Implications" (J. Bernsohn and H. J. Grossman, eds.) pp. 275–289, Academic Press, New York (1971).
16. A. Lajtha (ed.), "Handbook of Neurochemistry," Vols. 1–7, Plenum Press, New York (1969–1971). (Appropriate chapters.)
17. G. Schettler (ed.), "Lipids and Lipidoses," Springer-Verlag, New York (1967).
18. J. Eichberg, G. Hauser, and M. L. Karnovsky, in "The Structure and Function of Nervous Tissue" (G. H. Bourne, ed.) Vol. 3, pp. 185–287, Academic Press, New York (1969).
19. R. O. Brady, in "Neurosciences Research" (S. Ehrenpreis and O. C. Solnitzky, eds.) Vol. 2, pp. 301–315, Academic Press, New York (1969).
20. A. N. Davison, in "Applied Neurochemistry" (A. N. Davison and J. Dobbing, eds.) pp. 178–221, F. A. Davis, Philadelphia (1968).
21. W. Stoffel, Sphingolipids, *Ann. Rev. Biochem.* **40**:57–82 (1971).
22. P. Morell and P. Braun, Sphingolipid metabolism: Biosynthesis and metabolic degradation of sphingolipids not containing sialic acid, *J. Lipid Res.* **13**:293–310 (1972).
23. R. Ledeen and R. Yu, Structure and enzymatic degradation of sphingolipids, in "Lysosomes and Lysosomal Diseases" (H. G. Hers and F. van Hoof, eds.) Academic Press, New York (in press).
24. R. O. Brady and G. J. Koval, The enzymic synthesis of sphingosine, *J. Biol. Chem.* **233**: 26–31 (1958).
25. M. Sribney, Enzymatic synthesis of ceramide, *Biochim. Biophys. Acta* **125**:542–547 (1966).
26. S. Gatt, Enzymatic hydrolysis of sphingolipids, I. Hydrolysis and synthesis of ceramides by an enzyme from rat brain, *J. Biol. Chem.* **241**:3724–3730 (1966).
27. M. Sribney and E. P. Kennedy, The enzymatic synthesis of sphingomyelin, *J. Biol. Chem.* **233**:1315–1322 (1958).
28. R. O. Brady, R. M. Bradley, O. M. Young, and H. Kaller, An alternative pathway for the enzymatic synthesis of sphingomyelin, *J. Biol. Chem.* **240**:PC 3693–3694 (1965).
29. W. W. Cleland and E. P. Kennedy, The enzymatic synthesis of psychosine, *J. Biol. Chem.* **235**:45–51 (1960).
30. R. O. Brady, Studies on the total enzymatic synthesis of cerebrosides, *J. Biol. Chem.* **237**: PC 2416–2417 (1962).
31. R. O. Brady, in "Metabolism and Physiological Significance of Lipids" (R. M. C. Dawson and D. N. Rhodes, eds.) pp. 95–109, John Wiley, London (1964).
32. P. Morell and N. S. Radin, Synthesis of cerebroside by brain from uridine diphosphate galactose and ceramide containing hydroxy fatty acid, *Biochemistry* **8**:506–512 (1969).
33. P. Morell, E. Costantino-Ceccarini, and N. S. Radin, The biosynthesis by brain microsomes of cerebrosides containing nonhydroxy fatty acids, *Arch. Biochem. Biophys.* **141**: 738–748 (1970).

34. S. Hammarström, On the biosynthesis of cerebrosides containing nonhydroxy acids. 1. Mass spectrometric evidence for the psychosine pathway, *Biochem. Biophys. Res. Commun.* **45**:459–467 (1971).

35. S. Hammarström, On the biosynthesis of cerebrosides containing nonhydroxy acids. 2. Mass spectrometric evidence for the ceramide pathway, *Biochem. Biophys. Res. Commun.* **45**:468–475 (1971).

36. J. B. Hay and G. M. Gray, Glycosphingolipid biosynthesis in kidneys of normal C3H/He mice and of those with BP8 ascites tumours, *Biochem. Biophys. Res. Commun.* **38**:527–532 (1970).

37. L. Coles and G. M. Gray, The biosynthesis of digalactosylceramide in the kidney of the C57/Bl mouse, *Biochem. Biophys. Res. Commun.* **38**:520–526 (1970).

38. S. Basu, B. Kaufman, and S. Roseman, Enzymatic synthesis of ceramide-glucose and ceramide lactose by glycosyltransferases from embryonic chicken brain, *J. Biol. Chem.* **243**:5802–5804 (1968).

39. G. Hauser, The enzymatic synthesis of ceramide lactoside from ceramide glucoside and UDP-galactose, *Biochem. Biophys. Res. Commun.* **28**:502–509 (1967).

40. J. Hildebrand and G. Hauser, Biosynthesis of lactosylceramide and triglycosylceramide by galactosyltransferases from rat spleen, *J. Biol. Chem.* **244**:5170–5179 (1969).

41. B. Kaufman, S. Basu, and S. Roseman, *in* "Inborn Disorders of Sphingolipid Metabolism" (S. M. Aronson and B. W. Volk, eds.) pp. 193–213, Pergamon Press, Oxford (1967).

42. S. Handa and R. M. Burton, Biosynthesis of glycolipids: Incorporation of *N*-acetyl galactosamine by a rat brain particulate preparation, *Lipids* **4**:589–598 (1969).

43. M. C. M. Yip and J. A. Dain, The enzymic synthesis of ganglioside: I. Brain uridine diphosphate D-galactose: *N*-acetyl-galactosaminyl-galactosyl-glucosyl-ceramide galactosyl transferase, *Lipids* **4**:270–277 (1969).

44. S. Gatt, *in* "Chemistry and Metabolism of Sphingolipids" (C. C. Sweeley, ed.) pp. 235–249, North Holland, Amsterdam (1970).

45. E. Yavin and S. Gatt, Enzymatic hydrolysis of sphingolipids. VIII. Further purification and properties of rat brain ceramidase, *Biochemistry* **8**:1692–1698 (1969).

46. M. Heller and B. Shapiro, Enzymic hydrolysis of sphingomyelin by rat liver, *Biochem. J.* **98**:763–769 (1966).

47. J. N. Kanfer, O. M. Young, D. Shapiro, and R. O. Brady, The metabolism of sphingomyelin. I. Purification and properties of a sphingomyelin-cleaving enzyme from rat liver tissue, *J. Biol. Chem.* **241**:1081–1084 (1966).

48. Y. Barnholz, A. Roitman, and S. Gatt, Enzymatic hydrolysis of sphingolipids. II. Hydrolysis of sphingomyelin by an enzyme from rat brain, *J. Biol. Chem.* **241**:3731–3737 (1966).

49. P. B. Schneider and E. P. Kennedy, Sphingomyelinase in normal human spleens and in spleens from subjects with Niemann–Pick disease, *J. Lipid Res.* **8**:202–209 (1967).

50. N. J. Weinreb, R. O. Brady, and A. L. Tappel, The lysosomal localization of sphingolipid hydrolases, *Biochim. Biophys. Acta* **159**:141–146 (1968).

51. E. Mehl and H. Jatzkewitz, Eine Cerebrosidsulfatase aus Schweineniere, *Hoppe-Seylers Z. Physiol. Chem.* **339**:260–276 (1964).

52. E. Mehl and H. Jatzkewitz, Cerebroside 3-sulfate as a physiological substrate of arylsulfatase A, *Biochim. Biophys. Acta* **151**:619–627 (1968).

53. A. A. Farooqui and B. K. Bachhawat, The regional distribution, age dependent variation and species differences of brain arylsulfatases, *J. Neurochem.* **18**:635–646 (1971).

54. A. K. Hajra, D. M. Bowen, Y. Kishimoto, and N. S. Radin, Cerebroside galactosidase of brain, *J. Lipid Res.* **7**:379–386 (1966).

55. D. M. Bowen and N. S. Radin, Purification of cerebroside galactosidase from rat brain, *Biochim. Biophys. Acta* **152**:587–598 (1968).

56. D. M. Bowen and N. S. Radin, Properties of cerebroside galactosidase, *Biochim. Biophys. Acta* **152**:599–610 (1968).

57. D. M. Bowen and N. S. Radin, Cerebroside galactosidase: A method for determination and a comparison with other lysosomal enzymes in developing rat brain, *J. Neurochem.* **16:**501–511 (1969).

58. L. Svennerholm, Chromatographic separation of human brain gangliosides, *J. Neurochem.* **10:**613–623 (1963).

59. H. Wiegandt, Ganglioside, *Ergeb. Physiol. Biol. Chem. Exptl. Pharmakol.* **57:**190–222 (1966).

60. R. Öhman, A. Rosenberg, and L. Svennerholm, Human brain sialidase, *Biochemistry* **9:** 3774–3782 (1970).

61. C. L. Schengrund and A. Rosenberg, Intracellular location and properties of bovine brain sialidase, *J. Biol. Chem.* **245:**6196–6200 (1970).

62. R. Öhman, Subcellular fractionation of ganglioside sialidase from human brain, *J. Neurochem.* **18:**89–95 (1971).

63. S. Mahadevan, J. C. Nduaguba, and A. L. Tappel, Sialidase of rat liver and kidney, *J. Biol. Chem.* **242:**4409–4413 (1967).

64. S. Gatt and M. M. Rapport, Isolation of β-galactosidase and β-glucosidase from brain, *Biochim. Biophys. Acta* **113:**567–576 (1966).

65. S. Gatt, Enzymatic hydrolysis of sphingolipids. V. Hydrolysis of monosialoganglioside and hexosylceramides by rat brain β-galactosidase, *Biochim. Biophys. Acta* **137:**192–195 (1967).

66. Y. Z. Frohwein and S. Gatt, Enzymatic hydrolysis of sphingolipids. VI. Hydrolysis of ceramide glycosides by calf brain β-N-acetylhexosaminidase, *Biochemistry* **6:**2783–2787 (1967).

67. K. Sandhoff and W. Wässle, Anreicherung und Charakterisierung zweier Formen der menschlichen N-Acetyl-β-D-hexosaminidase, *Z. Physiol. Chem.* **352:**1119–1133 (1971).

68. K. Sandhoff and H. Jatzkewitz, A particle-bound sialyl lactosidoceramide splitting mammalian sialidase, *Biochim. Biophys. Acta* **141:**442–444 (1967).

69. Z. Leibowitz and S. Gatt, Enzymatic hydrolysis of sphingolipids. VII. Hydrolysis of gangliosides by a neuraminidase from calf brain, *Biochim. Biophys. Acta* **152:**136–143 (1968).

70. E. H. Kolodny, J. Kanfer, J. M. Quirk, and R. O. Brady, Properties of a particle-bound enzyme from rat intestine that cleaves sialic acid from Tay–Sachs ganglioside, *J. Biol. Chem.* **246:**1426–1431 (1971).

71. K. Sandhoff, H. Pilz, and H. Jatzkewitz, Über den enzymatischen Abbau von N-acetyl-neuraminsäurefreien Gangliosidresten (Ceramidoligosacchariden), *Z. Physiol. Chem.* **338:** 281–293 (1964).

72. R. O. Brady, R. M. Bradley, and E. Martensson, The metabolism of ceramide trihexosides. I. Purification and properties of an enzyme that cleaves the terminal galactose molecule of galactosylgalactosylglucosylceramide, *J. Biol. Chem.* **242:**1021–1026 (1967).

73. S. Gatt and M. M. Rapport, Enzymic hydrolysis of sphingolipids. Hydrolysis of ceramide lactoside by an enzyme from rat brain, *Biochem. J.* **101:**680–686 (1966).

74. N. S. Radin, L. Hof, R. M. Bradley, and R. O. Brady, Lactosylceramide galactosidase: Comparison with other sphingolipid hydrolases in developing rat brain, *Brain Res.* **14:** 497–505 (1969).

75. R. O. Brady, J. Kanfer, and D. Shapiro, The metabolism of glucocerebroside. I. Purification and properties of glucocerebroside-cleaving enzyme from spleen tissue, *J. Biol. Chem.* **240:**39–43 (1965).

76. S. Gatt, Enzymatic hydrolysis of sphingolipids. Hydrolysis of ceramide glucoside by an enzyme from ox brain, *Biochem. J.* **101:**687–691 (1966).

77. E. G. Lapetina, E. F. Soto, and E. deRobertis, Gangliosides and acetylcholinesterase in isolated membranes of the rat brain cortex, *Biochim. Biophys. Acta* **135:**33–43 (1967).

78. H. Wiegandt, The subcellular localization of gangliosides in the brain, *J. Neurochem.* **14:** 671–674 (1967).

79. L. Svennerholm, The distribution of lipids in the human nervous system—1. Analytical procedure, lipids of foetal and newborn brain, *J. Neurochem.* **11**:839–853 (1964).

80. T. Yamakawa and S. Suzuki, The chemistry of the lipids of posthemolytic residue or stroma of erythrocytes. III. Globoside, the sugar containing lipid of human blood stroma, *J. Biochem.* **39**:393–399 (1952).

81. A. C. Crocker, The cerebral defect in Tay–Sachs disease and Niemann–Pick disease, *J. Neurochem.* **7**:69–80 (1961).

82. A. C. Crocker and S. Farber, Niemann–Pick disease: A review of eighteen patients, *Medicine* **37**:1–95 (1958).

83. R. Lynn and R. D. Terry, Lipid histochemistry and electron microscopy in adult Niemann–Pick disease, *Am. J. Med.* **37**:987–994 (1964).

84. Y. Tanaka, G. Brecher, and D. S. Frederickson, Cellules de la maladie de Niemann–Pick et de quelques autres lipoidoses, *Nouv. Rev. Franc. Hematol.* **3**:5–12 (1963).

85. B. J. Wallace, L. Schneck, H. Kaplan, and B. W. Volk, Fine structure of cerebellum of children with lipidoses, *Arch. Pathol.* **80**:466–486 (1965).

86. S. Luse, *in* "Inborn Disorders of Sphingolipid Metabolism" (S. M. Aronson and B. W. Volk, eds.) pp. 93–105, Pergamon Press, Oxford (1966).

87. E. Klenk, Über die Natur der Phosphatide und anderer Lipoide des Gehirns und der Leber bei der Niemann-Pickschen Krankheit, *Z. Physiol. Chem.* **235**:24–36 (1937).

88. C. Tropp and B. Eckardt, Gehirn-Sphingomyelin bei Niemann-Pickscher Krankheit, *Z. Physiol. Chem.* **245**:163–174 (1936).

89. P. H. Teunissen and A. den Ouden, Nachtrag der Mitteilung: Beitrag zur Kenntnis der Chemie der Lipoidosis phosphatidica, *Z. Physiol. Chem.* **252**:271–279 (1938).

90. E. Chargaff, A study of the spleen in a case of Niemann–Pick disease, *J. Biol. Chem.* **130**:503–511 (1939).

91. J. N. Cumings, *in* "Cerebral Sphingolipidoses" (S. M. Aronson and B. W. Volk, eds.) pp. 171–178, Academic Press, New York (1962).

92. L. van Bogaert, F. Seitelberger, and G. W. F. Edgar, Études neuropathologiques et neurochimiques sur un cas de Niemann–Pick chez un jeune enfant, *Acta Neuropathol.* **3**:57–73 (1963).

93. S. Kamoshita, A. M. Aron, K. Suzuki, and K. Suzuki, Infantile Niemann–Pick disease: A chemical study with isolation and characterization of membranous cytoplasmic bodies and myelin, *Am. J. Dis. Child.* **117**:379–394 (1969).

94. M. Philippart L. Martin, J. J. Martin, and J. H. Menkes, Niemann–Pick disease: Morphologic and biochemical studies in the visceral form with late central nervous system involvement (Crocker's group C), *Arch. Neurol.* **20**:227–238 (1969).

95. H. Sobotka, E. Epstein, and L. Lichtenstein, The distribution of lipoid in a case of Niemann–Pick disease associated with amaurotic idiocy, *Arch. Pathol.* **10**:677–686 (1930).

96. H. Sobotka, D. Glick, M. Reiner, and L. R. Tuchman, The lipoids of spleen and liver in various types of lipoidoses, *Biochem. J.* **27**:2031–2034 (1933).

97. L. L. Uzman, The significance of the increase of nonspecific lipid components in primary lipoid-storage diseases, *Arch. Pathol.* **65**:331–339 (1958).

98. B. I. Ivemark, L. Svennerholm, C. Thorén, and R. Tunell, Niemann–Pick disease in infancy. Report of two siblings with clinical, histologic and chemical studies, *Acta Paediat.* **52**:391–404 (1963).

99. D. S. Frederickson, *in* "The Metabolic Basis of Inherited Disease," 2nd ed., (J. B. Stanbury, J. B. Wyngaarden, and D. S. Frederickson, eds.) pp. 586–617, McGraw-Hill, New York (1966).

100. R. M. Norman, R. M. Forrester, and A. H. Tingey, The juvenile form of Niemann–Pick disease, *Arch. Dis. Childh.* **42**:91–96. (1967).

101. G. Rouser, G. Kritchevsky, A. Yamamoto, A. G. Knudson, Jr., and G. Simon, Accumulation of a glycerophospholipid in classical Niemann–Pick disease, *Lipids* **3**:287–290 (1968).

102. P. N. Seng, H. Debuch, B. Witter, and H.-R. Wiedemann, Bis (monoacylglycerin) phos-phosäure-Vermehrung bei Sphingomyelinose (M. Niemann-Pick?), *Z. Physiol. Chem.* **352**:280–288 (1971).

103. C. W. Seiter and R. H. McCluer, Analysis of the structure of two gangliosides which ac-cumulate in the brain in Niemann–Pick disease, *J. Neurochem.* **17**:1525–1526 (1970).

104. A. C. Crocker and V. B. Mays, Sphingomyelin synthesis in Niemann–Pick disease, *Am. J. Clin. Nutr.* **9**:63–67 (1961).

105. R. O. Brady, J. N. Kanfer, M. B. Nock, and D. S. Frederickson, The metabolism of sphingo-myelin, II. Evidence of an enzymatic deficiency in Niemann–Pick disease, *Proc. Natl. Acad. Sci.* **55**:366–369 (1966).

106. P. B. Schneider and E. P. Kennedy, Sphingomyelinase in normal human spleens and in spleens from subjects with Niemann–Pick disease, *J. Lipid Res.* **8**:202–209 (1967).

107. J. P. Kampine, R. O. Brady, J. N. Kanfer, M. Feld, and D. Shapiro, Diagnosis of Gau-cher's disease and Niemann–Pick disease with small samples of venous blood, *Science* **155**: 86–88 (1967).

108. H. R. Sloan, B. W. Uhlendorf, J. N. Kanfer, R. O. Brady, and D. S. Frederickson, Defi-ciency of sphingomyelin-cleaving enzyme activity in tissue cultures derived from patients with Niemann–Pick disease, *Biochem. Biophys. Res. Commun.* **34**:582–588 (1969).

109. R. A. Snyder and R. O. Brady, The use of white cells as a source of diagnostic material for lipid storage diseases, *Clin. Chim. Acta* **25**:331–338 (1969).

110. C. J. Epstein, R. O. Brady, E. L. Schneider, R. M. Bradley, and D. Shapiro, *In utero* diagnosis of Niemann–Pick disease, *Am. J. Hum. Genet.* **23**: 533–535 (1971).

111. J. W. Callahan and M. Philippart, Phosphodiesterases (including sphingomyelinase) in Niemann–Pick disease types A and C, *Neurology* **21**:442 (1971).

112. D. S. Frederickson and H. R. Sloan, in "The Metabolic Basis of Inherited Disease," 3rd ed., (J. B. Stanbury, J. B. Wyngaarden, and D. S. Frederickson, eds.) pp. 730–759. McGraw-Hill, New York (1972).

113. R. G. Hibbs, V. J. Ferrans, P. R. Cipriano, and K. J. Tardiff, A histochemical and electron microscopic study of Gaucher cells, *Arch. Pathol.* **89**:137–153 (1970).

114. R. E. Lee, The fine structure of the cerebroside occurring in Gaucher's disease, *Proc. Natl. Acad. Sci.* **61**:484–489 (1968).

115. E. R. Fisher and R. Reidbord, Gaucher's disease: Pathogenetic considerations based on electron microscopic and histochemical observations, *Am. J. Pathol.* **41**:679–693 (1962).

116. B. W. Volk and B. J. Wallace, The liver in lipidoses; an electron microscopic and histo-chemical study, *Am. J. Pathol.* **49**:203–225 (1966).

117. S. W. Jordan, Electron microscopy of Gaucher cells, *Exptl. Molec. Pathol.* **3**:76–85 (1964).

118. B. Q. Banker, J. Q. Miller, and A. C. Crocker, in "Cerebral Sphingolipidoses" (S. M. Aronson and B. W. Volk, eds.) pp. 73–99. Academic Press, New York (1962).

119. R. M. Norman, H. Urich, and O. C. Lloyd, The neuropathology of infantile Gaucher's disease, *J. Pathol. Bacteriol.* **72**:121–131 (1956).

120. M. Adachi, B. J. Wallace, L. Schneck, and B. W. Volk, Fine structure of central nervous system in early infantile Gaucher's disease, *Arch. Pathol.* **83**:513–526 (1967).

121. K. Wakutani, H. Nakamura, H. Mori, H. Morihisa, and G. Ando, A case of infantile Gaucher's disease—Neuropathologic and electron microscopic observation, *Clin. Neurol.* **9**:261–270 (1969).

122. L. Svennerholm, in "Inborn Disorders of Sphingolipid Metabolism" (S. M. Aronson and B. W. Volk, eds.) pp. 169–186, Pergamon Press, Oxford (1967).

123. A. F. J. Maloney and J. N. Cumings, A case of juvenile Gaucher's disease with intra-neuronal lipid storage, *J. Neurol. Neurosurg. Psychiat.* **23**:207–213 (1960).

124. J. Montreuil, P. Bowanger, and E. Houcke, Chromatographie sur papier des constituants glucidiques des cérébrosides d'une rate de Gaucher, *Bull. Soc. Chim. Biol.* **35**:1125–1127 (1953).

125. L. Svennerholm, in "Brain Lipids and Lipoproteins, and the Leukodystrophies" (J. Folch-Pi and H. Bauer, eds.) pp. 104–119, Elsevier, Amsterdam (1963).

126. M Philippart and J. H. Menkes, Isolation and characterization of the principal cerebral glycolipids in the infantile and adult forms of Gaucher's disease, *J. Neuropathol. Exptl. Neurol.* **24**:389–400 (1967).

127. J. H. French, M. Brotz, and C. M. Poser, Lipid composition of the brain in infantile Gaucher's disease, *Neurology* **19**:81–86 (1969).

128. M. Philippart, B. Rosenstein, and J. H. Menkes, Isolation and characterization of the main splenic glycolipids in the normal organ and in Gaucher's disease: Evidence for the site of metabolic block, *J. Neuropathol. Exptl. Neurol.* **24**:290–303 (1965).

129. A Makita, C. Suzuki, and Z. Yosizawa, Glycol pids isolated from the spleen of Gaucher's disease, *Tohoku J. Exptl. Med.* **88**:277–288 (1966).

130. E. G. Trams and R. O. Brady, Cerebroside synthesis in Gaucher's disease, *J. Clin. Invest.* **39**:1546–1550 (1960).

131. R. O. Brady, J. N. Kanfer, and D. Shapiro, Metabolism of glucocerebrosides. II. Evidence of an enzymatic deficiency in Gaucher's disease, *Biochem. Biophys. Res. Commun.* **18**: 221–225 (1965).

132. R. O. Brady, J. N. Kanfer, R. M. Bradley, and D. Shapiro, Demonstration of a deficiency of glucocerebroside-cleaving enzyme in Gaucher's disease, *J. Clin. Invest.* **45**:1112–1115 (1966).

133. A. D. Patrick, A deficiency of glucocerebrosidase in Gaucher's disease, *Biochem. J.* **97**: 17C–18C (1965).

134. E. Beutler and W. Kühl, The diagnosis of the adult type of Gaucher's disease and its carrier state by demonstration of deficiency of β-glucosidase activity in peripheral blood leukocytes, *J. Lab. Clin. Med.* **76**:747–755 (1970).

135. R. O. Brady, Cerebral lipidoses, *Ann. Rev. Med.* **21**:317–334 (1970).

136. E. Beutler, W. Kühl, F. Trinidad, R. Teplitz, and H. Nadler, β-Glucosidase activity in fibroblasts from homozygotes and heterozygotes for Gaucher's disease, *Am. J. Hum. Genet.* **23**:62–66 (1971).

137. C. J. Epstein and R. O. Brady, personal communication.

138. P. A. Öckerman and P. Köhlin, Tissue acid hydrolase activities in Gaucher's disease, *Scand. J. Clin. Lab. Invest.* **22**:62–64 (1968).

139. B. Hagberg, H. Kolberg, P. Sourander, and H. O. Akesson, Infantile globoid cell leukodystrophy (Krabbe's disease). A clinical and genetic study of 32 Swedish cases 1953–1967, *Neuropädiatrie* **1**:74–88 (1969).

140. H. G. Dunn, B. D. Lake, C. L. Dolman, and J. Wilson, The neuropathy of Krabbe's infantile cerebral sclerosis (globoid cell leucodystrophy), *Brain* **92**:329–344 (1969).

141. B. J. Wallace, S. W. Aronson, and B. W. Volk, Histochemical and biochemical studies of globoid cell leukodystrophy (Krabbe's disease), *J. Neurochem.* **11**:367–376 (1963).

142. N. Allen and E. de Veyra, Microchemical and histochemical observations in a case of Krabbe's leukodystrophy, *J. Neuropathol. Exptl. Neurol.* **26**:456–474 (1967).

143. K. Suzuki and Y. Suzuki, in "The Metabolic Basis of Inherited Disease," 3rd ed., (J. B. Stanbury, J. B. Wyngaarden, and D. S. Frederickson, eds.) pp. 760–782. McGraw-Hill, New York (1972).

144. K. Suzuki and W. D. Grover, Krabbe's leukodystrophy (globoid cell leukodystrophy): An ultrastructural study, *Arch. Neurol.* **22**:385–396, (1970).

145. E. J. Yunis and R. E. Lee, The ultrastructure of globoid (Krabbe) leukodystrophy, *Lab. Invest.* **21**:415–419 (1969).

146. E. J. Yunis and R. E. Lee, Tubules of globoid leukodystrophy: A right-handed helix, *Science* **169**:64–66 (1970).

147. A. Bischoff and J. Ulrich, Peripheral neuropathy in globoid cell leukodystrophy (Krabbe's disease). Ultrastructural and histochemical findings, *Brain* **92**:861–870 (1969).

148. G. Lyon, L. Jardin, and J. Aicardi, Étude au microscope electronique d'un nerf périphérique dans un cas de leucodystrophie de Krabbe, *J. Neurol. Sci.* **12**:263–274 (1971).

149. J. H. Austin, Studies in globoid (Krabbe) leukodystrophy. I. The significance of lipid abnormalities in white matter in eight globoid and thirteen control patients, *Arch. Neurol.* **9**:207–221 (1963).

150. K. Suzuki and K. Suzuki, *in* "Lysosomes and Storage Diseases" (H. G. Hers and F. van Hoof, eds.) Academic Press, New York (in press).

151. Y. Eto and K. Suzuki, Brain sphingoglycolipids in Krabbe's globoid cell leukodystrophy, *J. Neurochem.* **18**:503–511 (1971).

152. J. M. Schibanoff, S. Kamoshita, and J. S. O'Brien, Tissue distribution of glycosphingolipids in a case of Fabry's disease, *J. Lipid Res.* **10**:515–520 (1969).

153. Y. Eto, K. Suzuki, and K. Suzuki, Globoid cell leukodystrophy (Krabbe's disease): Isolation of myelin with normal glycolipid composition, *J. Lipid Res.* **11**:473–479 (1970).

154. K. Suzuki and Y. Suzuki, Globoid cell leukodystrophy (Krabbe's disease): Deficiency of galactocerebroside β-galactosidase, *Proc. Natl. Acad. Sci.* **66**:302–309 (1970).

155. K. Suzuki, Y. Suzuki, and Y. Eto, *in* "Lipid Storage Diseases: Enzymatic Defects and Clinical Implications" (J. Bernsohn and H. J. Grossman, eds.) pp. 111–136, Academic Press, New York (1971).

156. J. Austin, K. Suzuki, D. Armstrong, R. O. Brady, B. K. Bachhawat, J. Schlenker, and D. Stumpf, Studies in globoid (Krabbe) leukodystrophy (GLD). V. Controlled enzymatic studies in ten human eases, *Arch. Neurol.* **23**:502–512 (1970).

157. Y. Suzuki and K. Suzuki, Krabbe's globoid cell leukodystrophy: Deficiency of galactocerebrosidase in serum, leukocytes and fibroblasts, *Science* **171**:73–75 (1971).

158. M. J. Malone, Deficiency in a degradative enzyme system in globoid leucodystrophy, *Abst. First Meeting Am. Soc. Neurochem., Albuquerque, N.M.*, p. 56 (1970).

159. K. Suzuki, E. Schneider, and C. J. Epstein, In utero diagnosis of globoid cell leukodystrophy (Krabbe's disease), *Biochem. Biophys. Res. Commun.* **45**:1363–1366 (1971).

160. K. Suzuki, Y. Suzuki, and T. Fletcher, Further studies of galactocerebroside β-galactosidase in globoid cell leukodystrophy, *in* "Proceedings of the Fourth International Symposium on Sphingolipids, Sphingolipidoses, and Allied Disorders" (B. W. Volk and S. M. Aronson, eds.) pp. 487–498. Plenum Press, New York (1972).

161. B. K. Bachhawat, J. Austin, and D. Armstrong, A cerebroside sulfotransferase deficiency in a human disorder of myelin, *Biochem. J.* **104**:15C–17C (1967).

162. J. H. Austin and D. Lehfeldt, Studies in globoid (Krabbe) leukodystrophy. III. Significance of experimentally-produced globoid-like elements in rat white matter and spleen, *J. Neuropathol. Exptl. Neurol.* **24**:265–289 (1965).

163. K. Suzuki, Ultrastructural study of experimental globoid cells, *Lab. Invest.* **23**:612–619 (1970).

164. K. Suzuki, Renal cerebroside in globoid cell leukodystrophy (Krabbe's disease), *Lipids* **6**: 433–436 (1971).

165. B. Hagberg, *in* "Brain Lipids and Lipoproteins, and the Leukodystrophies" (J. Folch-Pi and H. Bauer, eds.) pp. 134–146, Elsevier, Amsterdam (1963).

166. T. von Hirsch and J. Peiffer, Über histologische Methoden in der Differentialdiagnose von Leucodystrophien und Lipidosen, *Arch. Psychiat. Nervenkr.* **194**:88–104 (1955).

167. R. D. Terry, *in* "Lipid Storage Diseases: Enzymatic Defects and Clinical Implications" (J. Bernsohn and H. J. Grossman, eds.) pp. 3–25, Academic Press, New York (1971).

168. G. Aurebeck, K. Osterberg, M. Blaw, S. Chou, and E. Nelson, Electron microscopic observations on metachromatic leucodystrophy, *Arch. Neurol.* **11**:273–288 (1964).

169. A. Grégoire, O. Périer, and P. Dustin, Jr., Metachromatic leucodystrophy, an electron microscopic study, *J. Neuropathol. Exptl. Neurol.* **25**:617–636 (1966).

170. A. Résibois-Grégoire, Electron microscopic studies of metachromatic leucodystrophy. II. Compound nature of the inclusions, *Acta Neuropathol.* **9**:244–253 (1967).

171. H. de Webster, Schwann cell alterations in metachromatic leucodystrophy. Preliminary phase and electron microscopic observations, *J. Neuropathol. Exptl. Neurol.* **21**:534–554 (1962).

172. A. Résibois, Electron microscopic study of metachromatic leucodystrophy. III. Lysosomal nature of the inclusions, *Acta Neuropathol.* **13**:149–156 (1969).

173. K. Suzuki and K. Suzuki, *in* "Handbook of Neurochemistry" (A. Lajtha, ed.) Vol. 7, pp. 131–142. Plenum Press, New York (1972).

174. H. Jatzkewitz, Zwei Typen von Cerebrosid-schwefelsäureestern als sog. "Prälipoide" und Speichersubstanzen bei der Leukodystrophie, Typ Scholtz (metachromatische Form der diffusen Sklerose), *Z. Physiol. Chem.* **311**:279–282 (1958).

175. J. H. Austin, Metachromatic sulfatides in cerebral white matter and kidney, *Proc. Soc. Exptl. Biol.* **100**:361–364 (1959).

176. M. Malone, P. Stoffyn, and H. Moser, Structural studies on sulfatide in metachromatic leucodystrophy, *J. Neurochem.* **13**:1033–1036 (1966).

177. B. Hagberg, P. Sourander, L. Svennerholm, and H. Voss, Late infantile metachromatic leucodystrophy of the genetic type, *Acta Paediat.* **49**:135–153 (1960).

178. B. Hagberg, P. Sourander, and L. Svennerholm, Sulfatide lipidosis in childhood. Report of a case investigated during life and at autopsy, *Am. J. Dis. Child.* **104**:644–656 (1962).

179. H. Jatzkewitz, H. Pilz, and H. Holländer, Biochemische und vergleichende histochemische Untersuchungen in umschriebenen Gebieten des Gehirns bei Fallen von adulter und infantiler metachromatischer Leukodystrophie, *Acta Neuropathol.* **4**:75–89 (1964).

180. J. S. O'Brien and E. L. Sampson, Myelin membrane: A molecular abnormality, *Science* **150**:1613 (1965).

181. W. T. Norton, The variation in the chemical composition in myelin in disease and during development, *Charing Cross Hosp. Gazette* **8**:3–8 (1967).

182. K. Suzuki, K. Suzuki, and G. C. Chen, Metachromatic leucodystrophy: Isolation and chemical analysis of metachromatic granules, *Science* **151**:1231–1233 (1966).

183. K. Suzuki, K. Suzuki, and G. C. Chen, Isolation and chemical characterization of metachromatic granules from a brain with metachromatic leukodystrophy, *J. Neuropathol. Exptl. Neurol.* **26**:537–550 (1967).

184. J. Austin, D. Armstrong, S. Fouch, C. Mitchell, D. Stumpf, L. Shearer, and O. Briner, Metachromatic leukodystrophy. VIII. MLD in adults; diagnosis and pathogenesis, *Arch. Neurol.* **18**:225–240 (1968).

185. M. J. Malone and P. Stoffyn, Peripheral nerve glycolipids in metachromatic leukodystrophy, *Neurology* **17**:1033–1040 (1967).

186. E. Martensson, A. Percy, and L. Svennerholm, Kidney glycolipids in late infantile metachromatic leukodystrophy, *Acta Paediat. Scand.* **55**:1–9 (1966).

187. J. H. Austin, A. S. Balasubramanian, T. N. Pattabiraman, S. Saraswathi, D. K. Basu, and B. K. Bachhawat, A controlled study of enzymic activities in three human disorders of glycolipid metabolism, *J. Neurochem.* **10**:805–816 (1963).

188. J. Austin, D. Armstrong, and L. Shearer, Metachromatic form of diffuse cerebral sclerosis. V. The nature and significance of low sulfatase activity: A controlled study of brain, liver and kidney in four patients with metachromatic leukodystrophy (MLD), *Arch. Neurol.* **13**:593–614 (1965).

189. E. Mehl and H. Jatzkewitz, Evidence for the genetic block in metachromatic leucodystrophy (ML), *Biochem. Biophys. Res. Commun.* **19**:407–411 (1965).

190. J. Austin, D. McAfee, and L. Shearer, Metachromatic form of diffuse cerebral sclerosis. IV. Low sulfatase activity in the urine of nine living patients with metachromatic leukodystrophy (MLD), *Arch. Neurol.* **12**:447–455 (1965).

191. A. L. Percy and R. O. Brady, Metachromatic leukodystrophy: Diagnosis with samples of venous blood, *Science* **161**:594–595 (1968).

192. M. T. Porter, A. L. Fluharty, and H. Kihara, Metachromatic leukodystrophy: Arylsulfatase A deficiency in skin fibroblast cultures, *Proc. Natl. Acad. Sci.* **62**:887–891 (1969).

193. J. G. Leroy, J. Dumon, and J. Radermecker, Deficiency of arylsulfatase A in leucocytes and skin fibroblasts in juvenile metachromatic leucodystrophy, *Nature* **226**:553–554 (1970).

194. N. H. Bass, E. J. Witmer, and F. E. Dreifuss, A pedigree study of metachromatic leukodystrophy. Biochemical identification of the carrier state, *Neurology* **20**:52–62 (1970).

195. N. Taniguchi and I. Namba, Enzymatic abnormality of the carrier state in metachromatic leukodystrophy, *Clin. Chim. Acta* **29**:375–379 (1970).

196. F. Gabreëls, K. Lamers, J. Kok, M. Loonen, and E. Lommen, The biochemical differentiation between heterozygote carriers of metachromatic leucodystrophy and normal persons, *Neuropädiatrie* 2:461–469 (1971).

197. D. Stumpf and J. Austin, Metachromatic leukodystrophy (MLD). IX. Qualitative and quantitative differences in urinary arylsulfatase A in different forms of MLD, *Arch. Neurol.* 24:117–124 (1971).

198. M. Mossakowski, G. Mathieson, and J. N. Cumings, On the relationships of metachromatic leucodystrophy and amaurotic idiocy, *Brain* 84:585–604 (1961).

199. J. H. Austin, *in* "Medical Aspects of Mental Retardation" (C. H. Carter, ed.) pp. 768–812, Charles C. Thomas, Springfield, Ill. (1965).

200. M. Bischel, J. Austin, and M. Kemeny, Metachromatic leukodystrophy (MLD). VII. Elevated sulfated acid polysaccharide levels in urine and postmortem tissues, *Arch. Neurol.* 15:13–28 (1966).

201. F. Lüthy, J. Ulrich, F. Regli, and W. Isler, Amaurotic idiocy with metachromatic change in the white matter, *Proc. Fifth Internat. Congr. Neuropathol.*, pp. 125–130, Excerpta Medica Foundation, Amsterdam (1966).

202. S. Thieffry, G. Lyon, and P. Maroteaux, Leucodystrophie metachromatique (sulfatidose) et mucopolysaccharidose associées chez un même malade, *Rev. Neurol.* 114:193–200 (1966).

203. S. Thieffry, G. Lyon, and P. Maroteaux, Encéphalopathie métabolique associant une mucopolysaccharidose et une sulfatidose, *Arch. Franç. Pédiat.* 24:425–432 (1967).

204. S. Rampini, W. Isler, K. Baerlocher, A. Bischoff, J. Ulrich, and H. J. Plüss, Die Kombination von metachromatischer Leukodystrophie und Mukopolysaccharidose als selbständiges Krankheitzbild (Mukosulfatidose), *Helv. Paediat. Acta* 25:436–461 (1970).

205. J. V. Murphy, H. J. Wolfe, E. A. Balazs, and H. W. Moser, *in* "Lipid Storage Diseases: Enzymatic Defects and Clinical Implications" (J. Bernsohn and H. J. Grossman, eds.) pp. 67–110, Academic Press, New York (1971).

206. K. Suzuki, *in* "Inborn Disorders of Sphingolipid Metabolism" (S. M. Aronson and B. W. Volk, eds.) pp. 215–230, Pergamon Press, Oxford (1967).

207. K. Suzuki, *in* "Handbook of Neurochemistry" (A. Lajtha, ed.) pp. 17–32, Vol. 7, Plenum Press, New York (1972).

208. G. Dawson and A. O. Stein, Lactosyl ceramidosis: Catabolic enzyme defect of glycosphingolipid metabolism, *Science* 170:556–558 (1970).

209. G. Dawson and A. O. Stein, *in* "Lipid Storage Diseases: Enzymatic Defects and Clinical Implications" (J. Bernsohn and H. J. Grossman, eds.) pp. 183–201, Academic Press, New York (1971).

210. G. Dawson, R. Matalon, and A. O. Stein, Lactosylceramidosis: Lactosylceramide galactosyl hydrolase deficiency and accumulation of lactosylceramide in cultured skin fibroblasts, *J. Pediat.* 79:423–429 (1971).

211. Z. Hruban, J. R. Esterly, G. Dawson, and A. O. Stein, Ultrastructure of liver in lactosyl ceramidosis, *Proc. Twenty-ninth Ann. Meeting Electron Microscop. Soc. Am.* (C. J. Arceneaux, ed.)

212. W. Anderson, A case of "angiokeratoma," *Brit. J. Dermatol.* 10:113–117 (1898).

213. J. M. Opitz, F. C. Stiles, D. Wise, R. R. Race, R. Sanger, G. R. von Gemmingen, R. R. Kierland, E. G. Gross, and W. P. de Groot, The genetics of angiokeratoma corporis diffusum (Fabry's disease) and its linkage relations with the *Xg* locus, *Am. J. Hum. Genet.* 17:325–342 (1965).

214. W. Kahlke, *in* "Lipids and Lipidoses" (G. Schettler, ed.) pp. 332–351, Springer-Verlag, Berlin (1967).

215. B. Duperrat and G. Pluvinage, Angiokératose diffuse de Fabry avec hémiplégie, *Bull. Soc. Méd. Hôp.* 72:748–751 (1956).

216. D. Wise, H. J. Wallace, and E. H. Jellinek, Angiokeratoma corporis diffuum, *Quart. J. Med.* 31:177–206 (1962).

217. K. Scriba, Zur Pathogenese des Angiokeratoma corporis diffusum Fabry mit cardio-vasorenalem Symptomcomplex, *Verh. Deutsch. Ges. Pathol.* 34:221–226 (1950).

218. P. Fessas, M. M. Wintrobe, and G. E. Cartwright, Angiokeratoma corporis diffusum universale (Fabry), first American report of a rare disorder, *Arch. Int. Med.* **95**:469–481 (1955).

219. A. W. Pompen, M. M. Ruiter, and H. J. G. Wyers, Angiokeratoma corporis diffusum (universale) Fabry, as a sign of an unknown internal disease; two autopsy reports, *Acta Med. Scand.* **128**:234–255 (1947).

220. A. N. Rahman and R. Lindenberg, The neuropathology of hereditary distopic lipidosis, *Arch. Neurol.* **9**:373–385 (1963).

221. H. O. Christensen Lou and E. Reske-Nielsen, The central nervous system in Fabry's disease, *Arch. Neurol.* **25**:351–359 (1971).

222. J. R. Colley, D. L. Miller, M. S. R. Hutt, H. J. Wallace, and H. E. de Wardener, The renal lesion in angiokeratoma corporis diffusum, *Brit. Med. J.* **1**:1266–1268 (1958).

223. J. E. Bethune, P. L. Landrigan, and C. D. Chipman, Angiokeratoma corporis diffusum universale (Fabry's disease) in two brothers, *New Engl. J. Med.* **264**:1280–1285 (1961).

224. M. B. Curry and T. L. Fleisher, Angiokeratoma corporis diffusum. A case report, *J. Am. Med. Ass.* **175**:864–869 (1961).

225. A. Bischoff, V. Fierz, F. Regli, and I. Ulrich, Peripher-neurologische Störungen bei der Fabryschen Krankheit (Angiokerotoma corporis diffusum universale), *Klin. Wschr.* **46**: 666–671 (1968).

226. R. S. Kocen and P. K. Thomas, Peripheral nerve involvement in Fabry's disease, *Arch. Neurol.* **22**:81–88 (1970).

227. M. W. Hartley, R. E. Miller, H. J. Dempsy, and J. F. Caroll, Dysphospholipidosis in Fabry's disease: A light and electron microscopic study, *Ala. J. Med. Sci.* **1**:361–367 (1964).

228. P. Frost, Y. Tanaka, and G. L. Spaeth, Fabry's disease—glycolipid lipidosis: Histochemical and electronmicroscopic studies of two cases, *Am. J. Med.* **40**:618–627 (1966).

229. A. I. Rae, J. C. Lee, and J. Hopper, Jr., Clinical and electron-microscopic studies of a case of glycolipid lipidosis (Fabry's disease), *J. Clin. Pathol.* **20**:21–27 (1967).

230. J. D. Bagdade, F. Parker, P. O. Ways, T. E. Morgan, D. Lagunoff, and S. Eidelman, Fabry's disease: A correlative clinical, morphologic and biochemical study, *Lab. Invest.* **18**:681–688 (1968).

231. M. Tondeur and A. Résibois, Fabry's disease in children, an electron microscopic study, *Virchows Arch. Abt. B, Zellpathol.* **2**:239–254 (1969).

232. M. Ruiter, Histological investigation of the skin in angiokeratoma corporis diffusum with particular regard to the associated disturbance of phosphatide metabolism, *Dermatologia* **109**:272–286 (1954).

233. C. C. Sweeley and B. Klionsky, Fabry's disease: Classification as a sphingolipidosis and partial characterization of a novel glycolipid, *J. Biol. Chem.* **238**:3148–3150 (1963).

234. C. C. Sweeley and B. Klionsky, in "The Metabolic Basis of Inherited Disease," 2nd ed., (J. B. Stanbury, J. B. Wyngaarden, and D. S. Frederickson, eds.) pp. 618–632, McGraw-Hill, New York (1966).

235. H. O. Christensen Lou, A biochemical investigation of angiokeratoma corporis diffusum, *Acta Pathol. Microbiol. Scand.* **68**:332–342 (1966).

236. V. W. Steward and C. Hitschcock, Fabry's disease, *Pathol. Europ.* **3**:377–386 (1968).

237. T. Miyatake, A study on glycolipids in Fabry's disease, *Jap. J. Exptl. Med.* **39**:35–45 (1969).

238. S. Handa, T. Ariga, T. Miyatake, and T. Yamakawa, Presence of α-anomeric glycosidic configuration in the glycolipids accumulated in kidney with Fabry's disease, *J. Biochem.* **69**:625–627 (1971).

239. I. Bensaude, J. Callahan, and M. Philippart, Fabry's disease as an α-galactosidosis: Evidence for an α-configuration in trihexosyl ceramide, *Biochem. Biophys. Res. Commun.* **43**:913–918 (1971).

240. J. T. R. Clarke, L. S. Wolfe, and A. S. Perlin, Evidence for a terminal α-D-galactopyranosyl residue in galactosylgalactosylglucosylceramide from human kidney, *J. Biol. Chem.* **246**: 5563–5569 (1971).

241. H. Loeb, G. Jonniaux, M. Tondeur, P. Danis, P. E. Gregoire, and P. Wolf, Étude clinique, biochimique et ultrastructurelle de la maladie de Fabry chez l'enfant, *Helv. Paediat. Acta* **23**:269–286 (1968).

242. R. Matalon, A. Dorfman, G. Dawson, and C. C. Sweeley, Glycolipid and mucopolysaccharide abnormality in fibroblasts of Fabry's disease, *Science* **164**:1522–1523 (1969).

243. M. Philippart, L. Sarlieve, and A. Manacorda, Urinary glycolipids in Fabry's disease, *Pediatrics* **43**:201–206 (1969).

244. R. J. Desnick, G. Dawson, S. Desnick, C. C. Sweeley, and W. Krivit, Diagnosis of glycosphingolipidoses by urinary-sediment analysis, *New Engl. J. Med.* **284**:739–744 (1971).

245. R. O. Brady, A. E. Gal, R. M. Bradley, E. Martensson, A. L. Warshaw, and L. Loster, Enzymatic defect in Fabry's disease. Ceramidetrihexosidase deficiency, *New Engl. J. Med.* **276**:1163–1167 (1967).

246. J. A. Kint, Fabry's disease: Alpha-galactosidase deficiency, *Science* **167**:1268–1269 (1970).

247. G. Romeo and B. R. Migeon, Genetic inactivation of the α-galactosidase locus in carriers of Fabry's disease, *Science* **170**:180–181 (1970).

248. C. A. Mapes, R. L. Anderson, and C. C. Sweeley, Galactosylgalactosylglucosylceramide: Galactosyl hydrolase in normal human plasma and its absence in patients with Fabry's disease, *FEBS Letters* **7**:180–182 (1970).

249. R. O. Brady, B. W. Uhlendorf, and C. B. Jacobson, Fabry's disease: Antenatal detection, *Science* **172**:174–175 (1971).

250. L. A. Lockman, W. Krivit, and R. J. Desnick, Relief of the painful crises of Fabry's disease by diphenylhydantoin, *Neurology* **21**:423 (1971).

251. H. Jatzkewitz and K. Sandhoff, On a biochemically special form of infantile amaurotic idiocy, *Biochim. Biophys. Acta* **70**:354–356 (1963).

252. J. S. O'Brien, M. B. Stern, B. H. Landing, J. K. O'Brien, and G. N. Donnell, Generalized gangliosidosis. Another inborn error of ganglioside metabolism? *Am. J. Dis. Child.* **109**:338–346 (1965).

253. N. K. Gonatas and J. Gonatas, Ultrastructural and biochemical observations on a case of systemic late infantile lipidosis and its relationship to Tay–Sachs disease and gargoylism, *J. Neuropathol. Exptl. Neurol.* **24**:318–340 (1965).

254. B. H. Landing and J. H. Rubinstein, in "Cerebral Sphingolipidoses" (S. M. Aronson and B. W. Volk, eds.) pp. 1–13, Academic Press, New York (1962).

255. K. Suzuki, K. Suzuki, and G. C. Chen, Morphological, histochemical and biochemical studies on a case of systemic late infantile lipidosis (generalized gangliosidosis), *J. Neuropathol. Exptl. Neurol.* **27**:15–38 (1968).

256. K. Suzuki, K. Suzuki, and G. C. Chen, in "Cerebral Lipidoses II" (A. Nunes Vicente, P. Dustin, and A. Lowenthal, eds.) pp. 273–294, Présses Académiques Européenes, Brussels (1968).

257. R. Sacrez, J. G. Juif, J. M. Gigonnet, and J. E. Gruner, La maladie de Landing, ou idiotie amaurotique infantile précose avec gangliosidose généralisée de type G_{M1}, *Pédiatrie* **22**:143–162 (1967).

258. H. Roels, J. Quatacker, A. Kint, H. Vander Eecken, and L. Vrints, Generalized gangliosidosis-G_{M1} (Landing disease) II. Morphological study, *Europ. Neurol.* **3**:129–160 (1970).

259. C. R. Scott, D. Lagunoff, and B. F. Trump, Familial neurovisceral lipidosis, *J. Pediat.* **71**:357–366 (1967).

260. S. Takebayashi, D. B. von Bassewitz, and H. Themann, Feinstrukturelle Veränderungen der Niere bei generalisierter Gangliosidose G_{M1}. *Virchows Arch. Abt. B, Zellpathol.* **5**:301–313 (1970).

261. P. Hubain, E. Adam, A. Dewelle, G. Druez, J.-P. Farriaux, and A. Dupont, Étude d'une observation de gangliosidose à G_{M1}, *Helv. Paediat. Acta* **24**:337–351 (1969).

262. K. Suzuki, K. Suzuki, and S. Kamoshita, Chemical pathology of G_{M1}-gangliosidosis (generalized gangliosidosis), *J. Neuropathol. Exptl. Neurol.* **28**:25–73 (1969).

263. R. Ledeen, K. Salsman, J. Gonatas, and A. Taghavy, Structure comparison of the major monosialogangliosides from brains of normal human, gargoylism, and late infantile systemic lipidosis. Part I, *J. Neuropathol. Exptl. Neurol.* **24**:341–351 (1965).

264. K. Suzuki and G. C. Chen, Brain ceramide hexosides in Tay-Sachs disease and generalized gangliosidosis (G_{M1}-gangliosidosis), *J. Lipid Res.* **8**:105–113 (1967).

265. Y. Suzuki, A. C. Crocker, and K. Suzuki, G_{M1}-gangliosidosis: Correlation of clinical and biochemical data, *Arch. Neurol.* **24**:58–64 (1971).

266. K. Suzuki, Cerebral G_{M1}-gangliosidosis: Chemical pathology of visceral organs, *Science* **159**:1471–1472 (1968).

267. L. S. Wolfe, J. Callahan, J. S. Fawcett, F. Andermann, and C. R. Scriver, G_{M1}-gangliosidosis without chondrodystrophy or visceromegaly, *Neurology* **20**:23–44 (1970).

268. J. W. Callahan and L. S. Wolfe, Isolation and characterization of keratan sulfates from the liver of a patient with G_{M1}-gangliosidosis type I, *Biochim. Biophys. Acta* **215**:527–543 (1970).

269. P. Seringe, B. Plainfosse, E. Mautmann, J. Lorilloux, G. Calamy, J.-P. Berry, and J.-M. Watchi, Gangliosidose généralisée, du type Norman–Landing, à G_{M1}, *Ann. Pédiat.* **44**:685–704 (1968).

270. S. Okada and J. S. O'Brien, Generalized gangliosidosis: Beta-galactosidase deficiency, *Science* **160**:1002–1004 (1968).

271. G. Dacremont and J. A. Kint, G_{M1}-ganglioside accumulation and β-galactosidase deficiency in a case of G_{M1}-gangliosidosis (Landing disease), *Clin. Chim. Acta* **21**:421–425 (1968).

272. F. van Hoof and H. G. Hers, The abnormalities of lysosomal enzymes in mucopolysaccharidoses, *Europ. J. Biochem.* **7**:34–44 (1968).

273. R. O. Brady, J. S. O'Brien, R. M. Bradley, and A. E. Gal, Sphingolipid hydrolases in brain tissue of patients with generalized gangliosidosis, *Biochim. Biophys. Acta* **210**:193–195 (1970).

274. M. C. MacBrinn, S. Okada, M. W. Ho, C. C. Hu, and J. S. O'Brien, Generalized gangliosidosis: Impaired cleavage of galactose from a mucopolysaccharide and a glycoprotein, *Science* **163**:946–947 (1969).

275. C. Hooft, R. F. Vlietinck, G. Dacremont, and J. A. Kint, G_{M1}-gangliosidosis type II, *Europ. Neurol.* **4**:1–21 (1970).

276. H. S. Singer and I. A. Schafer, White cell β-galactosidase activity, *New Engl. J. Med.* **282**:571 (1970).

277. G. H. Thomas, β-D-Galactosidase in human urine: Deficiency in generalized gangliosidosis, *J. Lab. Clin. Med.* **74**:725–731 (1969).

278. H. R. Sloan, B. W. Uhlendorf, C. B. Jacobson, and D. S. Frederickson, β-Galactosidase in tissue culture derived from human skin and bone marrow: Enzyme defect in G_{M1}-gangliosidosis, *Pediat. Res.* **3**:532–537 (1969).

279. J. W. Callahan, L. Pinsky, and L. S. Wolfe, G_{M1}-gangliosidosis (type II): Studies on a fibroblast cell strain, *Biochem. Med.* **4**:295–316 (1970).

280. D. M. Derry, J. S. Fawcett, F. Andermann, and L. S. Wolfe, Late infantile systemic lipidosis, major monosialogangliosidosis, delineation of two types, *Neurology* **18**:340–348 (1968).

281. J. S. O'Brien, S. Okada, M. W. Ho, D. L. Fillerup, M. L. Veath, and K. Adams, Ganglioside storage diseases, *Fed. Proc.* **30**:956–969 (1971).

282. I. A. Schafer, Personal communication.

283. L. Pinsky, E. Powell, and J. Callahan, G_{M1}-gangliosidosis types 1 and 2: Enzymatic differences in cultured fibroblasts, *Nature* **228**:1093–1095 (1970).

284. S. M. Aronson, *in* "Tay–Sachs Disease" (B. W. Volk, ed.) pp. 118–153, Grune and Stratton, New York (1964).

285. B. W. Volk, *in* "Tay–Sachs Disease" (B. W. Volk, ed.) pp. 36–67, Grune and Stratton, New York (1964).

286. S. S. Lazarus, B. J. Wallace, and B. W. Volk, Neuronal enzyme alterations in Tay–Sachs disease, *Am. J. Pathol.* **41**:579–591 (1962).

287. R. D. Terry and M. Weiss, Studies in Tay–Sachs disease: II. Ultrastructure of cerebrum, *J. Neuropathol. Exptl. Neurol.* **22**:18–55 (1963).

288. B. J. Wallace, B. W. Volk, L. Schneck, and H. Kaplan, Fine structural localization of two hydrolytic enzymes in the cerebellum of children with lipidoses, *J. Neuropathol. Exptl. Neurol.* **25**:76–96 (1966).

289. M. Adachi, J. Torii, L. Schneck, and B. W. Volk, The fine structure of fetal Tay–Sachs disease, *Arch. Pathol.* **91**:48–54 (1971).

290. E. Klenk, Beiträge zur Chemie der Lipoidosen, Niemann-Picksche Krankheit und amaurotische Idiotie, *Z. Physiol. Chem.* **262**:128–143 (1939).

291. E. Klenk, Über die Ganglioside des Gehirns bei der infantilen amaurotischen Idiotie vom Typ Tay-Sachs, *Ber. Deutsch. Chem. Ges.* **75**:1632–1636 (1942).

292. L. Svennerholm, The chemical structure of normal brain and Tay–Sachs gangliosides, *Biochem. Biophys. Res. Commun.* **9**:436–441 (1962).

293. R. Ledeen and K. Salsman, Structure of the Tay–Sachs' ganglioside, I. *Biochemistry* **4**: 2225–2232 (1965).

294. O. Eeg-Olofsson, K. Kristensson, P. Sourander, and L. Svennerholm, Tay–Sachs disease. A generalized metabolic disorder, *Acta Paediat. Scand.* **55**:546–562 (1966).

295. K. Sandhoff, Variation of β-N-acetylhexosaminidase pattern in Tay–Sachs disease, *FEBS Letters* **4**:351–354 (1969).

296. S. Okada and J. S. O'Brien, Tay–Sachs disease: Generalized absence of a β-D-N-acetyl-hexosaminidase component, *Science* **165**:698–700 (1969).

297. D. Robinson and J. L. Stirling, N-acetyl-β-glucosaminidases in human spleen, *Biochem. J.* **107**:321–327 (1968).

298. E. H. Kolodny, R. O. Brady, and B. W. Volk, Demonstration of an alteration of ganglioside metabolism in Tay–Sachs disease, *Biochem. Biophys. Res. Commun.* **37**:526–531 (1969).

299. J. S. O'Brien, S. Okada, A. Chen, and D. L. Fillerup, Tay–Sachs disease: Detection of heterozygotes and homozygotes by serum hexosaminidase assay, *New Engl. J. Med.* **283**: 15–20 (1970).

300. Y. Suzuki, P. H. Berman, and K. Suzuki, Detection of Tay–Sachs disease heterozygotes by assay of hexosaminidase A in serum and leucocytes, *J. Pediat.* **78**:643–647 (1971).

301. L. Schneck, J. Friedland, C. Valenti, M. Adachi, D. Amsterdam, and B. W. Volk, Prenatal diagnosis of Tay–Sachs disease, *Lancet* **1**:582–584 (1970).

302. J. S. O'Brien, S. Okada, D. L. Fillerup, M. L. Veath, B. Adornato, P. H. Brenner, and J. G. Leroy, Tay–Sachs disease: Prenatal diagnosis, *Science* **172**:61–64 (1971).

303. K. Sandhoff, U. Andreae, and H. Jatzkewitz, Deficient hexosaminidase activity in an exceptional case of Tay–Sachs disease with additional storage of kidney globoside in visceral organs, *Life Sci.* **7**:283–288 (1968).

304. H. Bernheimer and F. Seitelberger, Über das Verhalten der Ganglioside im Gehirn bei 2 Fällen von spatinfantiler amaurotischer Idiotie, *Wiener. Klin. Wschr.* **80**:163–169 (1968).

305. B. W. Volk, M. Adachi, L. Schneck, A. Saifer, and W. Kleinberg, G_5-ganglioside variant of systemic late infantile lipidosis, *Arch. Pathol.* **87**:393–403 (1969).

306. K. Suzuki, K. Suzuki, I. Rapin, Y. Suzuki, and N. Ishii, Juvenile G_{M2}-gangliosidosis. Clinical variant of Tay–Sachs disease or a new disease, *Neurology* **20**:190–204 (1970).

307. J. H. Menkes, J. S. O'Brien, S. Okada, J. Grippo, J. M. Andrews, and P. A. Cancilla, Juvenile G_{M2}-gangliosidosis. Biochemical and ultrastructural studies on a new variant of Tay–Sachs disease, *Arch. Neurol.* **25**:14–22 (1971).

308. C. Klibansky, A. Saifer, N. I. Feldman, L. Schneck, and B. W. Volk, Cerebral lipids in a case of systemic G_{M2}-gangliosidosis of a late infantile type, *J. Neurochem.* **17**:339–346 (1970).

309. Y. Suzuki and K. Suzuki, Partial deficiency of hexosaminidase component A in juvenile G_{M2}-gangliosidosis, *Neurology* **20**:848–851 (1970).

310. L. Schneck, J. Friedland, M. Pourfar, A. Saifer, and B. W. Volk, Hexosaminidase activities in a case of systemic G_{M2}-gangliosidosis of late infantile type, *Proc. Soc. Exptl. Biol. Med.* **133**:997–998 (1970).

311. E. P. Young, R. B. Ellis, B. D. Lake, and A. D. Patrick, Tay–Sachs disease and related disorders: Fractionation of brain N-acetyl-β-hexosaminidase on DEAE-cellulose, *FEBS Letters* **9**:1–4 (1970).

312. S. Okada, M. L. Veath, and J. S. O'Brien, Juvenile G_{M2}-gangliosidosis: Partial deficiency of hexosaminidase A, *J. Pediat.* **77**:1063–1065 (1970).

313. K. Sandhoff, U. Andreae, and H. Jatzkewitz, *in* "Cerebral Lipidoses II" (A. Nunes Vicente, P. Dustin, and A. Lowenthal, eds.) pp. 164–171, Presses Académiques Européennes, Brussels (1968).

314. H. Pilz, D. Müller, K. Sandhoff, and V. ter Meulen, Tay-Sachssche Krankheit mit Hexosaminidase-Defekt, *Deutsch. Med. Wschr.* **39**:1833–1839 (1968).

315. Y. Suzuki, J. C. Jacob, K. Suzuki, K. M. Kutty, and K. Suzuki, G_{M2}-gangliosidosis with total hexosaminidase deficiency, *Neurology* **21**:313–328 (1971).

316. K. Sandhoff and H. Jatzkewitz, The chemical pathology of Tay–Sachs disease, *in* "Proceedings of the Fourth International Symposium on Sphingolipids, Sphingolipidoses, and Allied Disorders" (B. W. Volk and S. M. Aronson, eds.) pp. 305–319. Plenum Press, New York (1972).

317. S. Farber, A lipid metabolic disorder—disseminated "lipogranulomatosis"—a syndrome with similarity to, and important differences from, Niemann–Pick and Hand–Schüller–Christian disease, *Am. J. Dis. Child.* **84**:499–500 (1952).

318. S. Farber, J. Cohen, and L. Uzman, Lipogranulomatosis; a new lipo-glycoprotein storage disease, *J. Mount Sinai Hosp.* **24**:816–837 (1957).

319. H. W. Moser, A. L. Prensky, H. J. Wolfe, and N. P. Rosman, Farber's lipogranulomatosis. Report of a case and demonstration of an excess of free ceramide and ganglioside, *Am. J. Med.* **47**:869–890 (1969).

320. A. L. Prensky, G. Ferreira, S. Carr, and H. W. Moser, Ceramide and ganglioside accumulation in Farber's lipogranulomatosis, *Proc. Soc. Exptl. Biol. Med.* **126**:725–728 (1967).

321. K. Samuelsson and R. Zetterström, Ceramides in a patient with lipogranulomatosis (Farber's disease) with chronic course, *Scand. J. Clin. Lab. Invest.* **27**:393–405 (1971).

322. K. Samuelsson, R. Zetterström, and B. I. Ivemark, Studies on a case of lipogranulomatosis (Farber's disease) with protracted course, *in* "Proceedings of the Fourth International Symposium on Sphingolipids, Sphingolipidoses, and Allied Disorders" (B. W. Volk and S. M. Aronson, eds.) pp.533–548. Plenum Press, New York (1972).

323. P. Durand, C. Borrone, and G. Della Cella, Fucosidosis, *J. Pediat.* **75**:665–674 (1969).

324. H. Loeb, M. Tondeur, G. Jonniaux, S. Mockel-Pohl, and E. Vamos-Hurwitz, Biochemical and ultrastructural studies in a case of mucopolysaccharidosis "F" (fucosidosis), *Helv. Paediat. Acta* **24**:519–537 (1969).

325. G. Dawson and J. W. Spranger, Fucosidosis: A glycosphingolipidosis, *New Engl. J. Med.* **285**:122 (1971).

326. W. Zeman and P. Dyken, Neuronal ceroid-lipofuscinosis (Batten's disease): Relationship to amaurotic family idiocy? *Pediatrics* **44**:570–583 (1969).

327. H. G. Hers, Inborn lysosomal diseases, *Gastroenterology* **48**:625–633 (1965).

328. S. R. Korey and J. Gonatas, Separation of human brain gangliosides, *Life Sci.* **2**:296–302 (1963).

329. S. Hammarström, On the biosynthesis of cerebrosides: nonenzymatic N-acylation of psychosine by stearoyl coenzyme A, *FEBS Letters* **21**:259–263 (1972).

330. M. Sugita, J. T. Dulaney, and H. W. Moser, Ceramidase deficiency in Farber's disease (lipogranulomatosis), *Science* **178**:110–1102 (1972).

Chapter 2

DISORDERS OF THE CEREBRAL CIRCULATION

A. G. Waltz

Department of Neurology
University of Minnesota Medical School
Minneapolis, Minnesota

I. INTRODUCTION

The delivery of metabolic substrates to cerebral tissue depends on the flow of blood to and in the brain. To develop rational methods for treating patients with cerebral vascular disorders, including ischemic infarcts, subarachnoid and intracerebral hemorrhages, and other less common conditions, the mechanisms for the regulation of cerebral blood flow must be understood, and the ways in which the mechanisms are affected or modified by abnormal circumstances must be delineated. Complete interruption of the supply of blood to the brain, with cessation of delivery of oxygen, glucose, and other substrates, causes irreversible impairment of neuronal function and death of cells within 4–5 min. With the more ordinary kinds of ischemic cerebral vascular lesions, however, delivery of substrates to brain cells is decreased to a variable degree: in many regions of ischemic brain, blood flow is maintained (at a reduced level) through collateral channels. Although the time required for irreversible impairment of neuronal function under these circumstances is unknown, it may be as long as several hours. Thus therapeutic measures that can increase the delivery of metabolic substrates to cerebral tissue may prevent irreversible damage to neurons.

II. METHODS FOR STUDY OF THE CEREBRAL CIRCULATION

A. Cerebral Blood Flow and Cerebral Vascular Resistance

Because the brain is supplied by at least four arterial inputs, drained by at least two venous outputs, and encased in bone, methods used for the measurement of the flow of blood in other organs are not always suitable for brain.

1. Methods Using Diffusible Indicators

Most substances diffuse relatively slowly through extravascular tissue[1-3]; in brain, at least several minutes are required for an appreciable amount of an indicator to diffuse even a part of a millimeter.[4,5] However, distances between cerebral capillaries are small, particularly in gray matter where intercapillary distances range from about 35–40 μ to about 70–80 μ.[6,7] Thus diffusion times in brain tissue are short for substances that pass freely through the vascular endothelium. Concentrations of inert, nonmetabolized, freely diffusible substances in brain tissue generally depend on the delivery and removal of the molecules of the substance by the blood, rather than on the diffusibility of the molecules through the vascular endothelium or in extravascular tissue; the concentrations are "flow limited" rather than "diffusion limited." Therefore, freely diffusible indicator substances can be used to measure the volume flow of blood in cerebral tissue, by determining the absolute or relative concentrations of the substance in brain as a function of time and of the concentrations in the blood. The times vary with the methods, but most methods require measurements during a steady state of at least several minutes and are not applicable to situations where blood flow may be changing because of changes of cerebral metabolic activity or systemic factors.

In abnormal circumstances, both the velocity and the volume of the flow of blood in brain tissue may change. At high velocities of flow, as with hyperemia, the concentrations of diffusible indicators in brain tissue may be diffusion limited rather than flow limited, and calculated values for cerebral blood flow may be erroneous. With ischemia, blood may not flow through portions of the cerebral capillary network because of blockage of vessels by formed elements of the blood or by endothelial swelling[8-12]; thus effective cerebral intercapillary distances may increase, causing a functional block to diffusion into the tissue.

Most methods for measuring cerebral blood flow with diffusible indicators are based on the Fick principle, a restatement of the law of conservation of matter. The principle states simply that the amount of indicator taken up by an organ during a given period of time is equal to the amount brought to the organ by arterial blood minus the amount carried away by venous blood. In the first application of the Fick principle to the brain, Kety and Schmidt[13-15] used gaseous nitrous oxide as a diffusible indicator administered to human subjects by inhalation. The concentrations of nitrous oxide were measured in arterial and jugular venous blood. It was assumed that after 10 min of inhalation the concentrations of nitrous oxide in jugular venous blood were equal to the concentrations in cerebral tissue, and cerebral blood flow (CBF) was calculated from the equation

$$\text{CBF} = \frac{\lambda C_{v_{10}}}{(Ca - Cv)dt} \times 100$$

where λ is the blood–brain partition coefficient, reflecting the differences of solubility of the indicator in blood and brain tissue; $C_{v_{10}}$ is the concentration of the indicator in jugular venous blood at 10 min; Ca is the arterial blood concentration and Cv the jugular venous blood concentration, determined at intervals. Because all measured units are concentrations, CBF is expressed per unit of volume or weight of brain: values are given as the numbers of milliliters of blood flowing through each 100 g of brain tissue each minute. With this method, normal values for humans approximate 50–54 ml/100 g/min, an average for the volume flow of blood in the gray matter, white matter, and vascular tissue of all the portions of the brain taking up the indicator. To minimize errors caused by the contributions of extracerebral tissues to jugular venous blood and by the mixing of blood from each of the two cerebral hemispheres, the concentrations of the indicator must be determined for venous blood obtained from both jugular bulbs simultaneously. In addition, the values used for venous concentrations should be extrapolated to infinite time, and the partition coefficients should be corrected for the hematocrit of each subject.

This method also can be used with other indicators, such as hydrogen[16] and radioactive krypton-85,[17] or by measuring the arterial and jugular venous blood concentrations during the period of desaturation of the brain after the delivery of the indicator is discontinued.[18] In one modification of the method, the concentration of the indicator is measured in the brain itself, by the external detection of the radioactivity of antipyrine labeled with iodine-131, so that C_b is substituted for $C_{v_{10}}$ in the equation.[19] This modification provides a measure of total cerebral blood flow, with a normal value of approximately 1100 ml/min. If cerebral metabolism is constant, measurements of the differences between arterial and jugular venous concentrations of oxygen can be used to provide a minute-to-minute value for CBF.[15]

When a diffusible radioactive indicator (such as gaseous xenon-133 or H_2O^{15}) is injected rapidly into an artery supplying only the brain, a value for average blood flow in the portion of brain receiving the indicator can be determined by kinetic or stochastic analysis of the rate of change of the concentration of the indicator in the brain, measured by the external detection of its radio-activity.[20-23] With this method,

$$\text{CBF} = \frac{H_0 - H_t}{A_t}$$

where H_0 is the concentration of the indicator (or count rate) at the peak of the time–concentration function, H_t is the residual concentration (or count rate) after nearly complete clearance from brain tissue (usually after 10 min), and A_t is the integral of the concentration or its equivalent, which in practice is either the area under a recorded time–concentration curve or the total number of counts obtained during the time allowed for clearance. Half-time analysis of the first few minutes of the clearance function provides a value for average CBF quickly and simply[23,24]; values obtained by half-time analysis correlate well with values obtained by 10 min stochastic analysis and are useful in changing or stressful situations.

If external detection of radioactivity is used to determine the rate of change of concentration of an indicator, it must be injected into the internal (not external or common) carotid artery to avoid contamination of the record of clearance by contributions from extracerebral tissues. Corrections for recirculation of inert gases are unnecessary, because gases are cleared quickly from blood as it passes through the lungs. CBF can be measured from regions of brain if several scintillation detectors are used. However, close attention must be paid to physical and geometric factors if accurate values for regional CBF are desired.[25]

If the brain is exposed, an indicator that emits β-activity (such as krypton-85) can be used to measure regional blood flow of cerebral cortex.[26-28] If the brain is not exposed, cortical blood flow often can be determined by analyzing the clearance of an indicator that emits γ-activity (such as xenon-133) for two exponential components.[21] In normal brain, CBF values for regions of gray matter are greatly different from those for white matter[29-32]; thus the rate constants for the faster and slower of the two exponential components probably represent the rates of clearance of the indicator from gray matter and white matter, respectively.[21] CBF values for gray matter and white matter can be obtained with reasonable assurance by multiplying the rate constants by the appropriate partition coefficients.

The relative contributions of the two components to the clearance function also can be determined, and in normal brain these may reflect the anatomical amounts of gray and white matter taking up the indicator in the field of the detector. Unfortunately, under abnormal circumstances such as ischemia and hyperemia the usual bimodal distribution of rate constants for the clearance of an indicator from brain tissue no longer exists: some rate constants for white matter are similar to those for gray matter.[31-33] Thus in abnormal situations the values for CBF derived from the faster and slower exponential components of clearance will not necessarily reflect values for gray matter and white matter, and the relative contributions of the two components will not reflect the anatomical amounts of gray and white matter in the field of the detector.[24] With ischemia or dementia (associated with decreases of CBF), changes or differences of the calculated values for CBF or of the relative contributions of two exponential components will not signify changes in the amounts or the blood flow of gray matter or white matter,[34] but changes in rapidly and slowly cleared regions of brain regardless of anatomical constituency.

For studies in animals, in which the brain can be removed, sectioned, and placed against photographic film, measurements of regional CBF can be made by autoradiography (Fig. 1).[29-33] If an indicator emitting β-activity (such as antipyrine-C^{14}) is injected intravenously over a relatively short period of time, approximately 1–2 min, the concentration of the indicator in a given region of brain will depend on delivery by the blood flowing to the region. The regional

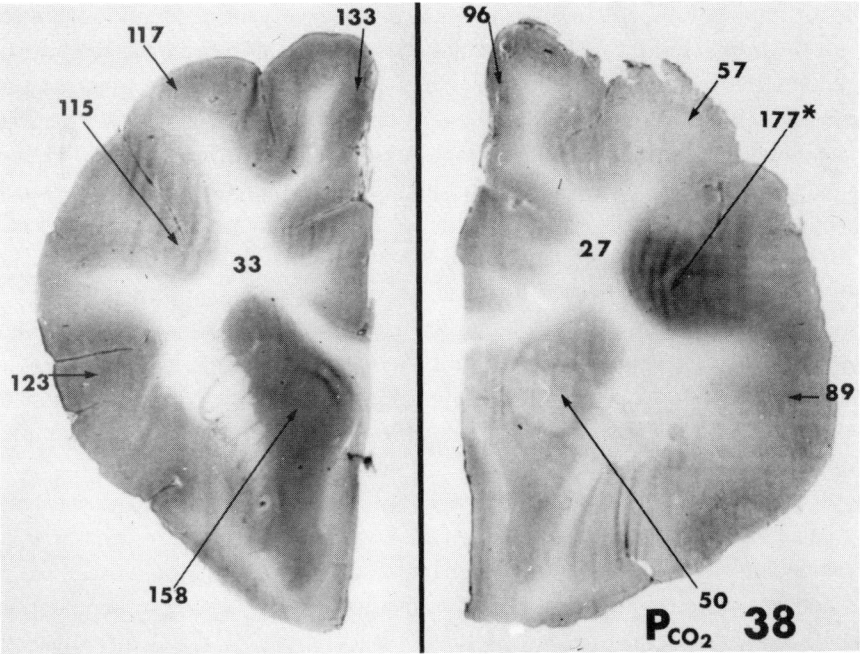

Fig. 1. Autoradiographs of the nonischemic (left) and the ischemic (right) hemispheres of the brain of a cat 5 hr after occlusion of the right middle cerebral artery. The densities of the images produced by antipyrine-C^{14} are directly related to regional CBF; the darker areas are those with higher CBF. Values for CBF are given in ml/100 g/min × 100. The lighter areas in the autoradiograph on the right are ischemic regions, primarily in the basal ganglia. The darker area indicated by an asterisk (CBF 177) is a region with reactive hyperemia.

concentrations of the radioactive indicator can be determined by densitometry, and CBF can be calculated by relating the concentrations of the indicator in brain to its concentration in blood, as it increases with time. This method allows accurate measurement of CBF in small regions in the depths of the brain, but because the brain must be sectioned only one set of measurements can be made for each animal. Regional values for CBF also can be obtained by analysis of the clearance of hydrogen gas[35] or by measurements of brain temperature.[36-39]

Xenon-133 can be given by inhalation so that carotid catheterization is unnecessary.[40-43] However, because the indicator is delivered to extracerebral structures and because relatively large amounts must be used, the clearance function must be corrected for arterial recirculation, and complex curve-fitting methods are needed for analysis. The potential usefulness of this method for study of patients with cerebral vascular disorders has not yet been realized in practice.

2. Methods Using Nondiffusible Indicators

Nondiffusible indicators do not leave blood vessels to enter cerebral tissue. Most applications of nondiffusible indicators have provided only qualitative information about the cerebral circulation, or indices of blood flow, rather than quantitative values for CBF.[44,45] Circulation times or transit times can be measured with dyes or with radio-opaque or radioactive indicators, but because the blood volume of the brain can not be determined conveniently, blood flow can not be calculated. The passage of nondiffusible radioactive indicators (such as technetium-99m) through major blood vessels in the neck and the head can be monitored with a scintillation camera or "gamma camera"; useful information can be obtained, but resolution is less than with angiography.[46-48]

Total blood flow through the brain can be determined with nondiffusible indicators by measuring the dilution of the indicator in jugular venous blood[49-50]; for accuracy and validity, dilution methods require injection of the indicator into at least one internal carotid artery and withdrawal of samples of blood from both jugular veins.

In theory, blood flow through a single cerebral blood vessel could be calculated from the time–concentration functions of an indicator, measured at two points along an unbranched segment of a vessel of known caliber. A videodensitometric application of this method, using radio-opaque materials and angiography, has proved useful for calculations of blood flow through coronary arteries[51]; applications to cerebral vessels have been limited.

3. Miscellaneous Methods

Electromagnetic flowmeters, ultrasonic flowmeters, and thermal probes can be used to measure blood flow in large blood vessels, such as the carotid arteries or the jugular veins. Because the regulation of cerebral blood flow is a function of regulatory arterial vessels distal to the large arteries and proximal to the large veins, measurements of flow in the blood vessels of the neck provide little information about regional CBF in the brain. Rheoencephalography (the measurement of the impedance of the head to an imposed electric current) and similar indirect methods that are related to pulsations or sounds within the cranial space have not proved to be of value for estimations of cerebral blood flow.

4. Cerebral Vascular Resistance

The resistance of the cerebral blood vessels to the flow of blood can be calculated mathematically with the equation

$$CVR = \frac{PP}{CBF}$$

where CVR is cerebral vascular resistance and PP is perfusion pressure. Perfusion pressure is commonly determined as mean systemic or carotid arterial blood pressure, less mean jugular venous blood pressure; an accurate determination of perfusion pressure would require knowledge of regional cerebral tissue pressure, or of intracranial pressure at least, and of the blood pressure in the surface arteries of the brain. For humans, CVR approximates 1.6 units, a unit being the pressure in millimeters of mercury required to force 1 ml of blood through 100 g of brain tissue in 1 min.[15]

Cerebral vascular resistance is a mathematical concept. Although changes of CVR frequently are related to changes of caliber of cerebral arterial vessels and regulatory arterioles, other factors—such as the viscosity of the blood flowing through the cerebral capillary network, and regional tissue pressure— also can influence CVR. Because CVR is a mathematical concept, changes of the calculated value should not be assumed to be due to changes of the caliber of cerebral blood vessels unless the blood vessels are observed directly.

B. Cerebral Blood Vessels

Much of the information available at present about the changes of caliber and the characteristics of the flow of blood in cerebral blood vessels is inferential because of the limitations inherent in the methods used for studies of the vasculature of the brain. As yet, only the vessels on the surface or at the base of the brain are available to study. Direct observations of cerebral blood vessels require removal of at least a portion of skull; however, with a craniectomy intracranial pressure and average brain tissue pressure are virtually the same as atmospheric pressure. Moreover, changes of intracranial pressure that occur when the skull is closed are prevented. Thus vascular caliber, pulsation profiles, and even the shapes of vessels (particularly venous vessels) may not be the same when the skull is open as when it is closed. Although many techniques for direct observation and photography of cerebral vessels have been developed, including fluorescein angiography[52] and high-speed microcinephotography,[53] interpretations of the results obtained with such techniques must be made with caution. For example, pulsatile displacements of cerebral arterial and venous vessels may not occur if the skull is intact and if vessel–tissue–atmosphere pressure gradients do not exist. Direct observation of vessels in the depths of the brain requires disruption of cerebral tissue with resulting impairment of the normal regulatory functions of the vasculature.

Surface and basal vessels differ in important respects from vessels in the depths of the brain. For example, basal and surface vessels are much more richly endowed with nerve fibers, adrenergic and cholinergic synaptic vesicles, and fluorescent catecholamine products than are penetrating cerebral vessels.[54–58] Additionally, arterial–arterial and venous–venous anastamoses are much more common between larger vessels (50–250 μ) on the surface of the brain than in the depths of the brain, where anastamoses are limited to vessels less than 20–30 μ in diameter.[7,59] Although all the blood that supplies the brain flows through the surface arterial network, it is not certain that changes of caliber of blood vessels deep in the brain are consonant with those of surface vessels or that the vascular processes involved in the regulation of the circulation in the depths of the brain are the same as those that have been observed in surface vessels. In animals with craniectomies, dissociations between changes of vascular caliber and changes of blood flow through tissue occur frequently,[60,61] suggesting that different mechanisms for regulating tissue perfusion may operate at different times or under different circumstances.

Methods that do not depend on the removal of a portion of skull also have serious limitations. Anatomical studies of the cerebral vasculature using microradiographic or histological techniques have provided valuable information, such as the lack of arterial–venous anastamoses in brain tissue,[7] but such techniques can not delineate the important dynamic responses of cerebral vessels to changing circumstances. Radiographic procedures using radio-opaque contrast media do not provide adequate resolution to allow study of the regulatory arterioles of the brain; magnification angiography does not resolve vessels as small as 20–25 μ in diameter. The imaging of cerebral blood vessels with radioactive nuclides and scintillation detectors (such as the "gamma camera") is possible only for large cervical or basal vessels. Thus, many concepts about the regulatory changes of cerebral vascular caliber are based on measurements of cerebral perfusion pressure and CBF with the skull intact and separate observations of the surface vessels of the brain with the skull removed. Changes of CVR, calculated from measurements of perfusion pressure and blood flow, have long been thought to be due to such changes of vascular caliber as those that are observed in surface vessels when systemic blood pressure is changed, but only recently have observations of vascular caliber and measurements of cerebral blood flow been made in the same experimental preparations. Caution must be exercised in attributing changes of CVR or CBF to changes of vascular caliber, unless blood vessels are observed directly at the same time as CBF is measured.

III. REGULATION OF THE CEREBRAL CIRCULATION

A. Regulatory Responses of Vascular Caliber and Blood Flow

The functional responses of cerebral blood vessels and the cerebral circulation can be expressed simplistically and teleologically as attempts to maintain constant, adequate supplies of glucose and oxygen to the brain and to remove excess CO_2 and other acidic metabolites from the brain. It is generally accepted, although not established, that vascular mechanisms that are intrinsic to the brain are responsible for the changes of vascular caliber, with or without changes of velocity or volume flow of blood, that occur in response to changes of such physical factors as perfusion pressure, systemic blood pressure, and intracranial pressure and such chemical factors as local decreases or increases of metabolites and products of metabolism. Thus it is thought that the brain is capable of "autoregulation" of its blood supply, defined broadly as "the capability of an organ to regulate its blood supply in accordance with its needs."[62] However, cerebral blood vessels react to changes of functional activity of the autonomic nervous system[60,63-66] and to circulating vasoactive agents[61,67-69] (although less than vessels of other organs), and most studies of the regulation of the supply of blood to the brain have not excluded the possibility of neurogenic and humoral influences.

1. Pressure–Flow Relationships

In a narrow sense, the term *autoregulation* has come to denote a particular type of pressure–flow relationship, in which the vessels of an organ have the intrinsic ability to maintain blood flow at relatively constant levels despite moderate changes of perfusion pressure. In normal brain, CBF is relatively stable despite alterations of mean systemic arterial blood pressure in the range of about 60–160 mm Hg (Fig. 2)[70-73]; moderate changes of intracranial[74] or venous pressure[75] likewise do not influence CBF appreciably. Because of the equation used for calculation of cerebral vascular resistance, changes of CVR must accompany changes of perfusion pressure if CBF remains stable. Changes of arte-

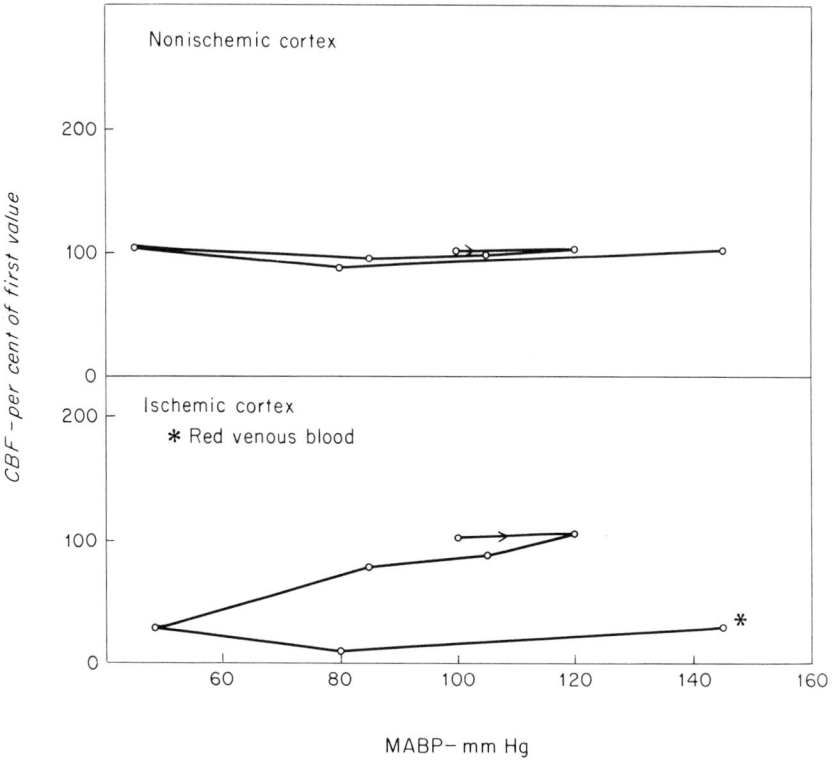

Fig. 2. Effects of successive changes of mean systemic arterial blood pressure, measured in the aorta, on cortical blood flow of the ischemic and the nonischemic cerebral hemispheres of a cat with one middle cerebral artery occluded. CBF was measured by the intra-arterial injection of krypton-85 and is expressed as a percent change from the first value obtained, which was set to 100%. Blood pressure was changed by the intravenous injection of sodium nitroprusside and phenylephrine. CBF of the nonischemic hemisphere did not change appreciably with changes of blood pressure, but CBF of the ischemic hemisphere decreased passively as CBF blood pressure decreased. Ischemia then was severe enough to prevent passive increases of when blood pressure was increased, perhaps because of blockage of the capillary network. Venous hyperoxemia (red venous blood) was noted with increased blood pressure, despite persistent ischemia. [Reproduced from Waltz,[73] p. 615, by permission of the publisher.]

rial caliber have been observed to be associated with changes of perfusion pressure: dilatation of arterial vessels accompanies a decrease of perfusion pressure, and constriction of arterial vessels accompanies an increase of perfusion pressure.[8,76–79] Thus changes of vascular caliber can account for the changes of CVR and the relative stability of CBF of normal brain during moderate changes of perfusion pressure.

There are several theories about the mechanisms responsible for cerebral pressure–flow relationships. According to the "myogenic" theory, vascular smooth muscle responds directly to the changes of tension of vessel walls that are produced by changes of intraluminal pressure; increases of intraluminal pressure cause increased muscular contraction and vasoconstriction. According to the "metabolic" theory, transient changes of blood flow through vessels occur when perfusion pressure changes; as a result of the transient changes of flow, there are changes of the local concentrations of acidic metabolites (such as CO_2), and vascular caliber is altered secondarily because of alterations of extravascular, extracellular pH.

Studies of the rich adrenergic (and probably cholinergic) innervation of the surface arterial vessels of the brain and of the autonomic influences on the cerebral vasculature have led to a reexamination of the "neurogenic" theory for the regulation of the cerebral circulation. Certain vasoactive agents, such as epinephrine[8] and serotonin,[80] can produce changes of arteriolar caliber, although intravascular injection may not be as effective as topical application because of the location of receptor sites on the outside of the vessels.[58,67] Thus a "humoral" theory also must be considered. Mechanisms for the regulation of CBF are discussed more fully in a later section.

CBF remains relatively constant despite changes of heart rate or other systemic factors, unless cardiac output is decreased appreciably by complete heart block. Under these circumstances, a subsequent increase of cardiac output by artificial pacing may cause an increase of CBF that is unrelated to a change of systemic arterial blood pressure, jugular venous blood pressure, or cerebral perfusion pressure.[81,82] Calculated CVR then must decrease; the decrease of CVR probably is not related to changes of caliber of the arterial vessels of the brain, because it is probable that dilatation of regulatory arterioles would have occurred previously in response to the decrease of blood flow. Rather, the apparent decrease of CVR is most likely a fallacious result of the mathematical calculation; the increase of CBF can be attributed to increases of the velocity of the flow of blood through the brain rather than to an increase of intravascular blood volume because of vasodilatation.

2. Relationships to Cerebral Functional Activity

In the waking state, such changes of the functional activity of the brain as intense concentration and the performance of mental arithmetic do not produce detectable increases of total CBF. Regional CBF, however, does change in response to changes of neuronal activity, as in the occipital cortex with stimulation of the retina by light[83] and in other cortical regions during appropriate activities.[84,85] In addition, alterations of cerebral function by sleep, anesthesia, dementia, lowering of the state of consciousness, or activation of cortical electrical activity can influence regional and total CBF and metabolism.[86-96] Presumably, neuronal metabolic activity produces local changes of the uptake of glucose and oxygen and the production of CO_2 and acidic metabolites, with changes of local tissue pH causing vascular regulatory mechanisms to operate to restore local concentrations of metabolites toward normal.

3. Effects of Arterial Carbon Dioxide Tension and Tissue pH

Changes of arterial carbon dioxide tension (Pa_{CO_2}) greatly influence CBF and the cerebral circulation.[28,77,86,97–100] When Pa_{CO_2} increases, CBF increases and calculated CVR decreases because of dilatation of cerebral arterial vessels (Figs. 3 and 4). Hyperventilation causes a reduction of CBF because of arterial constriction. CBF changes approximate 3–4 ml/g/min for each 1 torr change of Pa_{CO_2}, depending on initial values. Decreases of CBF caused by hyperventilation can be adequate to produce an increase of cerebral anaerobic carbohydrate metabolism to abnormal levels.[101]

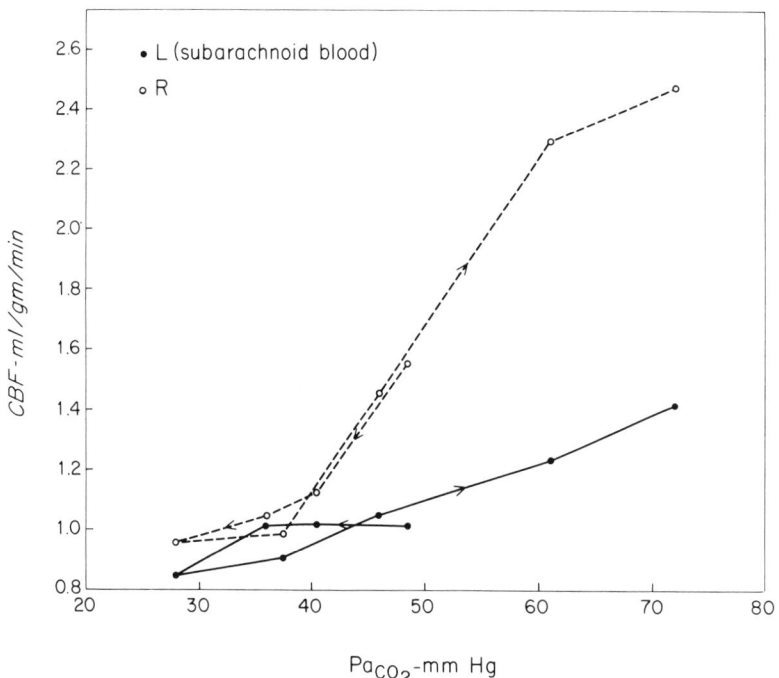

Fig. 3. Effects of successive changes of Pa_{CO_2} on cortical blood flow of a cat, measured by the intra-arterial injection of krypton-85. CBF of the exposed but otherwise normal hemisphere (open circles) responded normally to changes of Pa_{CO_2}. CBF of the hemisphere affected by surgically induced subarachnoid hemorrhage (closed circles) responded to a lesser degree. Any damage to cerebral tissue can result in an impaired— or paradoxical—response to changes of Pa_{CO_2}. [Reproduced from Waltz,[161] p. 68, by permission of the publisher.]

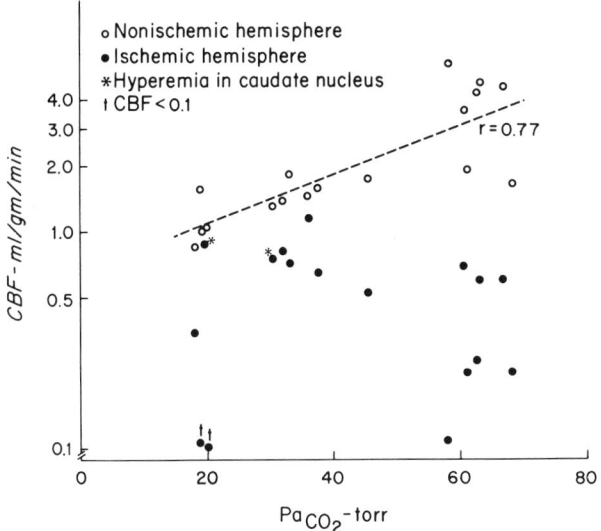

Fig. 4. Relationships between Pa_{CO_2} and regional CBF of the caudate nuclei, measured by autoradiography with antipyrine-C^{14}, in the ischemic and the nonischemic cerebral hemispheres of 17 cats with occlusion of one middle cerebral artery. Regional CBF was decreased on the side of the occluded artery and was less with greater Pa_{CO_2} (paradoxical reaction). Regions with hyperemia were found only in cats with normal or low Pa_{CO_2}.

The effects of Pa_{CO_2} on the cerebral circulation probably are mediated by changes of extravascular, extracellular pH.[102–105] Adaptation to changes of Pa_{CO_2} with reversion of CBF toward normal may occur after exposure of the brain to high or low Pa_{CO_2} for hours or days, as during the hyperventilation that occurs at high altitudes.[106] Adaptation of CBF to Pa_{CO_2} is accompanied by changes of cerebrospinal fluid pH, and presumably also extravascular, extracellular brain tissue pH, toward normal.

Changes of arterial pH in themselves have little influence on CBF.[107,108] Thus it is probable that the endothelium of cerebral blood vessels is more permeable to the CO_2 molecule than to the hydrogen ion. Because of this differential permeability, with increases of Pa_{CO_2} the CO_2 molecule diffuses into extracellular, extravascular spaces to cause secondary local acidosis and vasodilatation. Acetazolamide, a carbonic anhydrase inhibitor, causes increases of CBF that are related only in part to changes of Pa_{CO_2}.[61,109] Presumably, acetazolamide causes increases of extracellular, extravascular pH because of inhibition of the production of diffusible CO_2 and water from bicarbonate and hydrogen ions; extracellular tissue pH remains at lower levels than normal because of the excess of hydrogen ions.

4. Effects of Arterial Oxygen Tension

Despite a relative permeability of cerebral vascular endothelium to the oxygen molecule and a relatively rapid diffusion of oxygen in brain (largely because of diffusion through vascular endothelium), changes of arterial oxygen tension (Pa_{O_2}) are not as potent stimuli to the regulatory mechanisms of the cerebral circulation as are changes of Pa_{CO_2} or tissue pH.[110] Decreases of Pa_{O_2} cause dilatation of cerebral arterial vessels and increases of CBF only at levels near 40–50 torr, or when arterial oxygen saturation becomes less than 40–50%.[110,111] However, decreases of Pa_{O_2} may interfere with cerebral pressure–flow responses.[112] Increases of Pa_{O_2} may produce minimal cerebral arterial constriction and decreases of CBF[113]; more remarkable changes may occur with the administration of oxygen at high pressures in a hyperbaric chamber.[114]

B. Vascular, Neural, and Humoral Mechanisms for the Regulation of the Cerebral Circulation

As mentioned earlier, it has been thought that the brain is capable of autoregulation: that mechanisms intrinsic to cerebral tissue, and probably intrinsic to the blood vessels of cerebral tissue, are responsible for changes of vascular caliber that cause changes of CBF and regulate the cerebral circulation. For example, it is thought that the smooth muscle fibers in the walls of cerebral arterioles react directly to changes of extravascular, extracellular pH to cause dilatation or constriction of regulatory arterioles and changes of CBF in response to changes of functional activity or Pa_{CO_2}. Normal cerebral pressure–flow relationships are thought to be due to changes of arteriolar caliber that are caused either by direct responses of the smooth muscle of vessel walls to changes of intraluminal pressure or by changes of local concentrations of acidic metabolites during transient alterations of CBF that occur as immediate responses to changes of perfusion pressure. However, there is as yet no conclusive proof that the regulation of the cerebral circulation is an intrinsic function of the vascular tissue of the brain.

Numerous anatomical studies of the cerebral blood vessels have shown that arterial vessels on the surface and in the depths of the brain are invested by nerve fibers.[55] Neuromuscular contacts and synaptic vesicles have been described for arteries at the base and over the convexity of the brain, suggesting that vascular nerves are capable of efferent functional activity.[54,56] Synaptic vesicles often have granular characteristics that in other organs are thought to be typical of adrenergic or cholinergic structures; however, synaptic vesicles generally are limited to the adventitia of cerebral blood vessels. Adrenergic neurotransmitter substances, catecholamines and their products, have been demonstrated by fluorescence techniques to be abundant on arteries at the base of the brain, somewhat less abundant on smaller arterial vessels on the surface of the cortex, and relatively sparse on arterioles penetrating the brain.[57,58] The absence of abundant nerve fibers and fluorescent catecholamine products on parenchymal arterioles does not preclude a regulatory function for the autonomic nervous system; the entire brain is supplied by blood that passes through surface arterial vessels, and these are responsible for a major proportion of the total CVR.[115,116] Catecholamines cannot be demonstrated on ipsilateral cerebral vessels after unilateral sympathectomy.[57]

Although cerebral blood vessels are less responsive to vasoactive agents and neurogenic influences than vessels of other organs, adequate stimulation of sympathetic nerves can cause constriction of superficial cortical arterial blood vessels and decreases of blood flow in brain tissue or in cervical arteries and veins.[60,63-66] The responses are inconstant and somewhat capricious, and increases of blood flow from section of or damage to sympathetic nerves are even less consistent.[63,64] Stimulation of the parasympathetic fibers of the seventh cranial nerve may cause an ipsilateral increase of regional CBF if the nerve is sectioned before stimulation, or a decrease of CBF if the nerve is intact, presumably because of antidromic stimulation of brain stem structures.[117] Stimulation or section of sympathetic or parasympathetic nervous structures can modify the regulatory responses of the cerebral vasculature to changes of Pa_{CO_2} and perfusion pressure.[60,63] In addition, stimulation or section of autonomic structures can cause a dissociation between changes of caliber of cerebral arterial vessels and CBF; for example, decreases of cortical blood flow during stimulation of the cervical sympathetic trunk can be associated with decreases, increases, or no change of arterial caliber.[60] Interference with the integrity of brain stem structures by trauma of various sorts can impair the regulatory responses of the cerebral cortical vasculature to changes of perfusion pressure and Pa_{CO_2}, perhaps because of damage to autonomic structures.[64,118]

Many of the results of stimulation or destruction of autonomic structures may be due to effects on relatively large blood vessels in the neck and at the base of the brain.[60,119] In particular, the dissociation between changes of arterial caliber and of cortical blood flow may be related to constriction of cervical or basal vessels.[60] For example, after stimulation of a cervical sympathetic trunk a decrease of cortical blood flow unaccompanied by constriction of cortical arterial vessels may be due to constriction of cervical and basal cerebral blood vessels, with failure of secondary dilatation of surface vessels caused by the sympathetic stimulation overriding the regulatory mechanisms that would normally produce vascular dilatation in response to a decrease of perfusion pressure and CBF resulting from proximal vascular constriction. However, there is now abundant evidence that the autonomic nervous system can influence directly both CBF and the caliber of the surface vessels of the brain.[65,66]

As mentioned previously, vasoactive agents, including autonomic effector agents, can affect the caliber of cerebral blood vessels directly. Adrenergic agents are particularly effective in increasing CVR during hypotension, when cerebral arterial vessels presumably are dilated.[67] Blockade of β-adrenergic receptor sites with propranolol can interfere with the responses of the cerebral vasculature to the toxic agent barium chloride.[120] Blockade of α-adrenergic receptor sites with phenoxybenzamine inhibits the development of cerebral vascular spasm in response to the topical application of blood in experimental subarachnoid hemorrhage[58] and impairs the reactivity of basal cerebral blood vessels.[121]

The latency of the responses of the cerebral vasculature to increases of systemic blood pressure is different from that to decreases of pressure.[112] The differences in latency may be related to the methods used for changing perfusion pressure. Differences of the rate of change of pressure from the resting level may occur, depending on whether drugs are used to produce a rapid change or whether removal of blood is used to produce a slow-onset hypotension. The removal of blood, or cross-clamping the aorta to produce intracranial hypertension, may activate the renin–angiotensin system or cause release of other vasoactive agents.[122] Alternatively, the mechanisms underlying the cerebral vascular responses to increases of systemic blood pressure may differ from those related to decreases of pressure: cholinergic agents may have a particular role in mediating dilatation of cerebral blood vessels.[123]

Thus, there is no conclusive proof that intrinsic mechanisms are responsible for the regulatory responses of the cerebral vasculature, and there is abundant evidence that extrinsic neural and humoral influences can directly affect the blood vessels of the brain. In contrast, there is conclusive evidence that normal cerebral vascular pressure–flow relationships and normal responses to changes of Pa_{CO_2} can occur during blockade of β-adrenergic receptors with propranolol and in the absence of sympathetic innervation of cerebral arterial vessels,[124] although the resting values for blood flow may be greater after sympathectomy than before.[125]

When there is failure of the normal regulatory responses of the cerebral vasculature because of trauma, ischemia, or exposure of the surface of the brain to the atmosphere, there may be no obvious change of the size of the superficial cortical arterial vessels.[8] Thus, extrinsic influences, such as the autonomic nervous system, may be responsible for the "normal" or "standard" cerebral vascular caliber, determined by the "usual" state of contraction of vascular smooth muscle. Changes of vascular caliber and CBF in response to changes of such factors as perfusion pressure and extravascular tissue pH may be superimposed, and mediated or modulated either by extrinsic influences or by mechanisms intrinsic to brain tissue, or both.

IV. DYSFUNCTION OF REGULATORY MECHANISMS

Whatever the actual mechanisms for the regulation of cerebral vascular caliber and CBF may be, the regulatory processes frequently are impaired or abolished by anything that interferes with the integrity of the neural or vascular tissue of the brain.

A. Effects of Ischemia

The most common cause of deranged neuronal function in the brains of humans probably is ischemia, occurring regionally and related to atherosclerosis or other focal disease of blood vessels. Functional derangement of the vascular tissue of the brain also occurs with ischemia; as with neural tissue, the degree of impairment of function is related to the severity, extent, and acuteness of the ischemic process. Although naturally occurring cerebral ischemia, or strokes, can develop in animals, for most experimental studies of pathophysiology cerebral ischemia has been produced by occlusion of one or more major cervical or intracranial blood vessels. Occlusion of individual blood vessels in the neck or on the surface of the brain may not cause a cerebral infarct, because of the flow of blood through collateral channels, although some ischemia may develop.[8,52,126-129] Nearly all the results and interpretations of studies of experimental cerebral ischemia in animals have been confirmed in humans, either by angiography or by measurements of CBF.

1. Impairment of Vascular Reactivity

Occlusion of a middle cerebral artery of an animal produces a cerebral infarct on the same side as the occlusion, but the size and severity are different for different species.[126] CBF in the region of brain that is normally supplied by an occluded middle cerebral artery decreases (Fig. 4); the extent of the decrease is species related as well as variable in individual animals, and also is related to such systemic factors as blood pressure and Pa_{CO_2}.[130] In cats, decreases of CBF may range from 20 to 60%; in squirrel monkeys (*Saimiri sciureus*) CBF decreases may range from 40 to 90%.[27]

The ischemia resulting from occlusion of a middle cerebral artery in an animal almost invariably abolishes the regulatory responses of the cerebral blood vessels in the ischemic region, at least temporarily. For variable periods of time, the normal pressure–flow relationships ("autoregulation") are impaired: the decreased CBF in the ischemic region becomes pressure dependent (Fig.

2).[73,131] Changes of perfusion pressure, such as those caused by changes of systemic arterial blood pressure, will cause corresponding changes of CBF. Regulatory arterial blood vessels cannot respond to changes of perfusion pressure by dilatation or constriction to change CVR and maintain CBF at a relatively constant level.[8] Similarly, regulatory vessels do not respond to changes of Pa_{CO_2} by dilatation or constriction, and often there is no change of CBF when Pa_{CO_2} is changed.[28,33,131,132] At times, there may be a dissociation of regulatory dysfunctions: reactivity to changes of perfusion pressure may be impaired while reactivity to changes of Pa_{CO_2} is preserved, or vice versa.[133]

It has been postulated that the impairment or abolition of the regulatory responses of the cerebral blood vessels that occurs during ischemia is due to maximal vasodilatation[134] produced by decreases of tissue pH from accumulation of acidic metabolites such as carbon dioxide and lactic acid: because of local tissue acidosis, blood vessels are maximally dilated and cannot respond to such stimuli as changes of intraluminal pressure, intraluminal P_{CO_2}, or vasoactive agents. However, during ischemia the surface arterial vessels of the brain do not necessarily dilate; some vessels may, but others do not change size or even may constrict.[8,31,61] Therefore, ischemic impairment or abolition of regulatory responses is not due to generalized maximal vasodilatation. Impairment of responses to changes of Pa_{CO_2} may be related to decreases of tissue pH caused by ischemia, in that increases or decreases of Pa_{CO_2} may not produce appreciable additional changes of pH of severely acidotic tissue. However, impairment of this or other regulatory processes by ischemia may result from ischemic damage to the neural or muscular tissue responsible for changes of vascular caliber. Until more data are available, it is best to ascribe impaired cerebral vascular regulation nonspecifically to impaired vascular reactivity.

2. Paradoxical Reactions

Occasionally, changes of Pa_{CO_2} cause paradoxical reactions of CBF in ischemic brain: CBF may decrease when Pa_{CO_2} is increased (Fig. 4).[28,33,61,131,132] The presumed mechanisms for this paradoxical reaction are as follows: (a) Increases of CBF in normal brain, accompanied by increases of vascular volume, cause increases of intracranial pressure and tissue pressure, which in turn cause increases of local CVR.[28,135] Because ischemic regulatory arterial vessels are nonreactive, there is no compensatory dilatation in response to the increased CVR, and CBF decreases in ischemic regions of brain. (b) Paradoxical reactions occur in animals with craniectomies, in which increases of intracranial pressure cannot occur[28]; thus another mechanism also must be operative. If cardiac output and the cephalad flow of blood through cervical arteries remain the same or increase only slightly when Pa_{CO_2} increases, and if there is vasodilatation in nonischemic brain and no vasodilatation in ischemic brain, then a proportion of the blood flowing toward the head will be diverted to nonischemic brain tissue, or "shunted" from ischemic brain tissue. This phenomenon has been called the *intracerebral steal*,[136] but such redirections of the flow of blood must take place in the major anastomotic channels at the base and over the surface of the brain rather than within the brain, so that the term is inaccurate. Paradoxical reactions to papaverine have been described,[137,138] but other vasodilators do not produce vasodilatation in normal brain adequate to cause a paradoxical reaction.[137]

All of the impaired regulatory responses that have been demonstrated in experimental acute cerebral ischemia in animals also have been demonstrated in humans with cerebral vascular disorders, including paradoxical reactions to vasodilators.[34,136,138–148] In humans as well as animals, the degree of impairment of regulatory responses depends on the severity, extent, and acuteness of the ischemic process. When cerebral ischemia is relatively minor, as with occlusion of a surface branch of a middle cerebral artery in an animal, impairment of reactivity to changes of Pa_{CO_2} may not be adequate to cause a paradoxical reaction.[52,127] Similarly, impairment of regulatory responses is most severe shortly after the onset of cerebral ischemia; reactivity to changes of Pa_{CO_2}, for example, may revert toward normal several days after occlusion of a major cerebral artery.[28] In humans, most studies of CBF in patients with cerebral vascular disorders have been done days to weeks after an ictus, so that paradoxical reactions are uncommon.

3. Implications for Treatment of Acute Cerebral Ischemia

Dysfunction of vascular regulatory mechanisms has specific implications for the treatment of acute cerebral ischemia and infarction. For example, attempts to produce cerebral vasodilatation, as by inhalation of CO_2, may be useless or even potentially harmful for patients with acute, severe cerebral ischemia. On the other hand, CO_2 inhalation may be potentially beneficial (that is, it may cause vasodilatation and increases of CBF in ischemic regions) for patients who have acute but slight or moderate ischemia in small or scattered regions of brain. Less potent vasodilators may be potentially less harmful, but they also are potentially less useful.

Because of the paradoxical reactions of CBF that have been observed with increases of Pa_{CO_2}, hyperventilation to hypocapnia has been advocated for the treatment of acute cerebral ischemia.[149] However, paradoxical reactions in which CBF in ischemic regions increases with subnormal Pa_{CO_2} have not been observed in animals and only rarely in human[28,137,150,151]; hyperventilation to hypocapnia does not protect against cerebral infarction after occlusion of a middle cerebral artery.[152–154] The preliminary results of studies of hyperventilation for the treatment of strokes in humans have been disappointing.[155]

It is unlikely that any therapeutic procedure, including CO_2 inhalation, will be of benefit to patients several days after the onset of acute cerebral ischemia, regardless of effects on CBF: once irreversible impairment of neuronal function has occurred, no therapeutic measures are likely to be beneficial. Yet if there is a marginal zone around an ischemic zone, in which neuronal function is not irreversibly impaired but where CBF is low, measures to maintain or restore the flow of blood may be helpful. A rational choice of therapy for a given patient with cerebral ischemia will depend on measurements of regional blood flow in ischemic brain, or, better, measurements of neuronal function, both at rest and during tests of therapeutic procedures. Unfortunately, safe, simple, and reliable methods for measuring blood flow and neuronal function are not available for humans as yet.

B. Effects of Hypoxemia and Hyperoxemia

Slight to moderate hypoxemia causes impairment but not abolition of the regulatory responses of the cerebral vasculature.[110,112] Even at levels of Pa_{O_2} approaching 40–50 torr, CBF increases with increases of Pa_{CO_2}. Vasodilatation and increases of CBF caused by hypoxia appear to be mediated by decreases of local tissue pH, perhaps because of increases of local concentrations of acidic metabolites (such as lactate) resulting from increased rates of anaerobic metabolic processes. However, the decreases of tissue pH are not adequate to prevent additional vascular reactions to further decreases of tissue pH by increases of Pa_{CO_2}. Severe hypoxemia, such as that caused by respiratory arrest or ventilation with pure nitrogen, abolishes the usual regulatory responses of the cerebral vasculature.

Hyperoxemia produced by ventilation with 100% oxygen causes only slight and variable cerebral vasoconstriction and decreases of CBF. Regulatory responses are not affected by the degree of hyperoxemia that can be achieved by ventilation with oxygen at normal atmospheric pressure.[113] The effects of hyperbaric oxygenation have not been studied adequately; toxic reactions to hyperbaric hyperoxemia—including convulsive activity—may interfere with cerebral vascular regulatory responses.

C. Effects of Increased Intracranial Pressure

Increases of intracranial pressure decrease effective cerebral perfusion pressure and cause increases of tissue pressure that in turn cause increases of CVR if there is no compensatory dilatation of regulatory arterial vessels. Increases of intracranial pressure can be compensated for over a wide range of perfusion pressures, so that CBF usually remains virtually constant until intracranial pressure approaches diastolic blood pressure. Further increases of intracranial pressure then can cause increases of CVR to a point where focal cerebral ischemia may develop.[74]

D. Effects of Sustained Hypertension

Prolonged and severe increases of systemic arterial blood pressure, such as occur with accelerated hypertension in humans or experimental renal hypertension in animals, can cause headache, impairment of consciousness, focal neurological symptoms and signs such as seizures, weakness, and sensory disturbance, and eventual death: the clinical syndrome of hypertensive encephalopathy. Focal constriction of arterial vessels accompanies the manifestations of encephalopathy; constriction of cerebral vessels has been observed in animals and retinal vessels in humans.[156–159] In the past, it has been assumed that the constricted vessels were the abnormal ones and that vasoconstriction was responsible for the clinical findings. It is true that focal arterial and arteriolar constriction could be caused by an overreaction of regulatory vessels to the increased intraluminal pressure, perhaps because of hypersensitization by circulating vasoactive agents. Vasoconstriction itself then could cause focal cerebral ischemia with resulting impairment of neuronal function and the development of neurological symptoms and signs. However, extravasation of fluid and protein tracers can occur in the brain with increases of blood pressure,[160] and focal cerebral edema and ischemic changes usually are related to areas of brain supplied by arterial vessels that are not severely constricted.[159] Therefore, it seems likely that the focal arterial constriction is a normal regulatory response, to increase CVR and limit CBF and the transmission of elevated pressures distally, and that the nonconstricted vessels are the abnormal ones. A local failure of regulatory responses, perhaps because of an impairment of the ability of smooth muscle to maintain contraction, would allow segmental dilatation. Secondary extravasation of fluid, caused by the increased intraluminal pressure, and secondary ischemia, caused by focal edema, then would result in impaired neuronal function. Alternatively, cerebral vasoconstriction in hypertensive encephalopathy may be the result of the release of vasoconstricting agents from ischemic cerebral tissue near dilated, nonreactive vessels.

E. Effects of Miscellaneous Factors

Impairment of the regulatory responses of the cerebral vasculature may follow induced seizures.[133] As with ischemia, there may be a dissociation of the regulatory responses to changes of systemic blood pressure and to changes of Pa_{CO_2}.

Cerebral vascular regulation generally is preserved despite lowered levels of consciousness and decreases of functional cerebral activity. Surgical anesthesia, for example, ordinarily does not interfere with the regulatory responses of the cerebral vasculature unless a very low level of cerebral function is reached.[88]

Anything that interferes with the integrity of cerebral tissue, the cerebral vasculature, or neuronal function may interfere with cerebral vascular regulatory responses. For example, spontaneous or experimental subarachnoid hemorrhage may cause focal vasoconstriction and impaired regulation of CBF (Fig. 3).[161,162] Hypoglycemia will interfere with regulatory responses. Impaired reactivity and paradoxical reactions to changes of Pa_{CO_2} have been demonstrated with intracranial neoplasm, particularly if intracranial pressure is increased, presumably because of vasodilatation in neoplastic tissue at the expense of surrounding marginally ischemic tissue.[163] Cerebral trauma from head injury or surgical exposure also may interfere severely with the regulatory responses of the cerebral vasculature; paradoxical reactions to changes of Pa_{CO_2} have been demonstrated after head injury.[164] In assessments of experimental studies of cerebral vascular physiology and pathophysiology, one must be certain that results are not related to surgical trauma or to exposure of the brain rather than to the experimental variable under study.[165]

V. INTRAVASCULAR PHENOMENA

A. Normal Cerebral Circulation

Normally, the walls of cerebral arterial and venous vessels are not visualized by the usual methods for study of the cerebral circulation. The walls of small blood vessels are nearly transparent to reflected light; usually, a column of blood is seen rather than a vessel wall. Occasionally, when a vessel constricts and the thickness of the wall increases in proportion to the lumen, the wall may be seen.[8]

The flow of blood in cerebral arteries and arterioles is pulsatile, not continuous; the pulse is transmitted to venules, at least in animals with craniectomies.[53] The velocity of the flow of blood in cerebral vessels is greatest in arteries, least in the capillary network, and intermediate in venous vessels. Normally, formed elements of blood cannot be seen in arterial vessels on the surface of the brain, because of the velocity of flow. In certain species, notably cats, formed elements (particularly erythrocytes) can be seen moving in venous vessels ("particulate flow").[8]

Blood is thixotropic; that is, the viscosity of blood varies with its velocity. Viscosity is greatest where velocity is least: in precapillary and postcapillary vessels. Viscosity in capillaries, which can transport red cells serially only, is difficult to calculate.[53,166,167] Despite its thixotropic character, the flow of blood in large- and medium-size vessels is similar to that of Newtonian fluids which have a constant viscosity at different velocities of flow, or shear rates. The flow of blood in vessels is normally nonturbulent, but smooth, "laminar," or "streamlined." The laminae are of definite thickness and are composed largely of plasma near vessel walls and red cells centrally ("axial streaming"); the laminae are affected differentially by systemic influences.[167] Lamination persists when vessels branch or merge; in veins, contrasting laminae may remain separate and unmixed for considerable distances.[8] Axial streaming and laminar flow occur in vessels as small as precapillary arterioles. Near irregularities of vessels, such as atherosclerotic stenosis, or near vessels with abnormally high linear velocities, such as arterial–venous fistulas or shunts, turbulence may occur and energy may be expended as sound, causing bruits. Arterial–venous shunts and thoroughfare channels are not known to exist normally in the cerebral vasculature.[7,8,168]

The red blood cell concentration (hematocrit value) is less in blood vessels in the brain than it is in systemic and cervical blood vessels supplying the brain.[169,170] The lower hematocrit value in the brain is related to the velocity of the red cells, which is greater than that of plasma because most red cells travel centrally in the blood in the rapid axial stream.[171] The difference between the velocities of red cells and plasma is greater in smaller vessels than in larger ones; thus red cells move faster through the small vessels of the brain than does plasma, causing a lowering of the average hematocrit value. The lower red cell concentration of the blood tends to decrease blood viscosity, compensating in part for the thixotropic increase of viscosity of blood at lower velocities of flow.

The flow of blood in the arterial vessels of the surface of the brain usually is in only one direction. In the retina, and probably in the brain, all capillaries are always open: blood may flow through any capillary at any time. In large surface cerebral veins, the flow of blood is unidirectional, but in cortical venules (and, rarely, in small cortical arterioles) the flow of blood may stop or reverse, from changes of pressure gradients and vascular resistance.[8,53] These changes may be related to variations in the functional activity of regions of brain causing variations of the local concentrations of acidic metabolites and thus of vascular caliber and blood flow.

B. Abnormal Circulatory States

1. Changes Due to Ischemia

Experimental cerebral ischemia, produced in cats and monkeys by occlusion of a middle cerebral artery, causes changes in the surface vessels of the brain that may also occur in humans during the process of cerebral infarction.[8,31,52,127,129,172–175] Early changes that occur in association with ischemia include darkening of the color of venous blood, a decrease of the velocity of the flow of blood in venous vessels, pallor of the cortex (from constriction or decrease of the flow of blood through precapillary vessels and capillaries), and aggregation of the formed elements of the blood, producing "sludging." These changes occur chiefly in venous vessels but may occur in arterial vessels, particularly if the velocity of the flow of blood decreases severely. As mentioned earlier, arterial vessels do not always dilate despite the decrease of perfusion pressure and presumed decrease of local tissue pH that results from interference with the arterial supply to a region of brain. Although some vessels dilate, others do not change caliber, and some may constrict, perhaps because of the release of vasoactive agents from ischemic brain tissue.

If ischemia persists, blood may stagnate in venous vessels, and blood cells may aggregate in large clumps. Perivenous hemorrhages may develop, and plasma proteins (or radioactive indicators or dyes) may leak into cerebral tissue. White thrombi, composed largely of platelets, may form in venous vessels. Brain swelling or cerebral edema may be minimal or striking. The decreases of the velocity of flow and the aggregation of formed blood elements usually cause a decrease of blood flow in brain tissue. Presumably, there is irregular and patchy blockage of the capillary network, with obstruction to the flow of blood in some capillaries and collapse of other capillaries. Hyperemia and venous hyperoxemia will be described later.

2. Effects of Other Abnormal States

Local cerebral trauma and hemorrhage cause many of the same intravascular phenomena as ischemia, but vasoconstriction (spasm), hyperemia, and edema are more frequent and generally more severe.[58,175–182] In contrast to ischemia, the higher perfusion pressures and greater CBF may be responsible for extravasation of fluid and electrolytes. Cerebral edema that is caused by experimental trauma may be ameliorated by adrenal corticosteroids; evidence about the efficacy of steroids in reducing the edema caused by ischemia is not conclusive.

Pathological states of the blood, such as proteinopathies, hemoglobinopathies, erythrocythemia, and polycythemia, may cause aggregation of cellular elements of the blood in clumps or stacks (rouleaux) in the cerebral circulation.[8,183] Pathological states generally are associated with increases of the viscosity of the blood and secondarily of CVR.[167,183–186] Such increases of CVR may aggravate situations of marginal cerebral ischemia to cause infarction, or impairment of neuronal function. Anemia and other conditions that cause decreases of blood viscosity are associated with increases of the velocity of the flow of blood in cerebral blood vessels[187]; the increases of velocity may disrupt laminar flow. At very high velocities, turbulence may interfere with the delivery of oxygen and other substrates to neuronal tissue, and neuronal function may be affected.

3. Reversibility of Intravascular Phenomena

When an occluding clip is removed from a middle cerebral artery of an animal within several hours of occlusion, the surface vessels of the brain can be seen to dilate, the cortex may become reddish in color, and the velocity of the flow of blood increases in vessels that previously had very slow or virtually no flow.[12,188,189] Aggregation of the formed elements of the blood decreases; many clumps are swept from venous vessels, presumably into the systemic circulation. CBF may increase to preocclusion levels or may increase above normal, indicating reactive hyperemia to the preexisting ischemia. In humans, reactive hyperemia may occur after the release of an occluding clip from an internal carotid artery during endarterectomy for atherosclerotic stenosis.[146,148]

Studies with carbon black have shown that not all of the intravascular phenomena caused by ischemia are reversible.[9,11,190] If both the arterial input and the venous output of the brain are occluded for even a few minutes, blockage of the capillary network may not be reversible. Particles of carbon black will not fill many capillaries, particularly those in the depths of the brain, in the brain stem, and in the cerebellum: the "no-reflow" phenomenon. If only the arterial input to a region of brain is occluded, however, the failure of filling, or "no-reflow," is less widespread and less severe.[10,12,188,189,191,192] Moreover, the reversibility of ischemic intravascular phenomena, as measured by filling of the capillary network by carbon black, depends on the duration of the ischemic process and such systemic factors as blood pressure, blood volume, and blood osmolality.[193-196]

The reversibility of the intravascular phenomena caused by ischemia is of particular relevance to the treatment of cerebral vascular disorders. If the flow of blood cannot be restored to ischemic brain because of blockage of the capillary network, caused either by swelling of capillary endothelium[190] or by aggregation of formed elements of blood,[8] then attempts at surgical repair of an occluded or obstructed vessel would be fruitless. However, flow through cerebral vessels can be restored in animals despite occlusion of a major end artery of the brain for as long as 3–4 hr.[188,189] It is likely that irreversible neuronal damage, rather than irreversible intravascular phenomena, limits the time during which a surgical procedure will be beneficial to a patient with an acute embolic or thrombotic occlusion of a major cerebral arterial vessel.

VI. SECONDARY EFFECTS OF DISTURBED REGULATION

A. Hyperemia and Venous Hyperoxemia

Although incongruous, hyperemia (increases of the volume of flow of blood through tissue) and venous hyperoxemia (increases of oxygen saturation, manifested by reddening of the color of venous blood) occur in association with cerebral ischemia. Hyperemia and venous hyperoxemia have been demonstrated in cats and squirrel monkeys with occlusion of a middle cerebral artery,[8,28, 31–33,52,61,73,129,131,137,154,180,188,189,197] and hyperemia has been found in humans with acute cerebral infarcts.[139–141,151,198] Correlates of hyperemia, capillary "blushing" or increased capillary density and early filling of veins, frequently have been demonstrated by angiography in humans and animals with cerebral vascular disorders.[175,199–202] Hyperemia and venous hyperoxemia also occur in animals and humans in association with cerebral trauma or neoplasm and during the recovery phase after transient ischemia or hypoxia.[146,148,163,180,203–205] Hyperemia and venous hyperoxemia both have been included in the term *luxury perfusion*,[149] although hypermia and hyperoxemia cannot be equated.

1. Reactive Hyperemia

Hyperemia that develops after transient ischemia or hypoxia can be defined as "reactive" hyperemia. Reactive hyperemia of this type may be due to decreases of tissue pH caused by accumulation of acidic metabolites during the period of hypoxia or ischemia, with vasodilatation and increased CBF continuing even after restoration of the delivery of oxygen adequate for neuronal function. Alternatively, hypoxia or ischemia may cause impairment of the regulatory responses of the cerebral vasculature, so that the mechanisms that normally limit CBF to the needs of brain tissue are transiently impaired and inoperative. Reactive hyperemia frequently is associated with venous hyperoxia.

2. Hyperemia and Continuing Ischemia

Hyperemia in association with ongoing ischemia may have several causes.[32] If hyperemia occurs soon after the onset of acute ischemia, near the margin or periphery of the ischemic zone, it may be related to diffusion of acidic metabolites from ischemic regions to nonischemic regions of brain, with resultant decreases of local tissue pH and increases of regional CBF. Ischemic neuronal changes have been found in some hyperemic regions of the brains of cats shortly after occlusion of a middle cerebral artery; thus reactive hyperemia can develop during acute cerebral ischemia (Figs. 1 and 4).[32] Presumably, ischemia adequate to produce neuronal damage is followed by restoration of the flow of blood to the ischemic region, perhaps by increases of blood flow through collateral vascular channals or by the resolution of such intravascular phenomena as aggregation or "sludging."

Hyperemia also can occur at some distance from an ischemic zone. In experimental studies, such hyperemia may be related to surgical procedures or to increases of blood flow through collateral channels.

Late after the onset of cerebral ischemia, hyperemia can develop because of proliferation and hypertrophy of capillaries and other blood vessels in regions that once were ischemic.

3. Venous Hyperoxemia and Continuing Ischemia

Venous hyperoxemia, manifested by reddening of venous blood, can be found in relation to acute focal cerebral ischemia without associated hyperemia (Fig. 2).[28,73] In this situation, regional CBF is abnormally low, and the ischemic cerebral tissue is not using all of the decreased amount of oxygen that is being delivered to it. When venous hyperoxemia occurs without hyperemia, it is most likely the result of impairment of neuronal function: the metabolic activity of cerebral tissue is decreased, and oxygen is not utilized. An alternative explanation for venous hyperoxemia without hyperemia is blockage of the capillary network.[8,180] If only a few capillaries remain open in an ischemic region of brain, the velocity of the flow of blood through these open capillaries may be increased even though CBF of the block of tissue is decreased. Because of the increased velocity of flow, not all the available oxygen can be extracted from the blood and utilized by the surrounding cerebral tissue.

4. Hyperemia and Cerebral Infarction

The relationships between hyperemia and ischemia, and between hyperemia and such systemic factors as arterial blood pressure and Pa_{CO_2}, are unknown. It is possible that reactive hyperemia may be beneficial to ischemic brain: if CBF increases in a region that was previously ischemic, neuronal function may improve or the development or extension of an infarct may be prevented. However, if reactive hyperemia is caused by ischemic damage to regulatory arterial vessels so that there is merely a passive increase of CBF, it may have no influence on the processes of cerebral infarction. Finally, it is possible that reactive hyperemia may be detrimental to ischemic brain, by increasing the risk of hemorrhagic infraction or by causing increased cerebral edema and the extension of an ischemic infarct.

Reactive hyperemia occurring within the first day or two after the onset of cerebral ischemia is not associated with an increase of the extent or severity of cerebral edema, cerebral ischemia, or cerebral infarction.[32,33] Early hyperemia can be related either to increases of Pa_{CO_2} (produced before arterial occlusion)[130] or to decreases of Pa_{CO_2} (produced after arterial occlusion) (Fig. 4).[33] Hyperemia present 5–8 days after arterial occlusion can be related to increases of Pa_{CO_2}[154]; when acetazolamide is used to cause the increases of Pa_{CO_2}, the hyperemia then is associated with an increase of the extent and severity of edema, ischemia, and infarction.[206] Although no situations have yet been found in which hyperemia protects against the development of cerebral infarction, the exact relationships among hyperemia, edema, and infarction remain to be elucidated.

B. Cerebral Edema

Cerebral edema or brain swelling commonly accompanies any injury to cerebral tissue, including ischemia. Despite decreased CBF, ischemic cerebral edema may be massive enough to cause displacement of midline structures, as measured by angiography or echoencephalography, and to cause tentorial herniation with damage to the brain stem and death. Edema, rather than direct impairment of neuronal function, is the usual cause of death in animals with experimental occlusion of a middle cerebral artery.

The exact mechanisms underlying the development of ischemic cerebral edema are unknown, except that the cerebral vasculature apparently becomes more permeable to the passage of ions and molecules.[207] When cerebral edema is associated with neoplasm or head inujry, it may be modified by adrenal corticosteroids. However, in humans with cerebral neoplasms or head injury, CBF frequently is near normal or increased; with ischemia, the resolution of edema might be delayed or prevented because of decreased CBF even if the differential characteristics of endothelial permeability could be restored. Additional information about ischemic cerebral edema awaits the development of a suitable radioactive indicator or other means for assessment; indicators available at present are more useful for assessments of damage to the "blood–brain barrier" or disruption of cellular membranes than of the accumulation of water in brain tissue.

C. "Diaschisis" and Generalized Decreases of CBF

CBF may be decreased in regions remote from damaged brain tissue, whether the damage is caused by ischemia, trauma, neoplasm, or hemorrhage. Such decreases of blood flow have been ascribed to "diaschisis."[208] However, when decreases of CBF remote from injured brain are more than minimal, there usually is generalized atherosclerosis, secondary vasoconstriction (spasm), a generalized decrease of cerebral function manifested by dementia, or an altered level of consciousness that may be related to impaired function of the reticular activating system. In animals, there is no consistent decrease of cortical blood flow in the cerebral hemisphere opposite an occluded middle cerebral artery.[209]

D. Cardiac and Systemic Effects

Cardiac arrhythmias, alterations of pulse rate or systemic blood pressure, and alterations of respiratory patterns may be found in humans or animals with acute damage to brain tissue from ischemia, trauma, or other causes. Such effects may be due to impairment of neuronal function of brain stem centers, either primarily or secondary to cerebral edema and tentorial herniation, or may be contributing factors to cerebral ischemia.

VII. CONCLUSION

Regulatory mechanisms exist that adjust the supply of blood to cerebral tissue to meet the needs of the tissue for oxygen and glucose and for the removal of acidic metabolites. Constriction and dilatation of cerebral arterial vessels are basic components of these mechanisms. Regulatory functions may be intrinsic to the cerebral vasculature and be related to contraction or relaxation of the smooth muscle of the walls of blood vessels in response to such factors as intraluminal pressure and extravascular, extracellular tissue pH. Alternatively, the regulatory mechanisms may be dependent on the nerve fibers that are associated with arterial vessels. The nerve fibers may be affected directly by intraluminal pressure and tissue pH, but there is evidence that cerebral vascular regulatory responses can be modulated or modified by changes of function of the autonomic nervous system, from which the perivascular nerves arise. The regulatory mechanisms of the cerebral vasculature are affected by anything that interferes with the integrity of cerebral neural or vascular tissue, including ischemia, hypoxia, hypoglycemia, trauma, hemorrhage, and neoplasm; the impairment of regulatory mechanisms is nonspecific. Because the vasculature of abnormal brain does not respond to stimuli in the same way as the vasculature of normal brain, the treatment of abnormal cerebral circulatory states must depend on a knowledge of the reactivity of the blood vessels in the abnormal region for each patient; such knowledge unfortunately cannot be provided by the methods available at present for the study of the cerebral circulation.

REFERENCES

1. A. Krogh, The rate of diffusion of gases through animal tissues, with some remarks on the coefficient of invasion, *J. Physiol. (London)* **52**:391–408 (1918).
2. A. V. Hill, The diffusion of oxygen and lactic acid through tissues, *Proc. Roy. Soc. Ser. B* **104**:39–96 (1928).
3. F. J. W. Roughton, Diffusion and chemical reaction velocity in cylindrical and spherical systems of physiological interest, *Proc. Roy. Soc. Ser. B* **140**:203–229 (1952).
4. V. A. Levin, J. D. Fenstermacher, and C. S. Patlak, Sucrose and inulin space measurements of cerebral cortex in four mammalian species, *Am. J. Physiol.* **219**:1528–1533 (1970).
5. V. A. Levin, T. H. Milhorat, J. D. Fenstermacher, M. K. Hammock, and D. P. Rall, Physiological studies on the development of obstructive hydrocephalus in the monkey, *Neurology* **21**:238–246 (1971).
6. W. Lierse and E. Horstmann, Quantitative anatomy of the cerebral vascular bed with especial emphasis on homogeneity and inhomogeneity in small parts of the gray and white matter, *Acta Neurol. Scand. Suppl.* **14**:15–19 (1965).
7. R. L. de C. H. Saunders and M. A. Bell, X-ray microscopy and histochemistry of the human cerebral blood vessels, *J. Neurosurg.* **35**:128–140 (1971).
8. A. G. Waltz and T. M. Sundt, Jr., The microvasculature and microcirculation of the cerebral cortex after arterial occlusion, *Brain* **90**:681–696 (1967).
9. A. Ames, III, R. L. Wright, M. Kowada, J. M. Thurston, and G. Majno, Cerebral ischemia. II. The no-reflow phenomenon, *Am. J. Pathol.* **52**:437–453 (1968).
10. K. A. Hossmann and K. Sato, Effect of ischemia on the function of the sensorimotor cortex in cat, *Electroencephalog. Clin. Neurophysiol.* **30**:535–545 (1971).
11. N. R. Clendenon, N. Allen, T. Kamatsu, L. Liss, W. A. Gordon, and K. Heimberger, Biochemical alterations in the anoxic–ischemic lesion of rat brain, *Arch. Neurol. (Chicago)* **25**:432–448 (1971).
12. R. M. Crowell and Y. Olsson, Impaired microvascular filling after focal cerebral ischemia in monkeys, *J. Neurosurg.* **36**:303–309 (1972).
13. S. S. Kety and C. F. Schmidt, The determination of cerebral blood flow in man by the use of nitrous oxide in low concentrations, *Am. J. Physiol.* **143**:53–66 (1945).
14. S. S. Kety and C. F. Schmidt, The nitrous oxide method for the quantitative determination of cerebral blood flow in man: Theory, procedure and normal values, *J. Clin. Invest.* **27**:476–484 (1948).
15. S. S. Kety, Quantitative determination of cerebral blood flow in man, *Methods Med. Res.* **1**:204–217 (1948).
16. J. S. Meyer and Y. Shinohara, A method for measuring cerebral hemisphere blood flow and metabolism, *Stroke* **1**:419–431 (1970).
17. N. A. Lassen and O. Munck, The cerebral blood flow in man determined by the use of radioactive krypton, *Acta Physiol. Scand.* **33**:30–49 (1955).
18. L. C. McHenry, Jr., Quantitative cerebral blood flow determination: Application of a krypton[85] desaturation technique in man, *Neurology* **14**:785–793 (1964).
19. O. M. Reinmuth, P. Scheinberg, and B. Bourne, Total cerebral blood flow and metabolism, *Arch. Neurol. (Chicago)* **12**:49–66 (1965).
20. K. L. Zierler, Equations for measuring blood flow by external monitoring of radioisotopes, *Circ. Res.* **16**:309–321 (1965).
21. K. Hoedt-Rasmussen, Regional cerebral blood flow: The intra-arterial injection method, *Acta Neurol. Scand. Suppl.* **27**:1–81 (1967).
22. M. M. Ter-Pogossian, J. O. Eichling, D. O. Davis, M. J. Welch, and J. M. Metzger, The determination of regional cerebral blood flow by means of water labeled with radioactive oxygen 15, *Radiology* **93**:31–40 (1969).
23. O. B. Paulson, S. Cronqvist, J. Risberg, and F. I. Jeppesen, Regional cerebral blood flow: A comparison of 8-detector and 16-detector instrumentation, *J. Nucl. Med.* **10**:164–173 (1969).

24. A. G. Waltz, A. R. Wanek, and R. E. Anderson, Comparison of analytic methods for calculation of cerebral blood flow after intracarotid injection of ^{133}Xe, *J. Nucl. Med.* **13:** 66–72 (1972).

25. E. J. Potchen, D. O. Davis, T. Wharton, R. Hill, and J. M. Taveras, Regional cerebral blood flow in man. I. A study of the xenon-133 washout method, *Arch. Neurol. (Chicago)* **20:**378–383 (1969).

26. D. H. Ingvar and N. A. Lassen, Regional blood flow of the cerebral cortex determined by krypton[85], *Acta Physiol. Scand.* **54:**325–338 (1962).

27. A. G. Waltz, T. M. Sundt, Jr., and C. A. Owen, Jr., Effect of middle cerebral artery occlusion on cortical blood flow in animals, *Neurology* **16:**1185–1190 (1966).

28. A. G. Waltz, Effect of Pa_{CO_2} on blood flow and microvasculature of ischemic and nonischemic cerebral cortex, *Stroke* **1:**27–37 (1970).

29. M. Reivich, J. Jehle, L. Sokoloff, and S. S. Kety, Measurement of regional cerebral blood flow with antipyrine-^{14}C in awake cats, *J. Appl. Physiol.* **27:**296–300 (1969).

30. M. Reivich, R. Slater, and N. Sano, Further studies on experimental models of cerebral clearance curves, *in* "Cerebral Blood Flow: Clinical and Experimental Results" (M. Brock, C. Fieschi, D. H. Ingvar, N. A. Lassen, and K. Schürmann, eds.) pp. 8–10, Springer-Verlag, Berlin (1969).

31. R. D. G. Blair and A. G. Waltz, Regional cerebral blood flow during acute ischemia: Correlation of autoradiographic measurements with observations of cortical microcirculation, *Neurology* **20:**802–808 (1970).

32. T. Yamaguchi, A. G. Waltz, and H. Okazaki, Hyperemia and ischemia in experimental cerebral infarction: Correlation of histopathology and regional blood flow, *Neurology* **21:** 565–578 (1971).

33. T. Yamaguchi, F. Regli, and A. G. Waltz, Effect of Pa_{CO_2} on hyperemia and ischemia in experimental cerebral infarction, *Stroke* **2:**139–147 (1971).

34. J. E. Rees, J. W. D. Bull, G. H. DuBoulay, J. Marshall, R. W. Ross Russell, and L. Symon, The comparative analysis of isotope clearance curves in normal and ischemic brain, *Stroke* **2:**444–451 (1971).

35. C. Fieschi, L. Bozzao, A. Agnoli, M. Nardini, and A. Bartolini, The hydrogen method of measuring local blood flow in subcortical structures of the brain: Including a comparative study with the ^{14}C antipyrine method, *Exptl. Brain Res. (Berlin)* **7:**111–119 (1969).

36. R. Cooper, Local changes of intra-cerebral blood flow and oxygen in humans, *Med. Electron. Biol. Eng.* **1:**529–536 (1963).

37. E. Betz, D. H. Ingvar, N. A. Lassen, and F. W. Schmahl, Regional blood flow in the cerebral cortex, measured simultaneously by heat and inert gas clearance, *Acta Physiol. Scand.* **67:**1–9 (1966).

38. J. N. Hayward and M. A. Baker, A comparative study of the role of the cerebral arterial blood in the regulation of brain temperature in five mammals, *Brain Res.* **16:**417–440 (1969).

39. R. E. Anderson, A. G. Waltz, T. Yamaguchi, and R. D. Ostrom, Assessment of cerebral circulation (cortical blood flow) with an infrared microscope, *Stroke* **1:**100–103 (1970).

40. B. L. Mallett and N. Veall, The measurement of regional cerebral clearance rates in man using xenon-133 inhalation and extracranial recording, *Clin. Sci.* **29:**179–191 (1965).

41. K. B. Jensen, K. Høedt-Rasmussen, E. Sveinsdottir, B. M. Stewart, and N. A. Lassen, Cerebral blood flow evaluated by inhalation of ^{133}Xe and extracranial recording: A methodological study, *Clin. Sci.* **30:**485–494 (1966).

42. W. D. Obrist, H. K. Thompson, Jr., C. H. King, and H. S. Wang, Determination of regional cerebral blood flow by inhalation of 133-xenon, *Circ. Res.* **20:**124–135 (1967).

43. W. D. Obrist, H. K. Thompson, H. S. Wang, and S. Cronqvist, A simplified procedure for determining fast compartment rCBFs by [133]xenon inhalation, *in* "Brain and Blood Flow: Proceedings of the Fourth International Symposium on the Regulation of Cerebral ʔlood Flow" (R. W. Ross Russell, ed.) pp. 11–15, Pitman, London (1971).

44. W. H. Oldendorf and M. Kitano, Isotope study of brain blood turnover in vascular disease, *Arch. Neurol. (Chicago)* **12**:30–35 (1965).

45. J. O. Rowan, J. N. Cross, G. M. Tedeschi, and W. B. Jennett, Limitations of circulation time in the diagnosis of intracranial disease, *J. Neurol. Neurosurg. Psychiat.* **33**:739–744 (1970).

46. G. Burke and A. Halko, Cerebral blood flow studies with sodium pertechnetate Tc99*m* and the scintillation camera, *J. Am. Med. Ass.* **204**:319–324 (1968).

47. R. Janeway, G. Schweitzer, D. Addario, R. L. Witcofski, and C. D. Maynard, Precision analysis of intravenous rapid sequence scintiphotography: Further experience with the gamma camera, *in* "Brain and Blood Flow: Proceedings of the Fourth International Symposium on the Regulation of Cerebral Blood Flow" (R. W. Ross Russell, ed.) pp. 48–53, Pitman, London (1971).

48. D. C. Moses, A. E. James, Jr., H. W. Strauss, and H. N. Wagner, Jr., Regional cerebral blood flow estimation in the diagnosis of cerebrovascular disease, *J. Nucl. Med.* **13**:135–141 (1972).

49. G. Nylin, B. P. Silfverskiold, S. Löfstedt, O. Regnström, and S. Hedlund, Studies on cerebral blood flow in man, using radioactive-labeled erythrocytes, *Brain* **83**:293–335 (1960).

50. F. R. Hellinger, B. M. Bloor, and J. J. McCutcheon, Total cerebral blood flow and oxygen consumption using the dye-dilution method, *J. Neurosurg.* **19**:964–969 (1962).

51. H. C. Smith, R. L. Frye, D. E. Donald, G. D. Davis, J. R. Pluth, R. E. Sturm, and E. H. Wood, Roentgen videodensitometric measure of coronary blood flow, *Mayo Clin. Proc.* **46**:800–806 (1971).

52. Y. L. Yamamoto, K. M. Phillips, C. P. Hodge, and W. Feindel, Microregional blood flow changes in experimental cerebral ischemia: Effects of arterial carbon dioxide studied by fluorescein angiography and xenon^{133} clearance, *J. Neurosurg.* **35**:155–166 (1971).

53. W. I. Rosenblum, Erythrocyte velocity and a velocity pulse in minute blood vessels on the surface of the mouse brain, *Circ. Res.* **24**:887–892 (1969).

54. E. Nelson and M. Rennels, Neuromuscular contacts in intracranial arteries of the cat, *Science* **167**:301–302 (1970).

55. E. Nelson and M. Rennels, Innervation of intracranial arteries, *Brain* **93**:475–490 (1970).

56. T. Iwayama, J. B. Furness, and G. Burnstock, Dual adrenergic and cholinergic innervation of the cerebral arteries of the rat, *Circ. Res.* **26**:635–646 (1970).

57. K. C. Nielsen and C. Owman, Adrenergic innervation of pial arteries related to the circle of Willis in the cat, *Brain Res.* **6**:773–776, (1967).

58. R. A. R. Fraser, B. M. Stein, R. E. Barrett, and J. L. Pool, Noradrenergic mediation of experimental cerebrovascular spasm, *Stroke* **1**:356–362 (1970).

59. L. Alexander and T. J. Putnam, Pathological alterations of cerebral vascular patterns, *Res. Publ. Ass. Res. Nerv. Ment. Dis.* **18**:471–543 (1938).

60. S. Kobayashi, A. G. Waltz, and A. L. Rhoton, Effects of stimulation of cervical sympathetic nerves on cortical blood flow and vascular reactivity, *Neurology* **21**:297–302 (1971).

61. F. Regli, T. Yamaguchi, and A. G. Waltz, Responses of surface arteries and blood flow of ischemic and nonischemic cerebral cortex to aminophylline, ergotamine tartrate, and acetazolamide, *Stroke* **2**:461–470 (1971).

62. P. C. Johnson, Review of previous studies and current theories of autoregulation, *Circ. Res. 15*, Suppl. 1, 2–9 (1964).

63. I. M. James, R. A. Millar, and M. J. Purves, Observations on the extrinsic neural control of cerebral blood flow in the baboon, *Circ. Res.* **25**:77–93 (1969).

64. J. S. Meyer, T. Teraura, K. Sakamoto, and A. Kondo, Central neurogenic control of cerebral blood flow, *Neurology* **21**:247–262 (1971).

65. T. Yamaguchi and A. G. Waltz, Non-uniform response of regional cerebral blood flow to stimulation of cervical sympathetic nerve, *J. Neurol. Neurosurg. Psychiat.* **34**:602–606 (1971).

66. W. I. Rosenblum, Neurogenic control of cerebral circulation, *Stroke* **2**:429–439 (1971).

67. E. S. Gabrielyan and A. M. Harper, Effect of noradrenalin on regional cerebral blood flow depending on initial state of the mean arterial pressure, *Byull. Eksp. Biol. Med.* **69** (6): 9–11 (1970).

68. W. D. Anderson and W. G. Kubicek, Effects of betahistine HCl, nicotinic acid, and histamine on basilar flow in anesthetized dogs, *Stroke* **2**:409–415 (1971).

69. I. C. Denton, Jr., R. P. White, and J. T. Robertson, The effects of prostaglandins E_1, A_1, and $F_{2\alpha}$ on the cerebral circulation of dogs and monkeys, *J. Neurosurg.* **36**:34–42 (1972).

70. N. A. Lassen, Autoregulation of cerebral blood flow, *Circ. Res. 15*, Suppl. 1, 201–204 (1964).

71. A. M. Harper, Autoregulation of cerebral blood flow: Influence of the arterial blood pressure on the blood flow through the cerebral cortex, *J. Neurol. Neurosurg. Psychiat.* **29**: 398–403 (1966).

72. C. E. Rapela, H. D. Green, and A. B. Denison, Baroreceptor reflexes and autoregulation of cerebral blood flow in the dog, *Circ. Res.* **21**:559–568 (1967).

73. A. G. Waltz, Effect of blood pressure on blood flow in ischemic and in nonischemic cerebral cortex: The phenomena of autoregulation and luxury perfusion, *Neurology* **18**: 613–621 (1968).

74. N. N. Zwetnow, Effects of increased cerebrospinal fluid pressure on the blood flow and on the energy metabolism of the brain: An experimental study, *Acta Physiol. Scand. Suppl.* **339**:1–31 (1969).

75. B. Ekström-Jodal, Effect of increased venous pressure on cerebral blood flow in dogs, *Acta Physiol. Scand. Suppl.* **350**:51–61 (1970).

76. M. Fog, Cerebral circulation: II. Reaction of pial arteries to increase in blood pressure, *AMA Arch. Neurol. Psychiat.* **41**:260–268 (1939).

77. W. I. Rosenblum, Cerebral microcirculation: A review emphasizing the interrelationships of local blood flow and neuronal function, *Angiology* **16**:485–507 (1965).

78. R. W. Ross Russell, J. P. Simcock, I. M. S. Wilkinson, and C. C. Frears, The effect of blood pressure changes on the leptomeningeal circulation of the rabbit, *Brain* **93**:491–504 (1970).

79. L. Symon, Regional vascular reactivity in the middle cerebral arterial distribution: An experimental study in baboons, *J. Neurosurg.* **33**:532–541 (1970).

80. W. H. Bell, T. M. Sundt, Jr., and J. D. Nofzinger, The response of cortical vessels to serotonin in experimental cerebral infarction, *J. Neurosurg.* **26**:203–212 (1967).

81. I. A. Sulg, S. Cronqvist, H. Schüller, and D. H. Ingvar, The effect of intracardial pacemaker therapy on cerebral blood flow and electroencephalogram in patients with complete atrioventricular block, *Circulation* **39**:487–494 (1969).

82. W. Shapiro and N. P. S. Chawla, Observations on the regulation of cerebral blood flow in complete heart block, *Circulation* **40**:863–870 (1969).

83. L. Sokoloff, Local cerebral circulation at rest and during altered cerebral activity induced by anesthesia or visual stimulation, *in* "Regional Neurochemistry" (S. S. Kety and J. Elkes, eds.) pp. 107–117, Pergamon Press, New York (1961).

84. J. Risberg and D. H. Ingvar, Regional changes in cerebral blood volume during mental activity, *Exptl. Brain Res. (Berlin)* **5**:72–78 (1968).

85. J. Olesen, Contralateral focal increase of cerebral blood flow in man during arm work, *Brain* **94**:635–646 (1971).

86. N. A. Lassen, Cerebral blood flow and oxygen consumption in man, *Physiol. Rev.* **39**: 183–238 (1959).

87. M. Reivich, G. Isaacs, E. Evarts, and S. Kety, The effect of slow wave sleep and REM sleep on regional cerebral blood flow in cats, *J. Neurochem.* **15**:301–306 (1968).

88. J. D. Michenfelder, G. A. Gronert, and K. Rehder, Neuroanesthesia, *Anesthesiology* **30**: 65–100 (1969).

89. M. D. O'Brien and B. L. Mallett, Cerebral cortex perfusion rates in dementia, *J. Neurol. Neurosurg. Psychiat.* **33**:497–500 (1970).

90. W. D. Obrist, E. Chivian, S. Cronqvist, and D. H. Ingvar, Regional cerebral blood flow in senile and presenile dementia, *Neurology* **20**:315–322 (1970).

91. D. Simard, J. Olesen, O. B. Paulson, N. A. Lassen, and E. Skinhøj, Regional cerebral blood flow and its regulation in dementia, *Brain* **94**: 273–288 (1971).

92. T. W. Langfitt and N. F. Kassell, Cerebral vasodilatation produced by brain-stem stimulation: Neurogenic control vs. autoregulation, *Am. J. Physiol.* **215**:90–97 (1968).

93. M. Baldy-Moulinier and D. H. Ingvar, EEG frequency content related to regional blood flow of cerebral cortex in cat, *Exptl. Brain Res. (Berlin)* **5**:55–60 (1968).

94. J. W. Bean and N. E. Leatherman, Cerebral blood flow during convulsions, *Arch. Neurol. (Chicago)* **20**:396–405 (1969).

95. J. S. Meyer, F. Nomura, K. Sakamoto, and A. Kondo, Effect of stimulation of the brain-stem reticular formation on cerebral blood flow and oxygen consumption, *Electroencephalog. Clin. Neurophysiol.* **26**:125–132 (1969).

96. G. I. Mchedlishvili, D. H. Ingvar, D. G. Barmidze, and R. Ekberg, Blood flow and vascular behavior in the cerebral cortex related to strychnine-induced spike activity, *Exptl. Neurol.* **26**:411–423 (1970).

97. M. Reivich, Arterial pCO_2 and cerebral hemodynamics, *Am. J. Physiol.* **206**:25–35 (1964).

98. A. M. Harper and H. I. Glass, Effect of alterations in the arterial carbon dioxide tension on the blood flow through the cerebral cortex at normal and low arterial blood pressures, *J. Neurol. Neurosurg. Psychiat.* **28**:449–452 (1965).

99. W. Shapiro, A. J. Wasserman, and J. L. Patterson, Mechanism and pattern of human cerebrovascular regulation after rapid changes in blood CO_2 tension, *J. Clin. Invest.* **45**: 913–922 (1966).

100. J. Olesen, O. B. Paulson, and N. A. Lassen, Regional cerebral blood flow in man determined by the initial slope of the clearance of intra-arterially injected ^{133}Xe, *Stroke* **2**: 519–540 (1971).

101. S. C. Alexander, T. C. Smith, G. Strobel, G. W. Stephen, and H. Wollman, Cerebral carbohydrate metabolism of man during respiratory and metabolic alkalosis, *J. Appl. Physiol.* **24**:66–72 (1968).

102. J. W. Severinghaus and N. Lassen, Step hypocapnia to separate arterial from tissue PCO_2 in the regulation of cerebral blood flow, *Circ. Res.* **20**:272–278 (1967).

103. V. Fencl, J. R. Vale, and A. Brock, Respiration and cerebral blood flow in metabolic acidosis and alkalosis in humans, *J. Appl. Physiol.* **27**:67–76 (1969).

104. E. Skinhøj and O. B. Paulson, Carbon dioxide and cerebral circulatory control, *Arch. Neurol. (Chicago)* **20**:249–252 (1969).

105. M. Wahl, P. Deetjen, and K. Thurau, Micropuncture evaluation of the importance of perivascular pH for the arteriolar diameter on the brain surface, *Pflügers Arch.* **316**: 152–163 (1970).

106. J. W. Severinghaus, H. Chiodi, E. I. Eger, II, B. Brandstater, and T. F. Hornbein, Cerebral blood flow in man at high altitude: Role of cerebrospinal fluid pH in normalization of flow in chronic hypocapnia, *Circ. Res.* **19**:274–282 (1966).

107. A. M. Harper and R. A. Bell, The effect of metabolic acidosis and alkalosis on the blood flow through the cerebral cortex, *J. Neurol. Neurosurg. Psychiat.* **26**:341–344 (1963).

108. E. Betz and D. Heuser, Cerebral cortical blood flow during changes of acid–base equilibrium of the brain, *J. Appl. Physiol.* **23**:726–733 (1967).

109. S. Cotev, J. Lee, and J. W. Severinghaus, Effects of acetazolamide on cerebral blood flow and cerebral tissue PO_2, *Anesthesiology* **29**:471–477 (1968).

110. K. Kogure, P. Scheinberg, O. M. Reinmuth, M. Fujishima, and R. Busto, Mechanisms of cerebral vasodilatation in hypoxia, *J. Appl. Physiol.* **29**:223–229 (1970).

111. S. Shimojyo, P. Scheinberg, K. Kogure, and O. M. Reinmuth, The effects of graded hypoxia upon transient cerebral blood flow and oxygen consumption, *Neurology* **18**:127–133 (1968).

112. K. Kogure, P. Scheinberg, M. Fujishima, R. Busto, and O. M. Reinmuth, Effect of hypoxia on cerebral autoregulation, *Am. J. Physiol.* **219**:1393–1396 (1970).

113. F. Regli, T. Yamaguchi, and A. G. Waltz, Effects of inhalation of oxygen on blood flow and microvasculature of ischemic and nonischemic cerebral cortex, *Stroke* **1**:314–319 (1970).

114. J. W. Bean, J. Lignell, and J. Coulson, Regional cerebral blood flow, O_2, and EEG in exposures to O_2 at high pressures, *J. Appl. Physiol.* **31**:235–242 (1971).

115. E. Kanzow and D. Dieckhoff, On the location of the vascular resistance in the cerebral circulation, *in* "Cerebral Blood Flow: Clinical and Experimental Results" (M. Brock, C. Fieschi, D. H. Ingvar, N. A. Lassen, and K. Schürmann, eds.) pp. 96–97, Springer-Verlag, Berlin (1969).

116. H. M. Shapiro, D. D. Stromberg, D. R. Lee, and C. A. Wiedershielm, Dynamic pressures in the pial arterial microcirculation, *Am. J. Physiol.* **221**:279–283 (1971).

117. V. D. Salanga and A. G. Waltz, Regional cerebral blood flow during stimulation of seventh cranial nerve, *Stroke* **4**:213–217 (1973).

118. M. N. Shalit, O. M. Reinmuth, S. Shimojyo, and P. Scheinberg, Carbon dioxide and cerebral circulatory control. III. The effects of brain stem lesions, *Arch. Neurol. (Chicago)* **17**:342–353 (1967).

119. A. M. Harper, V. D. Deshmukh, J. O. Rowan, and W. B. Jennett, Studies on possible neurogenic influences on the cerebral circulation, *in* "Brain and Blood Flow: Proceedings of the Fourth International Symposium on the Regulation of Cerebral Blood Flow" (R. W. Ross Russell, ed.) pp. 182–186, Pitman, London (1971).

120. W. I. Rosenblum, Cerebral arteriolar spasm inhibited by β-adrenergic blocking agents, *Arch. Neurol. (Chicago)* **21**:296–302 (1969).

121. R. A. R. Fraser, B. M. Stein, and J. L. Pool, Adrenergic blockade of hypocapnic cerebral arterial constriction, *Stroke* **2**:219–231 (1971).

122. A. G. Waltz and T. Yamaguchi, Pressure–flow relationships of the cerebral vasculature: Autoregulatory responses to changes of perfusion pressure produced without drugs or hemorrhage, *Trans. Am. Neurol. Ass.* **95**:326–327 (1970).

123. G. I. Mchedlishvili and L. S. Nikolaishvili, Evidence of a cholinergic nervous mechanism mediating the autoregulatory dilatation of the cerebral blood vessels, *Pflügers Arch.* **315**:27–37 (1970).

124. A. G. Waltz, T. Yamaguchi, and F. Regli, Regulatory responses of the cerebral vasculature after sympathetic denervation, *Am. J. Physiol.* **221**:298–302 (1971).

125. M. J. Hernandez, M. E. Raichle, and H. L. Stone, The role of sympathetic innervation in cerebral blood flow autoregulation, *Panminerva Med.* **13**:173 (1971).

126. T. M. Sundt, Jr., and A. G. Waltz, Experimental cerebral infarction: Retro-orbital, extradural approach for occluding the middle cerebral artery, *Mayo Clin. Proc.* **41**:159–168 (1966).

127. K. Kogure, M. Fujishima, P. Scheinberg, and O. M. Reinmuth, Effects of changes in carbon dioxide pressure and arterial pressure on blood flow in ischemic regions of the brain in dogs, *Circ. Res.* **24**:557–565 (1969).

128. G. F. Molinari, Experimental cerebral infarction. I. Selective segmental occlusion of intracranial arteries in the dog. II. Clinicopathological model of deep cerebral infarction, *Stroke* **1**:224–244 (1970).

129. R. W. Ross Russell, The reactivity of the pial circulation of the rabbit to hypercapnia and the effect of vascular occlusion, *Brain* **94**:623–634 (1971).

130. J. H. Halsey and N. F. Capra, Physiological modification of immediate ischemia due to experimental middle cerebral occlusion—Its relevance to cerebral infarction, *Stroke* **2**:239–246 (1970).

131. J. H. Halsey, Jr., and L. C. Clark, Jr., Some regional circulatory abnormalities following experimental cerebral infarction, *Neurology* **20**:238–246 (1970).

132. B. W. Brawley, D. E. Strandness, and W. A. Kelly, The physiologic response to therapy in experimental cerebral ischemia, *Arch. Neurol. (Chicago)* **17**:180–187 (1967).

133. R. W. Brennan and F. Plum, Dissociation of autoregulation and chemical regulation in cerebral circulation following seizures, in "Cerebral Blood Flow: Clinical and Experimental Results" (M. Brock, C. Fieschi, D. H. Ingvar, N. A. Lassen, and K. Shürmann, eds.) pp. 218–222, Springer-Verlag, Berlin (1969).

134. J. F. Fogelson and R. W. Alman, Maximal dilatation of cerebral vessels, *Arch. Neurol. (Chicago)* **11**:303–309 (1964).

135. M. Brock, J. Beck, E. Markakis, and H. Dietz, Intracranial pressure gradients associated with experimental cerebral embolism, *Stroke* **3**:123–130 (1972).

136. N. A. Lassen and R. Palvölgyi, Cerebral steal during hypercapnia and the inverse reaction during hypocapnia observed by the ^{133}xenon technique in man, *Scand. J. Clin. Lab. Invest. Suppl.* **102**:XIII, D (1968).

137. F. Regli, T. Yamaguchi, and A. G. Waltz, Cerebral circulation: Effects of vasodilating drugs on blood flow and the microvasculature of ischemic and nonischemic cerebral cortex, *Arch. Neurol. (Chicago)* **24**:467–474 (1971).

138. J. Olesen and O. B. Paulson, The effect of intra-arterial papaverine on the regional cerebral blood flow in patients with stroke or intracranial tumor, *Stroke* **2**:148–159 (1971).

139. K. Høedt-Rasmussen, E. Skinhøj, O. Paulson, J. Ewald, J. K. Bjerrum, A. Fahrenkrug, and N. A. Lassen, Regional cerebral blood flow in acute apoplexy, *Arch. Neurol. (Chicago)* **17**:271–281 (1967).

140. C. Fieschi, A. Agnoli, N. Battistini, L. Bozzao, and M. Prencipe, Derangement of regional cerebral blood flow and of its regulatory mechanisms in acute cerebrovascular lesions, *Neurology* **18**:1166–1179 (1968).

141. O. B. Paulson, Regional cerebral blood flow in apoplexy due to occulsion of the middle cerebral artery, *Neurology* **20**:63–77 (1970).

142. O. B. Paulson, N. A. Lassen, and F. Skinhøj, Regional cerebral blood flow in apoplexy without arterial occlusion, *Neurology* **20**:125–138 (1970).

143. E. Skinhøj, K. Høedt-Rasmussen, O. B. Paulson, and N. A. Lassen, Regional cerebral blood flow and its autoregulation in patients with transient focal cerebral ischemic attacks, *Neurology* **20**:485–493 (1970).

144. M. L. Dyken, R. L. Campbell, and R. Frayser, Cerebral blood flow, oxygen utilization, and vascular reactivity: Internal carotid artery complete occlusion versus incomplete occlusion with infarction, *Neurology* **20**:1127–1132 (1970).

145. J. E. Rees, G. H. du Boulay, J. W. D. Bull, and J. Marshall, Regional cerebral blood flow in transient ischemic attacks, *Lancet* **2**:1210–1212 (1970).

146. G. Boysen, H. J. Ladegaard-Pedersen, H. Henriksen, J. Olesen, O. B. Paulson, and H. C. Engell, The effects of $PaCO_2$ on regional cerebral blood flow and internal carotid arterial pressure during carotid clamping, *Anesthesiology* **35**:286–300 (1971).

147. J. S. Meyer, Y. Fukuuchi, K. Shimazo, T. Ohuchi, and A. D. Ericsson, Effect of intravenous infusion of glycerol on hemispheric blood flow and metabolism in patients with acute cerebral infarction, *Stroke* **3**:168–180 (1972).

148. A. G. Waltz, T. M. Sundt, Jr., and J. D. Michenfelder, Cerebral blood flow during carotid endarterectomy, *Circulation* **45**:1091–1096 (1972).

149. N. A. Lassen, The luxury-perfusion syndrome and its possible relation to acute metabolic acidosis localized within the brain, *Lancet* **2**:1113–1115 (1966).

150. O. B. Paulson, J. Olesen, and M. S. Christensen, Restoration of autoregulation of cerebral blood flow by hypocapnia, *Neurology* **22**:286–293 (1972).

151. O. B. Paulson, Cerebral apoplexy (stroke): Pathogenesis, pathophysiology and therapy as illustrated by regional blood flow measurements in the brain, *Stroke* **2**:327–360 (1971).

152. M. Soloway, W. Nadel, M. S. Albin, and R. J. White, The effect of hyperventilation on subsequent cerebral infarction, *Anesthesiology* **29**:975–980 (1968).

153. M. Soloway, G. Moriarty, J. Fraser, and R. J. White, Effect of delayed hyperventilation on experimental cerebral infarction, *Neurology* **21**:479–485 (1971).

154. T. Yamaguchi, F. Regli, and A. G. Waltz, Effects of hyperventilation with and without carbon dioxide on experimental cerebral ischemia and infarction: Studies of regional cerebral blood flow and histopathology after occlusion of a middle cerebral artery in cats, *Brain* **95**:123–132 (1972).

155. M. S. Christensen, Stroke treated with prolonged hyperventilation, *in* "Brain and Blood Flow: Proceedings of the Fourth International Symposium on the Regulation of Cerebral Blood Flow" (R. W. Ross Russell, ed.) pp. 358–364, Pitman, London (1971).

156. F. B. Byrom, The pathogenesis of hypertensive encephalopathy and its relation to the malignant phase of hypertension: Experimental evidence from the hypertensive rat, *Lancet* **2**:201–211 (1954).

157. J. S. Meyer, A. G. Waltz, and F. Gotoh, Pathogenesis of cerebral vasospasm in hypertensive encephalopathy. II. The nature of increased irritability of smooth muscle of pial arterioles in renal hypertension, *Neurology* **10**:859–867 (1960).

158. R. Rodda and D. Denny-Brown, The cerebral arterioles in experimental hypertension. II. The development of arteriolonecrosis, *Am. J. Pathol.* **49**:365–371 (1966).

159. F. B. Byrom, "The Hypertensive Vascular Crisis," Grune and Stratton, New York (1969).

160. K. A. Hossmann and Y. Olsson, Influence of ischemia on the passage of protein tracers across capillaries in certain blood–brain barrier injuries, *Acta Neuropathol.* **18**:113–122 (1971).

161. A. G. Waltz, Regional cerebral blood flow: Responses to changes in arterial blood pressure and CO_2 tension, *in* "Cerebral Vascular Diseases: Transactions of the Sixth Conference" (J. F. Toole, R. G. Siekert, and J. P. Whisnant, eds.) pp. 66–70, Grune and Stratton, New York (1968).

162. T. Yamaguchi and A. G. Waltz, Effects of subarachnoid hemorrhage from puncture of the middle cerebral artery on blood flow and vasculature of the cerebral cortex in the cat, *J. Neurosurg.* **35**:664–671 (1971).

163. R. Palvölgyi, Regional cerebral blood flow in patients with intracranial tumors, *J. Neurosurg.* **31**:149–163 (1969).

164. E. Gordon, The effect of controlled ventilation on the clinical course of patients with severe traumatic brain injury, *in* "Brain and Blood Flow: Proceedings of the Fourth International Symposium on the Regulation of Cerebral Blood Flow" (R. W. Ross Russell ed.) pp. 365–369, Pitman, London (1971).

165. W. R. Hudgins and J. H. Garcia, The effect of electrocautery, atmospheric exposure, and surgical retraction on the permeability of the blood–brain barrier, *Stroke* **1**:375–380 (1970).

166. W. I. Rosenblum, Effect of blood pressure and blood viscosity on fluorescein transit time in the cerebral microcirculation in the mouse, *Circ. Res.* **27**:825–833 (1970).

167. W. I. Rosenblum, The differential effect of elevated blood viscosity on plasma and erythrocyte flow in the cerebral microcirculation of the mouse, *Microvasc. Res.* **2**:399–408 (1970).

168. T. Hasegawa, J. R. Ravens, and J. F. Toole, Precapillary arteriovenous anastamoses, *Arch. Neurol. (Chicago)* **16**:217–224 (1967).

169. O. A. Larsen and N. A. Lassen, Cerebral hematocrit in normal man, *J. Appl. Physiol.* **19**: 571–574 (1964).

170. V. A. Levin and J. I. Ausman, Relationship of peripheral venous hematocrit to brain hematocrit, *J. Appl. Physiol.* **26**:433–437 (1969).

171. W. I. Rosenblum, Ratio of red cell velocities near the vessel wall to velocities at the vessel center in cerebral microcirculation, and an apparent effect of blood viscosity on this ratio, *Microvasc. Res.* **4**:98–101 (1972).

172. D. Denny-Brown and J. S. Meyer, The cerebral collateral circulation. II. Production of cerebral infarction by ischemic anoxia and its reversibility in early stages, *Neurology* **7**: 567–579 (1957).

173. E. S. Gurdjian and L. M. Thomas, Human pial circulation, *Arch. Neurol. (Chicago)* **5**: 111–118 (1961).

174. F. C. A. Romanul and A. Abramowicz, Changes in brain and pial vessels in arterial border zones, *Arch. Neurol. (Chicago)* **11**:40–65 (1964).

175. P. Huber, H. Mogun, and R. Rivoir, The effect of pharmacologically increased blood pressure on the brain circulation: Angiographic investigation of arterial diameter and blood flow in patients with normal and pathologic arteriograms (preliminary report), *Neuroradiology* **3**:68–74 (1971).

176. L. Symon, An experimental study of traumatic cerebral vascular spasm, *J. Neurol. Neurosurg. Psychiat.* **30**:497–505 (1967).

177. J. Kapp, M. S. Mahaley, and G. L. Odom, Cerebral arterial spasm, *J. Neurosurg.* **29**: 331–356 (1968).

178. D. R. Smith, T. B. Ducker, and L. G. Kempe, Experimental *in vivo* microcirculatory dynamics in brain trauma, *J. Neurosurg.* **30**:664–672 (1969).

179. M. Brock, J. Risberg, and D. H. Ingvar, Effects of local trauma on cortical CBF studied by infrared thermography, *Brain Res.* **12**:238–241 (1969).

180. A. G. Waltz, Red venous blood: Occurrence and significance in ischemic and nonische-mic cerebral cortex, *J. Neurosurg.* **31**:141–148 (1969).

181. R. H. Wilkins and P. Levitt, Intracranial arterial spasm in the dog: A chronic experimental model, *J. Neurosurg.* **33**:260–269 (1970).

182. F. Echlin, Experimental vasospasm, acute and chronic, due to blood in the subarachnoid space, *J. Neurosurg.* **35**:646–656 (1971).

183. A. G. Waltz and J. S. Meyer, Effects of changes in composition of plasma on pial blood flow. II. High molecular weight substances, blood constituents, and tonicity, *Neurology* **9**: 815–825 (1959).

184. W. I. Rosenblum and R. M. Asofsky, Malfunction of cerebral microcirculation in macroglobulinemic mice, *Arch. Neurol. (Chicago)* **18**:151–159 (1968).

185. W. I. Rosenblum, Erythrocyte velocity and fluorescein transit time through the cerebral microcirculation in experimental polycythemia, *J. Neuropathol. Exptl. Neurol.* **31**:126–131 (1972).

186. V. A. Levin and D. D. Gilboe, Blood volume, hematocrit and pressure relationships in the isolated perfused dog brain, *Stroke* **1**:270–277 (1970).

187. W. I. Rosenblum, Effect of reduced hematocrit on erythrocyte velocity and fluorescein transit time in the cerebral microcirculation of the mouse, *Circ. Res.* **29**:96–103 (1971).

188. R. M. Crowell, Y. Olsson, I. Klatzo, and A. Ommaya, Temporary occlusion of the middle cerebral artery in the monkey: Clinical and pathological observations, *Stroke* **1**: 439–448 (1970).

189. T. M. Sundt, Jr., and A. G. Waltz, Cerebral ischemia and reactive hyperemia: Studies of cortical blood flow and microcirculation before, during and after temporary occlusion of the middle cerebral artery of squirrel monkeys, *Circ. Res.* **28**:426–433 (1971).

190. J. Chiang, M. Kowada, A. Ames, III, R. L. Wright, and G. Majno, Cerebral ischemia. IV. Vascular changes, *Am. J. Pathol.* **52**:455–476 (1968).

191. K. A. Hossmann, Cortical steady potential, impedance and excitability changes during and after total ischemia of cat brain, *Exptl. Neurol.* **32**:163–175 (1971).

192. J. H. Garcia, J. V. Cox, and W. R. Hudgins, Ultrastructure of the microvasculature in experimental cerebral infarction, *Acta Neuropathol.* **18**:273–285 (1971).

193. T. M. Sundt, Jr., A. G. Waltz, and G. P. Sayre, Experimental cerebral infarction: Modification by treatment with hemodiluting, hemoconcentrating, and dehydrating agents, *J. Neurosurg.* **26**:46–56 (1967).

194. R. C. Cantu, A. Ames, III, G. DiGiacinto, and J. Dixon, Hypotension: A major factor limiting recovery from cerebral ischemia, *J. Surg. Res.* **9**:525–529 (1969).

195. R. C. Cantu, A. Ames, J. Dixon, and G. DiGiacinto, Reversibility of experimental cerebrovascular obstruction induced by complete ischemia, *J. Neurosurg.* **31**:429–431 (1969).

196. Y. Olsson and K. A. Hossmann, The effect of intravascular saline perfusion on the sequelae of transient cerebral ischemia: Light and electron microscopical observations, *Acta Neuropathol.* **17**:68–79 (1971).

197. C. Fieschi, N. Battistini, and M. Nardini, Experimental cerebral infarction: Focal or perifocal reactive hyperemia and its relationship to red softening, *in* "Research on the Cerebral Circulation: Fourth International Salzburg Conference," (J. S. Meyer, M. Reivich, H. Lechner, and O. Eichhorn, eds.) pp. 17–23, Charles C. Thomas, Springfield, Ill. (1970).

198. M. E. Jaffe, L. C. McHenry, Jr., and H. I. Goldberg, Regional cerebral blood flow measurement with small probes. II. Application of the method, *Neurology* **20**:225–237 (1970).

199. S. Cronqvist and F. Laroche, Transitory hyperemia in focal cerebral vascular lesions studied by angiography and regional cerebral blood flow measurements, *Brit. J. Radiol.* **40**:270–274 (1967).

200. S. Cronqvist, Regional cerebral blood flow and angiography in apoplexy, *Acta Radiol. Diagn.* **7**:521–534 (1968).

201. J. M. Taveras, J. M. Gilson, D. O. Davis, B. Kilgore, and C. L. Rumbaugh, Angiography in cerebral infarction, *Radiology* **93**:549–558 (1969).

202. R. M. Crowell, Y. Olsson, and A. K. Ommaya, Angiographic and microangiographic observations in experimental cerebral infarction, *Neurology* **21**:710–719 (1971).

203. W. Feindel and P. Perot, Red cerebral veins: A report on arteriovenous shunts in tumors and cerebral scars, *J. Neurosurg.* **22**:315–325 (1965).

204. H. M. Shapiro, R. E. Myers, J. Segarra, and S. Sotsky, Posthypoxic jugular venous hyperoxia, *Neurology* **20**:791–801 (1970).

205. W. Feindel, Y. L. Yamamoto, and C. P. Hodge, Red cerebral veins and the cerebral steal syndrome: Evidence from fluorescein angiography and microregional blood flow by radioisotopes during excision of an angioma, *J. Neurosurg.* **35**:167–179 (1971).

206. F. Regli, T. Yamaguchi, and A. G. Waltz, Effects of acetazolamide on cerebral ischemia and infarction after experimental occlusion of middle cerebral artery, *Stroke* **2**:456–460 (1971).

207. F. Plum, J. B. Posner, and E. C. Alvord, Edema and necrosis in experimental cerebral infarction; effect of steroids in experimental cerebral infarction, *Arch. Neurol. (Chicago)* **9**:563–573 (1963).

208. J. S. Meyer, Y. Shinohara, T. Kanda, Y. Fukuuchi, A. D. Ericsson, and N. K. Kok, Diaschisis resulting from acute unilateral cerebral infarction: Quantitative evidence for man, *Arch. Neurol. (Chicago)* **23**:240–247 (1970).

209. A. G. Waltz, Cortical blood flow of opposite hemisphere after occlusion of middle cerebral artery, *Trans. Am. Neurol. Ass.* **92**:293 (1967).

Chapter 3

EFFECTS OF NARCOTIC ANALGESICS ON BRAIN FUNCTION

Doris H. Clouet

New York State Narcotic Addiction Control Commission
Testing and Research Laboratory
Brooklyn, New York

I. INTRODUCTION

Drugs which produce dependence have one characteristic in common: in effective doses, they induce alteration in behavior. This characteristic, however, does not define addictive drugs, since other drugs modify behavior without inducing addiction. The behavioral responses to addictive drugs do indicate the central nervous system as the target tissue, and nervous function as the site of drug intervention, during the development of dependence to chronic use of opiates or other drugs which produce dependence.

The chronic use of narcotic analgesics produces a type of dependence termed *drug dependence of the morphine type.*[1] The characteristics of this type of dependence will be described in the following sections. In order to specify which drugs are agonists in this drug category, it is necessary to assess each drug for its ability to sustain dependence and prevent withdrawal symptoms in animals chronically treated with morphine. The means of identifying narcotic antagonists is by testing the ability of a drug, in appropriate dosage, to block the effects of morphine.

There have been a number of recent reviews in which the structure and activity of the narcotic analgesics and antagonists are described and discussed.[2-4] It is sufficient for the purposes of this chapter to describe the drugs

briefly. Morphine is derived from the opium poppy plant, heroin is manufac-
tured by diacetylation of morphine and other agonists such as levorphanol, and
meperidine and methadone are either partially or completely synthetic (Fig. 1).
The almost pure antagonists are nalorphine and levallorphan, while pentazocine
and cyclazocine are mixed agonist/antagonists. Because levallorphan, nalorphine, and naloxone (N-allyl-7,8-dihydro-14-hydroxymorphinone) have N-
allyl groups, it was considered possible that there was a direct relationship
between agonist potency of a N-methyl compound and the antagonist potency
of its N-allyl derivative.[5] However, in some of the newer analgesic series, the
N-allyl and N-cyclopropylmethyl derivatives are potent agonists.[6] The structural requirements for a narcotic receptor include an anionic site for the nitrogen
atom, a flat surface for the phenol ring with a two-carbon cavity between, and
possible additional sites of cationic and lipophilic binding.[7,8] Accurate models
of the narcotic receptor can be constructed for homologous series of compounds,

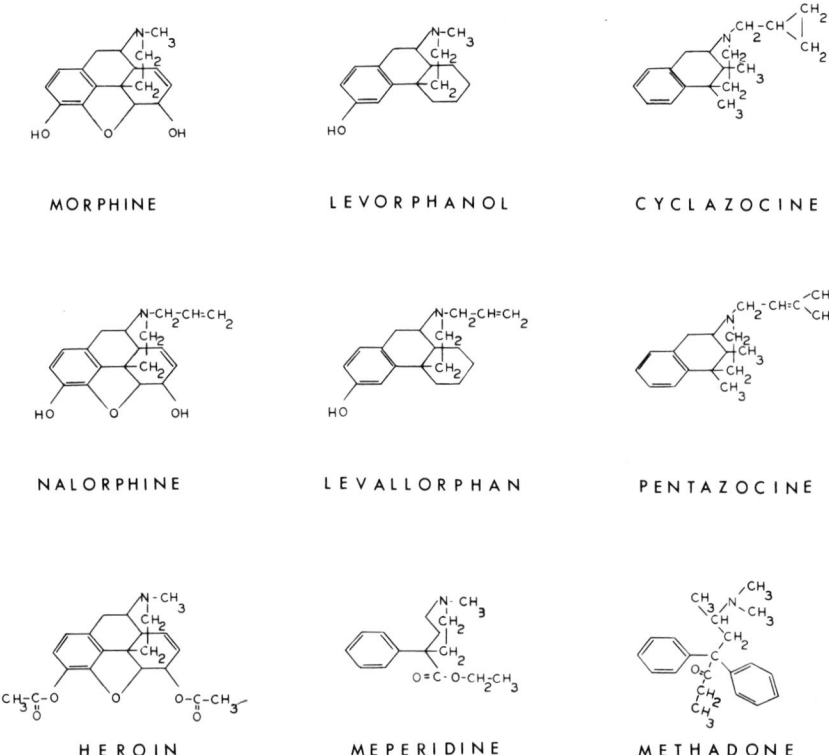

Fig. 1. Chemical structures of agonists: morphine, levorphanol, heroin, meperidine, and
methadone; the antagonists: nalorphine and leavllorphan; and the mixed agonists/an-
tagonists: cyclazocine and pentazocine.

but the structural requirements are to some extent dependent on the basic structure of the drugs.[8]

In the succeeding sections, the responses of the central nervous system to the administration of narcotic agonists and/or antagonists are described on several levels of examination ranging from electrophysiological recordings by scalp electrodes to neurotransmitter transport in the isolated nerve ending.

II. PHYSIOLOGICAL EFFECTS OF NARCOTIC ANALGESICS

A. Pharmacology of Narcotics

Narcotic analgesic agents, by definition, induce narcosis and analgesia. In man, the major clinical response to the administration of narcotic analgesics is a decreased awareness of pain (analgesia) with an associated drowsiness and mental clouding which do not necessarily lead to sleep. The depressant effects in the central nervous system are selective, affecting pain perception, emesis, temperature control, mental performance, and hypothalamic-mediated target tissue responses, without producing sleep or loss of motor coordination in therapeutic doses. In addition to the central nervous system depression, the drugs induce miosis, respiratory depression, and decreased motility of the lower gastrointestinal tract. In most species, there is no direct effect on blood pressure, although the increased levels of blood CO_2 due to respiratory depression lead eventually to a hypoxia-induced fall in blood pressure and a cerebral vasodilatation. The drugs also produce peripheral histamine release. In man, excitatory responses to the administration of narcotic analgesics include convulsions after high doses and, possibly, euphoria.

In experimental animals, the responses to the administration of narcotic analgesics are to some extent species determined. Rodents are depressed but are so insensitive that relatively high doses of the drugs are required for activity. The "Straub tail," which is a characteristic response in rodents to the administration of morphine, results from an increase in tone of the anal sphincter. Excitatory responses are more prominent in the dog, and especially in the cat, after drug administration. Among the excitatory effects are defecation, clonic convulsions, vomiting, and reflex irritability.

Not only are the pharmacological responses species dependent, but they are also drug dependent. Among the narcotic analgesics, the relative effectiveness varies for each pharmacological parameter. For example, morphine, methadone, and meperidine are active analgesic agents; codeine produces an antitussive response at a dose low enough to have little depressant effect; diphenoxylate acts mainly on the gastrointestinal tract, thus is an excellent antidiarrheal agent; heroin is the most effective euphoriant; methadone is more active than most of the other opiates when administered orally; and propoxyphene is a useful analgesic agent because its side-effects limit its abuse potential. A class of chemically related compounds, the antagonists, have little pharmacological activity other than an antagonism of the agonist effects of the narcotic analgesics, although some of the antagonists also have some agonist activity. Naloxone, a congener of oxymorphone, has least agonist activity of the antagonists.

It seems superfluous to point out that the pharmacological responses are also dose dependent and time dependent, since these parameters merit consideration whenever a response is evoked by a drug. However, time and dose are especially important for the rate of development of tolerance during chronic drug treatment. The state of tolerance, which is characterized by a decrease in most of the depressant effects with repeated doses of the drug, results in an increase in both the effective dose and the lethal dose of the drug. The rate of development of tolerance to any of the narcotic analgesics is influenced by the dose administered and by the interval between doses as related to the duration of action of the drug. One narcotic analgesic can be substituted for another in order to maintain tolerance, a phenomenon termed *cross-tolerance*. The inclusion of drugs in the category of addictive drugs depends on whether the drug has the ability to maintain tolerance in chronically morphine-treated animals.

The discontinuance of drug use in a tolerant individual produces a syndrome of withdrawal symptoms which may be characterized as a rebound hyperexcitability of the formerly depressed central nervous system with associated secondary responses. The severity of the withdrawal syndrome is maximal in man over a period of 24–72 hr after the last dose. However, it is evoked in minutes after the administration of a suitable dose of an antagonist to a tolerant man or animal. The symptoms arising upon drug withdrawal constitute the evidence of physical dependence on the drug.

Psychic dependence on narcotic analgesics, also called *habituation,* is evidenced by a compulsive chronic consumption of the drug. Habituation to the narcotic analgesics can be induced in animals by the use of various self-administration devices. In man, psychic dependence leads to behavioral patterns focused entirely on the drug. Such chronic involvement with drugs to the detriment of the drug users and to society is generally termed *drug abuse* or *addiction.*

When the effects of narcotic analgesics on brain function are discussed, a distinction must be made between studies of the mechanisms of action of narcotic analgesics (the way in which the pharmacological responses to an initial dose of the drug are produced) and studies of the mechanisms of addiction (the ways in which tolerance and dependence are produced). In the succeeding sections, the response to an initial dose of an opiate in man, and especially in experimental animals, will be described and related to the anatomical and biochemical sites of drug action and compared to the responses in tolerant subjects. There is no definitive evidence that the factors responsible for producing opiate dependence are, or are not, involved in the primary pharmacological effects of opiate administration.

B. Alterations of Lower Brain Stem Functions by Narcotics

The anatomical localization of the action of morphine and other opiates in the nervous system has been sought in studies of the effect of the drug on function at various levels of organization in the nervous system. The simplest element is a reflex arc in the spinal cord. The muscle stretch reflex, a monosynaptic arc, is not altered by the application of morphine, nor are other simple reflex arcs. Polysynaptic reflexes, however, are depressed by the administration of morphine to a spinal-transected animal. Thus a polyneuronal organization seems to be required for drug action. Tolerance to the depression of polysynaptic reflexes and withdrawal activity can be demonstrated in the spinal dog.[9]

Three major physiological activities are affected by the application of morphine in the decerebrate animal: (1) Respiration is depressed, tidal volume and respiratory frequency are decreased, and expiratory CO_2 concentrations are increased by the injection of morphine in the decerebrate cat.[10] (2) Emesis, which is controlled in the area postrema of the medulla, is evoked by local application of morphine, although larger doses depress the vomiting center. Tolerance develops to the depressant effect.[11] (3) In most species, changes in blood pressure cannot be separated from the secondary responses to elevated blood CO_2 concentrations or histamine release. However, in the decerebrate cat, there is a hypotensive response to morphine during assisted ventilation to prevent CO_2 buildup.[12] The effect of the development of tolerance on the responses in the decerebrate animal is not known.

C. Alterations of Subcortical Functions by Narcotics

The integrative functions of the hypothalamus and basal ganglia are depressed by the administration of morphine. The regulation of body temperature and the control of the hypothalamic–pituitary hormonal system are influenced by narcotic drugs. Tolerance and withdrawal are demonstrable in these modulating systems.[13]

In the decorticate animal, miosis is produced by morphine if the optic nerve is intact. The effect of morphine on the pupil reflex constriction seems to be induced centrally in the oculomotor nucleus, although it is modulated by peripheral input from the retina.[14]

A secondary "stress" response involving the adrenal hormones and the pituitary hormones is elicited by narcotic drugs if the sympathetic splanchnic pathway to the central nervous system is intact. Changes in blood glucose concentration, corticosteroid utilization, thyroid output, etc., are the result of the adrenal involvement.[15-17]

An almost complete picture of tolerance and withdrawal is obtained in the decorticate animal.[18]

D. Alterations of Cortical Functions by Narcotics

The administration of morphine and other opiates produces an alteration in the electrical activity of the cerebral cortex. The electroencephalographic response is actually the most sensitive indicator of a narcotic analgesic effect in man and animals. The changes in the recorded EEG patterns resemble sleep patterns. The changes, however, have not been correlated with the depression of cerebral cortical function evoked by narcotics. The general effect of morphine administration seems to be a depression of some of the integrative functions of the cortex involving pathways of reverberating loops to the reticular system and connections to the somatosensory pathways.[18] The psychic reaction component, which includes such parameters as mood, motivation, and learning, presumably is sited in the cortex. An understanding of the effect of narcotic analgesics on these behavioral components is as elusive as is the neurobiological basis of behavior itself.

The behavioral aspects of narcotic dependence have been examined in experimental animals using the classical procedures of operant conditioning. Self-administration of morphine and other opiates is effected through the use of an indwelling catheter, implanted in a vein or in the peritoneal cavity, which is connected to the drug solution by a signal-activated syringe and pump. Lever pressing in order to activate the injection procedure begins spontaneously in monkeys and rats and has been found to increase continually over a period of 30 days if the dose–lever-press ratio is kept fixed.[19,20] Morphine, methadone, and other narcotic drugs act as positive reinforcers, while the withdrawal syndrome acts as a negative avoidance reinforcer in tolerant animals. The avoidance response to withdrawal symptoms is shown in experiments in which the injection of an antagonist, nalorphine, in morphine-tolerant monkeys to induce withdrawal caused an increase in level-pressing activity.[19] The avoidance reinforcement has been separated from the positive reinforcer effect, however, in experiments in which the number of bar presses was directly proportional to the concentration of drug in the injection solution both when the drug concentration was increasing gradually and when the drug concentration was decreasing gradually. In the latter case, the monkeys were reinforced to self-inject morphine, while tolerance was lost so slowly that withdrawal symptoms were never apparent.[21]

Rats, too, self-administer morphine, responding to a fixed ratio with an increase in bar pressing over a period of days and assuming control of the press schedule when the ratios of dose–press are altered. There is an element of learning in the lever pressing, since rats cycled through a period of tolerance and withdrawal responded more readily in a second self-administration period than a second group of animals which had received the same amount of morphine by injection as the first group self-administered.[22] A second cycle of self-administration is begun more readily in the same physical surroundings in which the first cycle occurred.[23]

The elimination of the operant conditioning is not accomplished by punishment (a shock delivered simultaneously with the drug). Satiation with methadone by injection eliminates the lever press for morphine as does morphine satiation by injection. When the narcotic analgesic is removed from the injection solution, extinction occurs slowly over a period of months. The rate of unrewarded lever presses during the period in which physical symptoms of withdrawal are seen decreases rapidly to a low rate which then persists for a long time.[24] The responses in monkeys and rats bear a striking resemblance to addiction in man. It has been postulated that while the narcotic analgesics act as positive reinforcers in man, the pharmacological blockade by the drugs of the effects of adverse environment and sensory deprivation may be more important in reinforcing drug use.

III. METABOLISM OF NARCOTIC ANALGESICS IN THE NERVOUS SYSTEM

A. Penetration and Distribution in the Nervous System

The great variety in chemical characteristics of narcotic analgesic drugs makes difficult the task of discussing the metabolism and activity of the drugs as a class. Consequently, only those drugs which are in current use clinically or socially will be considered, with morphine as a prototype. Among the chemical requirements for activity are an aromatic moiety and a basic center which may be protonated at physiological pH. The pK_as for the common narcotics range from 6.5 to 9, a range which satisfies the latter requirement and makes it possible for the drugs to exist both as lipid-soluble and as ionic species at body pH. The property of lipid solubility enables the drugs to be absorbed and to enter the central nervous system without great difficulty, while the ionic form is held within the tissue by binding. After systemic drug administration, the amount of morphine, methadone, meperidine, cyclazocine, or codeine in the brain at peak concentration is not over 1% of the injected dose. After intracisternal or intraventricular injection, most of the dose is found in the central nervous system with the effective dose accordingly reduced.

The extra-CNS distribution of narcotics after a systemic route of administration is important only in that binding to plasma proteins, biotransformation in the liver, and excretion by the kidney alter the levels of the drugs in the CNS. In most species, morphine-C^{14} is found throughout the nervous system with only slightly higher levels in gray matter than in white matter after the administration of radiolabeled morphine.[25] There is, also, no special localization of methadone, dihydromorphine, levorphanol, cyclazocine, nalorphine, or pentazocine in the CNS. The distribution of narcotic analgesics in gross areas of the CNS seems to be related to the lipid solubility of the drug rather than to the pharmacological effects.[26] There are no regions within the blood–brain barrier in which the drug concentrations are especially high. In the dog, a significantly smaller amount of morphine was found in the brains from tolerant animals. However, in monkeys, rabbits, guinea pigs, and rats, there was no consistent difference in distribution of the drug between tolerant and nontolerant animals. A general conclusion may be reached from the many studies in which the gross anatomical distribution of narcotics in the CNS was examined both in tolerant and in nontolerant animals, namely, that "tolerance and physical dependence do not appear to be the result of an altered disposition . . . of narcotic drugs in the CNS."[27]

The intracellular disposition of narcotics in the CNS after centrifugal fractionation of the tissue is mainly in the soluble supernatant fraction of the cell. However, specific drug uptake has been found in particulate fractions.[28–31] In all areas of rat brain, dihydromorphine-H^3 and other labeled narcotics are accumulated in the synaptosomal fractions.[31] One fraction of synaptosomal membranes shows a specific uptake of methadone-H^3 after injection of the labeled drugs to rats by the ventricular route.[32] In addition, labeled drugs have been found in the nuclear and myelin fractions of brain. To date, there have been no studies in which the technique of radioautography–electron microscopy was used in an attempt to refine the tissue localization of narcotics.

B. Blood–Brain Barrier to Narcotics

The blood–brain barrier is an imperfect exclusion mechanism which resides in the glial sheath surrounding the small blood vessels and operates to decrease the penetration of ionic, hydrophilic compounds from blood to brain while increasing the penetration of lipid-soluble compounds through the lipoidal membranes. The barrier is permeable to the un-ionized forms of the lipid-

activity of the enzymes that catalyze the metabolism of morphine and other opiates is inducible by such inducer drugs as phenobarbital, although not by a type II inducer such as methylcholanthrene.[44] Type I binding to the microsomal cytochrome system is exhibited by morphine and methadone.[45] Further evidence that narcotics bind to cytochrome P_{450} in type I binding is that phenobarbital induces the demethylation of morphine and meperidine. However, while codeine N-demethylase activity is induced by phenobarbital, codeine O-demethylase activity is not.

There is some indirect evidence that glucuronide formation in the liver is catalyzed by the same liver drug-metabolizing enzymes, namely, that the glucuronidation of O-aminophenol, sulfadimethoxine, and bilirubin is inducible by either phenobarbital or polyhydrocarbons. However, phenobarbital does not induce the glucuronidation of steroids, and in male rats chronic morphine treatment decreases the activity of N-demethylase but not that of estradiol glucuronyl transferase.

There are other metabolic transformations in liver that have not been established as oxidative reactions requiring the microsomal cytochrome, namely, the hydrolysis of meperidine and normeperidine, the N-methylation of methadone to N-methylmethadone, and the deacetylation of heroin.[40]

A number of metabolic reactions have been found to occur in the CNS: the N-dealkylation (though not necessarily oxidative) of morphine, codeine, nalorphine, and meperidine; the remethylation of normorphine to morphine; and the hydrolysis of heroin. Heroin is hydrolyzed in many tissues, including kidney, lung, and blood as well as liver and brain. The conjugation of narcotics also occurs in the kidney.[46]

Urinary catabolites vary in amount and number, depending on species, drug, dose, tolerance state, period of urine collection, etc.[47] In heroin-dependent man, the main urinary catabolites are morphine and conjugated morphine. In brain, also, there is little unchanged heroin. The monodeacetylated product, 6-monoacetylmorphine, accumulates rapidly, and the dideacetylated compound, morphine, is formed more slowly in the CNS. It is postulated that morphine is the active form of the drug and that heroin and monoacetylmorphine function as carriers into the CNS.[33]

Codeine is both O-demethylated to morphine and N-demethylated to norcodeine. There is some evidence that morphine is the active form of codeine, namely, that the relative potency of codeine compared to morphine is much less by the intraventricular route of administration than after systemic administration.

Morphine is N-demethylated to normorphine, and both compounds are conjugated for excretion. While normorphine is less active than morphine after parenteral administration, it is equal in potency when given intercerebrally.

C. Biotransformations of Narcotics

The importance of the physiological transformations of narcotic analgesics in the body for the effects of the drugs on brain function is directly related to the metabolic alterations in blood levels, and subsequently in brain levels, of the active form of the drug. The effective brain level of an opiate is increased whenever an inactive form of a narcotic is converted to an active form. The brain levels of an active drug are more commonly decreased by drug metabolism. The active drug is usually converted into a less active or an inactive form by catabolism, and a lipid-soluble drug is usually converted to a more water-soluble form and is thereby more readily excreted.

Narcotic analgesic drugs belong in the general category of foreign organic bases and vary widely in the type and number of substituents capable of reacting chemically.[39,40] Common to all is a piperidine, or quasi-piperidine, ring with a nitrogen that is cationic at physiological pH. Alkyl groups on the ring nitrogen such as the methyl, allyl, phenethyl, and cyclopropylmethyl groups are removed through an oxidative N-alkylation reaction. Alkyl groups on the hydroxyl oxygen of codeine or ethylmorphine are removed by oxidative O-dealkylation. The two acetyl groups in heroin are removed successively by hydrolysis. Other esters such as meperidine are also hydrolyzed. A major pathway in the biotransformation of narcotic drugs is conjugation of the drug with glucuronic, acetic, or sulfuric acids. The glucuronides of compounds with a free hydroxyl, carboxyl, amino, or sulfhydryl group are formed enzymatically by the transfer of glucuronic acid from uridine diphosphate glucuronic acid. In addition, there are oxidative hydroxylations and ring closures in the metabolism of some of the narcotics.

The major site of metabolism of the opiates and other narcotics is the liver. The liver microsomal enzyme system called *the drug-metabolizing enzymes* catalyzes many of the biotransformations of narcotic analgesics. These reactions are oxidative but require TPNH as well as oxygen and the microsomal cytochrome respiratory system.[41] There is evidence that drugs bind to a component of the cytochrome P_{450} system in two types of binding: type I to the lipoprotein of the apoenzyme and type II as a ligand of iron in the heme of P450, inducing a change in configuration manifested by a shift in the wavelength of the ultraviolet absorption.[42] The N- and O-dealkylations, hydroxylations, and oxidation–reduction reactions of narcotic analgesics are catalyzed by this system of enzymes. There is evidence that the microsomal cytochrome respiratory chain is common to all of the drug-metabolizing enzyme systems.

A characteristic of the drug-metabolizing enzymes of the liver is that the activity is inducible by the administration of single or multiple doses of inducer drugs.[43] Morphine and other narcotics are not inducer drugs. However, the

soluble narcotic drugs but excludes such ionized compounds as the quaternary base, N-methylmorphine, which has little activity when administered systemically. For most narcotic analgesics, a blood–brain barrier exists only in a relative sense: the rate of penetration of one narcotic drug surpasses that of another when similarly presented to an animal or man. However, a real barrier to the uptake of morphine seems to exist in some species. The development of a barrier is indicated by a decrease in the brain–blood drug ratio with increasing age from birth to adult or by a decrease in toxicity with increasing age. In rats, morphine becomes less toxic and accumulates at a lower concentration in brain as the rat grows from 8 to 32 days old. Heroin and meperidine do not have this age-dependent change in toxicity, so presumably the barrier is not as effective in limiting the uptake of these drugs into the nervous system. A barrier to codeine uptake in the CNS is also found, but it operates less effectively than that to morphine when measured in same group of rats.[33] In man, also, a barrier to morphine seems to develop, since in the human infant morphine is much more potent than meperidine in doses which are equally active in adult man.[34]

Another age-dependent factor must be assessed before ascribing changes in the toxicity with age of neuroactive drugs to the development of a barrier in the CNS. The catabolic activity of the drug-metabolizing enzymes of liver also increases with age, so that drugs are more effectively biotransformed in the adult than in the newborn. Therefore, in the adult an increase in dose is necessary in order to maintain the blood level of the active form of the drug.[35]

When morphine-C^{14} is administered subcutaneously to rats, 45 min before sacrifice, radioautographs of the central nervous system show a large accumulation in areas outside the blood–brain barrier, in the choroid plexus and in the ventricular ependyma.[36] The uptake of narcotics in the choroid plexus has the characteristics of an active transport system, accumulating all of the narcotics tested against a concentration gradient in an energy-requiring process. There is no age-dependent parameter to drug transport in the choroid plexus, which is probably the route of egress from the cerebrospinal fluid (CSF) to blood. When labeled narcotics are perfused through the CSF, they appear in blood as rapidly and to the same extent as in brain tissue.[37]

A placental barrier can be demonstrated for certain narcotic analgesics. When dihydromorphine and etorphine are injected intramuscularly into pregnant rats, etorphine reaches fetal blood and brain at a much faster rate than dihydromorphine.[38] However, the placental barrier is also only a partial exclusion mechanism, since most drugs appear in fetal circulation after injection into pregnant animals. In man, the placental barrier is permeable to narcotics. The human infant of a heroin-dependent mother is born addicted and enters a withdrawal state soon after birth.

However, there is some evidence that normorphine may be remethylated to morphine in brain.[46]

Methadone is transformed by *N*-demethylation to a nor derivative which then is transformed by ring closure to a pyrrolidine compound. This metabolite has been isolated from human urine. Unchanged methadone is excreted via the bile.[46]

Since Rossbach speculated, in 1880, that the explanation of tolerance to morphine may reside in changes either in the rate of elimination of the drugs during chronic use or in the tissue levels in tolerant subjects, hundreds of investigators have compared the tissue distribution and the rate of elimination of the drugs and catabolites in tolerant and nontolerant animals. The general conclusion is summarized by Mulé as follows: "It is concluded that neither distribution nor metabolism of morphine provides any further insight into the mechanism of tolerance development."[27]

IV. BIOCHEMICAL RESPONSES TO NARCOTICS

A. General Effects

A major effect of the administration of narcotic analgesics to man or animals is a depression of respiration. Oxygen consumption decreases, as do both the tidal volume and the respiratory rate.[48] However, the reflection of these physiological changes in biochemical parameters in the central nervous system has been difficult to demonstrate. While morphine has a definite effect on cerebral glucose metabolism, most of the changes are seen when isolated tissue preparations are incubated in the presence of high concentrations of narcotic analgesics.[49] The respiration of isolated cerebral cortical slices is unaffected by the presence of the drugs. However, the stimulated respiration of cortex slices is inhibited by methadone, meperidine, morphine, and other narcotic drugs.[50,51] The utilization of glucose for oxidation through the citric acid cycle intermediates is decreased in brain slices incubated in the presence of narcotic drugs.[52] These effects are also seen *in vivo,* although there are several factors which have limited the observation of changes in brain respiration in the living animal: (1) The *time factor.* The fact that the insult to brain homeostasis induced by a drug is biphasic, first a response, then a counterresponse (recovery or adaptation), means that both the response and the counterresponse must be examined during a limited time period. (2) The *dose factor.* The ratio of depressant to stimulant effects resulting from drug administration is dose dependent (as well as drug and species dependent). In addition, secondary responses mediated by such mechanisms as histamine-induced vasodilatation or CO_2-induced hypertension contribute to the overall oxygen consumption of the central nervous system. (3) The *specificity factor*. Other categories of drugs also affect the utilization of glucose in the brain. For example, the administration of morphine or pentobarbital to rats produces an increase in the concentrations of glucose and ATP in brain.[53] Although the effects are similar, the morphine response is reversible by nalorphine while the pentobarbital response is not,[53] demonstrating a certain specificity in the mechanism for producing the effect at least at a regulatory site. These factors are also limiting in studies of most of the metabolic systems of the brain, of course, and are not the only causes of discrepancies in the results in similar experiments. For example, the administration of 20 mg/kg of morphine to rats resulted in a decrease in the conversion of glucose-C^{14} to aspartate-C^{14} and Gaba-C^{14} in the brains of rats killed 30 min after the injection of morphine,[54] while the administration of 60 mg/kg of morphine to rats produced no change in the conversion of leucine-C^{14} to labeled aspartate, Gaba, or glutamate in the brains of rats killed 2 hr after the injection.[55] In addition to the differences in time and dose in the two studies, another relevant factor which may be operating is that there are at least two pools of glutamate in brain: one labeled predominantly from glucose (the "pyruvate" pool) and the other from acetate and other 2-carbon intermediates such as those derived from leucine (the "acetate" pool).[56,57] The possibility exists that only one pool of glutamate is affected by the administration of morphine. This possibility is especially intriguing because the differences seen in the turnover of citric acid cycle intermediates in the glutamine pools in brain have been ascribed either to cellular (glial *vs*. neuronal) or to subcellular (mitochondrial *vs*. cytoplasmic) localization of the compartments.[58]

In order to be detectable, a change in the rate of synthesis of macromolecules in the brains of intact morphine-treated animals must represent a major change in the rate of biosynthesis of proteins or nucleic acids in the brain. When rats are injected with a fairly large dose of morphine (60 mg/kg), the rate of protein synthesis in whole brain, as measured by the incorporation of a labeled amino acid into protein *in vivo,* is inhibited transiently over a 1–4 hr period after the injection and recovers to over 150% of the control concentration by 24 hr.[59] The same dose of morphine also has a biphasic effect on the synthesis of RNA in rat brain, an initial inhibition followed by an increased incorporation of labeled precursor into nuclear RNA.[60] It is possible that these gross effects on the rates of biosynthesis of protein and RNA are secondary effects resulting from a decreased oxygenation of the tissue or a general inhibition of biosynthesis by a stepdown in cerebral function during the depressed state produced by the narcotic analgesics. In rodents, hypothermia is among the depressant effects after the administration of morphine. A relationship between depressed body temperature and depressed rate of protein synthesis in brain after the adminstration of morphine was eliminated, however, by experiments in which rats kept normothermic at an ambient temperature of 30°C had the same inhibition of protein synthesis in brain as morphine-treated rats kept at room temperature (22°C).[61] In simple systems, narcotic analgesics inhibit both protein and RNA synthesis. When bacterial cells or mammalian cells in tissue culture are incubated in the presence of various opiates, protein synthesis is inhibited.[62,63] This inhibition has been associated with an initial decrease in concentration of ATP followed by a decreased incorporation of precursors into protein and nucleic acids.[64] In Sindbis virus growing on monolayers of chick embryo cells, levorphanol produces both cytotoxic effects in the fibroblasts and inhibition of RNA and protein synthesis in both the virus and the fibroblasts.[65] Because the drug was cytotoxic at $\frac{1}{45}$ the concentration required to inhibit the synthesis of RNA and protein, the drug effect was ascribed to a primary membrane disruption prior to the disruption of the biosynthetic processes.[65] The antagonist, levallorphan, has also been shown to be bactericidal. The lethal effect of this drug is primarily on the plasma membrane, indicated by permeability changes and lysis of spheroblasts.[66]

Attempts to develop bacterial variants with a requirement for an opiate have not been successful,[67] although earlier reports describe the development of tolerance to narcotic drugs in cultures of tissue explants or mammalian cells.[68,69] Studies of the effect of tolerance on the biosynthesis of RNA and protein in animal brain are limited to experiments in which rats were injected once daily for 5 days. Although tolerance to morphine injection was shown on the fifth day in several pharmacological parameters, in one biochemical parameter, the rate of turnover of brain catecholamines,[70] tolerance did not develop, in contrast to other regimens of chronic drug administration. The increased rate of biosynthesis of nuclear RNA (2 hr after the injection) and the inhibition of protein synthesis (1.5 hr after the injection) in the brains after five daily injections were magnifications of the responses after the first injection of morphine.[59,71] These results do not exclude the possibility that tolerance in these biochemical parameters would develop after a suitable regimen of drug treatment.

The need for macromolecular synthesis during the development of tolerance to chronic opiate treatment is suggested by a number of studies in which the effects of the administration of inhibitors of RNA or protein synthesis were examined. The administration of an inhibitor of RNA synthesis, actinomycin D, results in a delay in the development of tolerance to the analgesic effects of chronic morphine treatment in mice.[72] Another inhibitor of RNA synthesis, 8-azaguanine, produces the same effect: tolerance to chronic morphine administration is less in mice also receiving the inhibitor.[73,74] Tolerance may be divided into *short-term* tolerance, a hyposensitivity to the drug occurring 2–5 hr after the administration of the drug, and *long-term* tolerance, occurring after

6 hr or later. The administration of actinomycin D blocks the development of long-term tolerance but not that found immediately after drug use.[75] However, when morphine, meperidine, or heroin is infused intravenously into rats, the development of tolerance during the infusion is prevented by the simultaneous infusion of actinomycin D, although established tolerance is not reversed by the inhibitor.[76] Other inhibitors of RNA synthesis, 6-mercaptopurine and 5-fluorouracil, also reduce the level of tolerance developed during morphine infusion.[77,78] Thus the evidence tends to indicate that synthesis of RNA is required for the development of tolerance to narcotic drugs.

Inhibition of RNA synthesis affects the synthesis of new messenger RNA and preribosomal units, so that protein synthesis is also inhibited in time. Whether the inhibition of protein synthesis in brain without interference with RNA turnover also affects the rate of development of tolerance is the subject of several experimental studies. The development of tolerance to the production of lenticular opacity in mice during chronic levorphanol treatment is significantly depressed when an inhibitor of protein synthesis, puromycin, is administered simultaneously with levorphanol.[75,79] The administration of another inhibitor, cycloheximide,[80,81] or of a suppressor of the immune response, cyclophosphoramide,[82] also decreases the level of tolerance to morphine arising during multiple injection schedules. When protein synthesis in brain is reduced by 40% by the administration of cycloheximide or puromycin, the development of acute tolerance to morphine infusion is also reduced.[77] These latter studies suggest that the action of inhibitors of RNA synthesis on the development of tolerance is affected through the subsequent inhibition of protein synthesis. However, further studies correlating biochemical effects in the nervous system with pharmacological responses are necessary before a firm conclusion about the requirement of protein synthesis in brain for the development of tolerance may be reached.

In order to distinguish between direct effects of the administration, or the addition in *in vitro* experiments, of narcotic drugs on lipid metabolism in the nervous system and indirect effects consequent to changes in energy transformation, oxygen uptake, or ATP pool size, these parameters have been examined at the same time as phospholipid synthesis. As already described, the oxygen uptake in a brain slice does not change in the presence of morphine. In the same experiments, the incorporation of P^{32} into phospholipid was increased 40%, with the largest increase in phosphoinositides and phosphatidic acids.[83] That these changes in the rate of synthesis of phospholipids reflect relative changes in the rate of synthesis of individual phospholipid classes is shown by the studies of Mulé,[84] who examined the time curves for the incorporation of P^{32}, glycerol-C^{14}, or choline-C^{14} into the phospholipids of guinea pig cerebral cortex slices. In the presence of morphine, there is more synthesis of phosphatidic acids and phosphatidyl inositide and less synthesis of phosphorylcholine. Mulé[85] suggests that the 1,2-diglyceride becomes limiting in brain slice experiments. When morphine and P^{32} are given to guinea pigs *in vivo,* all of the classes of phospholipids in brain show an increase in new synthesis, suggesting that in the whole animal the diglyceride concentration is not limiting.[85] In a model system consisting of an organic and an aqueous phase with the movement of morphine from the organic to the aqueous phase as an indicator of binding and transport of the drug, both calcium ions and phospholipids are required, with phosphatidic acids isolated from brain the most effective in transporting the drug into the aqueous phase.[86]

In the isolated guinea pig ileum preparation, morphine in very low doses (0.01 μg/ml) inhibits the effect of prostaglandin PGE_1.[87,88] Nalorphine and etonitazene have similar effects.

B. Pituitary–Adrenal Effects

Many of the responses to the administration of narcotic analgesics are mediated via the pituitary–adrenal hormone system. The neural control of the pituitary gland is transmitted through hypothalamic pathways, with the pituitary stalk the direct connection between the hypothalamic nuclei and the anterior pituitary.[89] The transmitter substances produced in the hypothalamus, called *releasing factors,* are carried in the hypophyseal portal system as blood-borne hormones and elicit the release of the specific tropic hormones from the pituitary gland.[89] Several of the releasing factors have been isolated and chemically identified as peptides.[90–92] Studies on the effect of opiate administration on releasing factors and hypophyseal tropins employ target tissue response as an assay of drug effects. The involvement of the pituitary has been shown by the loss of drug response in hypophysectomized animals and the involvement of the hypothalamus by the loss of response when small areas in hypothalamic nuclei are ablated. The acute administration of morphine to rats produces a striking reduction in adrenal ascorbic acid concentrations, which have an inverse relation to adrenocorticotropin (ACTH) levels.[93] Hypophysectomy abolishes this response to morphine and to the active isomer of methadone. The antagonist, nalorphine, abolishes the effect of morphine on adrenal stimulation.[93] Lesions of the anterior median eminence of the hypothalamus also abolish the adrenal cortical response to morphine, indicating that the neural control of the pituitary is located in, or passes through, areas of the hypothalamus.[94]

By the intrahypothalamic injection of very small doses of narcotic analgesics, the localization of some sites of specific pharmacological responses to the opiates has been made in hypothalamic nuclei.[13] In rats, the injection of 5–50 μg of morphine in various areas of the hypothalamus produces an increase in plasma corticosterone and a decrease in adrenal ascorbic acid only when the drug is introduced into an area in the middle region of the hypothalamus.[95] Similarly, hypothermia is induced by the microinjection of morphine into an area in the anterior hypothalamus.[96]

In rats, a single injection of morphine produces an increase in the amounts of plasma and urinary corticosteroids.[97] Tolerance develops to these effects.[98] In man, the first injection of an opiate has little effect on the urinary excretion of corticosteroids,[99] possibly because the dose of drug injected is much less active pharmacologically in man. In tolerant man, however, there is a reduction in the amounts of both plasma and urinary hydroxycorticosteroids.[99] The administration of morphine also blocks the diurnal rise in corticosteroid secretion in the rat[100] and in man.[101]

The uptake and release of thyroidal I^{131} are indices of the secretion of thyrotropin (TSH), an increase in I^{131} utilization in the thyroid reflecting an increase in TSH secretion in the anterior pituitary.[102] A single injection of morphine, meperidine, and many other narcotic analgesics to mice results in the release of I^{131} from the thyroid gland.[102] The administration of the same drugs for 5 days results in a loss of response, indicating that there is tolerance in this biochemical parameter.[102] Other tropic hormones, the follicle-stimulating hormone (FSH) and the luteinizing hormone (LH), are also secreted in response to the administration of opiates.[103] Since the control of the secretion of tropic hormones may result in either more or less tropins from the pituitary, as some of the hypothalamic factors are inhibitory, the responses to opiate administration are complex. It is evident, however, that opiates influence pituitary function by their action on the CNS, definitely in the hypothalamus and possibly elsewhere.[13]

The hormones of the posterior pituitary are also affected by the administration of narcotics. The antidiuretic hormone is excreted after a single injection of opiates, and tolerance develops to this effect upon chronic administration in rats, chicken, and man.[47]

C. Role of the Biogenic Amines and Acetylcholine

The biogenic amines and acetylcholine have long been considered to be implicated in the mechanism of action of the narcotic analgesic drugs. In 1936, Bernheim and Bernheim [104] described the effects of analgesics on cholinesterase, and in 1939 Slaughter and Munsell[105] suggested that cholinergic mechanisms in the CNS may be inhibited by morphine and other narcotic analgesics. The involvement of the catecholamines of the CNS in the action of morphine was shown by Vogt[106] in 1954, when she demonstrated a reduction of "sympathin" in the midbrain and hypothalamus of cats after an injection of 30 mg/kg of morphine. Since these early reports, there have been many studies in which changes in the amounts of norepinephrine, dopamine,[107] and acetylcholine[108] have been found in various areas of the brains of animals treated with morphine either acutely or chronically.[107]

The need for an adequate concentration of norepinephrine in the brain in order to effect pharmacological responses to morphine is shown in experiments in which the concentrations of the brain amines are manipulated by pretreatment with other drugs which alter these amounts. The prior treatment of animals with reserpine to deplete brain catecholamines antagonizes the pharmacological responses to morphine administration.[109] The length of time of pretreatment with reserpine dictates the effect on the responses to morphine, with the simultaneous administration of both drugs leading to a potentiation of the response.[110] Since reserpine also depletes brain dopamine and serotonin as well as norepinephrine, the need for norepinephrine is not established in these studies. However, when α-methyltyrosine, a specific inhibitor of tyrosine hydroxylase, the first enzyme in the norepinephrine biosynthetic pathway, is administered, the effect of morphine in producing analgesia is also antagonized,[111] implicating the catecholamines in this pharmacological response. In morphine-tolerant animals, there is no longer a depletion of brain catecholamines, which indeed are sometimes above control concentrations after an injection of morphine.[112,113] The possibility that an increased rate of biosynthesis of brain dopamine and norepinephrine occurs during the development of tolerance to morphine has been suggested as an explanation for the lack of effect in tolerant animals.[112,113] When the turnover of brain catecholamines is measured in $vivo$ by injecting the labeled precursor tyrosine-C^{14}, and measuring the rate of its incorporation into dopamine and norepinephrine, the effect of morphine is more pronounced on biosynthesis of dopamine than on that of norepinephrine.[70,114]

Administered acutely, morphine depletes brain dopamine as well as norepinephrine.[115] In most species, the chronic administration of morphine results in a loss of this dopamine-depleting effect.[116] The rate of biosynthesis of dopamine in areas of rat brain is increased after a single injection of morphine,[70] presumably as a result of the prior depletion of the amine. The dose of analgesic administered determines both the extent of the resynthesis and the time over which it occurs.[114] This effect of morphine is more pronounced in the brains of rats treated for 5 or 10 days with a single daily injection of morphine,[70] although tolerance in this biochemical parameter develops if the animals are injected four times a day[114] or are implanted with morphine pellets subcutaneously.[117]

A single dose of morphine produces no effect on the amount of serotonin in the brain.[118,119] There are a number of studies which nevertheless implicate this biogenic amine in the action of morphine. A monoamine oxidase inhibitor maintains a high level of serotonin in brain by preventing its catabolism by oxidation. If a MAO inhibitor is administered to mice, the toxic effects of high doses of morphine, meperidine, phenazocine, and pentazocine are potentiated.[120] The administration of p-chlorophenylalanine, (p-CPA), an inhibitor of serotonin biosynthesis, antagonizes the effect of morphine on analgesia.[121] In tolerant animals, there are no changes in brain serotonin concentrations.[122] However, an increased turnover of serotonin in the brains of tolerant animals has been found in several studies,[122,123] while in other studies no change has been found in the turnover rate of serotonin.[124] When p-CPA is administered to mice, the rate of development of tolerance and physical dependence is diminished.[122] Thus the involvement of serotonin in the mechanism of action or in the mechanism of addiction is not yet decided.

The concentrations of acetylcholine and its synthesizing and hydrolyzing enzyme systems, choline acetylase and acetylcholinesterase, have been examined in the nervous system after the administration of narcotic drugs. At very low doses, there is no change in the concentration of acetylcholine in the brain, but at higher doses (100 mg/kg) a transient increase in acetylcholine is observed in the brains of nontolerant mice, with a maximum 30 min after the injection of the drug.[125–128] During the development of tolerance to chronic drug administration, the increase in acetylcholine is no longer seen. In withdrawal induced either by discontinuing the drug or by nalorphine injection, the concentration of acetylcholine in brain increases sharply, peaking 39–46 hr after the last injection of morphine.[128] The activity of choline acetylase in one area of brain, the caudate nucleus, reflects the changes in acetylcholine concentration: 1 hr after the administration of a single dose of morphine, the activity of the enzyme is lower, returning to normal upon subsequent injections of morphine.[129] No change has been seen in the concentration of a precursor of acetylcholine, acetyl-CoA, at the time that the changes in acetylcholine and choline acetylase are found.[130–132] It has been suggested that the increase in bound acetylcholine in brain seen following the acute administration of morphine is due to a decreased liberation of acetylcholine from its storage site.[126] The inhibitory effect of morphine on acetylcholine release has also been seen in isolated preparations of preganglionic and superior cervical ganglion, and in perfused whole brain of the cat.[131] In brain cortex slices, the addition of morphine inhibits the release of preloaded acetylcholine from the slice, a reaction dependent on an adequate level of calcium ions.[133,134] It is likely that morphine competes with acetylcholine for the releasing mechanism, since the binding of acetylcholine has been shown to be competitively inhibited by morphine and other narcotic drugs both in brain cortex slices and in synaptic vesicle–rich fractions of brain.[135,136]

The hydrolysis of acetylcholine by brain acetylcholinesterase is inhibited *in vitro* in the presence of morphine and other agonists and antagonists.[137,138] Erythrocyte cholinesterase is also inhibited by narcotic drugs, including the newly introduced, highly active compounds.[139,140] The competitive nature of the interaction of narcotic drugs and acetylcholine for sites on serum cholinesterase has been shown to be limited to a single site, the hydrolytic site.[141,142] In spite of this strong inhibitory effect of narcotic drugs on the activity of acetylcholinesterase *in vitro*, no effect of morphine administration on the activity of brain acetylcholinesterase *in vivo* has been shown in either tolerant or nontolerant animals.[126,137] It is possible that cholinesterase is present in the nervous system in such excess that local inhibition is masked.

Thus it has been demonstrated that morphine and other narcotic drugs compete with acetylcholine at two metabolic sites: the binding and the hydrolytic sites. The relative importance of the interference in acetylcholine metabolism to the spectrum of pharmacological responses is difficult to assess. Maynert[125,143] suggests the interesting hypothesis that acetylcholine is involved in the excitatory responses, while catecholamines are related to the depressant effects of morphine.

D. Localization of Drug Responses

Some aspects of the localization of drug responses have been discussed in Section II. The anatomical sites in the nervous system where the neural actions of narcotic analgesics are located are described by the logic of exclusion.[18] If the action of morphine occurs after the ablation of an area of the CNS, then that area is excluded from the possible sites of action of the drug. The results of such studies suggest that morphine has an effect on any polysynaptic neural pathway and that the more complex responses depend on functioning higher centers only for the expression, or lack of expression, of the more complex pharmacological responses.

On the regional level, functional areas in the hypothalamus have been explored using microablation and microinjection techniques to see if pharmacological responses can be related to specific areas in the hypothalamus.[13] As discussed in Section IVC, specific hypothalamic nuclei can be related to certain pharmacological responses. The possibility exists that the hypothalamic nuclei thus identified are merely integrative pathways and that projections to cortical and other subcortical regions lead to the center for the response. These further pathways are not pertinent to this discussion if the actions of opiates on these responses are due only to intervention at the hypothalamic level.

Early studies on the effect of narcotic analgesics on single-cell systems suggested that phenomena resembling tolerance and dependence can be produced in tissue explants[69] or in fibroblasts in culture.[144] More recently, only toxic effects have been produced when cells were grown in culture medium containing opiates.[145] Noteboom and Mueller[146] have found the same relative order of toxicity of morphine and related drugs in mice and in HeLa cells in culture. Drugs with more toxicity than morphine in HeLa cell cultures, such as levorphanol and levallorphan, are also more active in inhibiting cell growth in bacterial systems.[147] The major effects of levorphanol in growing cultures of *Escherichia coli* are on macromolecular synthesis. In the presence of 1.8×10^{-3} M levorphanol, the rate of biosynthesis of both RNA and DNA is inhibited.[62] A smaller inhibition of protein biosynthesis is also found at this drug concentration.[62] Similar effects are seen on protein and RNA synthesis in *Staphylococcus aureus*.[148] At lower concentrations of levorphanol, 2×10^{-4} M, the flux of putrescine and spermidine through *E. coli* membranes is inhibited.[149] Additional evidence suggesting membrane damage in cells grown in culture in the presence of levorphanol comes from experiments in which the first effect of adding the drug was a leakage of ATP through the cell membrane.[64] The importance of such findings is that these relatively specific metabolic alterations induced by opiates in single cells seem to be reflections of alterations found in the CNS of higher organisms after drug treatment.

On the biochemical level, sites of opiate action in the CNS may be found,as the studies described in earlier sections demonstrate. The problem has been to separate from the abundance of prospective sites the one, or several, sites which are primary effective sites arising directly from interaction of drug and tissue component. On the one hand, interactions between narcotic analgesics and proteins, nucleic acids, phospholipids, or inorganic ions can be shown in isolated systems, and, on the other hand, biochemical responses may be demonstrated in the CNS of animals treated with the drugs. The missing link is the "initial site(s)" of action of the drug in the nervous system.

The activity of morphine and other narcotic analgesics in simple polysynaptic reflex arcs suggests that the neuronal synapse is an opiate tissue target. The function of each affected synapse would then determine which pharmacological effects would be expressed. If some component of the neuronal membrane system is the primary site of opiate action, it need not follow that the same receptors are directly involved in the development of tolerance to and dependence on chronic drug use. There are two other possibilities of mechanisms for biochemical adaptation to chronic drug exposure: (1) the adaptation is in secondary reactions which are consequent to the initial biochemical lesion, or (2) the adaptation occurs as a result of a drug–receptor interaction not directly related to the initial lesion. These possibilities are simplified versions of the various hypotheses of the mechanisms for tolerance and physical dependence, which will be discussed in the next section.

Several studies of the distribution in the CNS of labeled narcotic analgesics injected into animals have suggested that some of the drug is particulate bound.[28-30] We have recently studied the uptake of labeled narcotics into the nerve-ending particles of rat brain.[31] When homogenates of whole brain, or of brain areas, are incubated with dihydromorphine-H^3 or other labeled opiates, the largest amount of drug is found in the synaptosomal fraction. The localization is in the same fraction in the brains of rats killed 30 or 60 min after the injection of the labeled drug in pharmacologically active doses.[150] However, the injected drug is bound more tightly than the drug incubated with tissue *in vitro* and remains, in part, with the synaptosomal membrane when the nerve-ending particles are ruptured hypo-osmotically.[150] Recent studies have shown a localization of methadone-H^3 in a synaptosomal membrane fraction after injection of the drug into the CSF.[32] This specific localization was blocked by administering naloxone either into the CSF or subcutaneously.[32] Neuronal membrane binding has also been reported *in vitro* in studies in which nonspecific binding is excluded by preincubating with inactive isomers of the drug.[151] The functional significance of such binding is suggested both by the inhibition by morphine of catecholamine uptake into synaptosomes[152] and by the leakage of ATP and amino acids from cells in tissue culture.[64]

E. Tolerance to Biochemical Responses

The phenomenon of tolerance is defined by the relationship between the dose of a drug and the response to the drug in a tissue or organism. Tolerance is demonstrated by a decreased response to repeated exposure to a certain dose of a drug or by the need to increase the dose in order to obtain a response equal to that produced by the initial dose of drug. In animals and man, the term *tolerance* is used for two phenomena, which are differentiated by the time period needed to produce the condition into *short-term* and *long-term* tolerance. A decreased response to a drug is produced after exposure of the animal or tissue to the drug for a period of minutes or hours. In dogs, tolerance to morphine in some pharmacological parameters was produced within 8 hr by the continuous infusion of 3 mg/kg/hr.[153] In rats, the continuous infusion of 7.5 mg/kg/hr produced analgesic tolerance within 3 or 4 hr.[76] In mice, the two kinds of tolerance to opiates were demonstrated successively, with the short-term stage lasting 8 hr and the long-term at least 3 weeks,[75] using the cataractogenic effect as a measure of tolerance. Short-term tolerance to opiates may be closely related to the tachyphylaxis found after the chronic administration of sympathomimetic amines such as ephedrine and amphetamine.[154] Repeated intravenous doses of these amines at 10 min intervals resulted in a reduction in both intensity and duration of the hypertensive response in dogs after the fourth or fifth injection. Tachyphylaxis is mainly due to the depletion of neuronal biogenic amines which are needed for the pharmacological response,[155] although some sympathomimetic amines such as tyramine also enter catecholamine granular stores and serve as false transmitters.[156] Acute tolerance to acetylcholine can be demonstrated in nerve muscle preparations and is also attributed to insufficient time for recovery between consecutive doses of the drug.[157]

Short-term tolerance to opiates has been produced in isolated tissue preparations by repeated applications of morphine to the tissue.[158] The output of acetylcholine in response to nerve stimulation in the guinea pig ileum is diminished in the presence of morphine, an effect reversed by the direct application of acetylcholine.[159] Similarly, the response to excitation of isolated innervated nictitating membrane is reduced by morphine and reversed by the addition of norepinephrine.[160] Thus in isolated nerve–muscle preparations, tachyphylaxis to the inhibition of nervous transmission by morphine can be ascribed to the inability of the tissue to replenish its functional neurohormonal pool.

In the intact animal, the continuous intravenous infusion of morphine to dogs[153] or morphine, heroin, meperidine, and etorphine to rats[76] produced short-term tolerance in the analgesic response in 8 and 4 hr, respectively. The effects of opiates on neurotransmitter concentrations and turnover, described in Section IVC, suggest that short-term tolerance in intact animals may also be related to insufficient time for homeostatic recovery in one or more neurotransmitter systems before the application of another dose of drug.

Long-term tolerance, induced by repeated injections of opiates or by morphine pellet implantation, is manifested by diminished or absent responses to the drugs in many pharmacological parameters and also by adaptation in biochemical systems. Among the biochemical responses in which tolerance occurs are the depletion of biogenic amines in the CNS and of adrenal steroids, as well as the synthesis and release of pituitary–hypothalamic hormones.[161] In the CNS of tolerant animals, there is no longer an increased turnover of brain amines,[162] nor a reduction in acetylcholine release,[126,128] nor a fall in Ca^{2+} concentration,[134] after the administration of morphine to the animals. There are few examples of a lack of adaptation to chronic opiate administration in a biochemical parameter. One such example is the distribution and localization of radiolabeled opiates in brain. In tolerant animals, the concentrations and sites of localization of dihydromorphine-H^3, cyclazocine-H^3, and morphine-C^{14} are not significantly different from those in naive animals.[25] Similarly, the subcellular localization of drugs in the nerve-ending particles of rat brain is not changed when the rats are treated chronically with morphine.[117] That this lack of effect in tolerant animals may be ascribed to a major component of nonspecific drug localization is suggested by the results of recent experiments in which the binding of methadone-H^3, to one fraction of synaptosomal membranes was reversed by naloxone and decreased in nerve-ending preparations from tolerant animals.[32]

Behaviorally morphine and other opiates serve as reinforcers for the opiate-dependent animal,[163] so that the animals will self-administer opiates when given an opportunity to do so. Self-administration techniques have been developed for the monkey[19] and the rat[20] and used to assess the variables of reinforcement and withdrawal from the use of drugs which induce dependence.[21,22] However, there are few studies comparing neurochemical effects after self-administration of opiates to those induced by drug administration according to the experimenter's schedule. The role of motivational conditions for drug use has thus been assessed mainly in its psychological aspects.

V. ELECTROENCEPHALOGRAPHY

A. Electroencephalography in Experimental Animals

The effects of narcotic analgesics on the EEG of animals were described by Wikler and Altschul[164] in 1950. Small doses produced high-voltage slow activity, but moderate doses had little effect. The dose effect has been described in subsequent studies. At low doses, there are changes in the EEG indicative of a blocked arousal response and the elimination of rapid eye movement (REM) sleep. At moderate higher doses of the drugs, there is a normal arousal and at high doses a convulsive seizure–type pattern.[165,166] Similar changes are seen in the EEG patterns of fetal guinea pig when the pregnant female is injected with meperidine.[167] In the rat, the response to an initial dose of morphine is a complete loss of paradoxical sleep and the appearance of "slow bursts."[168] Tolerance develops to these effects. After a single dose of 8 mg/kg morphine in rats, tolerance persists for 7 days in the high-voltage slow wave response.[169] The application of morphine or levorphanol to the surface of the brain also produces the high-voltage slow wave typical of the response to the systemic administration of opiates.[170] The use of EEG recordings, an extremely sensitive method of measuring narcotic drug response, is limited only by the relative lack of specificity of the response and the technical difficulty of measuring EEG patterns in large numbers of experimental animals.

B. Electroencephalography in Man

The EEG patterns from scalp recordings in man during addiction cycles in stabilized addicts, in post-addicts, and in untreated controls were first described by Andrews[171] in 1943. The abundance of high-alpha waves was noted in the stabilized addicts. When the dosage was kept constant, the EEG patterns returned quickly to the preaddiction state. Theta activity was also found after a single dose of heroin, morphine, or methadone. An early correlation between

EEG and behavior was observed by Wikler,[172] who associated desynchronization with anxiety or stimulation and synchronization with depression.

The use of FM tape-recordings of EEG records for digital computer analysis has refined the evaluation of scalp responses.[173] In a number of studies, Fink and his colleagues have examined the effects of heroin and other opiates and antagonists on the EEG in addicted man.[174] A dose of 15–25 mg of heroin given intravenously during a 2 min injection period produced a two-stage response. The *early response* was an increase in α-voltage and a decrease in α-frequency.[174] The *late response* was a fragmentation of the α-rhythms and an increase in θ-activity. Some subjects showed paroxysmal activity at this dose of heroin.[174] Methadone also produces a decrease in α-frequency and an increase in θ-activity.[174] The effects disappear during chronic methadone treatment. In the same laboratory, the effect of a single dose of heroin on EEG patterns of addicts maintained on 100 mg/day of methadone was examined. The heroin response was characterized in these individuals as increased θ-activity.[175] The effects of heroin are blocked by the administration of the antagonist, naloxone, which does not elicit responses when administered alone.[175] The partial antagonist, cyclazocine, also blocks the EEG effects of an acute dose of heroin. However, the blocking effect was less in post-addicts receiving cyclazocine chronically.[174] Thus the changes in scalp potentials reflect the behavioral effects produced by the agonists and antagonists, which are both in turn reflections of the neurochemical responses in the CNS.

VI. THEORIES ON THE MECHANISM OF ADDICTION TO NARCOTICS

A general hypothesis proposed by Himmelsbach[177] in 1943 is termed the *homeostatic counteradaptive theory* of morphine tolerance and physical dependence. In this theory, the effect of opiates on hypothalamic centers disturbs homeostasis and leads to physiological adjustments in order to attain a new level of homeostasis. When drug use is discontinued, the new equilibrium is disturbed, as indicated by the withdrawal syndrome.

Newer theories have suggested some mechanisms whereby this biochemical adaptation to opiates is accomplished. Shuster[178] has suggested, in his *enzyme induction hypothesis*, that opiates act both at the neurotransmitter synaptic level and at the enzyme induction level. Goldstein and Goldstein[179] have suggested a similar hypothesis in which the opiates act to inhibit the synthesis of some of the enzymes which catalyze the biosynthesis of neurotransmitters, resulting in lower concentrations of neurotransmitters, which in turn would induce the synthesis of the inhibited enzyme. Another hypothesis, proposed by Collier,[180] suggests that it is not the neurotransmitter system which adapts but the receptor itself, which is transformed from "silent" to "active" by the continued presence of drugs. Another theory involving the receptor has been proposed by Jaffe and Sharpless,[181] called the *disuse supersensitivity theory*, in which central receptors in the synapses are subject to pharmacological blockade by chronic drug exposure, leading to a state of disuse hyperexcitability in the postsynaptic elements which overrespond when the blockade is lifted by drug withdrawal.

The possibility that transferable factors are synthesized in tolerant animals derives from analogy to the antigen–antibody reaction. A morphine–protein complex would act as an antigen and stimulate the production of antibodies to the drug. There have been several attempts to isolate serum factors, presumably containing antibodies, which transfer tolerance when injected into recipient animals.[182] The results have not been conclusive, since there was great variation in "potentiation" or "attenuation" by serum, particularly when different animal species were compared. Likewise, there have been attempts to demonstrate tissue factors which transfer a tolerance to recipient animals.[183] The injection of extracts prepared from the brains of tolerant animals into nontolerant animals has resulted in the transfer of partial tolerance to the recipient animals. However, these findings have not been reproduced in other laboratories in which the same phenomenon was studied.[183] One must conclude that the question of the transfer of tolerance is not decided. However, antibodies to opiates have been produced by administering a morphine–protein complex to rabbits using the usual immunological procedures and isolating antibodies which have a specificity for morphine and closely related compounds.[184]

VII. SUMMARY AND CONCLUSIONS

The ubiquitous role played by narcotic analgesics in altering metabolism in single cells, in inhibiting simple polysynaptic reflexes in the spinal dog, in affecting the regulation of body temperature in the decorticate dog, and in modulating behavior in intact animals and man may be attributed to a wide distribution of receptor sites for drug activity in both the central and peripheral nervous systems. Because the narcotic drugs are large ionizable organic molecules, they enter the nervous system readily and are distributed throughout the brain and in peripheral nerve. The localization of the drugs in gross anatomical or subcellular areas of the CNS seems to be more dependent on physicochemical characteristics of each drug than on the relative pharamacological action of the drug. In homologous series of drugs, correlations have been made between narcotic activity and one or more physicochemical characteristics, such as partition coefficients or dimensions of lipophilic sites. Comparisons of the "analgesic receptor(s)" as defined by structural modifications in various families of narcotics suggest that the "receptor" may exist in more than one conformation, with varying numbers of binding sites. This suggestion would offer a chemical basis for the varying pharmacological effects produced by administration of the various types of narcotic analgesics.

On the subcellular level, the localization of radiolabeled narcotic drugs after administration in active doses to animals in the nerve-ending fractions isolated from the CNS, and especially in the membranes obtained from ruptured nerve-endings, may offer a clue to the localization of the "receptor." If the synaptic membrane is the site of drug action, and synaptic transmission is the function which is disturbed, then secondary effects arising from these primary lesions could involve the biogenic amines and acetylcholine in the CNS and elsewhere and the hypothalamic–pituitary–adrenal system, which in turn would produce alterations in the neurohormonal modulation of the CNS. This progression would lead to a new equilibrium in neuronal functioning established to counter the drug effects. It is possible to include most physiological and biochemical responses to drug administration in this scheme of events. The development of tolerance to and dependence on chronic drug use could then be considered as adaptive responses to the effects of the drug through mechanisms as yet unknown.

REFERENCES

1. WHO Expert Committee on Drug Dependence, Drug dependence: Its significance and characteristics, *Bull. World Health Org.* **37**: (1965).
2. N. B. Eddy, The relation of chemical structure to analgesic action, *J. Am. Pharm. Ass.* **39**: 245–251 (1950).
3. A. F. Casy, The structure of narcotic analgesic drugs, *in* "Narcotic Drugs: Biochemical Pharmacology" (D. H. Clouet, ed.) Plenum Press, New York (1971).
4. P. S. Portoghese, Relationship between stereostructure and pharmacological activities, *Ann. Rev. Pharmacol.* **10**:51–76 (1970).
5. S. Archer and L. S. Harris, Narcotic antagonists, *Progr. Drug Res.* **8**:261–269 (1965).
6. L. S. Harris, Structure–activity relationships, *in* "Narcotic Drugs: Biochemical Pharmacology" (D. H. Clouet, ed.) Plenum Press, New York (1971).
7. A. H. Beckett and A. F. Casy, Synthetic analgesics: Stereochemical considerations, *J. Pharm. Pharmacol.* **6**:986–1001 (1954).
8. J. W. Lewis, K. W. Bentley, and A. Cowan, Narcotic analgesics and antagonists, *Ann. Rev. Pharmacol.* **11**:241–270 (1971).
9. A. Wikler, Sites and mechanisms of action of morphine and related drugs in the central nervous system, *Pharmacol. Rev.* **2**:435–506 (1950).
10. J. Florez, L. E. McCarthy, and H. L. Borison, A comparative study in the cat of the respiratory effects of morphine, *J. Pharmacol. Exptl. Therap.* **163**:448–455 (1968).
11. H. L. Borison and S. C. Wang, Physiology and pharmacology of vomiting, *Pharmacol. Rev.* **5**:193–225 (1953).
12. C. F. Schmidt and A. E. Livingston, The action of morphine in mammalian circulation, *J. Pharmacol. Exptl. Therap.* **47**:411–441 (1933).

13. R. George, The effects of narcotic analgesics on the hypothalamus: Pituitary gland, *in* "Narcotic Drugs: Biochemical Pharmacology" (D. H. Clouet, ed.) Plenum Press, New York (1971).

14. F. D. McCrea, G. S. Eadie, and J. E. Morgan, The mechanism of morphine miosis, *J. Pharmacol. Exptl. Therap.* **74**:239–246 (1942).

15. E. Mills and S. C. Wang, Liberation of antidiuretic hormone: Pharmacologic blockade of ascending pathways, *Am. J. Physiol.* **207**:1405–1410 (1964).

16. C. M. Brooks, R. A. Goodwin, and H. N. Willard, The effect of various brain lesions on morphine-induced hyperglycemia and excitement in the cat, *Am. J. Physiol.* **133**:226–227 (1941).

17. K. E. Moore, L. E. McCarthy, and H. L. Borison, Blood glucose and brain catecholamine levels in the cat following the injection of morphine into the cerebrospinal fluid, *J. Pharmacol. Exptl. Therap.* **148**:169–175 (1965).

18. H. L. Borison, Site of action of narcotics in the nervous system, *in* "Narcotic Drugs: Biochemical Pharmacology" (D. H. Clouet, ed.) Plenum Press, New York (1971).

19. T. Thompson and C. R. Schuster, Morphine self-administration, food-reinforced and avoidance behaviors in rhesus monkey, *Psychopharmacologia* **5**:87–94 (1964).

20. J. R. Weeks, Experimental morphine addiction: Method for automatic intravenous injection in unrestrained rats, *Science* **138**:143–144 (1962).

21. C. R. Schuster and T. Thompson, Self-administration of and behavioral dependence on drugs, *Ann. Rev. Pharmacol.* **9**:483–502 (1969).

22. J. R. Weeks and R. J. Collins, Factors affecting voluntary morphine intake in self-maintained addicted rats, *Psychopharmacologia* **6**:267–279 (1964).

23. T. Thompson and W. Ostlund, Susceptibility to readdiction as a function of the addiction and withdrawal environment, *J. Comp. Physiol. Psychol.* **59**:388–392 (1965).

24. T. Thompson and R. Pickens, Drug self-administration and conditioning, *in* "Scientific Basis of Drug Dependence" (H. Steinberg, ed.) J. and A. Churchill, London (1969).

25. S. J. Mulé, Physiological disposition of narcotic agonists and antagonists, *in* "Narcotic Drugs: Biochemical Pharmacology" (D. H. Clouet, ed.) Plenum Press, New York (1971).

26. A. Herz and H. J. Teschemacher, activities and sites of antinociceptive action of morphine-like analgesics, *in* "Advances in Drug Research" (N. J. Harper and A. B. Simmonds, eds.) Vol. 6, Academic Press, New York (1971).

27. S. J. Mulé, The relationship of the disposition and metabolism of morphine in the CNS to tolerance, *in* "Scientific Basis of Drug Dependence" (H. Steinberg, ed.) J. and A. Churchill, London (1969).

28. H. Kaneto and L. B. Mellett, The intracellular binding of *N*-methyl C^{14}-morphine in brain tissue of the rat, *The Pharmacologist* **2**:98 (1960).

29. D. Van Praag and E. J. Simon, Studies on the intracellular distribution and tissue binding of dihydromorphine-H^3 in the rat, *Proc. Soc. Exptl. Biol. Med.* **122**:6–16 (1966).

30. S. J. Mulé, C. M. Redman, and J. W. Flesher, Intracellular disposition of H^3-morphine in the brain and liver of non-tolerant and tolerant guinea-pigs, *J. Pharmacol. Exptl. Therap.* **157**:459–471 (1967).

31. D. H. Clouet and N. Williams, The binding of narcotic analgesics in synaptosomal and other particulate fractions, *The Pharmacologist* **13**:676 (1971).

32. S. J. Mulé, G. A. Casella, and D. H. Clouet, Localization of levo-H^3-methadone in synaptic membranes of rat brain, reported to the *Committee on Problems of Drug Dependence* (Natl. Acad. Sci-Natl. Research Council) pp. 322–339 (1972).

33. E. L. Way, Brain uptake of morphine: Pharmacologic implications, *Fed. Proc.* **26**:1115–1118 (1967).

34. W. L. Way, E. C. Costley, and E. L. Way, Respiratory sensitivity of the newborn infant to meperidine and morphine, *Clin. Pharmacol. Therap.* **6**:454–461 (1965).

35. A. H. Conney and J. J. Burns, Factors influencing drug metabolism, *Advan. Pharmacol.* **1**:31–58 (1962).

36. C. C. Hug, Transport of narcotic analgesics by choroid plexus and kidney tissue, *Biochem. Pharmacol.* **16**:345–359 (1967).

37. C. C. Hug, Transport of narcotic analgesics in the central nervous system, *in* "Narcotic Drugs: Biochemical Pharmacology" (D. H. Clouet, ed.) Plenum Press, New York (1971).

38. G. F. Blane and H. E. Dobbs, Distribution of H^3-labelled etorphine (M99) and dihydromorphine in pregnant rats at term, *Brit. J. Pharmacol.* **30**:166–172 (1967).

39. E. L. Way and T. K. Adler, The biological disposition of morphine and its surrogates, *Bull. World Health Org.* **25–27**:3–117 (1962).

40. E. L. Way and T. K. Adler, The pharmacologic implications of the fate of morphine and its surrogates, *Pharmacol. Rev.* **12**:383–446 (1960).

41. J. R. Gillette, Factors affecting drug metabolism, *Ann. N. Y. Acad. Sci.* **179**:43–67 (1971).

42. K. C. Liebman, A. G. Hildebrandt, and R. W. Estabrook, Spectrophotometric studies of interactions between various substrates in their binding to microsomal cytochrome P-450, *Biochem. Biophys. Res. Commun.* **36**:789–794 (1969).

43. A. H. Conney, Enzyme induction and drug toxicity, *in* "Drugs and Enzymes" (B. B. Brodie and J. R. Gillette, eds.) Pergamon Press, New York (1965).

44. D. H. Clouet and M. Ratner, The effect of altering liver microsomal *N*-demethylase activity on the development of tolerance to morphine in rats, *J. Pharmacol. Exptl. Therap.* **144**:362–372 (1964).

45. J. B. Schenkman, H. Remmer, and R. W. Estabrook, Spectral studies of drug interaction with hepatic microsomal cytochrome, *Molec. Pharmacol.* **3**:113–123 (1967).

46. J. T. Scrafani and D. H. Clouet, Biotransformations of narcotic analgesics, *in* "Narcotic Drugs: Biochemical Pharmacology" (D. H. Clouet, ed.) Plenum Press, New York: (1971).

47. J. M. Fujimoto, Sites of action of narcotic analgesics in the kidney, *in* "Narcotic Drugs: Biochemical Pharmacology" (D. H. Clouet, ed.) Plenum Press, New York (1971).

48. R. B. Nelson and H. W. Elliott, A comparison of some central effects of morphine, morphinone and thebaine in rats and mice, *J. Pharmacol. Exptl. Therap.* **155**:516–520 (1967).

49. A. E. Takemori, Intermediary metabolism: Effects by narcotic drugs, *in* "Narcotic Drugs: Biochemical Pharmacology" (D. H. Clouet, ed.) Plenum Press, New York (1971).

50. A. E. Takemori, Cellular adaptation to morphine in rats, *Science* **133**:1018–1019 (1961).

51. H. McIlwain, Actions of haloperidol, meperidine and related compounds on the excitability and ion content of isolated cerebral tissues, *Biochem. Pharmacol.* **13**:523–529 (1964).

52. A. E. Takemori, Effect of central depressant agents on cerebral G-6-P dehydrogenase activity of rats, *J. Neurochem.* **12**:407–415 (1965).

53. P. W. Dodge and A. E. Takemori, Changes in rat cerebral glycolytic intermediates *in vivo* after treatment with morphine, nalorphine or pentobarbital, *Biochem. Pharmacol.* **18**:1873–1882 (1969).

54. H. S. Bachelard and J. R. Lindsay, Effects of neurotropic drugs on glucose metabolism in rat brain *in vivo*, *Biochem. Pharmacol.* **15**:1053–1058 (1966).

55. D. H. Clouet and A. Neidle, The effect of the administration of morphine on the transport and metabolism of intracisternally administered leucine in the rat, *J. Neurochem.* **17**:1069–1074 (1970).

56. S. Berl and D. P. Purpura, Regional development of glutamic acid compartmentation in immature brain, *J. Neurochem.* **13**:293–304 (1966).

57. R. M. O'Neal and R. E. Koeppe, Precursors *in vivo* of glutamate, aspartate and their derivatives of rat brain, *J. Neurochem.* **13**:835–847 (1966).

58. R. Balazs, Y. Machiyama, B. J. Hammond, T. Julian, and D. Richter, The operation of the gamma-aminobutyrate bypass of the tricarboxylic acid cycle in brain tissue *in vitro*, *Biochem. J.* **116**:445–467 (1970).

59. D. H. Clouet and M. Ratner, The effect of morphine administration on the incorporation of C^{14}-leucine into the protein of rat brain *in vivo*, *Brain Res.* **4**:33–43 (1967).

60. D. H. Clouet, The effect of morphine administration on protein and RNA synthesis in rat brain, *in* "Drug Abuse: Social and Psychopharmacological Aspects" (J. O. Cole and J. R. Wittenborn, eds.) Charles C. Thomas, Springfield, Ill. (1969).

61. D. H. Clouet, The effects of drugs upon protein synthesis, *in* "Protein Metabolism in the Nervous System" (A. Lajtha, ed.) Plenum Press, New York (1970).
62. E. J. Simon and D. Van Praag, Inhibition of RNA synthesis in *E. coli* by levorphanol, *Proc. Natl. Acad. Sci.* **51**:877–883 (1964).
63. W. D. Noteboom and G. C. Mueller, Inhibition of cell growth and the synthesis of RNA and protein in Hela cells by morphinans and related compounds, *Molec. Pharmacol.* **2**: 534–542 (1966).
64. R. Greene and B. Magasanik, The mode of action of levallorphan as an inhibitor of cell growth, *Molec. Pharmacol.* **3**:453–472. (1967).
65. T. Rossman, F. F. Becker, and J. Vilchek, An investigation into the mechanism of cyto-toxicity of levorphanol, *Molec. Pharmacol.* **7**:480–483 (1971).
66. P. L. Boquet, M. A. Devynck, H. Aurelle, and P. Fromageot, On the bacterialcidal action of levallorphan: Irreversible alterations of the plasma membrane, *Europ. J. Biochem.* **21**: 536–541 (1971).
67. E. J. Simon, Inhibition of the synthesis of RNA in *E. coli* by the narcotic drug levorphanol, *Nature* **198**:794–795 (1963).
68. G. Corssen and I. A. Skora, "Addiction" reactions in cultural human cells, *J. Am. Med. Ass.* **187**:328–332 (1964).
69. K. Sanjo, Experimentelle Untersuchungen über die Gewöhnung der Irisepithelkulturen an Morphin, *Folia Pharmacol. Jap.* **17**:219–229 (1934).
70. D. H. Clouet and M. Ratner, Catecholamine biosynthesis in brains of rats treated with morphine, *Science* **168**:854–856 (1970).
71. D. H. Clouet, The effects of narcotic analgesics on protein and RNA metabolism, *in* "Narcotic Drugs: Biochemical Pharmacology" (D. H. Clouet, ed.) Plenum Press, New York (1971).
72. M. Cohen, A. S. Keats, W. Krivoy, and G. Ungar, Effect of actinomycin D on morphine tolerance, *Proc. Soc. Exptl. Biol. Med.* **119**:381–383 (1965).
73. J. Yamamoto, R. Inoki, Y. Tamari, and K. Iwatsubo, Inhibitory effect of 8-azaguanine on the development of tolerance in the analgesic action of morphine, *Jap. J. Pharmacol.* **17**:140–142 (1967).
74. M. T. Spoerlein and J. Scrafani, Effects of time and 8-azaguanine on the development of morphine tolerance, *Life Sci.* **6**:1549–1564 (1967).
75. A. A. Smith, M. Karmin, and J. Gavitt, Tolerance to the lenticular effects of opiates, *J. Pharmacol. Exptl. Therap.* **156**:85–91 (1967).
76. B. M. Cox, M. Ginsburg, and O. H. Osman, Acute tolerance to narcotic analgesics in rats, *Brit. J. Pharmacol.* **33**:245–256 (1968).
77. B. M. Cox and O. H. Osman, The role of protein synthesis inhibition in the prevention of morphine tolerance, *Brit. J. Pharmacol.* **35**:373 (1969).
78. B. M. Cox and M. Ginsburg, Is there a relationship between protein synthesis and toler-ance to analgesic drugs?, *in* "Scientific Basis of Drug Dependence" (H. Steinberg, ed.) J. and A. Churchill, London (1969).
79. A. A. Smith, M. Karmin, and J. Gavitt, Blocking effect of puromycin, ethanol and chloro-form on the development of tolerance to an opiate, *Biochem. Pharmacol.* **151**:1877–1879 (1966).
80. H. H. Loh, F. Shen, and E. L. Way, Effects of cycloheximide on the development of mor-phine tolerance and physical dependence, *Biochem. Pharmacol.* **18**:2711–2718 (1969).
81. M. P. Feinberg and J. Cochin, Effect of weekly doses of cycloheximide on tolerance to morphine in the rat, *The Pharmacologist* **11**:256 (1969).
82. M. P. Feinberg and J. Cochin, Effect of cyclophosphoramide on tolerance to morphine, *The Pharmacologist* **10**:188 (1968).
83. M. Brossard and J. H. Quastel, Effect of morphine and tofranil on the incorporation of P^{32} into phospholipids of slices, *Biochem. Pharmacol.* **12**:766–768 (1963).
84. S. J. Mulé, Effect of morphine and nalorphine on the metabolism of phospholipids in guinea-pig cerebral cortex slices, *J. Pharmacol. Exptl. Therap.* **154**:370–383 (1966).

85. S. J. Mulé, Morphine and the incorporation of P^{32} into brain phospholipids of nontolerant, tolerant and abstinent guinea-pigs, *J. Pharmacol. Exptl. Therap.* **156**:92–100 (1967).

86. S. J. Mulé, Inhibition of phospholipid facilitated Ca^{++} transport by CNS acting drugs, *Biochem. Pharmacol.* **18**:339–346 (1969).

87. R. Jaques, Morphine as an inhibitor of prostaglandin in isolated guinea-pig intestine, *Experientia* **25**:1059–1060 (1969).

88. J. Sanner, Prostaglandin inhibition with a dibenzoxazepine hydrazide derivative and morphine, *Ann. N.Y. Acad. Sci.* **180**:396–406 (1971).

89. J. D. Green and G. W. Harris, The neurovascular link between the neurohypophysis and adenohypophysis, *J. Endocrinol.* **5**:136–146 (1947).

90. R. Guillemin, The adenohypophysis and its hypothalamic control, *Ann. Rev. Physiol* **29**:313–348 (1967).

91. A. V. Schalley, A. Arimura, C. Y. Bowers, A. J. Kastin, S. Sawano, and T. W. Redding, Hypothalamic neurohormones regulating anterior pituitary function, *Recent Progr. Horm. Res.* **24**:497–588 (1968).

92. R. Burgus and R. Guillemin, Hypothalamic releasing factors, *Ann. Rev. Biochem.* **39**:499–526 (1970).

93. R. George and E. L. Way, Studies on the mechanism of pituitary–adrenal activation by morphine, *Brit. J. Pharmacol.* **10**:260–264 (1955).

94. R. George and E. L. Way, The role of the hypothalamus in pituitary–adrenal activation and antidiuresis by morphine, *J. Pharmacol. Exptl. Therap.* **125**:111–115 (1959).

95. V. J. Lotti, N. Kokka, and R. George, Pituitary–adrenal activation by intra-hypothalamic micro-injection of morphine, *Neuroendocrinology* **4**:326–332 (1969).

96. V. J. Lotti, P. Lomax, and R. George, Temperature response in the rat following intracerebral micro-injection of morphine, *J. Pharmacol. Exptl. Therap.* **150**:135–139 (1965).

97. O. Nikodijevic and R. P. Maickel, Some effects of morphine in pituitary–adrenocortical function in the rat, *Biochem. Pharmacol.* **16**:2137–2142 (1967).

98. E. Paroli and P. Melchiorri, Urinary excretion of hydroxysteroids and aldosterone in rats during a cycle of treatment with morphine, *Biochem. Pharmacol.* **6**:1–17 (1961).

99. A. J. Eisenmann, H. T. Fraser, and J. W. Brooks, Urinary excretion and plasma levels of 17-hydroxycorticosteroids during a cycle of addiction to morphine, *J. Pharmacol. Exptl. Therap.* **132**: 226–231 (1961).

100. M. G. Slusher and B. Browning, Morphine inhibition of plasma corticosteroid levels in chronic catherized rats, *Am. J. Physiol.* **200**:1032–1034 (1961).

101. R. K. McDonald, F. T. Evans, V. K. Weise, and R. W. Patrick, Effects of morphine and nalorphine on plasma hydroxycorticosteroid levels in man, *J. Pharmacol. Exptl. Therap.* **125**:241–247 (1959).

102. T. W. Redding, C. Y. Bowers, and A. V. Schalley, Effects of morphine and other narcotics on thyroid function in mice, *Acta Endocrinol.* **51**:391–399 (1966).

103. E. G. Rennels, Effect of morphine on pituitary cytology and gonadotrophic levels in the rat, *Texas Rep. Biol. Med.* **19**:646–657 (1961).

104. F. Bernheim and M. L. C. Bernheim, Action of drugs on the cholinesterase of brain, *J. Pharmacol. Exptl. Therap.* **57**:427–436 (1936).

105. D. Slaughter and D. W. Munsell, New aspects of morphine action: Cholinergic effects on pain, *J. Pharmacol. Exptl. Therap.* **66**:33–36 (1939).

106. M. Vogt, Concentration of sympathin in different parts of the CNS under normal conditions and after the administration of drugs, *J. Physiol.* **123**:451–481 (1954).

107. E. L. Way and F. H. Shen, The effects of narcotic analgesic drugs on catecholamines and 5-hydroxytryptamine, *in* "Narcotic Drugs: Biochemical Pharmacology" (D. H. Clouet, ed.) Plenum Press, New York (1971).

108. M. Weinstock, The effects of narcotic analgesic drugs on acetylcholine and cholinesterases, *in* "Narcotic Drugs: Biochemical Pharmacology" (D. H. Clouet, ed.) Plenum Press, New York (1971).

109. J. A. Schneider, Reserpine antagonism of morphine analgesia in mice, *Proc. Soc. Exptl. Biol. Med.* **87**:614–615 (1954).
110. W. L. Dewey, L. S. Harris, J. F. Howes, and J. A. Nuite, The effect of various neurohumoral modulators on the activity of morphine and the narcotic antagonists in the tail-flick and phenylquinone tests, *J. Pharmacol. Exptl. Therap.* **175**:435–442 (1970).
111. R. A. Verri, F. G. Graef, and A. P. Corrado, Antagonism of morphine analgesia by reserpine and alpha-methyl tyrosine and the role played by catecholamines in morphine analgesic action, *J. Pharm. Pharmacol.* **19**:264–265 (1967).
112. L. M. Gunne, Catecholamines and 5-hydroxytryptamine in morphine tolerance and withdrawal, *Acta Physiol. Scand.* **58**:1–91 (1963).
113. E. W. Maynert and G. I. Klingman, Tolerance to morphine: Effects on catecholamines in brain and adrenal gland, *J. Pharmacol. Exptl. Therap.* **143**:285–295 (1964).
114. C. B. Smith, J. E. Villarreal, J. H. Bednarczyk, and M. I. Sheldon, Tolerance to morphine-induced increases in C^{14}-catecholamine synthesis in mouse brain, *Science* **170**:1106–1113 (1970).
115. H. Tagaki and N. Nakama, Effect of morphine and nalorphine on the content of dopamine in mouse brain, *Jap. J. Pharmacol.* **16**:482–483 (1966).
116. L. M. Gunne and J. Johnson, Effects of morphine intoxication in brain catecholamine neurons, *Europ. J. Pharmacol.* **5**:338–342 (1969).
117. D. H. Clouet and M. Ratner, The biosynthesis of catecholamines in the brains of morphine-treated rats, Reported to the Committee on Problems of Drug Dependence (1970).
118. B. B. Brodie, P. A. Shore, and A. Pletscher, Serotonin-releasing activity limited to rauwolfia alkaloids with tranquilizing action, *Science* **123**:992–993 (1956).
119. E. W. Maynert, G. I. Klingman, and H. K. Taki, Tolerance to morphine: Lack of effect on 5-OH tryptamine and GABA, *J. Pharmacol. Exptl. Therap.* **135**:296–299 (1962).
120. K. J. Rogers and J. A. Thornton, The interaction between monoamine oxidase inhibitors and narcotic analgesics in mice, *Brit. J. Pharmacol.* **36**:470–480 (1969).
121. S. S. Tenon, Antagonism of the analgesic effect of morphine and other drugs by *p*-chlorophenylalanine, a serotonin depletor, *Psychopharmacologia* **12**:278–285 (1968).
122. E. L. Way, H. H. Loh, and F. H. Shen, Morphine tolerance, physical dependence and the synthesis of brain serotonin, *Science* **162**:1290–1292 (1968).
123. D. R. Haubrich and D. E. Blake, Effect of acute and chronic administration of morphine on the metabolism of brain serotonin in rats, *Fed. Proc.* **28**:793 (1969).
124. D. L. Cheney and A. Goldstein, Narcotic tolerance and dependence: Lack of relationship with serotonin turnover in the brain, *Science* **171**:1169–1170 (1971).
125. E. W. Maynert, Analgesic drugs and brain neurotransmitters. I. Effects of morphine on acetylcholine and certain other transmitters, *Arch. Biol. Med. Exptl.* **4**:136–137 (1967).
126. K. Hano, H. Kaneto, T. Kakunaga, and N. Moribayashi, The administration of morphine and changes in acetylcholine metabolism in mouse brain, *Biochem. Pharmacol.* **13**:441–447 (1964).
127. J. F. Howes, L. S. Harris, W. L. Dewey, and C. A. Voyda, Brain acetylcholine levels and inhibition of the tail-flick reflex in mice, *J. Pharmacol. Exptl. Therap.* **169**:23–28 (1969).
128. W. A. Large and A. S. Milton, The effect of acute and chronic morphine administration on brain acetylcholine levels, *Brit. J. Pharmacol.* **38**:451 P (1970).
129. K. Datta, L. Thal, and I. Wajda, The effect of morphine on choline acetyltransferase levels in the caudate nucleus of the rat, *Brit. J. Pharmacol.* **41**:84–92 (1971).
130. R. W. Morris, Effects of drugs on the biosynthesis of acetylcholinesterase: Pentobarbital, morphine and morphinan derivatives, *Arch. Int. Pharmacodyn. Therap.* **133**:236–243 (1961).
131. D. Beleslin and R. L. Polak, Depression by morphine and chloralose of acetylcholine release from cat's brain, *J. Physiol.* **177**:411–419 (1965).
132. J. Schuberth, J. Sollenberg, A. Sundwall, and B. Sörbo, Acetyl CoA in brain: The effect of centrally active drugs, insulin coma and hypoxia, *J. Neurochem.* **13**:819–822 (1966).

133. T. Shikimi, H. Kaneto, and K. Hano, Effect of morphine on the liberation of acetylcholine from the mouse cerebral cortex slices in relation to Ca^{++} concentration in the medium, *Jap. J. Pharmacol.* **17**:136–137 (1967).

134. M. Sharkawi and M. P. Shulman, Inhibition by morphine of the release of acetylcholine-^{14}C from rat brain cortex slices, *J. Pharm. Pharmacol.* **21**:546–547 (1969).

135. J. Schuberth and A. Sundwall, Effects of some drugs on the uptake of acetylcholine in cortex slices of rat brain, *J. Neurochem.* **14**:807–812 (1967).

136. K. Kuriyama, E. Roberts, and J. Vos, Some characteristics of the binding of GABA and AcCH to a synaptic vesicle fraction from mouse brain, *Brain Res.* **9**:231–252 (1968).

137. W. Schaumann, A. hypothesis of cholinergic mechanism for the action of morphine, *N.-S. Arch. Parmakol. Exptl. Pathol.* **237**:229–240 (1959).

138. W. L. Dewey and L. S. Harris, Narcotic-antagonist analgesics, effects on brain cholinesterases, *The Pharmacologist* **9**:230 (1967).

139. A. C. Lane, I. R. Macfarlane, and A. McCoubrey, Inhibition of cholinesterases by complex derivatives of morphine, *Biochem. Pharmacol.* **15**:122–123 (1966).

140. G. E. Hein and K. Powell, Evaluation of kinetic constants for mixed inhibitors of cholinesterase, *Biochem. Pharmacol.* **16**:567–573 (1966).

141. M. J. Ettinger and A. Gero, Interactions of narcotics and their antagonists with human serum esterase, *Arch. Int. Pharmacodyn. Therap.* **164**:96–110 (1966).

142. M. J. Ettinger and A. Gero, Nature of the antagonism between narcotics and antagonists on human serum esterase, *Arch. Int. Pharmacodyn. Therap.* **164**:111–119 (1966).

143. E. W. Maynert, Some aspects of the comparative pharmacology of morphine, *Fed. Proc.* **26**:1111–1114 (1967).

144. M. Sasaki, Studies on the phenomenon of abstinence of the morphinized culture *in vitro*, *Jap. J. Med. Sci.* **9**:34–61 (1936).

145. E. J. Simon, Effects of narcotic analgesics in single cell, in "Narcotic Drugs: Biochemical Pharmacology" (D. H. Clouet, ed.) Plenum Press, New York (1971).

146. W. D. Noteboom and G. C. Mueller, Inhibition of cell growth and the synthesis of RNA and protein in Hela cells by morphinans and related compounds, *Molec. Pharmacol.* **5**: 38–48 (1969).

147. E. J. Simon, Inhibition of bacterial growth by drugs of the morphine series, *Science* **144**: 543–544 (1964).

148. E. F. Gale, Effects of diacetylmorphine and related morphinans on some biochemical activities of *S. aureus*, *Molec. Pharmacol.* **6**:128–133 (1970).

149. E. J. Simon, L. Schapira, and N. Wurster, Effect of levorphanol on cell membranes, *Bull. N.Y. Acad. Med.* **45**:500 (1969).

150. D. H. Clouet and N. Williams, Localization in brain particulate fractions of narcotic analgesic drugs administered intracisternally to rats, *Biochem. Pharmacol.* **22**:1283–1294 (1973).

151. A. Goldstein, L. I. Lowney, and B. K. Pal, Stereospecific and nonspecific interactions of the morphine congener, levorphanol, in subcellular fractions of mouse brain, *Proc. Natl. Acad. Sci.* **68**:1742–1749 (1971).

152. D. H. Clouet and N. Williams, The effect of morphine on the uptake and release of neurotransmitters by isolated synaptosomes, *Abst. Commun. Fifth Internat. Congr. Pharmacol.* (July 1972).

153. W. R. Martin and C. G. Eades, Demonstration of tolerance and physical dependence in the dog following short-term infusion of morphine, *J. Pharmacol. Exptl. Therap.* **133**: 262–270 (1961).

154. M. D. Day and M. J. Rand, Tachyphylaxis to some sympathomimetic amines in relation to monoamine oxidase, *Brit. J. Pharmacol.* **21**:84–89 (1963).

155. U. Trendelenburg, A. Muskus, W. W. Fleming, and B. G. Alonso de la Sierra, Modification by reserpine of the action of some sympathomimetic amines in spinal cats, *J. Pharmacol. Exptl. Therap.* **138**:170–172 (1962).

156. I. J. Kopin, J. E. Fleisher, J. M. Musacchio, W. D. Horst, and V. K. Weise, False neuro-transmitters and the mechanism of sympathetic blockade by monoamine oxidase inhibitors, *J. Pharmacol. Exptl. Therap.* **147**:186–190 (1965).

157. W. D. M. Paton, Transmission and block in autonomic ganglia, *Pharmacol. Rev.* **6**:59–64 (1954).

158. W. Schaumann, The inhibition by morphine of the release of acetylcholine from the intestine of the guinea-pig, *Brit. J. Pharmacol.* **12**:115–118 (1957).

159. W. D. M. Paton, The action of morphine and related substances on contraction and acetylcholine output, *Brit. J. Pharmacol.* **12**:119–124 (1957).

160. A. B. Carnie, H. W. Kosterlitz, and D. W. Taylor, Effect of morphine on some sympathetically innervated effectors, *Brit. J. Pharmacol.* **17**:539–551 (1961).

161. D. H. Clouet, The alteration of brain metabolism by narcotic drugs, *in* "Handbook of Neurochemistry" (A. Lajtha, ed.) Vol. 6, Plenum Press, New York (1971).

162. C. B. Smith, M. I. Sheldon, J. H. Bednarczyk, and J. E. Villarreal, Morphine-induced increases in the incorporation of C^{14}-tyrosine into C^{14}-dopamine and C^{-14}norepinephrine in the mouse brain: Antagonism by naloxone and tolerance, *J. Pharmacol. Exptl. Therap.* **180**:547–557 (1972).

163. C. P. Headlee, H. W. Coppock, and J. R. Nichols, Apparatus and technique involved in a laboratory method of detecting addictiveness of drugs, *J. Am. Pharm. Ass.* **44**:229–231 (1955).

164. A. Wikler and S. Altschul, Effects of methadone and morphine on the electroencephalogram of the dog, *J. Pharmacol. Exptl. Therap.* **98**:437–446 (1950).

165. A. S. De Carolis and V. G. Longo, The effects of morphine and related drugs on the electrical activity of brain: Their relationship with analgesic action, *Arch. Biol. Med. Exptl.* **4**:24–28 (1967).

166. S. D. Echols and R. E. Jewett, Effects of morphine on the sleep of cats, *The Pharmacologist* **11**:254 (1969).

167. W. A. Bleyer and M. G. Rosen, Meperidine-induced changes in the maternal and fetal EEGs of the guinea-pig, *Electroencephalog. Clin. Neurophysiol.* **24**:249–258 (1968).

168. N. Khazan, J. R. Weeks, and L. A. Schroeder, Electroencephalographic, electromyographic and behavioral correlates during a cycle of self-maintained morphine addiction in the rat, *J. Pharmacol. Exptl. Therap.* **155**:521–531 (1967).

169. J. M. Nelson and C. Kornetsky, Single dose tolerance to morphine sulfate, *The Pharmacologist* **10**:188 (1968).

170. H. Gangloff and M. Monnier, The topical action of morphine, levorphanol and levallorphan on the unanesthetized rabbit's brain, *J. Pharmacol. Exptl. Therap.* **121**:629–636 (1957).

171. H. L. Andrews, Changes in the EEG during a cycle of morphine addiction, *Psychosom. Med.* **3**:399–409 (1943).

172. A. Wikler, "Opiate Addiction," Charles C. Thomas, Springfield, Ill. (1953).

173. M. Matousek, J. Volavka, J. Roubicek, and Z. Roth, EEG frequency analysis related to age in normal adults, *Electroencephalog. Clin. Neurophysiol.* **23**:162–167 (1967).

174. M. Fink, A. Zaks, J. Volavka, and J. Roubicek, Electrophysiological studies in man, *in* "Narcotic Drugs: Biochemical Pharmacology" (D H. Clouet, ed.) Plenum Press, New York (1971).

175. J. Roubicek, J. Volavka, A. Zaks, and M. Fink, Electrographic effects of chronic administration of methadone, *Neuropharmacol. Psychopharmacol.* (in press).

176. M. Fink, T. Itil, A. Zaks, and A. M. Freedman, EEG Patterns of cyclazocine, a narcotic antagonist, *in* "Neurophysiological and Behavioral Aspects of Psychotropic Drugs" (A. G. Karczmar and W. P. Koella, eds.) Charles C. Thomas, Springfield, Ill. (1969).

177. C. K. Himmelsbach, With reference to physical dependence, *Fed. Proc.* **2**:201 (1943).

178. L. Shuster, Repression and de-repression of enzyme synthesis as a possible explanation of some aspects of drug action, *Nature* **189**:314–315 (1961).

179. D. B. Goldstein and A. Goldstein, Possible role of enzyme inhibition and repression in drug tolerance and addiction, *Biochem. Pharmacol.* **8**:48–49 (1961).
180. H. O. J. Collier, Supersensitivity and dependence, *Nature* **220**:228–231 (1968).
181. J. Jaffe and S. K. Sharpless, The rapid development of physical dependence on barbiturates, *J. Pharmacol. Exptl. Therap.* **150**:140–145 (1965).
182. L. Shuster, Tolerance and physical dependence, *in* "Narcotic Drugs: Biochemical Pharmacology" (D. H. Clouet, ed.) Plenum Press, New York (1971).
183. G. Ungar and L. Galvan, Conditions of transfer of morphine tolerance by brain extracts, *Proc. Soc. Exptl. Biol. Med.* **130**:287–290 (1969).
184. S. Spector and C. W. Parker, Morphine: Radioimmunoassay, *Science* **168**:1347–1348 (1970).

Chapter 4

GENETIC DISORDERS OF BRAIN DEVELOPMENT: ANIMAL MODELS

Norbert N. Herschkowitz

Department of Pediatrics
University of Berne
Berne, Switzerland

I. INTRODUCTION

The normal functioning of the central nervous system depends on the integrated and coordinated action of an immense number of neurons. The integrated cellular pattern is the result of a sequence of developmental steps: *proliferation* of cells, *migration* to determined locations, *differentiation* to specific functions, and *cell death*. These processes depend on the genetically controlled selective synthesis and degradation of proteins at specific times during development. Genetic mutations or exogenous factors which affect genetic control can, therefore, interfere with normal brain development and may result in severe brain dysfunction.

Multidisciplinary sequential studies during brain development are necessary for the investigation of these pathogenetic mechanisms. Animal models are very useful for the following reasons: many mutations in animals are known which lead to disorders of brain development resembling those observed in man, sequential studies can be performed beginning at the embryonic stage, the animals can be kept under specified conditions, and the normal littermates of the inbred strains can be used as controls. The timing of brain development in regard to prenatal and postnatal life is species specific[1] and has to be considered carefully when one compares developmental disorders in man to those in other mammalian species.

The comparison of behavior in humans and other mammals is extremely difficult. Even relatively simple phenomena such as tremor can depend on different mechanisms in different species.

The effect of a mutation on development depends partly on the structure of the genome; therefore, current hypotheses about the eukaryotic genome will be discussed first.

Since mutations of the nervous system have been studied extensively in the mouse, the normal and abnormal development of the brain will be discussed with special emphasis on this animal.

II. CHROMOSOMAL ORGANIZATION IN EUKARYOTES

There is some evidence that in the chromosomes of eukaryotes, as opposed to those of bacteria, most or all genes are represented not only once, but in multiples (up to several tens of thousands) which form a gene family.[2,3] One member of the family, the master gene, specifies the nucleotide sequence of the slave genes. "The one gene–one protein theory is to be modified to read one gene family one protein type."[4] A possible consequence of this theory is that more mRNA can be synthesized not only by gene activation, i.e., increased synthesis of mRNA from one template, but also by gene amplification, the formation of more slave genes which serve as templates for mRNA.[2]

An implication of this theory for brain development is the following: the neuronal network shows definite patterns of connectivity.[5] The number of gene families would not be sufficient if single gene products were specific for each single connection, since there are at least 10^{12} of these connections in man. If, however, a code were used arranging, e.g., ten different gene products in ordered sets of 12, then 10^{12} unique arrangements coding for specific connections would result.[4]

Not only is the protein structure under genetic control but also to a certain extent the rate of protein synthesis and degradation, the time during development at which specific protein synthesis occurs in specific cells, and the localization and distribution of proteins within the cell.

For convenience, Paigen[6] divides the genes into four classes: structural, regulatory, architectural, and temporal genes.

A. Structural Genes

Structure genes code the primary amino acid sequence of the proteins and influence, therefore, their secondary, tertiary, and quaternary structures.

Changes in the amino acid sequence may alter the susceptibility to physical denaturation of the protein, electrophoretic mobility, kinetic constants, and immunological properties. The best indication of changes in protein structure is abnormal thermostability. The probability that a random amino acid substitution will change thermostability is about 50%, which is higher than for any other test.[6] The final proof of a structural mutation, however, is detection of an altered peptide sequence.

B. Regulatory Genes

Mutations of regulatory genes, which would confirm the latter's presence, have to meet at least two criteria: (1) a change in the rate of synthesis or degradation of a normal enzyme and (2) a different locus than the structural locus of this enzyme. None of the known mutations in mammals strictly meets these two criteria.

C. Architectural Genes

Cells are probably assembled by self-organizing processes in which architectural genes determine the intracellular site of the proteins. An example of a possible mutation of this type is the omission of endoplasmic β-glucuronidase in the eg^o mouse.[7] Normal mice show approximately equal amounts of β-glucuronidase activity in the endoplasmic reticulum and in the lysosomes. The two enzymes appear to be identical, since no catalytic or physical differences can be found between them. The eg and the β-glucuronidase structural gene loci interact but are not linked genetically. In mice homozygous for the eg^0 allele, the β-glucuronidase is found only in the lysosomes; the endoplasmic reticulum contains no enzyme activity.

D. Temporal Genes

There may exist a genetic region next to the structural gene which responds to intracellular signals and determines the "turning on" and turning off" of structural gene activity. These signals may occur at fixed times during development and may lead to an inherent program of time-dependent, selective, genetic expression. It can be hypothesized that due to a mutation of a "temporal gene," the age-dependent onset of synthesis of one or of a group of enzymes would not occur and the enzyme activity would therefore remain low. The residual enzyme protein would have normal catalytic properties and a normal amino acid sequence. The locus of mutation would be different from the structural gene locus.

III. DEVELOPMENT OF THE CNS

A. Formation of the Neural Tube

Around the sixth gestational day in the mouse, the formation of the neural plate from ectoderm is induced by the underlying mesodermal archenteric roof. During the following days, the edges of the plate elevate and unite by continuous folding to form the neural tube (see Fig. 1), around the ninth gestational day. The neurulation is probably caused by factors intrinsic to the neural plate: medially directed cell movements cause a narrowing of the neural folds, and changes in cell adhesion may cause the curvature. At this time, cells move away from the closure into the crevice between the tube and overlying epidermis and form the neural crest. From this newly formed neural crest, which later becomes paired and segmented, the melanocytes, the craniospinal and sympathetic ganglia, and the adrenal medulla are formed. It is of special interest that most Schwann cells which form the peripheral myelin originate in the neural crest.[8] Shortly after the formation of the neural tube, the cells adjacent to the lumen begin to proliferate.

B. Cell Proliferation

There is evidence suggesting the existence of an inductive influence—possibly a mitotic stimulation factor—from the underlying mesoderm to cells in the neural tube.[9] In the neural tube, sheets of columnar epithelial cells stretch from the ventricular surface to the outside of the organ and form the ventric-

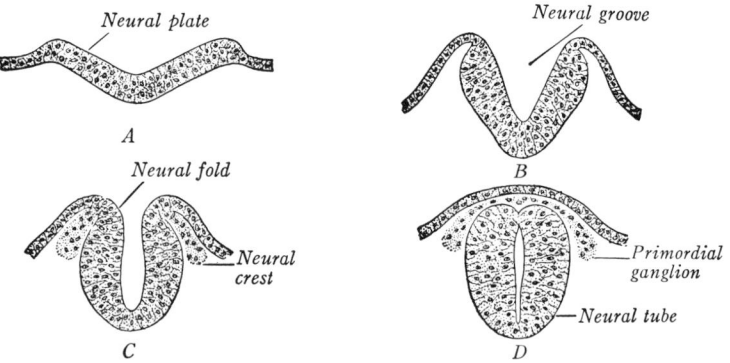

Fig. 1. Neurulation. Origin of the neural tube and neural crest, illustrated by transverse sections from early human embryos. [From Arey,[116] p. 413, by permission of the publisher.]

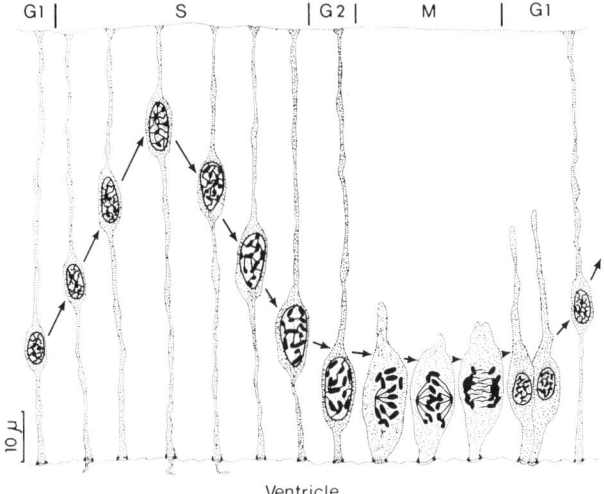

Fig. 2. Intermitotic "migration." Diagram of a section of neural tube of the chick embryo, showing the intermitotic migration of a single neuroepithelial germinal cell at approximately half-hour intervals throughout the mitotic cycle. [From Jacobson,[117] p. 7, by permission of the publisher.]

ular zone. During mitosis, the ventricular cells become round, and the nuclei move to the ventricular surface with a to-and-fro movement (see Fig. 2).[10] It is not known how many times the cells divide or what controls the number of the cell cycles.

An immense number of cells are formed within a few days, and slight alterations in the time and rate of formation will have enormous effects on the final number of cells.[11] This final number, however, is not the result of cell generation alone but of a balance between cell formation and cell death. Cell death is a normal regulating factor in determining the size of a cell population.[12] The rate of cell death may be influenced by a retrograde flow of metabolites from the periphery to the cell body.[13]

The ventricular cells are the progenitors of the neurons and the macroglia. This zone exists in all regions of the CNS, but at different times, and eventually disappears.

In the mouse, proliferation in the cortex takes place between the tenth and eighteenth gestational days and ends around birth. An approximate "timetable" of brain development is shown in Fig. 3. A second area of proliferation is the subventricular zone (see Fig. 4), which is formed around the thirteenth day of gestation and persists well after birth. The proliferative activity of the subventricular cells continues, therefore, postnatally.[14] Subventricular cells differ from ventricular cells with respect to shape and size. According to Altman,[15] they are the progenitors of the glia cells and of the microneurons.

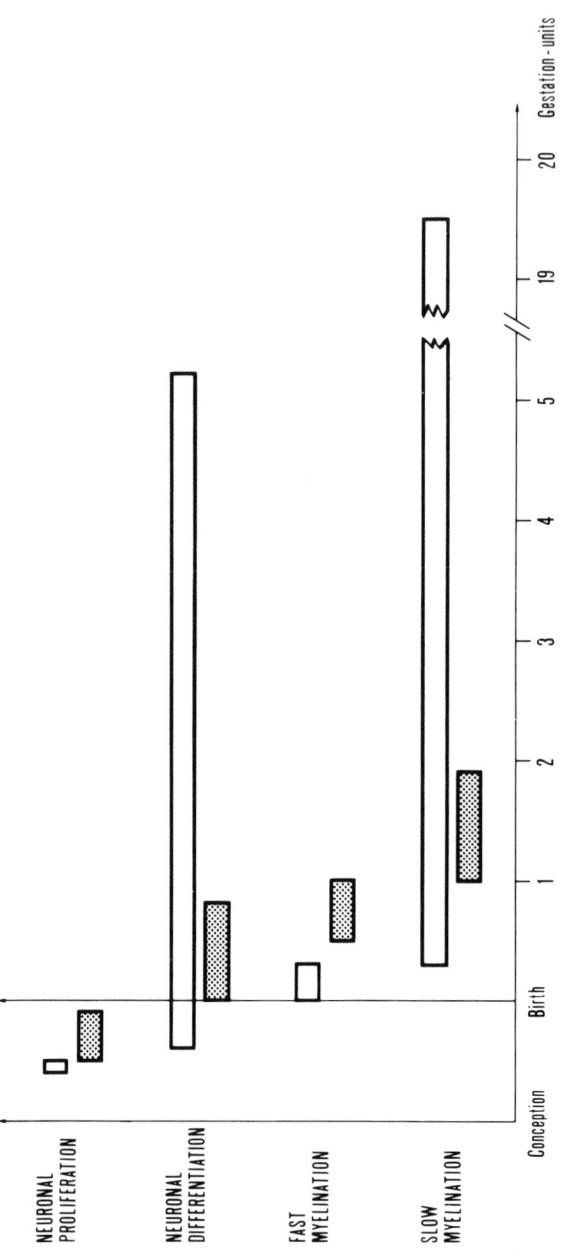

Fig. 3. Timetable of brain development in man and mice. Values are averages; 1 gestation unit in man equals 270 days, and 1 gestation unit in mouse equals 20 days. Open bar, man; dotted bar, mouse.

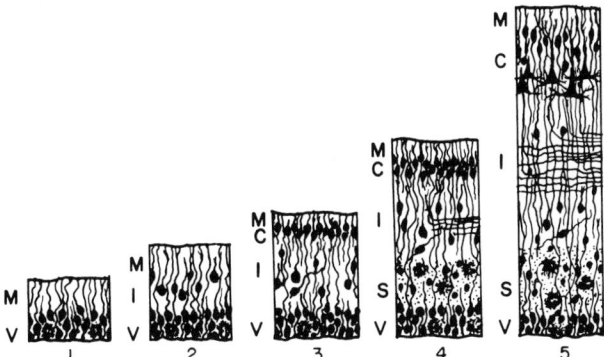

Fig. 4. Five stages in the development of the vertebral central nervous system. C, cortical plate; I, intermediate zone; M, marginal zone; S, subventricular zone; V, ventricular zone. [From Angervine et al.,[14] p. 258, by permission of the publisher.]

The proliferation of ventricular cells in the human cerebellar cortex slows down markedly at the end of the thirteenth gestational week, and it can be assumed that the entire Purkinje cell population has been established by that time. An external granular layer is observed around the eleventh gestational week, and the cells of this layer divide at a high rate throughout the first postnatal year.[16]

1. Chemical Markers of Cell Proliferation

If it is assumed that most of the brain cells are of diploid character and that the DNA concentration of cells is constant irrespective of the cell type and size, then the amount of DNA can be used as a measure of cell number. Lapham,[17] however, has shown that in the cerebellar cortex in man, Purkinje neurons are of tetraploid character. It is thus possible that this rule does not apply to certain areas in brain. Estimation of DNA content in whole brain is of limited value, since the proliferation of cells differs vastly from one brain area to another. Another shortcoming of this method is the fact that it does not allow a distinction between the types of cells which are in the proliferation phase. Nevertheless, under certain circumstances this method reveals important facts.

Dobbing and Sands[18] analyzed DNA of whole brain of human embryos and fetuses from the tenth week of gestation until after birth and found two distinct periods of sharp increase in DNA in the whole brain, one between 15 and 20 weeks of gestation, the other starting around 25 weeks of gestation and probably ending in the second year of postnatal life. The second phase is due to a predominantly glial division.

C. Cell Migration

After an unknown number of divisions, the cells stop dividing and migrate out from their area of generation. Migration of cells can be observed during the whole period when cell division occurs.

Experiments by DeLong[19] give some insight into the mechanisms which regulate cell migration and the formation of cellular layers. Cells from fetal mouse isocortex and hippocampus were disaggregated with trypsin and allowed to reaggregate in liquid cell cultures in rotating tubes for a period of 6–8 days. The cells from 17- to 18-day-old fetuses containing large hippocampal pyramidal cells showed *in vitro* cluster formation and separation from smaller cells. If cells were taken from $18\frac{1}{2}$-day-old fetuses, the pyramidal cells formed a specific internal organization showing three specific features: parallel alignment, uniform polar orientation, and formation of a discrete lamina plate with the characteristic curvature of the hippocampus comparable to that *in vivo* (see Fig. 5). If the cells were taken from embryos younger than 17 days or older than $18\frac{1}{2}$ days, no specific cellular organization could be observed. The segregation processes observed in these aggregates seem to be equivalent to the migration observed in neural histogenesis. It can be assumed that the cells aggregating together have a greater affinity for their own type than for other cells because of surface affinities between the cells of one population. The parallel alignment of the cells within the homogeneous cell population requires asymmetrical cell surface properties which make the sides of the cells mutually attractive but not the ends. The uniform polar orientation of the elongated cells may be due to patches on the bases which have an affinity for similar patches on the bases of other cells but not for the apices of these cells. Thus it can be assumed that the process of cell migration in the brain is due to specific properties of the cell membrane and that these properties are formed at specific periods during development.

Weiss and Taylor[20] have shown that complete organs can be reconstituted *in vitro* from single-cell suspension of chicken embryos when taken at an advanced stage of differentiation and that this self-organization takes place without external inductive instructions.

D. Differentiation of Neurons

The growth and branching of the dendrites begin after the proliferation of the cells has ceased. In the mouse cortex, this period extends from birth to 14 days postnatally. Formation of synaptic junctions starts around the seventeenth postnatal day, at which time the neurons have a "mature" appearance. In man, differentiation starts around the twenty-fifth postnatal week and continues until about the fourth postnatal year.

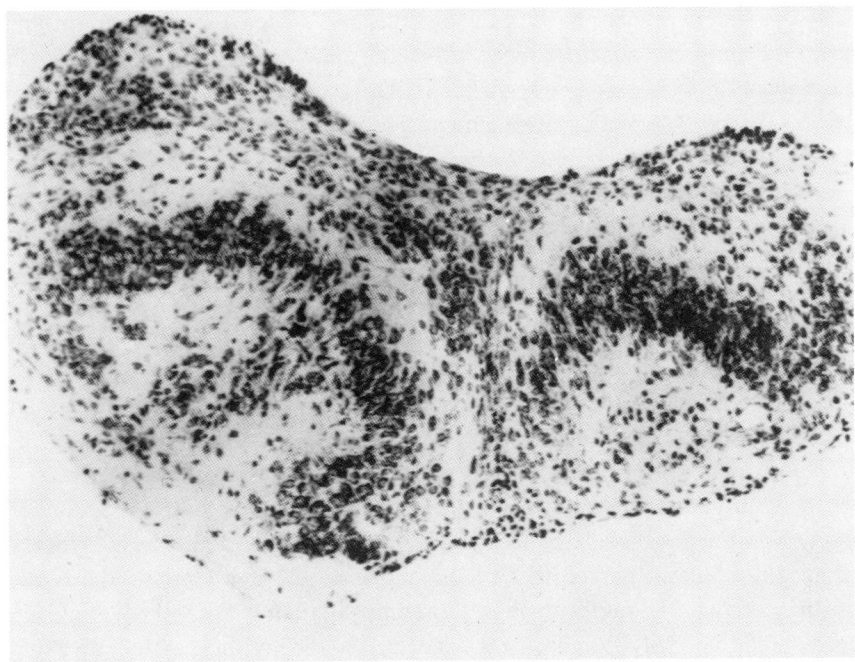

Fig. 5. Low-power view of an aggregate of hippocampal cells dissociated at 18.5 days of gestation and maintained in rotating culture 7 days. The hippocampal pyramidal cells have organized into two curved laminar formations, each similar to the normal configuration of Ammon's horn, with apical processes directed toward the concave border. The pyramidal population is completely segregated from other cell populations, and this segregation is emphasized by the characteristic cell-poor area around the Ammon's horn formation. [From DeLong,[19] p. 568, by permission of the publisher.]

The time of onset of differentiation varies among cell types and different brain areas. There is no direct relationship between the time of formation of a cell and the onset of its differentiation.[16] It is not known what mechanisms initiate the differentiation of neurons. The hypothesis that the migration of granule cell neurons in the vicinity of Purkinje cells influences the differentiation of the Purkinje cells does not seem to be justified in view of the fact that human Purkinje cells show multibranched apical dendrites when only very few granule cells have passed them.[16]

1. Chemical Markers of Neuronal Differentiation

Since gangliosides are present in neurons[21] and particularly in the membranes of the synaptic junctions, their presence can be used as a marker for neurons and for neuronal differentiation. Developmental profiles of gangliosides in human and rat brain have been studied, and three different periods of early brain development have been suggested.[22] Period 1 lasts in the rat until birth and in man until the twenty-fifth fetal week and is characterized by the multiplication of neurons and glial cells. Only a moderate increase in gangliosides is found. The predominant gangliosides are GM_1 and GT_1. The beginning of GD_{1A}-ganglioside synthesis is observed.

The duration of period 2 is from birth to the tenth postnatal day in the rat and in man from the twenty-fifth fetal week to term. This period is characterized by multiplication of glial cells and microneurons and differentiation of dendrites and axons combined with the establishment of neural connections. In this period a first phase of rapid increase in ganglioside concentration is observed, mainly because of an increase in GD_{1A}.

Period 3 lasts in the rat from the tenth to the twentieth postnatal day and in man from birth to 8 months of age. Neuronal connections are further extended, and myelination begins. In this period, a second phase of rapid increase in the ganglioside concentration is observed. The GD_{1A} is still a predominant fraction in the ganglioside pattern and seems to be stabilized. The biosynthesis of gangliosides is observed to be most active around the tenth to the twelfth postnatal day in the rat.[23,24] In interpreting ganglioside data, the fact must be considered that GM_1-ganglioside is connected to the myelin sheath or an axonal membrane closely associated with myelin.[25] Norton and Podulso[26] have also shown that gangliosides are normal constituents of glial cells. A sharp increase of sodium- and potassium-activated ATPase in rat brain between the tenth and the twentieth day[27] may be related to the observed increase in the amount of dendrites and nerve endings. The high activity of glutamic acid decarboxylase and other enzymes involved in dicarboxylic acid metabolism suggests that these enzymes play an active role during the growth period of the dendrites and axons.[28]

From a functional point of view, it is interesting to note that the appearance of a mature pattern of dendrites in the mouse cortex at 17 days of age corresponds to the maturation of the electrocorticogram and the attainment of the adult type of muscular control.[29] However, since many different processes occur simultaneously during this period, it is difficult to establish a causal relationship. A possible step in this direction is the investigation of the correlation between the development of brain structure, metabolism, and functions.

E. Differentiation of Oligodendrocytes: Myelination

An extensive review of the morphological and biochemical aspects of myelination has been made by Davison and Peters.[30]

In the mouse CNS, fast myelination starts around the tenth postnatal day, reaches a maximum at 20 days, and then slows down and is essentially finished at about 40 days. In man, the period of fast myelination extends from birth until the third month; then the process slows down and is completed at about 15 years. At the beginning of myelination, each axon is partially surrounded by a cell process. The process then extends until it completely encircles the axon; then together with the plasma membranes of the opposed lips of the process it forms a mesaxon. The mesaxon then becomes elongated and forms a loose spiral. Some outer leaflets of the opposed plasma membranes close together and form the intraperiod line (see Fig. 6).

At a later stage of myelination, the cytoplasm disappears from between the turns, which causes the apposition of the cytoplasmic leaflets of the plasma membranes in the spiral and thereby the formation of a compact myelin spiral.[31,32] The later slow myelination consists of the addition of lamellae to the spiral.

The oligodendrocytes seem to be the myelin-forming cells. This conclusion is based on observations which show that these cells are in continuity with the myelin sheath and that in the optic nerve of the 7-day-old rat oligodendrocytes appear prior to myelination.[30] Also, prior to the histological appearance of

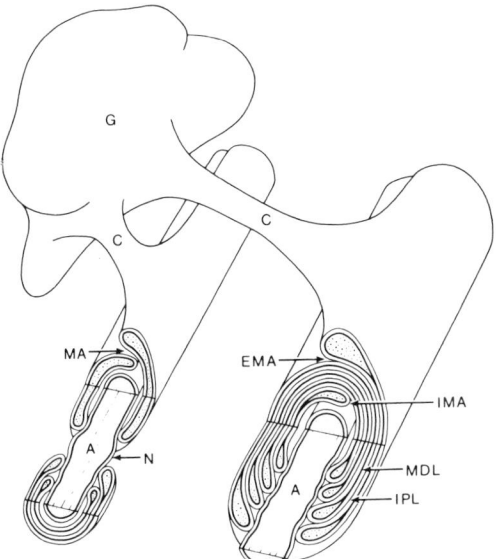

Fig. 6. Diagram illustrating myelination of axons in the central nervous system. Myelination is further advanced on the axon at the right than on the axon at the left. A, axon; C, cytoplasmic process of oligodendrocyte; EMA, external mesaxon; IMA, internal mesaxon; IPL, intraperiod line; MA, mesaxon; MDL, major dense line; N, node; G, oligodendroglial cell. [From Jacobson,[117] p. 176, by permission of the publisher.]

myelin, the proliferation of presumably oligodendroglial cells can be observed, as indicated by increase of the DNA content of whole brain[33,18] and thymidine-H^3 incorporation into the cell nuclei.[34]

It is not yet known what induces the myelination of the axon or what the function of the axon itself is in this process. Although there is a strong indication that the oligodendrocytes are the cells responsible for myelin formation, final proof is still lacking.

1. Chemical Markers of Myelination

Characteristic components of the myelin membrane are the sphingolipids cerebroside and sulfatide, cholesterol, proteolipid protein, and basic proteins.[35,36] Histologically mature myelin is found in the rat brain after 15 days of age. A myelin-like fraction can be isolated from 15-day-old rat brain, which, in contrast to mature brain, does not contain cerebrosides and basic proteins.[37] The synthesis of cerebrosides in brain increases postnatally and reaches a maximum at 11 days. The newly synthesized cerebroside contains C_{24}-chain fatty acids. The synthesis and elongation of the fatty acids by rat brain microsomes reach a maximum at 15 days of age. In human brain, no basic protein is present in the first postnatal month[38]—a similarity to the lack of basic protein in the early myelin of the rat.

Sulfatide synthesis in the rat brain begins at about 8 days of age, reaching a maximum at 20 days.[39-41] The lipid is synthesized in the endoplasmic reticulum. Sulfatide is bound to proteins and by this means forms as water-soluble lipoprotein, which transports the glycolipid from the endoplasmic reticulum to the myelin, where sulfatide is finally incorporated (see Fig. 7).[42] When puromycin is added, the incorporation of newly synthesized proteins into the myelin membrane ceases but not that of sulfatide, which suggsest an independent incorporation of sulfatide and protein into myelin.

The synthesis of glycolipids, fatty acids, and basic proteins increases almost synchronously,[44] which suggests a common mechanism responsible for this multiple, time-dependent, simultaneous, increase in activity.

There is some evidence that myelination is under the hormonal control of thyroxin,[45] cortisone,[46] and estradiol.[47] However, myelination and the synthesis of myelin lipids also occur in tissue cultures in vitro independent of hormonal influences at the same time these processes occur in vivo (see Fig. 8).[43]

IV. GENETIC DISORDERS OF BRAIN DEVELOPMENT

Many mutations are known which affect the central nervous system. The mouse brain has been studied most intensively in this respect and will therefore

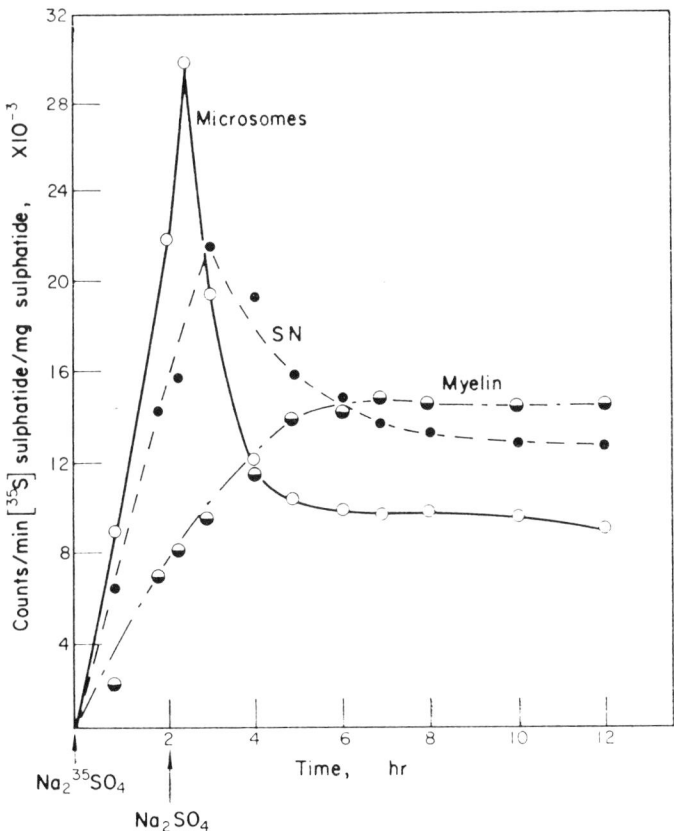

Fig. 7. Turnover of sulfatide-S[35] in subcellular fractions. Animals received an intraperitoneal injection of 2 μc $Na_2S^{35}O_4$ per gram of body weight and after 2 hr a second intraperitoneal injection of 0.5 ml of 7% Na_2SO_4. SN, myelin, and microsomes were separated, and the specific activity of sulfatide-S[35] was determined. [From Herschkowitz et al.,[42] p. 1186, by permission of the publisher.]

be discussed in detail. Since brain development is regulated by the coordinated and synchronized activity of metabolic events, mutations will frequently interfere with the normal growth and maturation of the nervous system. The primary effect of the mutation is, however, altered by the subsequent development, and this can make its recognition extremely difficult.

A. Abnormal Induction of the Neural Tube

In a recessive mutation t^o, the homozygous mouse embryos show no induction of the neural plate, which may be due to a generalized defect in the organization of ectoderm caused by lack of inducing mesodermal material. The

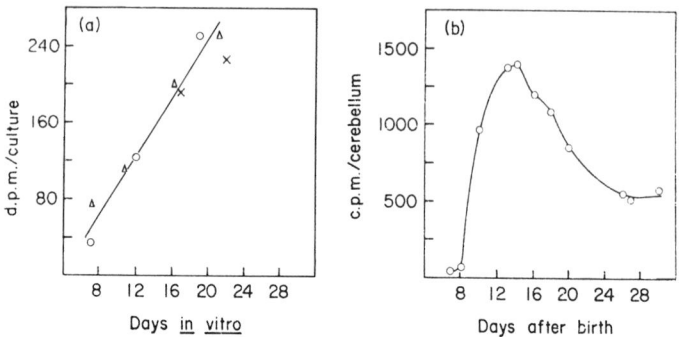

Fig. 8. Incorporation of sulfate into sulfatide in myelinating cultures in whole rat cerebellum. (a) In culture; (b) *in vivo*. [From Silberberg *et al.*,[43] p. 13, by permission of the publisher.]

homozygous animals die at 6 days of gestation. In a dominant mutation *T*, irregularities and partial absence of the neural tube are observed. The abnormalities of the underlying mesoderm suggest a defect in the inductive activity of this material.[48] *T/T* animals die at 10 days of gestation.

B. Abnormal Formation of the Neural Crest

Homozygotes of the dominant Kinky (*Fu^{ki}*) mutation show duplication of the neural folds and other organs. They die at 9 days of gestation. There seem to be allelisms between Kinky and Fused. Homozygote Fused (*Fu*) show the abnormalities of irregular folding of the neural tube (see Fig. 9) with ventral and caudal outgrowths together with abnormalities of the urogenital system. The underlying mesodermal material seems to be normal, which suggests that the mutation affects directly the mechanisms of self-organization of the neural crests.[48]

C. Abnormal Folding of the Neural Tube

In the same linkage group are two mutations which show incomplete folding of the neural tube leading to dystrophic disorders. Loop-tail (*Lp*) homozygotes show a failure of fusion of the neural folds at $9\frac{1}{2}$ days of gestation. At that stage, the neural plate and the bases of the folds seem to be normal. A day later, the first signs of degeneration in the neural plate are visible, and hemorrhagic areas and collapsed hemispheres are found.[49] The neural tube may remain open from the mesencephalon to the shortened tail. The animals die shortly before birth. Heterozygous animals show no pathological changes in the nervous system but may have a twisted tail, instability of gait, and wobbling of the head.

Splotch (*Sp*) homozygotes show rachischisis and myeloschisis in the lumbosacral region and, in half of the cases, in the hindbrain. Again, no folding of the neural tube at 9½ days of gestation is observed. Additionally, however, there is excessive growth of neural tissue on either side of the dorsal midline.[50] There is a reduction or absence of the dorsal root ganglia and their derivatives and an abnormal tail. The affected animals show a complete failure to produce hair pigment, probably because of a defect of the pigment-producing melanophores originating from the neural crest. The animals die on the thirteenth gestational day. In the heterozygous mouse, only the pigment anomaly related to the defect in the neural crest is present.

D. Abnormal Induction of Sensory Organs

There is some evidence suggesting that the neural tube has an inductive function in the formation of the sensory organs.[51] In the 9-day-old homozygous Kreisler mouse (*kr*) embryo, the otic pit appears separated from the neural folds of the rhombencephalon (see Fig. 10). The normally occurring fifth neuromere of the rhombencephalon is missing. In the fourth neuromere, there is extensive degeneration of cells, which suggests an anomaly of the neural tube. The sub-

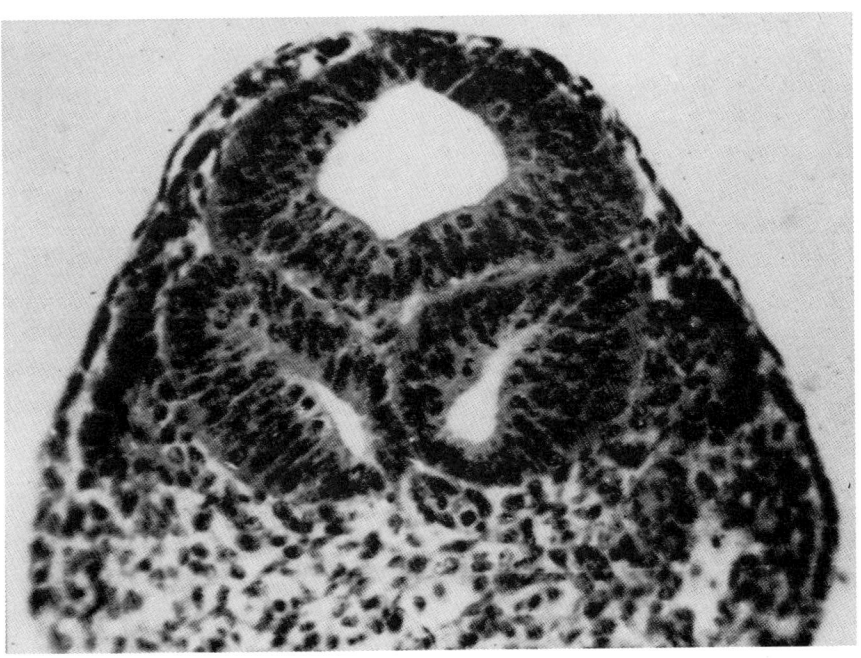

Fig. 9. Abnormal neurulation. Cross-sections in fused (*Fu/Fu*) embryo of 13 days showing three neural tubes. [From Glueckson-Waelsch,[48] p. 380, by permission of the publisher.]

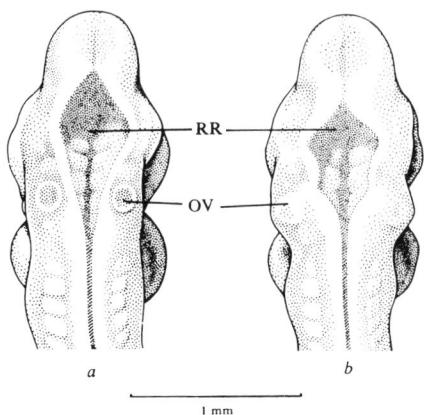

Fig. 10. Head development of the Kreisler embryo. Dorsal views of the heads of a 9-day Kreisler embryo (b) and its normal littermate (a). OV, otic vesicle; RR, roof of the rhombencephalon. [From Deol,[51] p. 484, by permission of the publisher.]

sequent development of the inner ear in *kr/kr* is extremely variable and does not depend on the mutant gene itself but on the separation of the otic vesicle from the neural tube. Neuromeres are considered to be formed by localized high mitotic activity in the neural tube. In the Kreisler mouse, there seems to be a primary defect in the proliferative activity of the cells of this region with secondary effects on further ear development. In the inner ear, the endolymphatic duct can be lacking, and the ear can show defective semicircular canals and a rudimentary cochlea. Clinically, in the second postnatal week the animals crawl in circles, fail to right themselves when placed on the back, are deaf, and show vertical head tossing.

In Dreher (*dr*) mice at the ninth gestational day, the lateral folds of the neural plate fuse abnormally. The roof of the rhombencephalon terminates before the fourth neuromere at the tenth day and does not reach the vicinity of the otic vesicle as in normal animals.[52] The development of the endolymphatic duct is retarded in the otic vesicle, which leads to secondary abnormalities in the development of the inner ear with similar effects to those in the Kreisler mouse.[53] Deol[52] suggests that a defect in the inductive function of the neural tube in specific regions may be the common cause of genetically determined malformations of, for example, the eye, the skull, and the axial skeleton.

The Fidget (*fi*) gene causes anomalies of the ear and eye.[54] The inner ear in this mutation seems to develop normally until the twelfth gestational day. The endolymphatic duct is formed; however, the semicircular canals do not develop. A secondary consequence of the abnormal auditory capsule may be the abnormal shape and position of the parafloccular lobe of the cerebellum. At the tenth gestational day, the lens vesicle is already distinctly smaller than in normals, leading secondarily to reduction of eye size and absence of lachrymal glands. Clinically, the affected animals show head shaking and severe lack of motor coordination 3–4 days postnatally. They are not deaf. As suggested by Deol,[52] a mutation affecting the neural tube could be the common cause for the diverse defects in this mutation.

In the Eyeless (*ey*) mutant, an optic vesicle has been formed by the tenth embryonic day. This does not, however, grow into the optic cup and does not reach the surface ectoderm. The inductive contact necessary to form a lens is therefore missing. Structures such as the lateral geniculate body and the ophthalmic division of cranial nerve V still look embryonic at birth. Cranial nerves III, IV, and VI form but become smaller at the age of 6 months. Again, this mutation may be a defect of cell proliferation caused by lack of induction by the neural tube.

E. Abnormal Migration and Alignment of Cortical Cells

In normal mice, during the second postnatal week there is high mitotic activity of granule cells in the external granular layer with subsequent migration inward to form the granular layer. In the Weaver (*wv*) homozygotes, this migration does not take place and the granule cells degenerate. The picture of this disease resembles that of neonatal viral encephalitis[55] or neonatal X-irradiation.[56] In all three diseases, the same population of cells seems to be affected, confirming the known fact that proliferating cells are particularly susceptible to exogenous noxious agents. At about 8 days postnatally, the affected animals show an instability of gait, and they topple over to either side. The hindlimbs remain abducted and extended. Superimposed on slow swaying movements is a fine rapid tremor of the trunk and extremities. The limbs are poorly coordinated. The mice sometimes leap more than 10 inches into the air when agitated. Tonic–clonic seizures are rare. The symptoms do not progress with age.

In the monozygous Reeler (*rl*) mouse, the cortical cells seem to proliferate and migrate normally; however, they show an abnormal histological organization due to a failure to align into the characteristic brain structures.

Histological analysis of the homozygous Reeler mutant prior to weaning reveals abnormalities of the organization and cytoarchitecture of the cerebral cortex, the hippocampus, and the cerebellum.[57] At the time of weaning, the cerebellar cortex shows abnormal lamination; in particular, no molecular layer is visible. The layer directly below the surface of the cerebral cortex is crowded with different cells of varying size and shape. Pyramidal and granular cells are present. They do not, however, form the normal structures. In the hippocampus of *rl/rl*, no lamination is visible, and pyramidal cells and granule cells are dispersed in the fascia dentata.[58] In the cerebellum, the abnormalities are found to be the most extensive and to involve all the areas. The number of Purkinje cells and their dendritic branching seem to be reduced. In the granular layer and in the white matter, an excessive number of larger neurons are visible where normally none of these cells are seen. These cells show a positive reaction to specific cholinesterase,[59] while normally only neurons in the deep nuclei of the white matter exhibit a comparable intense staining reaction.

The abnormalities of the cerebral cortex and hippocampus are already visible at birth, which suggests that the mutant gene affects development prenatally. Histological studies of embryos at different stages show normal cell proliferation. The cells seem to proliferate normally around the twelfth day of gestation, but an abnormal cellular distribution is found in the cortex on the sixteenth embryonic day. The cerebral isocortex probably contains the normal number of cells; however, the arrangement of the cells is disturbed in all areas. Principally similar disturbances are found in the hippocampus. In the cerebellum, the abnormal pattern is observed postnatally.

DeLong and Sidman[60] studied the reaggregation of dissociated cortical and hippocampal cells of embryonic Reeler brains using rotating cell cultures. If dissociated isocortex cells were allowed to reaggregate from fetuses between $17\frac{1}{2}$ and $18\frac{1}{2}$ days old, an aggregation of neurons was observed (see Fig. 11).

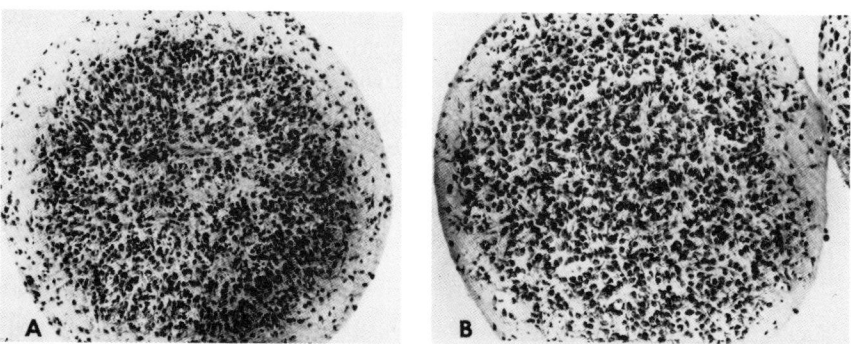

Fig. 11. Cell aggregation. A, Section through the center of an aggregate of cells dissociated from the isocortex of a normal fetus at 18.5 days of gestation and aggregated in rotating culture for 6 days. The young cortical neurons have formed a concentric middle layer, with a distinct outer margin and a less distinct but nevertheless definite inner margin. A central area containing cells and fibers without organization is present. The young cortical neurons are seen to be aligned radially, with their apical processes directed outward toward the peripheral cell-poor zone. B, Section through the center of an aggregate of isocortical cells for a Reeler littermate of the fetus of A. Dissociation and culture conditions were the same for both. A peripheral cell-poor zone is present but somewhat less well demarcated from the deeper cellular zone than in the normal. The concentric cortical plate extends to the center of the aggregate and is composed of cells with processes oriented randomly. [From DeLong and Sidman,[60] p. 586, by permission of the publisher.]

These cells, however, were randomly oriented and did not show the typical alignment of normal cell bodies and apical processes. The formation in the culture resembled that seen histologically in the brain of the mutant. In normals and in mutants when fetuses of less than $17\frac{1}{2}$ days of age were used, an aggregation of cells was observed but no alignment.[19] Reaggregates from both types were difficult to distinguish.

The same phenomenon was observed when brains from 19-day-old fetuses were investigated: the cells aggregated but did not show the internal organization within the homogeneous group. Again, there was no significant difference between mutants and normals. The same observations were made with cerebellar cells 3–4 days postnatally.

From these experiments, it can be concluded that aggregation of cells and alignment of cells within the aggregate are two different processes and that the latter is mainly affected in the Reeler mutant. Aggregation in cell culture is the clustering of cells which possess surface affinities. This process is equivalent to migration *in vivo*. The alignment is due to asymmetrical properties of the cell surface which lead to specific orientation within the homogeneous cell population (equivalent to the histogenetic process observed *in vivo*). The surface properties, which the isocortex and hippocampal cells attain during the short interval between $17\frac{1}{2}$ and $18\frac{1}{2}$ days of gestation, and the cerebellar cells, at 3–4 days postnatally, are missing in the Reeler mouse.

Clinically, homozygotes can be recognized about the eighteenth postnatal day. The homozygous animal cannot keep its hindquarters upright and sways slowly from side to side, sometimes falling over. It rights itself immediately but will fall again. There is a slight tremor of the foot when it is off the ground. When the mouse is excited, a more generalized tremor is sometimes observed. Such mice are weak and unable to climb an inclined wire mesh. Reelers are less active than normals, and it is difficult to induce them to run. They pay little attention to disturbances in and around the cage.

F. Disorders of Myelin Formation

The Quaking mouse (qk) is an autosomal recessive mutation, which shows a severe lack of myelin in the central nervous system, with no signs of primary demyelination. Neurons and especially the axons look ultrastructurally normal.[61] Characteristic myelin glycolipids such as cerebrosides and sulfatides are concomitantly reduced in the brain.[62]

A decreased activity of galactosylsphingosine transferase *in vitro* is found in the Quaking mouse. A double reciprocal plot of substrate concentration and enzyme activity gives practically the same K_m values for both control and Quaking brain. The reaction of this enzyme involves the transfer of galactose from UDP-galactose to sphingosine, which results in the synthesis of galactosyl-sphingosine (psychosine).[63] Polyphosphoinositol concentrations in Quaking brain are low, and the ability to incorporate inorganic P^{32} *in vivo* into these compounds is substantially reduced.[64] Long-chain fatty acids of cerebrosides and sphingomyelin in isolated myelin are reduced because of a block in chain elongation.[65] The analysis of fatty acids from brain of adult Quaking mice reveals a pattern similar to that of 12-day-old normal mice.

Proteolipid protein and basic proteins are also affected by the mutation.[66] In purified myelin of adult Quaking mice, qualitatively the same proteins are found as in normals, but high molecular weight proteins are present in greater percentage, and the ratio of proteolipid protein to basic protein is lowered. The myelin proteins of adult Quaking mice resemble those of 10-day-old normal mice, suggesting that there is an arrest in the development of myelin at an early phase in the Quaking mouse. These results suggest that the myelin defect is due to a defect in the formation of myelin and not only to a secondary degradation; however, both mechanisms could be responsible.

It is of interest that in brain of the Quaking mouse the activity of some lysosomal enzymes is lowered. Cerebroside galactosidase activity is 17% lower than normal, which does not account for the low cerebroside content and is probably a feedback response due to the low cerebroside concentration in the Quaking brain.[67] α-Mannosidase activity is only 30% of the normal value. This may be of importance with respect to the finding that mannose-containing glycoproteins have been isolated from brain tissue; Kurtz and Kanfer[68] speculate that there may be specific localizations of such linkages in myelin macromolecules.

The disease is recognized in the second postnatal week by an unsteady gait and a marked tremor, particularly in the caudal part of the trunk. The symptoms are fully expressed by about 3 weeks of age. Frequently, tonic seizures occur, which can also be elicited by touching. The number of seizures can be reduced by diphenylhydantoin.

Jimpy (*jp*) is a sex-linked recessive mutation. The reduction of myelin in the CNS is more severe than in the Quaking mouse. Occasionally, only one or two lamellae of myelin are wrapped around an axon.[61] The peripheral myelin has a normal appearance. In the nonmyelinated axons of the CNS, honeycomb-like tubular structures are frequently found. In normal littermates, these structures are found only in the myelinated axons.[69,70] Jimpy brain contains diminished amounts of cholesterol, phospholipid, and galactolipid, especially cerebrosides. The normally occurring increase of the content of cholesterol and galactolipids cannot be observed.[71,72] Hogan *et al.*[73] suggest that the fact that cerebroside increases until around the tenth postnatal day and then fails to increase further speaks against a primary defect in cerebroside synthesis. The synthesis of cerebroside and sulfatide is, however, decreased. *In vivo* incorporation of galactose-C^{14} into cerebroside and sulfatide is low but not that of galactose-C^{14} into gangliosides.

The activity of ceramide galactosyl transferase *in vitro* is only 10% of that in normal brain microsomes. This enzyme is involved in cerebroside synthesis.[75] As in Quaking brain, *in vitro* activity of galactosylsphingosine transferase activity is low but shows a normal K_m value, which indicates a normal protein structure of the enzyme. The activity of this enzyme normally increases sharply in the brain of 8-day-old animals, reaches a maximum at about 17 days, and declines sharply after the twentieth day. This age-dependent activity increase takes place in the brains of Jimpy and Quaking but only at a very reduced level.

There is, therefore, in both mutations not a primary defect in cerebroside synthesis but rather an inability to *increase* cerebroside synthesis after the eighth postnatal day, shortly before myelination normally starts.

It will be of special importance to investigate whether the increase in activity of a number of enzymes around the eighth postnatal day is caused by the increased synthesis of the same enzyme protein or by the synthesis of new isoenzymes. The comparison of the physical–chemical characteristics of the enzymes around birth and after the eighth postnatal day will help clarify this important question.

The metabolism of sulfatide in the Jimpy mutant was studied by Herschkowitz *et al.*[76] In the central nervous system, the amount of the glycolipids cerebroside and sulfatide is diminished. In the peripheral nervous system, no abnormality is found—this parallels the histological and ultrastructural findings. $S^{35}O_4$ incorporation into sulfatide *in vivo* and *in vitro* is only about 20% of normal. When homogenates of normal and Jimpy brain are mixed, no inhibition of enzyme activity is observed. Sulfatide synthesis in the normal brain increases sharply after the eighth postnatal day. This increase is not observed in Jimpy. The residual activity of galactocerebroside sulfate transferase in Jimpy brain was compared to that in normal brain, and no differences were found in heat inactivation, pH optimum, or K_m for galactocerebroside and PAPS (phosphoadeno sinephosphosulfate). These results indicate that the enzyme protein in Jimpy brain is of a normal structure,[77] but only amino acid sequencing of the isolated protein will answer this question.

At 15–16 days postpartum, there is a marked deficiency of long-chain fatty acids of cerebrosides and sulfatides. The defect is more severe than in the Quaking mouse. (The maturation of the normal brain is characterized by the progressive increase in the long-chain fatty acids of these sphingolipids.) The lack of long-chain fatty acids could reflect either a primary defect in fatty acid elongation or a secondary manifestation of a failure in myelin formation.[78]

The myelination in cerebellar explants of Jimpy mice was studied in tissue culture conditions by Wolf and Holden.[79] In practically all Jimpy explants, no myelin is formed, in contrast to the normals. All other developmental features, however, were judged to be normal. When cerebellar tissues from normals and Jimpy are grown on the same slide in a common medium, no mutual effects can be observed: the Jimpy does not myelinate, and the normal does not stop myelination. No diffusible factors seem to be involved in the defective myelination in Jimpy brain.

Clinically, the animals can be recognized at 10 days of age. Ataxia and tremor, which are present at about 8 days and then disappear in normals, increase in the affected males. From about 18 days on, generalized convulsions occur, and the animals usually die during a tonic–clonic seizure at about 20–25 days. This disorder resembles Pelizaeus–Merzbacher disease in man and is certainly an excellent model for studying the pathogenetic mechanisms of leukodystrophies.

Summarizing the findings in Jimpy, we see that several enzymes which are involved in the synthesis of myelin lipids do not show the characteristic increase in activity several days before myelination can be observed histologically. The remaining activity, which shows normal physical and catalytic properties, leads to the assumption that the synthesized enzyme protein is, however, of a normal structure. The tissue culture experiments indicate that the defect is cell bound and not diffusible. These results can be explained by at least three possible mechanisms:

1. Mutation in a temporal gene which affects the normally synchronous activation or amplification of a group of genes around the eighth postnatal day.
2. Mutation in a structural gene which codes for a protein necessary for the activation of the affected enzymes.
3. A mutation which leads to a general arrest in the development of the oligodendrocytes. (A defect in brain cells other than the oligodendrocytes has not yet been ruled out.)

The Quaking mouse, which has similar features to the Jimpy, but less pronounced, is a different mutation (autosomal instead of sex linked). The important findings of the defect in myelin proteins in Quaking mice must still be investigated in Jimpy.

G. Metabolic Disorders Which May Act on Different Levels of Development

Shambling (*shm*) is a recessive mutation which, in the homozygous state, affects several organs such as brain, spinal cord, heart, kidney, and skeletal system. The mice are smaller than normal. Their hindlimbs are weak and stiff, and they show striking ataxia. In contrast to normals, the animals are bad swimmers and begin to tremble and shake violently (sometimes for over 30 min) when taken out of the water.[80] The first symptoms can be observed at about 2–3 weeks postnatally. The animals usually die of cardiac decompensation. Multiple deposits are found in the brain, lumbar spinal cord, heart, and kidneys. The number of plaques increases in older animals. The material is basophilic alcian blue and PAS positive.[81] In the spinal cord, the deposits are found mainly in the dorsal roots, only rarely in the ventral roots. Eccentric nuclei, karyolysis, chromatolysis, clumping of Nissl substance, and vacuolization are observed in the neurons of the dorsal root ganglia and anterior horns. In the brain, deposits are found in the medulla and brain stem, the pia mater, the proximal blood vessels, and the plexus choroidea. Plaques are also found in the myelinated motor and sensory nerves. In the myocardium, the muscle fibers show vacuolization, hyalinization, and necrosis before changes in the nervous system are observed. Coronaries and other arteries are affected, also in heterozygotes. The skeletal system shows periostial proliferations, calcific deposits, and cortical erosions. Plaques are found in the glomerular tufts and in the descending tubules in the kidneys of some of the Shambling homozygotes.

The chemical nature of the accumulated material is not known. It is chloroform–methanol soluble and insoluble in acetone. The material is totally resistant to β-glucuronidase and hyaluronidase digestion but is partially removed by trypsin. The PAS-reacting material remains after β-glucuronidase and partially disappears upon trypsinization. These properties indicate that the accumulated substance is a mucopolysaccharide–lipid–protein complex. Further work is needed to define the substance accumulated and the corresponding metabolic defect.

Dilute-lethal (d^l) is a recessive mutation. Clinical signs appear at the tenth postnatal day with clonic seizures and opisthotonus. The animals die at about 3 weeks.[82] In homozygotes, blood phenylalanine is increased and serum proteins are decreased. Epinephrine and norepinephrine are increased in the adrenal glands, and the urinary output of epinephrine is increased.[83] In the tectospinal, spinocerebellar, and vestibulospinal tracts, histochemical signs indicate a severe demyelination shortly following the normal onset of myelination.[84] The demyelination occurs *in foci*.

Phenylalanine-pyruvate transaminase activity in liver, which in the wild allele *D* increases sharply after the sixth postnatal day and reaches a peak at the fourteenth day, does not show this increase in the homozygote d^l/d^l.[85] In liver homogenate from normals, phenylalanine hydroxylase activity increases sharply around the seventeenth postnatal day; this increase is also not observed in the homozygous Dilute-lethal. If, however, the supernatant of the homogenate is assayed after centrifugation at $15,000 \times g$ for 30 min, a normal activity of phenylalanine hydroxylase can be found. This finding suggests that the low activity of phenylalanine hydroxylase in the liver homogenate is due to an inhibitor located in the particulate fraction and removed by the centrifuga-tion.[86] In control mice, this inhibitor is also present from the tenth to the seventeenth postnatal day, but its inhibitory activity decreases thereafter to almost zero; in the homozygous Dilute-lethal, the inhibitor activity con-tinues to increase after this age. The inhibitor is found in the liver, brain, kidney, and heart. Hydroxylation of phenylalanine to tyrosine is coupled with the oxidation of the cofactor dihydrobiopterin. The same cofactor is used in the hydroxylation of tryptophan in serotonin synthesis. The inhibitor seems to interfere with the oxidation of the cofactor.[87]

Although this mutation shows similarities to the human disease, the differ-ences must also be emphasized; in man, the phenylalanine values in blood are much higher than in the homozygous mouse, and phenylpyruvate transaminase activity is normal.

H. Increased Death Rate of Cells

The homozygous Staggerer (sg) mutation is characterized by smallness of the cerebellum. Histologically, the number of granule cells in the cerebellar cortex is greatly reduced.[88] The postnatal proliferation, migration, and align-ment of the cells seem to be normal in sg/sg. Since the number of cells, however, is reduced in the mutant mice, it can be assumed that the death rate of the cells is increased.

The animals show major difficulties in locomotion which are pronounced in the third postnatal week. Sometimes the beginning of motor activity is ac-companied by a mild tremor. There is a group of mutants in which degenerative processes occur postnatally in the neuroepithelial elements of the inner ear. To this group belong such mutants as Shaker-1 (sh-1), Shaker-2 (sh-2), Jerker (je), Pirouette (pi), Spinner (sr), Waltzer (v), and Whirler (wi).[89] All the diseases are genetically and pathologically distinct, but they have in common a normal development of the inner ear until around the tenth postnatal day, after which degeneration of the apparently normally formed structures begins.

In Pirouette, for example, the organ of Corti develops normally until the twelfth postnatal day, after which degeneration of the hair cells becomes ap-parent. Clinically, at around the tenth postnatal day the animals show disturbed locomotion, later along with whirling in tight circles, there is some vertical head tossing. According to Deol,[90] the circular running and head shaking imply a disturbance in the central nervous system and not in the inner ear alone. In Shaker-1, the clinical and pathological changes occur later than in Pirouette but are essentially of a similar type. Again, not all the clinical symptoms can be explained by inner ear degeneration alone, and it will be important to look for defects in the brain itself.

I. Genetic Mutations Affecting Brain Metabolism

Numerous metabolic disorders of the central nervous system have been described in mammals other than mice. In many cases, the genetic etiology of the disease is not certain. Some examples will be discussed in which an inborn error of metabolism can be assumed. The pathogenesis of most of these diseases is not yet clear; therefore, it is best at the current stage of knowledge to discuss them according to their pathological, anatomical, and chemical appearance.

1. Globoid Cell Leukodystrophy

Fankhauser *et al.*[91] described for the first time a spontaneous leukodystrophy in two unrelated young dogs, a male and a female. The animals showed the first symptoms between the second and the third postnatal month, beginning with ataxia. The progression of the symptoms was slow. In one dog, paresis developed, which spread from caudal to cranial; in the other animal, there were disturbances of coordination and blindness. The behavior of the animals remained unaffected for a long period. Euthanasia was performed around 6 months of age, probably shortly before natural death would have occurred. The brains of the animals showed widespread and severe demyelination with axonal degeneration especially in the corpus callosum, centrum ovale, internal capsule, cerebellum, spinal cord, optic and trigeminal nerves, and spinal roots. The gray matter was essentially normal. The capillaries in the affected white matter were surrounded by large, round, or polygonal, sometimes multinucleated cells which were PAS positive. These cells were morphologically very similar to the globoid cells in human Krabbe's leukodystrophy. No other organ was affected.

The familial pattern of the disease and the possible autosomal recessive character was stressed by the occurrence of this disease in three cairn terriers[92] which were from different litters of the same dam with father and son as sires. A sex-linked or sex-limited inheritance was suggested for a similar disease in cats.[93] Jortner and Jonas[94] observed a decreased number of oligodendroglia in the severely affected regions of the brain. The decrease of axons in the involved regions was considered to be secondary to the loss of myelin and the extensive cellular reaction to it. Two cairn terriers, products of inbreeding, with a similar disease and pathology were described by Fletcher *et al.*[95] The globoid cells contained material which was PAS positive, not sudanophilic, and not metachromatic. It was thought that this material was a cerebroside–protein complex derived from the breakdown of myelin.

There are many similarities in Krabbe's disease to the described leukodystrophy in dog: in both, there are regions of intensive demyelination accompanied by glial and mesenchymal tissue response, and the morphological and histochemical properties of the globoid cells are very similar. In the human disease, an accumulation of galactocerebroside in the globoid cells has been found,[96] probably caused by the significant decrease of galactocerebroside-β-galactosidase.[97]

2. Metachromatic Leukodystrophy

A probably recessive inherited form of leukodystrophy occurs in inbred minks.[98-100] The first symptoms of the disease are observed around the age of 40 days. There is a universal tremor which is most pronounced in the head. Paresis of the hind legs progresses to paralysis, with concomitant incontinence of urine and feces during the final stage of the disease. The duration of the disease is about 2–4 months. At the onset of the disease, an increase of cells—probably astroglia—in the brain and spinal cord is visible. In this early stage, myelin already shows signs of degeneration. In areas in which demyelination is present, a perivascular, intracellular accumulation of cholesterol esters occurs. There is moderate metachromatic staining with toluidine blue and cresyl violet of granules in macrophages and also in extracellular sites. These granules are PAS positive. In the later stages of the disease, demyelination and vacuolization of the white matter progress. The proliferation of astrocytes increases. Macrophages around the vessels contain more sudanophilic granules, which stain metachromatically with toluidine blue. During the terminal stages, the number of sudanophilic and metachromatic granules is reduced and some insufficient remyelination can be observed.

In brain of normal littermates, cerebrosides and sulfatides increase sharply between 50 and 150 days postnatally. In the affected minks, the content of the glycolipids at the fiftieth day is like that in the normals. However, the increase of the glycolipid content per weight unit of brain is thereafter much reduced, and after the nintieth day the total content actually decreases to about 25% of the normal value. The ratio of cerebroside to sulfatide remains at about 4 : 1, as in normal brain.

Determination of the specific activity of sulfatide after sulfate-S^{35} injection reveals the specific activity of sulfatide-S^{35} to be four to five times higher than normal, and this remains almost constant between 36 and 204 hr after injection of the label, while in normals the specific activity has dropped by 10% at 204 hr. The increased activity is found in myelin, mitochondria, and microsomes. In contrast, C^{14}-labeling of the cerebrosides shows decreased synthesis of this lipid; the specific activity during the following days drops at the same rate in test and control animals, which indicates that the turnover of these lipids is not affected.

Interpretation of these results is difficult. Andersen and Palludan[100] suggest that the pathway of sulfatide synthesis in the mink is different from that proposed in man[39,40,101] and that the defect is an inhibition of acylation of psychosine and psychosine sulfate:

$$\text{Sphingosine} \xrightarrow{\text{UDP-galactose}} \text{psychosine} \xrightarrow{\text{Acyl-CoA}} \text{cerebrosides}$$

$$\downarrow \text{PAPS}$$

$$\text{psychosine-SO}_4 \xrightarrow{\text{Acyl-CoA}} \text{sulfatide}$$

The finding of decreased amounts of sulfatide in mink brain makes this metachromatic leukodystrophy quite different from that known in human infants, where sulfatide accumulates because of decreased activity of the sulfatide-degrading enzymes galactocerebroside sulfatase or arylsulfatase A.[102,103]

3. GM₂-Gangliosidosis

A sex-linked recessive mutation in German shorthair pointers has been described by Karbe and Schiefer.[104] Six males from three closely related litters were affected by probably the same disease. All affected dogs were males and were derived indirectly through dams from one female. The transmission is assumed to be sex-linked recessive. At the age of 6 months, the animals showed nervousness and a decreased ability for training. At about 9–12 months, progressive ataxia developed, and seizures occurred occasionally. Vision was partially impaired. The dogs lost contact with their surroundings to various degrees and were considered to be "dummies."

Histologically, there was an accumulation of granular material in most of the neurons of the CNS, associated with cell degeneration and necrosis. In advanced stages, the nuclei were pushed to the periphery of the enlarged cells. The lesions were most prominent in the gyrus orbitalis and gyrus preoreus of the rhinencephalon, in layers III and IV of the telencephalon, in the hippocampus, and in the spinal cord. In the perivascular spaces, there were macrophages with foamy material. Histochemically, the material stained black with sudan black and was PAS positive. The material was not birefringent in polarized light. There was a loss of neurons in the retina, and the few remaining ganglion cells contained a sudan black–positive and PAS-positive material in their cytoplasm. The other organs (liver, spleen, lymph nodes, kidney, heart, lung, and intestine) were normal. There was mild demyelination in the spinal cord.

The chemical nature of the accumulated material from the brain of one affected German shorthair pointer was investigated.[105] In the cortex, the total ganglioside content was found to be five times as high as in normals, which was mainly because of the accumulation of GM₂-gangliosides. The composition of this ganglioside was the same as in humans. It was resistant to neuraminidase and could be degraded to the corresponding hexosylceramide by partial hydrolysis. There was an additional accumulation of some GM₃-ganglioside, trihexosylceramide, and lactosylceramide. Since the material was stored in formalin, which may cause ganglioside degradation, the authors used control material which had been stored in formalin for the same length of time. Thus the finding of the slight GM3 accumulation might be unspecific. Accumulation of the asialo derivatives of the GM2- and GM3-gangliosides is also known in the human disease.

From these findings, it can be concluded that the disease observed in the dog is comparable to GM2-gangliosidosis in man. It must, however, be considered that in man the known GM2-gangliosidosis is not sex-linked recessive. It will be interesting to specify the enzymatic defect of this disease.

4. Mucopolysaccharidosis

Snorter-dwarf in cattle is an autosomal recessive mutation which occurs in Hereford and Angus cattle. Characteristic of these dwarf animals is the heavy, labored breathing from which the name *Snorter* is derived. An extensive investigation of this mutation was undertaken in 1960 by Lorincz.[106] At birth, the calves seem to be normal. Symptoms are observed at 2–4 months of age, when regression starts. Most animals die at the age of 1 year. The head of the affected animals is short, broad, and low set, and the neck is shorter and wider than in normal animals. The eyes are widely spaced, the nose is short, and the nostrils are upturned. The lips do not meet. This appearance corresponds to the "gargoyle" faces in some forms of human mucopolysaccharidosis. The long bones are shortened, the lumbar vertebrae abnormal. Internal hydrocephalus, enlarged tongue, rough, thickened skin, and abdominal hernias are often present.

In urine collected over 24 hr, estimation of crude mucopolysaccharide was 25–50 mg per liter in the affected animals compared to 5 mg per liter in normals. In the affected animals, the major mucopolysaccharide fractions were hyaluronic acid and chondroitin sulfate B. In the liver of these animals, the same mucopolysaccharides were accumulated.

This disease shows many similarities to human mucopolysaccharidosis and could be an excellent model in which to study the pathogenesis of the disease and to attempt a treatment of this inborn error of metabolism. Neufeld and Fratantoni[107] were able to show that the metabolic defect in mucopolysaccharidosis is the impairment of the degradation of mucopolysaccharides and that this defect can be corrected in cultured fibroblasts from patients with this disease. It would be very interesting to apply this knowledge in this animal model.

V. PREVENTION OF BRAIN DYSFUNCTION

Is it possible to prevent a genetically determined alteration of normal brain development which might cause brain dysfunction? Eugenic procedures such as selective breeding in animals or artificial insemination might prevent the disease completely. This is possible when the type of transmission of the mutation is known and when carriers of the mutant gene can be recognized. Once conception has taken place, a correction of the biochemical effect of the mutation is still possible and can perhaps prevent abnormal brain development, but only when the correction is applied before irreversible damage has occurred.

Correction can take place at different levels of the biochemical pathway of genetic expression, e.g., at the level of (1) transcription, (2) messenger RNA, (3) synthesized enzyme protein, or (4) the enzyme action itself. Transcription can be corrected by inserting or transducing a normally functioning gene into the cells.[108-110] Induction of enzyme activity can be achieved by several means.[111] Tomkins et al.[112] speculate that in the case of induction of tyrosine amino-transferase by dexamethasone the site of action might be messenger RNA. Substitution of lacking enzymes in cultured cells is possible and might lead to correction of the impaired metabolism.[113] Additional possibilities exist in the substitution of a missing end product where there is a deficiency of the key enzyme in a synthetic pathway. Where there is lack of a transferring or degrading enzyme, accumulation of toxic substances can be prevented by the avoidance or removal of the responsible precursors.[114]

Correction of the altered metabolism has to take place before or during the period when this metabolic step is a key regulator of the development of brain structures. Theoretically, a metabolic disorder which affects neuronal proliferation must be treated in man during the first half of the pregnancy, when the rate of proliferation is very high. A disorder affecting myelination must be treated at the latest between birth and the fourth postnatal month, since it is at this time that fast myelination is taking place. (see Fig. 1). Since much of brain development occurs in many species before birth, it is necessary to recognize the inborn error of metabolism prenatally and start the correction at the appropriate time. The possibilities for studying the metabolism in cultured amniotic cells of the embryo or fetus make possible in many cases the necessary early recognition.[115]

Animals with mutations of the CNS can be used to study the effects of the correction of the impaired metabolism on brain development. This is a necessary step toward the prevention of genetically caused brain dysfunction.

REFERENCES

1. A. N. Davison and J. Dobbing, The developing brain, in "Applied Neurochemistry" (A. N. Davison and J. Dobbing, eds.) pp. 253–286, Blackwell Scientific Publications, Oxford (1968).
2. H. G. Callan, The organisation of genetic units in chromosomes, J. Cell. Sci. 2:1–7 (1967).

3. R. J. Britten and D. E. Kohne, Repeated sequences in DNA, *Science* **161**:529–540 (1968).
4. C. A. Thomas, Jr., The theory of the master gene, *in* "The Neurosciences, Second Study Program" (F. O. Schmitt, ed.) pp. 973–998, Rockefeller University Press, New York (1970).
5. M. Jacobson, Development of specific neuronal connections, *Science* **163**:543–547 (1969).
6. K. Paigen, The genetics of enzyme realisation, *in* "Enzyme Synthesis and Degradation in Mammalian Systems" (M. Rechcigl, ed.) pp. 1–46, S. Karger, Basel and New York (1971).
7. R. Ganschow and K. Paigen, Glucuronidase phenotypes of inbred mouse strains, *Genetics* **59**:335–349 (1968).
8. J. A. Weston, An autoradiographic analysis of the migration and localisation of trunk neural crest cells in the chick, *Develop. Biol.* **6**:279–310 (1963).
9. B. Källén, Overgrowth malformation and neoplasia in embryonic brain, *Continia Neurol.* **22**:40–60 (1962).
10. L. J. Stensaas and S. S. Stensaas, An electron microscope study of cells in the matrix and intermediate laminae of the cerebral hemisphere of the 45 mm rabbit embryo, *Z. Zellforsch. Mikroskop. Anat.* **91**:341–365 (1968).
11. R. L. Sidman, Cell proliferation, migration and interaction in the developing mammalian central nervous system, *in* "The Neurosciences, Second Study Program" (F. O. Schmitt, ed.) pp. 100–107, Rockefeller University Press, New York (1970).
12. A. Glücksmann, Cell death in normal vertebrate ontogeny, *Biol. Rev. (Cambridge)* **26**: 59–86 (1951).
13. W. E. Watson, Centripetal passage of labelled molecules along mammalian motor axons, *J. Physiol. (London)* **196**:122P–123P (1968).
14. J. B. Angevine, Jr., D. Bodian, A. J. Coulombre, M. U. Edds, Jr., V. Hamburger, M. Jacobson, K. M. Lyser, M. C. Prestige, R. L. Sidman, S. Varon, and P. A. Wers, Embryonic vertebrate central nervous system: Revised terminology, *Anat. Rec.* **166**:257–262 (1970).
15. J. Altman, Autoradiographic and histological studies of postnatal neurogenesis, *J. Comp. Neurol.* **128**:431–474 (1966).
16. P. Rakic and R. L. Sidman, Histogenesis of cortical layers in human cerebellum, particularly the lamina dissecans, *J. Comp. Neurol.* **139**:473–500 (1970).
17. L. W. Lapham, Tetraploid DNA content of Purkinje neurones of human cerebellar cortex, *Science* **159**:310–312 (1968).
18. J. Dobbing and J. Sands, Timing of neuroblast multiplication in developing human brain, *Nature* **226**:639–640 (1970).
19. G. R. DeLong, Histogenesis of fetal mouse isocortex and hippocampus in reaggregating cell cultures, *Develop. Biol.* **22**:563–583 (1970).
20. P. Weiss and A. C. Taylor, Reconstruction of complete organs from single cell suspensions of chick embryos in advanced stages of differentiation, *Proc. Natl. Acad. Sci.* **46**:1177–1185 (1960).
21. J. A. Lowden and L. S. Wolfe, Studies on brain gangliosides. 3. Evidence for the location of gangliosides specifically in neurones, *Can. J. Biochem.* **42**:1587–1703 (1964).
22. M. T. Vanier, M. Holm, R. Oehman, and L. Svennerholm, Developmental profiles of gangliosides in human and rat brain, *J. Neurochem.* **18**:581–592 (1971).
23. R. M. Burton, L. Garcia-Bunnel, M. Golden, and Y. McBride Baltour, Incorporation of radioactivity of D-glycosamine-(1-^{14}C), D-glucose-(1-^{14}C), D-galactose-(1-^{14}C) and DL-serine-(3-^{14}C) into rat brain glycolipids, *Biochemistry* **2**:580–585 (1963).
24. H. S. Maker and G. Hauser, Incorporation of glucose carbon into gangliosides and cerebrosides by slices of developing rat brain, *J. Neurochem.* **14**:457–464 (1967).
25. K. Suzuki, G. E. Podulso, and S. E. Podulso, Further evidence for a specific ganglioside fraction closely associated with myelin, *Biochim. Biophys. Acta* **152**:576–586 (1968).
26. W. T. Norton and S. E. Podulso, Neuronal perikarya and astroglia of rat brain: Chemical composition during myelination, *J. Lipid Res.* **12**:84–90 (1971).

27. F. E. Samson and D. J. Quinn, Na⁺-K⁺-activated ATPase in rat brain development, *J. Neurochem.* **14**:421–427 (1967).

28. S. M. Bayer and W. C. McMurray, The metabolism of amino acids in developing rat brain, *J. Neurochem.* **14**:695–706 (1967).

29. T. Kobayashi, O. Inman, W. Buno, and H. E. Himwhich, A multidisciplinary study of changes in mouse brain with age, *Recent Advan. Biol. Psychiat.* **5**:293–308 (1963).

30. A. N. Davison and A. Peters, "Myelination," pp. 80–143, Chales C. Thomas, Springfield, Ill. (1970).

31. S. A. Luse, Formation of myelin in the central nervous system of mice and rats as studied with electron microscope, *J. Biophys. Biochem. Cytol.* **2**:777–783 (1956).

32. A. Peters, The formation and structure of myelin sheaths in the central nervous system, *J. Biophys. Biochem. Cytol.* **8**:431–446 (1960).

33. J. P. M. Bensted, J. Dobbing, R. S. Morgan, R. T. W. Reid, and G. Payling Wright, Neuroglial development and myelination in the spinal cord of the chick embryo, *J. Embryol. Exptl. Morphol.* **5**:428–437 (1957).

34. J. Schonbach, K. H. Hu, and R. L. Friede, Cellular and chemical changes during myelination: Histological, autoradiographic, histochemical and biochemical data on myelination in the pyramidal tract and corpus callosum in rat, *J. Comp. Neurol.* **134**:21–38 (1968).

35. M. L. Cuzner and A. N. Davison, The lipid composition of rat brain myelin and subcellular fractions during development, *Biochem. J.* **106**:29–34 (1968).

36. N. H. Bass and H. H. Hess, A comparison of cerebrosides, proteolipid, proteins and cholesterol as indices of myelin in the architecture of rat cerebrum, *J. Neurochem.* **16**:731–750 (1969).

37. H. C. Agrawal, N. L. Banitz, A. H. Bone, A. N. Davison, R. F. Mitchell, and M. Spohn, The identity of a myelin-like fraction isolated from developing brain, *Biochem. J.* **120**:635–642 (1970).

38. E. R. Einstein and J. Csejtev, Proteins in the developing human brain, *Trans. Am. Neurol. Ass.* **1966**:218 (1966).

39. A. S. Balasubramanian and B. K. Bacchawat, Studies on enzymic synthesis of cerebroside sulfate from 3′-phosphoadenosine-5′-phosphosulfate, *Indian J. Biochem.* **2**:212–216 (1965).

40. G. M. McKhann and W. Ho, The *in vivo* and *in vitro* synthesis of sulfatides during development, *J. Neurochem.* **14**:717–724 (1967).

41. H. P. Chase, J. Dorsey, and G. M. McKhann, The effect of malnutrition on synthesis of myelin lipid, *Pediatrics* **40**:551–559 (1967).

42. N. Herschkowitz, G. M. McKhann, S. Saxena, and E. M. Shooter, Characterisation of sulfatide containing lipoproteins in rat brain, *J. Neurochem.* **15**:1181–1188 (1968).

43. D. Silberberg, J. A. Benjamins, N. Herschkowitz, and G. M. McKhann, Incorporation of radioactive sulphate into sulphatide during myelination in cultures of rat cerebellum, *J. Neurochem.* **19**:11–18 (1972).

44. E. R. Einstein, K. B. Balal, and J. Csejtev, Biochemical maturation of the central nervous system. II. Protein and proteolytic enzyme changes, *Brain Res.* **18**:35–49 (1970).

45. B. Balazs, B. W. L. Brooksbank, A. N. Davison, J. T Earys, and D. A. Wilson, The effect of neonatal thyroidectomy on myelination in the rat brain, *Brain Res.* **15**:219–232 (1969).

46. J. de Vellis and D. Inglish, Hormonal control of glycerophosphate dehydrogenase in the rat brain, *J. Neurochem.* **15**:1061–1070 (1968).

47. J. J. Curry and L. M. Heim, Brain myelination after neonatal administration of oestradiol, *Nature* **209**:915–916 (1966).

48. S. Glueksohn-Waelsch, Genetic factors and the development of the nervous system, *in* "Biochemistry of the Developing Nervous System" (H. Waelsch, ed.) pp. 375–396, Academic Press, New York (1955).

49. L. J. Smith and K. F. Stein, Axial elongation in the mouse and its retardation in homozygous loop tail mice, *J. Embryol. Exptl. Morphol.* **10**:73–87 (1962).

50. R. Auerbach, Analysis of the developmental effects of a lethal mutation in the house mouse, *J. Exptl. Zool.* **127**:305–324 (1954).
51. M. S. Deol, The abnormalities of the inner ear in Kreisler mice, *J. Embryol. Exptl. Morphol.* **12**:475–490 (1964).
52. M. S. Deol, The origin of the abnormalities of the inner ear in Dreher mice, *J. Embryol. Exptl. Morphol.* **12**:727–733 (1964).
53. H. Fischer, Die Embryogenese der Innenohrmissbildungen bei dem spontan mutierten Dreherstamm der Hausmaus, *Z. Mikroskop. Anat. Forsch.* **64**:476–497 (1958).
54. G. M. Truslove, The anatomy and development of the Fidget mouse, *J. Genet.* **54**:64–86 (1956).
55. L. Kilham and G. Margolis, Cerebellar disease in cats, induced by inoculation of rat virus, *Science* **148**:244–246 (1965).
56. R. Schmidt, Die postnatale Genese der Kleinhirndefekte röntgenbestrahlter Hausmäuse, *Z. Hirnforsch.* **5**:164–209 (1962).
57. U. Hamburgh, Analysis of the postnatal development effect of "Reeler," a neurological mutation in mice, *Develop. Biol.* **8**:165–185 (1963).
58. H. Meier and W. G. Hoag, The neuropathology of "Reeler," a neuromuscular mutation in mice, *J. Neuropathol. Exptl. Neurol.* **21**:649–654 (1962).
59. G. B. Koelle, The histochemical identification of acetyl cholinesterase in cholinergic, adrenergic and sensory neurones, *J. Pharmacol. Exptl. Therap.* **114**:167–184 (1955).
60. R. DeLong and R. Sidman, Alignment defect of reaggregating cells in cultures of developing brains of Reeler mutant mice, *Develop. Biol.* **22**:584–600 (1970).
61. R. L. Sidman, M. M. Dickie, and S. H. Appel, Mutant mice (Quaking and Jimpy) with deficient myelination in the central nervous system, *Science* **144**:309–311 (1964).
62. N. A. Baumann, C. M. Jacque, S. A. Pollet, and M. L. Harpin, Fatty acid and lipid composition of the brain of a myelin deficient mutant, the "Quaking" mouse, *Europ. J. Biochem.* **4**:340–344 (1968).
63. N. Neskovic, J. L. Nussbaum, and P. Mandl, A study of glycolipid metabolism in myelination disorder of Jimpy and Quaking mice, *Brain Res.* **21**:39–53 (1970).
64. G. Hauser, J. Eichberg, and S. Jacobs, Polyphosphoinositide levels and biosynthesis in Quaking mouse brain, *Biochem. Biophys. Res. Commun.* **43**:1072–1080 (1971).
65. N. A. Baumann, M. L. Harpin, and J. M. Bourré, Long chain fatty acid formation: Key step in myelination studied in mutant mice, *Nature* **227**:960–961 (1970).
66. P. Morell, W. T. Norton, and S. Greenfield, Isolation and characterisation of myelin protein form adult Quaking mice and its similarity to "early myelin" protein of young controls, *Abst. Third Internat. Meeting Internat. Soc. Neurochem.*, p. 417 (1971).
67. D. M. Bowen and N. S. Radin, Hydrolase activities in brain of neurological mutants: Cerebroside galactosidase, nitrophenyl galactoside hydrolase, nitrophenyl glucoside hydrolase and sulfatase, *J. Neurochem.* **16**:457–460 (1969).
68. D. J. Kurtz and J. N. Kanfer, Cerebral acid hydrolase activities: Comparison in "Quaking" and normal mice, *Science* **168**:259–260 (1970).
69. A. Hirano, D. S. Sax, and H. M. Zimmermann, The fine structure of the cerebella of Jimpy mice and their "normal" litter mates, *J. Neuropathol. Exptl. Neurol.* **28**:388–400 (1969).
70. J. Torii, M. Adachi, and B. W. Volk. Histochemical and ultrastructural studies of inherited leucodystrophy in mice, *J. Neuropathol. Exptl. Neurol.* **30**:278–289 (1971).
71. J. L. Nussbaum, N. Neskovic, and P. Mandel, A study of lipid components in brain of the "Jimpy" mouse, a mutant with myelin deficiency, *J. Neurochem.* **16**:927–934 (1969).
72. C. Galli and D. Re C. Galli, Cerebroside and sulfatide deficiency in the brain of "Jimpy" mice, a mutant strain of mice exhibiting neurological symptoms, *Nature* **220**:165–166 (1968).
73. E. L. Hogan, K. C. Joseph, and G. Schmidt, Composition of cerebral lipids in murine sudanophilic leucodystrophy. The Jimpy mutant, *J. Neurochem.* **17**:75–83 (1970).
74. C. Galli, G. M. Kneebone, and R. Paoletti, An inborn error of cerebroside biosynthesis as the molecular defect of the Jimpy mouse brain, *Life Sci.* **8**:911–918 (1969).

75. M. N. Neskovic, J. L. Nussbaum, and P. Mandel, Enzymatic deficiency in neurological mutants, brain uridine diphosphate galactose : ceramide galactosyl transferase in Jimpy mouse, *FEBS Letters* **8**:213–216 (1970).

76. N. Herschkowitz, F. Vassella, and A. Bischoff, Myelin differences in the central and peripheral nervous system in the Jimpy mouse, *J. Neurochem.* **18**:1361–1363 (1971).

77. J. M. Matthieu, U. Schneider, and N. Herschkowitz, *In vitro* synthesis of sulfatides in a myelin deficient mutant, the Jimpy mouse, *Brain Research* **42**:433–439 (1972).

78. K. J. Joseph and E. L. Hogan, Fatty acid composition of cerebrosides, sulfatides and ceramides in murine sudanophilic leucodystrophy: The Jimpy mutant, *J. Neurochem.* **18**: 1639–1645 (1971).

79. M. K. Wolf and A. B. Holden, Tissue culture analysis of the inherited defect of central nervous system myelination in Jimpy mice, *J. Neuropathol. Exptl. Neurol.* **28**:195–213 (1969).

80. E. L. Green, Shambling, a neurological mutant of the mouse, *J. Hered.* **58**:65–68 (1967).

81. H. Meier, Pathological findings in Shambling hereditary neuropathy of mice, *J. Neuropathol. Exptl. Neurol.* **26**:620–633 (1967).

82. A. G. Searle, A lethal allele of Dilute in the house mouse, *Heredity* **6**:395–401 (1952).

83. C. H. Doolittle and H. Rauch, Epinephrine and norepinephrine levels in Dilute lethal mice, *Biochem. Biophys. Res. Commun.* **18**:43–47 (1964).

84. D. E. Kelton and H. Rauch, Myelination and myelin degeneration in the central nervous system of Dilute lethal mice, *Exptl. Neurol.* **6**:252–262 (1962).

85. D. L. Coleman, Phenylalanine hydroxylase activity in Dilute and non-Dilute strains of mice, *Arch. Biochem. Biophys.* **69**:562–568 (1962).

86. H. Rauch and M. T. Yost, Phenylalanine metabolism in Dilute lethal mice, *Genetics* **48**: 1487–1495 (1963).

87. S. Kaufman, The structure of the phenylalanine hydroxylation cofactor, *Proc. Natl. Acad. Sci.* **50**:1085–1093 (1963).

88. R. L. Sidman, P. W. Lane, and M. W. Dickie, Staggerer, a new mutation in the mouse affecting the cerebellum, *Science* **137**:610–612 (1962).

89. H. Grüneberg, "The Genetics of the Mouse," 2nd ed., Martinus Nijhoff, The Hague (1952).

90. M. S. Deol, The anatomy and development of the mutants Pirouette, Shaker-1 and Waltzer in the mouse, *J. Genet.* **52**:562–588 (1954).

91. R. Fankhauser, H. Luginbühl, and U. J. Hartley, Leukodystrophie vom Typus Krabbe beim Hund, *Schweiz. Arch. Tierheilkunde* **105**:198–207 (1963).

92. R. S. Hirth and S. W. Nielsen, A. familial canine globoid cell leucodystrophy (Krabbe type), *J. Small Anim. Pract.* **8**:569–575 (1967).

93. K. H. Johnson, Globoid leucodystrophy in the cat, *J. Am. Vet. Med. Ass.* **157**:2057–2064 (1970).

94. B. S. Jortner and A. M. Jonas, The neuropathology of globoid cell leucodystrophy in the dog, *Acta Neuropathol.* **10**:171–182 (1968).

95. T. F. Fletcher, H. J. Kurtz, and D. G. Low, Globoid cell leucodystrophy (Krabbe type) in the dog, *J. Am. Vet. Med. Ass.* **149**:165–172 (1966).

96. K. Suzuki and Y. Suzuki, Globoid cell leucodystrophy (Krabbe's disease): Deficiency of galacto ccrebroside-β-galactosidase, *Proc. Natl. Acad. Sci.* **66**: 302–309 (1970).

97. Y. Suzuki and K. Suzuki, Krabbe's globoid cell leucodystrophy: Deficiency of galacto cerebrosidase in serum, leucocytes and fibroblasts, *Science* **171**:73–75 (1971).

98. N. R. Brander and B. Palludan, Leucoencephalopathy in mink, *Acta Vet. Scand.* **6**:41–51 (1965).

99. H. A. Andersen, Leucodystrophy in mink, *Acta Neuropathol.* **7**:297–304 (1967).

100. H. A. Andersen and B. Palludan, Leucodystrophy in mink, *Acta Neuropathol.* **11**:347–360 (1968).

101. N. S. Radin, F. Martin, and J. R. Brown, Galactolipid metabolism, *J. Biol. Chem.* **224**: 499–507, (1957).

102. J. Austin, D. McAfee, D. Armstrong, M. O'Rouske, L. Shearer, and B. Bacchawat, Abnormal sulphatase activities in two human diseases, *Biochem. J.* **93**:15c (1964).

103. E. Mehl and H. Jatzkewitz, Evidence for the genetic block in metachromatic leucodystrophy, *Biochem. Biophys. Res. Commun.* **19**:407–411 (1965).

104. E. Karbe and B. Schiefer, Familial amaurotic idiocy in male German shorthair pointers, *Pathol. Vet.* **4**:223–232 (1967).

105. H. Bernheimer and E. Karbe, Morphologische und neurochemische Untersuchungen von zwei Formen der amaurotischen Idiotie des Hundes: Nachweis einer GM_2-Gangliosidose, *Acta Neuropathol.* **16**:243–261 (1970).

106. A. E. Lorincz, Heritable disorders of acid mucopolysaccharide metabolism in humans and snorter dwarf cattle, *Ann. N.Y. Acad. Sci.* **91**:644–658 (1960/61).

107. E. F. Neufeld and J. C. Fratantoni, Inborn errors of mucopolysaccharide metabolism, *Science* **169**:141–146 (1970).

108. D. Rabovsky, Gene insertion into mammalian cells, *Science* **174**:933–934 (1971).

109. W. Mumyon, E. Kraiselbrud, D. Davis, and J. Mann, Transfer of thymidine kinase to thymidine kinaseless L cells by infection with ultraviolet irradiated herpes simplex virus, *J. Virol.* **7**: 813–820 (1971).

110. C. E. Merill, M. R. Geier, and J. C. Petricciani, Bacterial virus gene expression in human cells, *Nature* **233**:398–400 (1971).

111. I. M. Arias, L. M. Gartner, M. Cohen, J. B. Ezzer, and A. J. Levi, Chronic nonhaemolytic unconjugated hyperbilirubinaemia with glucuronyl transferase deficiency, *Am. J. Med.* **47**:395–409 (1969).

112. G. M. Tomkins, T. D. Gelehrter, D. Gramer, D. Martin, H. H. Samuels, and E. B. Thompson, Control of specific gene expression in higher organisms, *Science* **166**:1474–1480 (1969).

113. U. Wiesmann, E. Rossi, and N. Herschkowitz, Treatment of metachromatic leucodystrophy in fibroblasts by enzyme replacement, *New Engl. J. Med.* **284**:672 (1971).

114. H. Bickel, Recent advances in the early detection and treatment of inborn errors with brain damage, *Neuropaediatrie* **1**:1–11 (1969).

115. A. Milunski, J. W. Littlefield, J. N. Kanfer, E. H. Kolodny, V. E. Shih, and L. Atkins, Prenatal genetic diagnosis, *New Engl. J. Med.* **283**:1370–1381 (1970).

116. L. Arey, "Developmental Anatomy," Saunders, Philadelphia (1946).

117. M. Jacobson, "Developmental Neurology," Holt, Rinehart and Winston, New York (1970).

Chapter 5

EXPERIMENTAL ALLERGIC ENCEPHALOMYELITIS

Marian W. Kies

Section on Myelin Chemistry, Laboratory of Cerebral Metabolism
National Institute of Mental Health
Bethesda, Maryland

I. INTRODUCTION

This chapter is a series of discussions based on research in experimental demyelination. The discussions more or less follow the development of research on experimental allergic encephalomyelitis (EAE) as it evolved from our studies and those of others in the field. It began quite simply with the desire to clarify the etiology and pathogenesis of EAE; it appeared that this information was not only necessary but would be sufficient to permit the more difficult analysis of the etiology and pathogenesis of the human illness, multiple sclerosis. Although we were not committed to the thesis that the two diseases were one and the same entity, it seemed to us that they must share some basic mechanism.

It is one of the ironies of scientific research that an increase in knowledge increases rather than decreases the number of questions to be answered. Even though a great deal has been learned about induction of the experimental disease, there is as yet no clear understanding of the initiating event(s?) in multiple sclerosis. Immunological studies which have failed to establish a common "antigen" have convinced us that the problem is more complex than was realized. Likewise, our ability to use encephalitogen for *prevention* of EAE prior to sensitization or for *therapy* after development of neurological signs has not provided the necessary information for prevention or treatment of human illness.

II. HISTORICAL BACKGROUND

A. Clinical "Accidents" Related to Injections of CNS Tissue

Occurrence of demyelination after the injection of CNS material was first observed in humans treated with rabies vaccine derived from rabbit brain and attenuated rabies virus. There was some evidence that the encephalomyelitis was somehow related to the CNS material present in the virus preparation.[1] In attempting to prove that the encephalomyelitis was caused by the nervous tissue rather than the attenuated virus, Rivers *et al.*[2,3] injected monkeys repeatedly with CNS tissue and, after a period of several months, succeeded in inducing neurological signs and CNS lesions experimentally. Although the implication of their observation was clear, the exploitation of this experimental approach to the study of demyelinating diseases was not feasible until it was shown by Kabat *et al.*[4] and Morgan[5] that Freund's mycobacterial adjuvant[6,7] enhanced the encephalitogenicity of whole CNS tissue to such an extent that EAE could be induced experimentally by a single injection. This observation set off a chain reaction which has continued to this day. Literally hundreds of papers have been published on the substance in CNS tissue responsible for the induction of EAE, not only the identification and purification of encephalitogenic protein but also its structure, immunological activity, significance in human disease, and possible *in vivo* function (reviewed by Alvord[8]).

B. Induction of EAE in Laboratory Animals

Early studies on EAE were carried out in several species of animals, including dogs, cats, rats, mice, rabbits, guinea pigs, monkeys, chickens, and pigeons. Occasionally, there were reports of attempts to induce EAE in cows, sheep, and hamsters. Hamsters appear to be unique in their resistance to this disease, although this may only reflect failure on the part of investigators to locate a susceptible strain, as was the case with rats and mice. At present, the animals most commonly employed for investigation of EAE are guinea pigs, rats, rabbits, and monkeys.

Variable susceptibility to EAE is common among rabbits and monkeys, some individuals being much more resistant to sensitization than others. This different range of susceptibility has also been observed among different strains

of rats and guinea pigs. Not all strains of small laboratory animals are susceptible. Of those commercially available, Hartley guinea pigs and Lewis rats
are probably the most widely used for studies of EAE. We use noninbred guinea
pigs randomly bred in a closed colony at NIH, a stock originally derived from
a group of mixed-color females and Hartley males. For studies involving cell
transfer, strain 13 guinea pigs inbred for histocompatibility are used.[9] Strain
2 guinea pigs, also histocompatible, are relatively resistant to induction of EAE.
Lewis rats, the most susceptible of the rat strains studied by Levine and
Wenk,[10] have also been inbred for histocompatibility and provide an excellent
tool for both bioassay and cell transfer studies.

III. IDENTIFICATION OF SPECIFIC ENCEPHALITOGENIC COMPONENTS OF CNS TISSUE

A. Early Studies

During the first decade after publication of the classic papers of Kabat
et al.,[4] Morgan,[5] and Freund *et al.*,[7] much attention was paid to the possible
encephalitogenicity of CNS lipids.[11] It soon became evident that lipids, although they possessed a unique immunological activity as haptens,[12,13] could
not by themselves induce EAE. In fact, it was found that most of the encephalitogenic activity in brain or spinal cord could be obtained in aqueous
solution[14] by treatment of the CNS residue with dilute acid after proteolipids and lipids had been removed with chloroform–methanol by the technique
of Folch and Lees.[15]

In addition to the acid-soluble encephalitogenic fraction which was obtained from chloroform–methanol treated CNS tissue, a collagen-like protein which
had some encephalitogenic activity was isolated in collaboration with Roboz-
Einstein *et al.*[16] from bovine spinal cord acetone powder. The collagen-like
protein (so named because of its amino acid composition) was judged to be
homogeneous by its behavior in the ultracentrifuge. Since the specific activity
of this fraction was much less than that of the acid-soluble protein, it was
concluded[17] that the biological activity of collagen-like protein was probably
related to contamination by a small amount of encephalitogenic protein. The
specific activity of the latter was such that amounts easily detected by bioassay
could not have been detected by ultracentrifugal analysis.

Once it had been established that the encephalitogen was a water-soluble
protein,[18] the problem of etiology and pathogenesis of EAE could be investigated by a variety of immunological techniques which were not feasible with
whole CNS tissue as the antigen.

B. Bioassay

The importance of and inherent difficulties in bioassay are not always recognized by investigators who utilize this technique to monitor isolation of a biologically active compound. *Bioassay* is defined in Webster's Third New International Dictionary as the "determination of the relative effective strength of a substance by comparing its effect on a test organism with that of a standard preparation." It is easy to lose sight of the fact that most bioassays yield an estimation rather than an absolute value and that personal judgment is involved. Thus the need for critical appraisal cannot be over emphasized.

Of the several species of animals available for testing encephalitogenic activity of CNS fractions, guinea pigs appeared to be the best choice at the time we developed the bioassay in collaboration with Alvord.[19] In the intervening years, rabbits have also been used successfully for bioassay by Waksman *et al.*[20] and Kibler *et al.*[21] as well as Lewis rats by Levine.[22] Each species has its own peculiar advantages, and, as it developed later, it was lucky that different investigators had studied encephalitogenic activity in different species. Whichever species one uses, the test requires large numbers of animals and careful attention to details of experimental conditions—preparation of the emulsion, concentration and type of mycobacterial adjuvant, route of injection, observation of animals, and standardization of records, to list a few. Histological ratings are essential for quantitative studies and should be made on coded specimens without prior knowledge of clinical results.

Successful isolation and identification of a biologically active compound depend on the ability of the bioassay to discriminate levels of activity of various fractions; i.e., there must be some means of estimating *specific activity* (units/mg dry wt). It is obvious that a fraction which induces EAE in only one of five animals is less active than a second fraction (tested at the same level) which induces EAE in all five of the animals. However, one cannot conclude that the two fractions differ by a factor of 5 in their specific activities. The number of animals showing clinical signs, the intensity of the signs, and the intensity and number of histological lesions all have to be taken into account in evaluating the reaction induced by a given weight of sample.[19]

In actual practice, five guinea pigs are injected with each dose of fraction tested; each animal is assigned an appropriate score based on the clinical and histological observations (see Table 1). The mean score for the group is the disease index for the particular dose of encephalitogenic fraction injected. A plot of the mean scores *vs.* weights of fraction tested (Fig. 1) yields a curve which can be used to calculate the specific activity, provided at least one of the points on the curve is near the 50% effective dose. The latter (ED_{50}) is a convenient "unit" of activity. Normally, it takes two or three bioassays to establish

**TABLE 1. BIOASSAY OF ENCEPHALITOGENIC ACTIVITY OF CNS
TISSUE FRACTIONS: ASSIGNMENT OF SCORE TO INDIVIDUAL
GUINEA PIGS**

Histology	Clinical			
	0	$+^a$	K^b	D^c
0	0	1	2	3
$\pm$	1	4	6	7
$+, ++$	4	8	9	10

^{a}Clinical signs include weight loss of more than 50 g in 4–7 days, fecal impaction, drooling,
marked loss of muscle tone, dirty, ruffled fur, and hind leg weakness.
^{b}K, killed when severe weight loss and paralysis are sufficient to warrant termination of experiment.
^{c}D, death, rated 1 point higher than K on the assumption that the disease reaction is more severe.
Animals with lipemic serum at time of sacrifice are scored the same as D.

specific activity of a fraction—more if the investigator over- or underestimates the activity of his preparation and fails to choose an amount suitable for bioassay.

C. Isolation of the Encephalitogen and its Characterization as Myelin Basic Protein

Search for an unknown biologically active compound is frequently like a treasure hunt—some clues are so obvious that they are hard to believe. Such a clue was the extremely high concentration of the encephalitogen in white matter. Preparations 50–100 times as actives as whole cord were readily obtained,[14,18] but attempts to increase the specific activity by a factor greater than this were unsuccessful. This increase in specific activity seemed insignificant to us in comparison to the thousandfold increases frequently obtained in purification of other biologically active proteins.

Although myelin had been implicated as the probable source of encephalitogenicity in CNS tissue, the protein was not isolated directly from purified myelin prior to our report in 1961 at the Fourth International Congress of Neuropathology.[23] This study, described in detail by Laatsch *et al.*,[24] provided a much-needed reference for our other purified fractions. On the basis of their specific encephalitogenic activities, they were as good as the protein isolated directly from myelin. Unfortunately, bioassay cannot provide the fine discrimination needed to establish purity or homogeneity, since small amounts of impurity do not affect the specific activity. Paper electrophoresis and ultracentrifugation likewise lack the sensitivity needed to establish homogeneity. With the advent of acrylamide gel electrophoresis, we could confirm that the encephalitogenic proteins that we had isolated from whole CNS tissue were comparable in purity to basic protein isolated directly from myelin (Fig. 2).

Bioassay of Bovine Myelin Basic Protein

µg Inj.	Clin.	DO$_{50}$	Hist.	D.I.
0.5	1/5		2/5	1.6
1.0	1/5		3/5	1.8-2.6
2.0	4/5	19 d.	5/5	6.2-8.2
4.0	3/5	16 d.	4/5	4.4-6.2
8.0	5/5	17 d.	5/5	9.4
50	5/5	14 d.	5/5	8.8-10.0

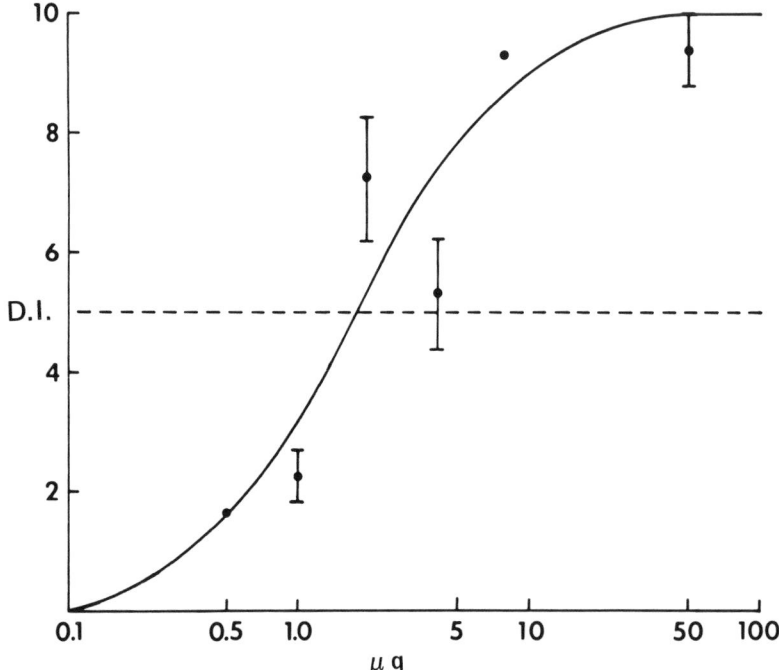

Fig. 1. A representative bioassay curve. The intersection of the curve with the dotted line indicates that the ED$_{50}$ (50% effective dose) of this preparation is approximately 2 µg. The spread in values arises from the fact that histological ratings of some specimens were $0 - \pm$ or $\pm - +$, rather than 0, $\pm$, or $+$.

Despite a claim to the contrary, encephalitogenic protein preparations as pure as those described more recently[25] have been available since 1960.[26]

There are certain theoretical advantages to isolation of the protein directly from myelin (there should be no contamination with basic proteins from other particulate cell fractions). However, it is much simpler to obtain large amounts of purified myelin basic protein from whole CNS tissue. By means of a procedure recently developed by Deibler *et al.*,[27] relatively pure myelin basic protein can be obtained even from CNS tissues low in white matter (e.g., rat brain) without resorting to the use of column fractionation.

Since it is highly basic and located in myelin, we chose the name *myelin basic protein* for the protein we isolated and characterized rather than the more cumbersome term *encephalitogenic protein*. Trivial terms such as A1 protein, EP, EBP, BEP, ELF, and PEProt serve no purpose other than to obscure the fact that these preparations are the same as the myelin basic protein originally described.[24]

On the basis of relative specific activity as well as the dry weight of basic protein actually extracted from myelin by 0.1 N HCl, we estimated its concentration to be approximately 30% of the total protein of myelin.[28] This figure has since been confirmed by several other investigators.[25,29,30] The high concentration of the encephalitogen in myelin explains our inability to increase the specific activity of the protein relative to that of whole cord more than a hundredfold; i.e., this protein accounts for roughly 1% of the dry weight of whole tissue. By our current preparative procedure,[27] about 0.5 g of myelin basic protein is obtained from 100 g of whole fresh guinea pig cord, or 1.5 g/100 g of lyophilized cord. This amount of pure myelin would contain approximately 1.8 g basic protein. Since cord is predominantly white matter, the amount actually obtained

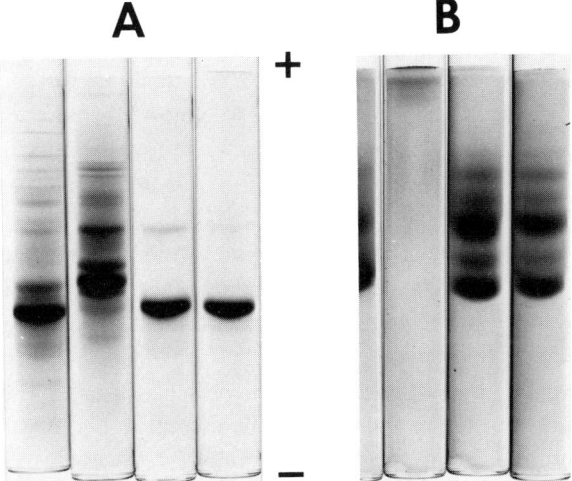

Fig. 2. Electrophoresis of brain and liver proteins on acrylamide gel at pH 2.4 (A) and at pH 10.6 (B). A (from left to right), pH 3.0 extract (80 μg) of chloroform–methanol (CM) treated guinea pig brain; pH 3.0 extract (100 μg) of CM-treated guinea pig liver; guinea pig myelin basic protein (25 μg) isolated from the extract of brain; and guinea pig myelin basic protein (25 μg) isolated from CNS myelin. B, Same sequence as A but with 200, 200, 75, and 75 μg, respectively, of protein applied to the gels. See Deibler *et al.*[27] for details of electrophoresis.

is in good agreement with the amount estimated. The ED_{50} of lypohilized whole spinal cord is about 0.1 mg, whereas the ED_{50} of a good preparation of myelin basic protein should be about 1 μg.

The isolation of the encephalitogen from purified myelin demonstrated for the first time the direct relationship between the "antigen" and its "target organ" which had been proposed earlier.[4] It was also the first demonstration of the presence in a highly specialized cell membrane of a water-soluble protein possessing some of the unusual characteristics of histones, yet distinctly different from the latter. It is remarkable that a tissue constituent amounting to approximately 1 % of the dry weight of white matter had not been recognized previously except by virtue of its biological activity. In 1965, Martenson and LeBaron[31] observed that the major constituent of a fraction obtained from bovine white matter by dilute acid treatment was a basic protein easily separated from the other acid-soluble proteins by chromatography. Their attention was drawn to it by its high concentration rather than by its biological activity, but its similarity to myelin basic protein was quickly recognized and the two proteins were found to be identical.[32]

1. Electrophoretic Analyses

Electrophoresis in acrylamide gel is one of the best techniques currently available for characterization of proteins and protein mixtures. The method, originally developed for serum proteins,[33] was modified for basic proteins by Reisfeld et al.[34] Their technique was used in almost all studies on myelin basic protein for a time and is still being used by some investigators. Unfortunately, it does not provide good resolution of all the proteins present in dilute acid extracts of CNS tissue.

In order to achieve maximum resolution of the proteins in crude preparations of myelin basic proteins, we have adopted other electrophoretic conditions for routine examination of these fractions. The technique is described in detail by Martenson et al.[35] At pH 2.4, all of the proteins in the pH 3.0 extract of CNS tissue migrate into 5 % acid gels in the presence of urea. The latter insures the dissociation of complexes which may be present in the mixture of proteins. Similar conditions have been used successfully for the analysis of histones by Johns.[36] For protein staining, amido black (buffalo black B) is used. It has the advantage of producing a characteristic color with highly basic proteins. When the gels are viewed with incandescent light, the blue-green color of myelin basic protein is easily identified in a mixture of CNS proteins, most of which are grayish-blue (Figs. 3 and 4).[37]

Electrophoresis in gels containing 8 M urea and pH 10.6 buffer has provided additional information regarding myelin basic protein. This technique can be

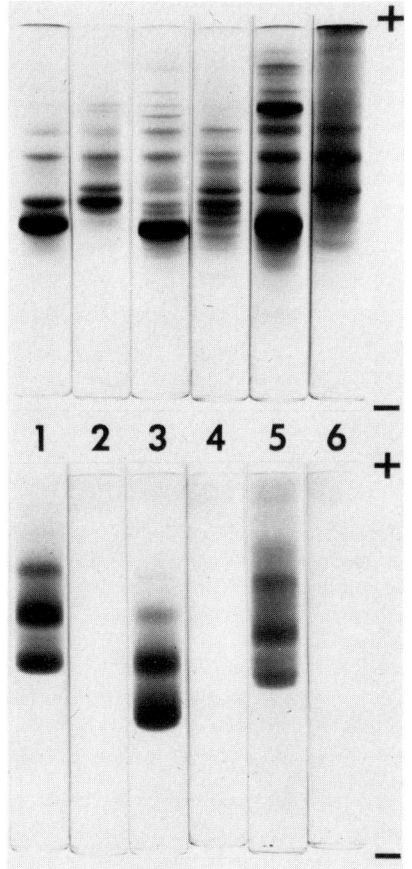

Fig. 3. Electrophoresis of brain and liver proteins from guinea pig, chicken, and turtle. Extracts were prepared as in Fig. 2. Upper gels run at pH 2.4: (1) guinea pig brain, (2) guinea pig liver, (3) chicken brain, (4) chicken liver, (5) turtle brain, and (6) turtle liver (1–4, 100 μg protein; 5 and 6, 200 μg protein). Lower gels run at pH 10.6: amounts applied to alkaline gels were double the amounts applied to acid gels.

used to examine myelin basic protein in pH 3.0 extracts without further purification, since most of the other proteins in the dilute acid extract migrate toward the anode at high pH and do not affect the electrophoretic pattern of myelin basic protein. In contrast to the single discrete band which characterizes this protein BP in low pH electrophoretograms, several bands are observed at high pH. This phenomenon was first reported by Martenson and Gaitonde[38] in a study of a highly basic protein from bovine white matter. It was subsequently shown by Martenson et al.[39] that the multiple components of myelin basic protein detected at high pH are neither artifacts nor impurities. Their behavior on gel chromatography suggests that they have the same molecular size. They can be separated by chromatography on carboxymethylcellulose, each retaining its original mobility on repeated electrophoretic analysis (Figs. 5 and 6). All the components have essentially the same encephalitogenic activity and identical amino acid compositions (Table 2).

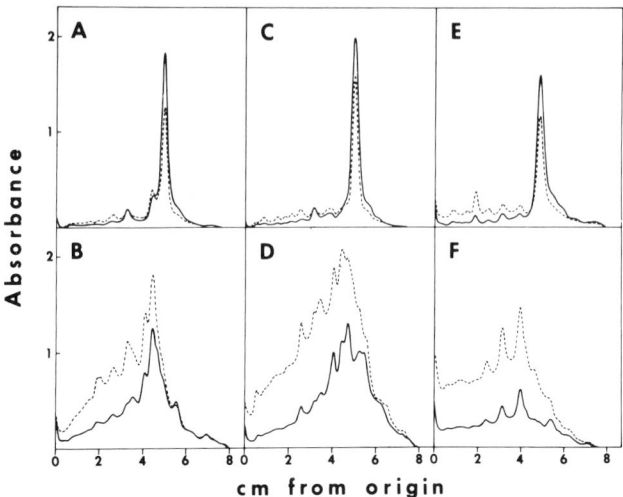

Fig. 4. Densitometric tracings of acid gels similar to those shown in Fig. 3. Solid lines, absorbance at 690 nm; dotted lines, absorbance at 580 nm. A, 50 μg guinea pig brain extract; B, 300 μg guinea pig liver extract; C, 50 μg chicken brain extract; D, 400 μg chicken liver extract; E, 100 μg turtle brain extract; and F, 400 μg turtle liver extract.

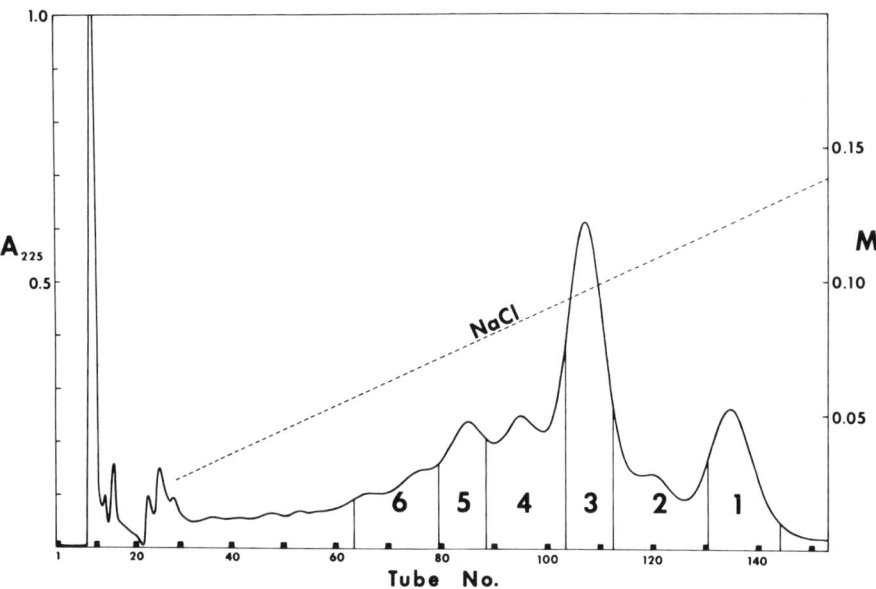

Fig. 5. Chromatographic separation of components of guinea pig myelin basic protein on carboxymethylcellulose. Eluting solvent, 0.01 M glycine–NaOH, pH 10.6, 2 M urea; NaCl gradient shown by broken line. Components were eluted in order of increasing electrophoretic mobility on acrylamide gel at pH 10.6.

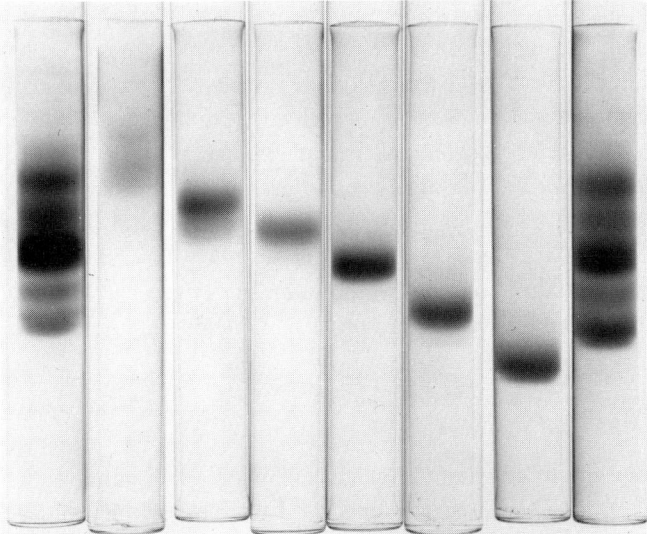

Fig. 6a. Electrophoresis of guinea pig myelin basic protein and its separated components on acrylamide gel at pH 10.6. Gel on left, 100 μg total protein; next 6 gels, 25 μg of each of the components in order of their elution from the column (i.e., 6, 5, 4, 3, 2, 1). Gel on right, reconstituted mixture of components 1–5.

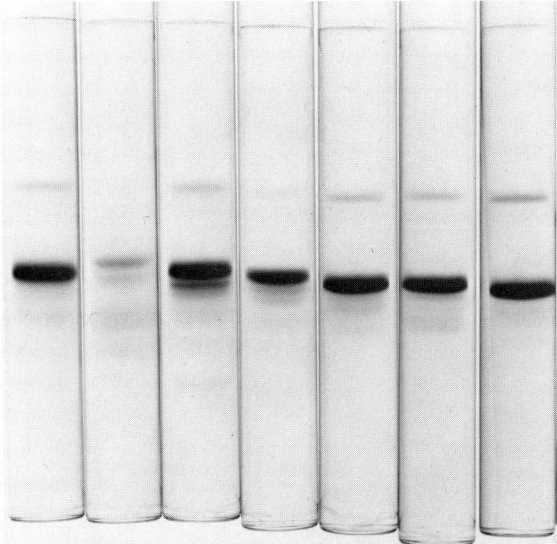

Fig. 6b. Electrophoresis at pH 2.4 of guinea pig myelin basic protein and its components. Gel on right, 25 μg total protein; next 6 gels, 25 μg of each component arranged as in 6a.

Although heterogeneity among proteins, particularly highly charged ones, is fairly common, the pattern of heterogeneity observed in myelin basic proteins is unique and has not yet been adequately explained. Variable methylation of a single arginine residue has been suggested by Baldwin and Carnegie[40] as one explanation for heterogeneity, but this phenomenon alone cannot account for heterogeneity in guinea pig myelin basic protein. With a new technique which permits analytical resolution of the basic amino acids and their methyl derivatives on the total acid hydrolysate, Deibler et al.[41] have shown that all of the components contain monomethylarginine and symmetrical dimethylarginine in an approximately 4:1 ratio. No other methylated amino acids were detected (Table 3).

2. Immunochemical Analyses

In addition to acrylamide gel electrophoresis, immunodiffusion or immunoelectrophoresis has been widely used for analysis of protein purity. The value of immunochemical techniques depends largely on the antigenic characteristics of the protein and its contaminants. A contaminant of low antigenicity accompanying a highly antigenic protein may not be detected at all by immunological methods. On the other hand, there is no doubt that immunochemical

TABLE 2. AMINO ACID ANALYSIS OF COMPONENTS OF GUINEA PIG MYELIN BASIC PROTEIN

Amino acid	Component[a]					
	1	2	3	4	5	6
Tryptophan	0.3	0.5	0.6	0.5	0.3	0.4
Lysine[b]	13.0	13.0	13.0	13.0	13.0	13.0
Histidine	8.5	8.7	9.5	8.6	9.2	8.5
Arginine	19.0	19.6	19.1	16.8	17.6	16.8
Aspartic acid	10.2	10.0	11.2	10.5	11.4	10.4
Threonine	7.3	7.0	7.4	6.8	7.5	6.6
Serine	17.3	12.4	18.0	16.2	17.4	14.8
Glutamic acid	10.4	10.1	10.7	10.4	10.7	10.4
Proline	11.9	11.0	11.5	11.0	11.1	10.7
Glycine[c]	23.0	23.0	23.0	23.0	23.0	23.0
Alanine	12.9	13.0	13.1	11.9	13.3	11.6
Valine	2.0	1.9	2.0	2.1	2.0	2.0
Methionine	2.0	2.0	2.1	2.0	2.1	1.8
Isoleucine	3.3	3.3	3.4	3.3	3.5	3.1
Leucine	7.3	7.4	7.5	7.2	7.5	6.6
Tyrosine	3.8	3.8	4.0	3.7	4.0	3.5
Phenylalanine	9.0	8.9	9.5	9.0	9.4	8.9

[a]Numbered according to relative electrophoretic mobility at pH 10.6: 1, most cathodic; 6, least cathodic
[b]Normalized to 13 moles per mole of protein.
[c]Normalized to 23 moles per mole of protein.

TABLE 3. PERCENTAGE[a] OF ARGININE (A), MONOMETHYLARGININE
(MMA), AND SYMMETRICAL DIMETHYLARGININE (SDMA) AT
RESIDUE "107" OF GUINEA PIG MYELIN BASIC PROTEIN

Component	% A	% MMA	% SDMA	% MMA + SDMA	MMA/DMA
1	57.0	34.4	8.6	43.0	4.0
2	59.7	32.2	8.1	40.3	4.0
3	58.9	33.3	7.8	41.1	4.2
4	31.2	55.8	13.0	68.8	4.3
5	52.0	38.0	10.0	48.0	3.8
6	33.8	51.3	14.7	66.2	3.5
Total BP	53.8	37.7	8.5	46.2	4.5

[a]Calculated on the basis of Lys as 13.0.

techniques are more sensitive than gel electrophoresis for detecting traces of highly antigenic impurities in a protein of low antigenicity, such as myelin basic protein. The relatively low antigenicity of this protein in rabbits has hindered production of good precipitating antisera for immunochemical analyses. This, coupled with the unavoidable loss of some rabbits by death from EAE, probably accounts for the infrequent use of immunochemical techniques for the analysis of purity of myelin basic proteins.

An early encephalitogenic preparation described by Roboz-Einstein *et al.*[42] as bovine far-cathodic protein was used successfully by Rauch and Raffel[43] to induce precipitating antibody in a rabbit by repeated injections (a total of 15 mg) over a period of 6 months. The single arc obtained with this antiserum and the "far-cathodic protein" were cited as proof of purity of the latter. The antibody was shown by immunofluorescent techniques to bind specifically to myelin and gave immunological reactions of identity with an encephalitogenic protein from human CNS tissue as well as with another preparation of bovine encephalitogen which Nakao *et al.*[44] designated fraction G. Their attempts to prepare antisera to fraction G were unsuccessful.[45,46] It should be noted that the "far-cathodic protein" differed from fraction G as well as from all other preparations of myelin basic protein in that it had a distinctly different amino acid composition—in particular, an appreciable amount of half-cystine.

Eylar *et al.*[25] have also cited the production of a single precipitin arc in immunoelectrophoresis as proof of homogeneity of myelin basic protein which they isolated from bovine cord. Their experimental details are insufficient for a critical evaluation of their immunochemical data, but in the light of our experience with the production of antibody to myelin basic proteins in rabbits, we doubt that they exploited the technique adequately to rule out the possibility of species-specific contaminants in their preparations.

The most extensive study to date of the induction of precipitating antibodies to myelin basic proteins by immunization of rabbits has been carried out by Hruby et al.[47] In view of the low antigenicity of myelin basic protein, they used a large amount of antigen in the initial injection (12 mg in Freund's complete adjuvant) followed by 6 mg in incomplete adjuvant twice weekly for 3 weeks. The multiple injections of myelin basic protein in incomplete adjuvant reduced the probability that EAE would develop during the primary immunization period but did not prevent its unpredictable onset later. This immunization schedule is very effective for the detection of minute amounts of highly antigenic impurities contaminating a protein of low antigenicity. Therefore, it was not surprising that they found impurities in almost every preparation tested regardless of the level of impurities detected by acrylamide gel electrophoresis at low pH. The contaminants most frequently found—i.e., the most antigenic—were species-specific proteins which were negatively charged or neutral when electrophoresis was carried out in agarose, pH 8.6.

We have recently induced the formation of antibody to myelin basic protein in rabbits with lower doses of guinea pig myelin basic protein (5–10 mg). In addition to antibody to myelin basic protein, the sera contain antibody to an impurity which is judged to be somewhat less basic than myelin basic protein by agarose electrophoresis in 0.1 M ammonium acetate, pH 7.0. This antibody also forms a precipitin line with a crude acid-soluble protein fraction of guinea pig liver.[48] It is possible to induce antibody specific to myelin basic protein by immunization of rabbits with homologous myelin basic protein, but preparation of good precipitating antisera is more readily achieved with heterologous myelin basic proteins. Regardless of the species of myelin basic protein used for the immunization, the antibody will react with other mammalian myelin basic proteins as well as with certain of the submammalian myelin basic proteins.[8,47,48]

Guinea pigs rarely form precipitating antibodies to myelin basic proteins. For this reason, guinea pig antisera have little value for analysis of antigenic purity, although antibody to basic protein can be detected in guinea pig sera by radioimmunoassay by a modification of the Farr technique.[49] Such an assay developed by Lisak et al.[50] involves incubation of I^{125}-labeled myelin basic protein with "immune" sera. Soluble antigen–antibody complexes are subsequently precipitated along with other globulins by rabbit antibody to guinea pig globulins. Under appropriately controlled conditions, immune sera bind significantly more I^{125} label than normal guinea pig serum. The assay is quantitative if successive dilutions of antiserum are tested. Antibody may also be detected qualitatively by radioautography of immunoelectrophoretic patterns developed with the appropriate antiglobulin serum.

It should be noted that this type of immunoassay fails to discriminate between antibody to impurity and antibody to myelin basic protein if the same heterologous antigen is used for both immunization and immunoassay. Any impurity present in the protein preparation will be iodinated as well as the myelin basic protein. If the contaminant is more antigenic than myelin basic protein, the amount of antibody to impurity in the globulin fraction will be disproportionately higher than the amount of impurity actually present. Thus the effect of an impurity on the "specific" binding of the serum to iodinated myelin basic protein depends not only on the amount of impurity but also on its tyrosine content and its antigenicity.

D. Structural Studies

The amino acid sequences of human and bovine myelin basic proteins as well as the sequences of certain regions of proteins from other species have been reported by three different groups: Carnegie and coworkers, University of Melbourne; Eylar, Hashim, and coworkers, formerly at the Salk Institute; and Kibler, Shapira, and coworkers, Emory University.

Many errors were made in the course of the studies, and although several of these have been recognized, it is possible that a few residues are still incorrectly assigned. A chronological review of the publications on this subject clearly shows that credit for development of the sequence must be shared and cannot honestly be claimed by a single individual or group (Table 4).

The protein sequenced by Carnegie was prepared from human brain by the method of Kies[51]; Eylar's was prepared from bovine cord by a similar series of extractions.[25] The basic technique was essentially the same for both preparations: CNS tissue was treated with chloroform–methanol to remove lipids; myelin basic protein was then obtained in soluble form by dilute acid treatment of the chloroform–methanol treated residue. The encephalitogen sequenced by Kibler et al.[21] was prepared by dilute acid extraction of CNS tissue previously treated with acetone–ether to remove lipids. It was very much smaller than the protein obtained after chloroform–methanol treatment, a difference which was shown to be related to the different solvents used in their preparation.[52] Chloroform–methanol destroys most if not all of the catheptic activity of whole CNS tissue, whereas at least 90% of the original catheptic activity is retained in the residue after removal of lipids with cold acetone–ether.[53]

Kibler and Shapira[54] reported the amino acid composition, tryptic peptide map, Phe as the N-terminal residue, and a molecular weight of approximately 5000 for this 45-residue encephalitogen. Eylar and Hashim[55] isolated a peptide from peptic digests of bovine myelin basic protein which contained the single tryptophan. This peptide was encephalitogenic in guinea pigs. It later developed that the sequence as reported was incorrect. Meanwhile, Carnegie reported the correct sequence of the Trp-containing tryptic peptide in human basic protein,

TABLE 4. DEVELOPMENT OF AMINO ACID SEQUENCE OF
MYELIN BASIC PROTEINS

Authors and dates[a]	Species source	Contribution
1. Kibler and Shapira S, June 1967 P, Jan. 1968	Bovine	Amino acid analysis; tryptic peptide map; N-terminal Phe approximately 4000 mol wt; active in rabbits
2. Eylar and Hashim S, Aug. 1968 P, Oct. 1968	Bovine	Sequence of Trp-containing peptic peptide active in guinea pigs: Ser-Arg-Phe-Gly-Ser-Trp-Gly-Ala-Glu-Gly-Gln-Ser-Pro-Phe-Gly-Lys
3. Eylar and Hashim S, Dec. 1968 P, April 1969	Bovine	Protein cleaved with N-bromosuccinimide lost activity; peptide following Trp cleavage had sequence underlined above
4. Carnegie S, Oct. 1968 P, Jan 1969	Human	Sequence of N-terminal CNBr fragment: Ac-Ala-Ser-Gln-Lys-Arg-Pro-Ser-Gln-Arg-His-Gly-Ser-Lys-Tyr-Leu-Ala-Thr-Ala-Ser-Thr-HSer
5. Hashim and Eylar S, Feb. 1969 P, Mar. 1969	Bovine	Amino acid analysis of N-terminal CNBr fragment duplicated sequence above, including His-Gly, except for additional Ala in 3-position; acknowledged discussion with Carnegie; C-terminal sequence (Lys, Ala,Ile)-Leu-Val-His-Phe-Met-Ala-Arg-Arg-OH incorrect
6. Kibler et al. S, Feb. 1969 P, May 1969	Bovine	Sequence of 5000 mol wt encephalitogen; first 22 of 45 residues correct; last 23 contained ambiguous tryptic peptides (T6-T7-T8 should have been T8-T7-T6); T6 sequence incorrect
7. Carnegie S, June 1969 P, Aug. 1969	Human	Sequence of Trp-containing tryptic peptide: Phe-Ser-Trp-Gly-Ala-Glu-Gly-Gln-Arg-Pro-Gly-Phe-Gly-Tyr-Gly-Gly-Arg; first correct sequence of guinea pig "recognition site," not recognized by author as active
8. Eylar and Hashim P, Sept. 1969	Bovine	Sequence of Trp-containing peptide partially corrected: Lys replaced Arg (see 7); sequence of 45-residue peptide duplicated first 22 residues reported in (6) above; changed order of T6-T7-T8 to T7-T6-T8 but retained error in T6
9. Hashim and Eylar S, June 1969 P, Dec. 1969	Bovine	Correct sequence of N-terminal CNBr fragment (no His-Gly); bovine C-terminal sequence His-Phe-Met-Ala-Arg-Arg-OH incorrect
10. Eylar et al. S, March 1970 P, June 1970	Bovine	Correct sequence of Trp-containing tryptic peptide; reported synthetic peptides to be slightly less active than T27
11. Lennon et al. S, May 1970	Human	Confirmed encephalitogenic activity of Trp-containing peptic peptides; note added in

[a]S, Submitted; P, published; R, revised. Complete references, which follow, are numbered as in the table.

TABLE 4. (*Continued*)

Authors and dates	Species source	Contribution
P, Nov. 1970		proof confirmed activity of Eylar's synthetic peptide: Ser---Trp-----Arg (11 residues)
12. Eylar S, Aug. 1970 P, Nov. 1970	Human and bovine	Sequence of two myelin basic proteins, no experimental details; reported possible modification of Arg 107 (no data); information in (13) was available to author prior to submission of manuscript
13. Carnegie S, June 1970 P, Jan. 1971	Human	Sequence of myelin basic protein almost complete; reported discovery of methylation of Arg-107 as possible source of microheterogeneity
14. Westall *et al.* S, May 1970 R, July 1970 P, Jan. 1971	Synthetic	Based on activity of synthetic peptides concluded ( )-( )-Trp-( )-( )-( )-( )-Gln—Lys (or Arg) required for activity
15. Baldwin and Carnegie S, Oct. 1970 R, Dec. 1970 P, Feb. 1971	Human	Reported enyzmatic methylation of Arg-107 by guinea pig CNS
16. Brostoff and Eylar S, Jan. 1971 P, May 1971	Bovine	Reported discovery of methylated Arg-107; acknowledged discussion with Carnegie but did not cite (13) or (15) above
17. Eylar *et al.* S, Nov. 18, 1970 P, May 1971	Bovine	Sequence of peptic peptide identical to 45-residue polypeptide fragment isolated and sequenced by Kibler (21)
18. Burnett and Eylar S, Nov. 1970 P, May 1971	Bovine	Oxidation of Trp did not destroy activity; cleavage at Trp destroyed most but not all activity (amino-terminal piece slightly active in guinea pigs)
19. Carnegie S, Dec. 1970 P, June 1971	Human	Sequence (some experimental details); corrected some data in (13)
20. Baldwin and Carnegie S, Dec. 1970 P, June 1971	Human	Reported methylation of Arg-107, giving more experimental details than (13) and (15) above
21. Shapira *et al.* S, Nov. 17, 1970 P, July 1971	Bovine, guinea pig, human, rabbit, monkey	Experimental details for sequence of 45-residue polypeptide; paper submitted to *J. Biol. Chem.* 1 day ahead of (17); two papers, almost identical, not reviewed and published simultaneously
22. Shapira *et al.* S, Apr. 1971 P, Aug. 1971	Synthetic	Synthetic peptide containing sequence Thr-Thr-His-Tyr-Gly-Ser-Leu-Pro-Gln-Lys moderately active in rabbits; 45-residue polypeptide active in rabbits and monkeys, mildly active in guinea pigs
23. Eylar *et al.* S, Apr. 1971 P, Sept. 1971	Bovine and human	"Final" structure of two species of myelin basic protein; experimental details still incomplete

1. R. F. Kibler and R. Shapira, Isolation and properties of an encephalitogenic protein from bovine, rabbit, and human central nervous system tissue, *J. Biol. Chem.* **243**:281–286 (1968).
2. E. H. Eylar and G. A. Hashim, Allergic encephalomyelitis: The structure of the encephalitogenic determinant, *Proc. Natl. Acad. Sci.* **61**:644–650 (1968).
3. E. H. Eylar and G. A. Hashim, Allergic encephalomyelitis: Cleavage of the C-tryptophyl bond in the encephalitogenic basic protein from bovine myelin, *Arch. Biochem. Biophys.* **131**:215–222 (1969).
4. P. R. Carnegie, N-terminal sequence of an encephalitogenic protein from human myelin, *Biochem. J.* **111**:240–242 (1969).
5. G. A. Hashim and E. H. Eylar, The structure of the terminal regions of the encephalitogenic A1 protein, *Biochem. Biophys. Res. Commun.* **34**:770–776 (1969).
6. R. F. Kibler, R. Shapira, S. McKneally, J. Jenkins, P. Selden, and F. Chou, Encephalitogenic protein: Structure, *Science* **164**:577–580 (1969).
7. P. R. Carnegie, Digestion of an Arg-Pro bond by trypsin in the encephalitogenic basic protein of human myelin, *Nature* **223**:958–959 (1969).
8. E. H. Eylar and G. A. Hashim, The structure of the encephalitogenic protein of myelin, *Second Internat. Meeting Internat. Soc. Neurochem., Milan.* (R. Paoletti, R. Famuagalli, and C. Galli, eds.) pp. 53–54 (1969).
9. G. A. Hashim and E. H. Eylar, The structure of the terminal regions of the encephalitogenic basic protein from bovine myelin, *Arch. Biochem. Biophys.* **135**:321–333 (1969).
10. E. H. Eylar J. Caccam, J. Jackson, F. C. Westall, and A. B. Robinson, Experimental allergic encephalomyelitis: Synthesis of disease-inducing site of the basic protein, *Science* **168**:1220–1223 (1970).
11. V. A. Lennon, A. V. Wilks, and P. R. Carnegie, Immunologic properties of the main encephalitogenic peptide from the basic protein of human myelin, *J. Immunol.* **105**:1223–1229 (1970).
12. E. H. Eylar, Amino acid sequence of the basic protein of the myelin membrane, *Proc. Natl. Acad. Sci.* **67**:1425–1431 (1970).
13. P. R. Carnegie, Properties, structure and possible neuroreceptor role of the encephalitogenic protein of human brain, *Nature* **229**:25–28 (1971).
14. F. C. Westall, A. B. Robinson, J. Caccam, J. Jackson, and E. H. Eylar, Essential chemical requirements for induction of allergic encephalomyelitis, *Nature* **229**:22–24 (1971).
15. G. S. Baldwin and P. R. Carnegie, Specific enzymatic methylation of an arginine in the experimental allergic encephalomyelitis protein from human myelin, *Science* **171**:579–581 (1971).
16. S. Brostoff and E. H. Eylar, Localization of methylated arginine in the A1 protein from myelin, *Proc. Natl. Acad. Sci.* **68**:765–769 (1971).
17. E. H. Eylar, F. C. Westall, and S. Brostoff, Allergic encephalomyelitis. An encephalitogenic peptide derived from the basic protein of myelin, *J. Biol. Chem.* **246**:3418–3424 (1971).
18. P. R. Burnett and E. H. Eylar, Allergic encephalomyelitis. Oxidation and cleavage of the single ryptophan residue of the A1 protein from bovine and human myelin, *J. Biol. Chem.* **246**:3425–3430 (1971).
19. P. R. Carnegie, Amino acid sequence of the encephalitogenic basic protein from human myelin, *Biochem. J.* **123**:57–67 (1971).
20. G. S. Baldwin and P. R. Carnegie, Isolation and partial characterization of methylated arginines from the encephalitogenic basic protein of myelin, *Biochem. J.* **123**:69–74 (1971).
21. R. Shapira, S. S. McKneally, F. Chou, and R. F. Kibler, Encephalitogenic fragment of myelin basic protein. Amino acid sequence of bovine, rabbit, guinea pig, monkey, and human fragments, *J. Biol. Chem.* **246**:4630–4640 (1971).
22. R. Shapira, F. C.-H. Chou, S. McKneally, E. Urban, and R. F. Kibler, Biological activity and synthesis of an encephalitogenic determinant, *Science* **173**:736–738 (1971).
23. E. H. Eylar, S. Brostoff, G. Hashim, J. Caccam, and P. Burnett, Basic A1 protein of the myelin membrane. The complete amino acid sequence, *J. Biol. Chem.* **246**:5770–5784 (1971).

Phe-Ser-Trp-Gly-Ala-Glu-Gly-Gln-Arg-Pro-Gly-Phe-Gly-Tyr-Gly-Gly-Arg, but found it to be inactive in guinea pigs.[56] Eylar and Hashim[57] partially modified their reported sequence of the bovine Trp-containing peptide to conform to that reported for the human—in place of the Arg in human,[56] Lys was found in the bovine.

In 1969, Kibler *et al.*[58] reported the complete sequence of their 45-residue encephalitogen. The first 22 residues were ordered correctly, but the last 23 were incorrect, primarily because of ambiguities in the order of the tryptic peptides. The order should have been T8, T7, T6, not T6, T7, T8 as reported.

Eylar and Hashim[57] later reported the same sequence for a peptic peptide of the whole protein except that the order of the ambiguous tryptic peptides was different (T7, T6, T8) but still not correct. The correct sequence, as reported by Shapira et al.,[59] is Phe-Gly-Ser-Asp-Arg-Gly-Ala-Pro-Lys-Arg-Gly-Ser-Gly-Lys-Asp-Gly-His-His-Ala-Ala-Arg-Thr-Thr-His-Tyr-Gly-Ser-Leu-Pro-Gln-Lys-Ala-Gln-Gly-His-Arg-Pro-Gln-Asp-Glu-Asn-Pro-Val-Val-His. They found that the tryptic peptides of this fragment could also be isolated from the tryptic digest of the larger myelin basic protein, with Phe-Phe-Gly-Ser-Asp-Arg in place of the N-terminal peptide of the fragment and Ala-Gln-Gly-His-Arg-Pro-Gln-Asp-Glu-Asn-Pro-Val-Val-His-Phe-Phe-Lys in place of the C-terminal peptide. These results support the assumption that the 45-residue fragment is produced by the action of an acid brain proteinase with high affinity for Phe-Phe-bonds.

Meanwhile, Carnegie[60] reported the sequence of the 21-residue N-terminal piece obtained by cyanogen bromide treatment of human myelin basic protein; the corresponding bovine sequence was later reported by Hashim and Eylar.[61] The C-terminal region offered somewhat more difficulty. Three incorrect versions were published[61-63] before the final sequence was reported, without experimental details, in November, 1970.[64] Despite the errors, the race for priority made the information more readily available to other investigators for structural studies of antigenicity and encephalitogenicity. Much more is known about the encephalitogenic than about the antigenic sites. The latter have not been completely characterized but it is apparent that at least three separate regions of myelin basic protein bind specifically to antibody induced in guinea pigs by homologous protein (Driscoll and Kies, unpublished). At least two and possibly three separate regions of the molecule also react with rabbit antibody induced with whole heterologous protein (Hruby and Alvord, unpublished).

E. Species Relationships

Our concept of the role of species-specificity in the induction of EAE was profoundly influenced by the identification of the encephalitogen. The apparent lack of species-specificity had led to the assumption that a lipid rather than a protein was responsible for disease induction. When the encephalitogen was finally identified as a protein[14] it was assumed that that part of the molecule which was responsible for its encephalitogenicity was the same regardless of the species from which the protein was derived. It was further assumed that this site was encephalitogenic in all species. Therefore, it was not surprising to find that the whole protein was not essential for encephalitogenic activity:

1. Limited papain digestion of guinea pig myelin basic protein yielded trichloroacetic acid–soluble fragments which were as active in guinea pigs as the original protein.[65]
2. Either serum or brain proteinases digested bovine and guinea pig myelin basic proteins to fragments smaller than the original proteins without apparent loss of activity.[44,52,66]
3. A polypeptide of about 5000 daltons was as active in rabbits as the whole protein.[58]

The relationships existing among these peptide fragments were not clear, however, until the amino acid sequence of the entire protein was determined. Through the efforts of several investigators, the entire sequence of human and bovine myelin basic proteins has been established as well as the substitutions and deletions in a 45-residue fragment of the bovine protein which occur in the corresponding portions of the human, monkey, rabbit, and guinea pig proteins (see Table 4 and Fig. 7). Partial sequence information is also available on the porcine myelin basic protein.[67]

The fortuitous use by several investigators of different species for bioassays of these proteins and peptides soon provided evidence for multiple encephalitogenic sites which were different in structure, with different degrees of encephalitogenicity for different species. Although there were reports in the literature of striking differences in susceptibility to EAE in various species, systematic analyses had been largely confined to the guinea pig, which developed EAE following sensitization to myelin basic protein from several species with almost equal facility. The observation of Levine and Wenk[10] that the Lewis rat was very susceptible to EAE stimulated widespread use of this strain for bioassay, and it soon became evident that there were significant differences among the mammalian myelin basic proteins which had not been detected by bioassays in guinea pigs. The difference between guinea pig and bovine myelin basic proteins is an example: whereas 5–10 μg of either protein induced maximal disease in the guinea pig and the same amount of guinea pig protein induced maximal disease in the rat, ten times as much bovine myelin basic protein was required to produce even mild EAE in the rat[68] (Table 5). Intermediate degrees of encephalitogenicity were evident in rabbit, human, and monkey myelin basic proteins in the rat, whereas all were essentially equal when tested in the guinea pig (Table 6). Further differences between the rat and the guinea pig were revealed when submammalian myelin basic proteins (chicken, turtle, frog) were compared: these were mildly encephalitogenic at 50 μg in rats but inactive even at 250 μg in guinea pigs[69] (Table 7). Fish myelin basic proteins obtained from dusky shark and carp were inactive in both species. Agrawal et al.[70] found myelin basic protein from spiny dogfish to be inactive in guinea pigs. Taken all together, these differences indicate (1) that there must be multiple encepha-

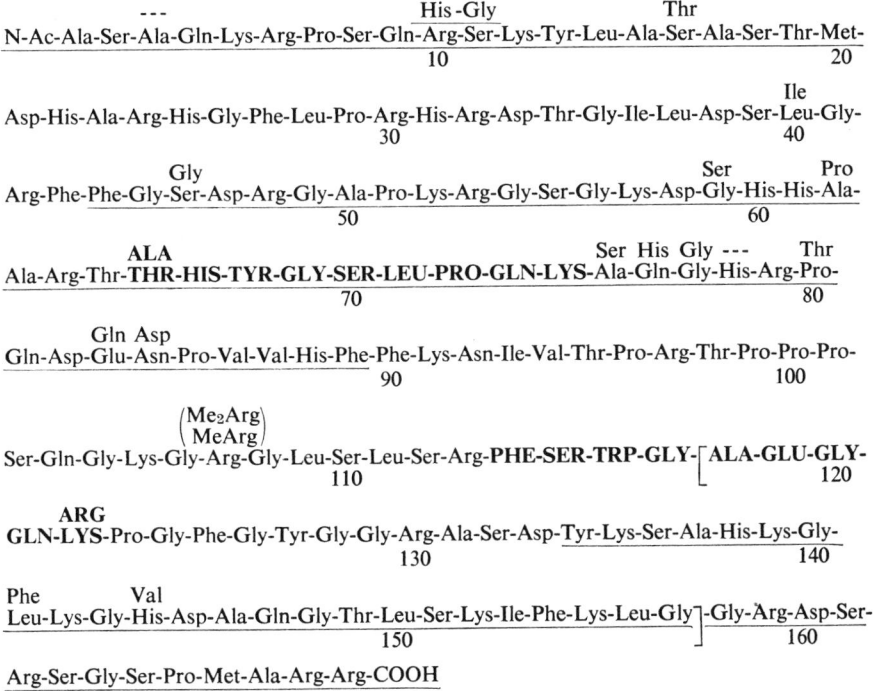

```
                 ---                      His -Gly              Thr
N-Ac-Ala-Ser-Ala-Gln-Lys-Arg-Pro-Ser-Gln-Arg-Ser-Lys-Tyr-Leu-Ala-Ser-Ala-Ser-Thr-Met-
                          10                                                    20

                                                                      Ile
Asp-His-Ala-Arg-His-Gly-Phe-Leu-Pro-Arg-His-Arg-Asp-Thr-Gly-Ile-Leu-Asp-Ser-Leu-Gly-
                          30                                            40

         Gly                                               Ser         Pro
Arg-Phe-Phe-Gly-Ser-Asp-Arg-Gly-Ala-Pro-Lys-Arg-Gly-Ser-Gly-Lys-Asp-Gly-His-His-Ala-
                50                                          60

     ALA                                          Ser His Gly  ---     Thr
Ala-Arg-Thr-THR-HIS-TYR-GLY-SER-LEU-PRO-GLN-LYS-Ala-Gln-Gly-His-Arg-Pro-
                70                                                     80

         Gln Asp
Gln-Asp-Glu-Asn-Pro-Val-Val-His-Phe-Phe-Lys-Asn-Ile-Val-Thr-Pro-Arg-Thr-Pro-Pro-Pro-
                          90                                      100

              (Me₂Arg\
              \ MeArg/
Ser-Gln-Gly-Lys-Gly-Arg-Gly-Leu-Ser-Leu-Ser-Arg-PHE-SER-TRP-GLY-[ALA-GLU-GLY-
                          110                                120

     ARG
GLN-LYS-Pro-Gly-Phe-Gly-Tyr-Gly-Gly-Arg-Ala-Ser-Asp-Tyr-Lys-Ser-Ala-His-Lys-Gly-
                130                                              140

Phe          Val
Leu-Lys-Gly-His-Asp-Ala-Gln-Gly-Thr-Leu-Ser-Lys-Ile-Phe-Lys-Leu-Gly]-Gly-Arg-Asp-Ser-
                          150                                     160

Arg-Ser-Gly-Ser-Pro-Met-Ala-Arg-Arg-COOH
```

Fig. 7. Amino acid sequence of bovine myelin basic protein (footnote 23, Table 4). Substitutions, additions, and deletions in human myelin basic protein are shown above the line (footnote 19, Table 4). Sequences referred to in text are identified as follows: sequence A, residues 114–122 (capital letters); sequence B, residues 66–74 (capital letters); fragment CB-1, residues 1–20 (underlined); 45-residue catheptic fragment sequenced by Shapira *et al.* (footnote 21, Table 4), residues 44–89 (underlined); peptide reported to be encephalitogenic in monkeys, residues 134–170 (underlined)[82]; sequence deleted from rat S, residues 118–157 (or 117–156) (in brackets).[74]

litogenic determinants, at least one for the guinea pig and another for the rat, and (2) that relatively minor modifications of each may account for the quantitative variations in degree of encephalitogenicity in any one species.

Of particular interest is the unique characteristic of rat myelin, which contains two basic proteins.[29,71–73] The isolation and separation of the two with an analysis of how they differ have been carried out by Martenson *et al.*[35,74] For purposes of discussion, the two proteins are referred to as rat L and rat S. Rat L is similar in most respects to the myelin basic proteins of other mammalian species, whereas rat S differs from rat L by an internal deletion of about 40 amino acids between Trp-116 and Gly-157 (Figs. 7 and 8). The Trp-containing encephalitogenic determinant is significantly modified in rat S, and rat S has relatively little encephalitogenic activity in the guinea pig[35] (see

TABLE 5. BIOASSAY OF MYELIN BASIC PROTEINS IN TWO SPECIES

Source of myelin basic protein	Guinea pigs		Lewis rats	
	μg	D.I.[a]	μg	D.I.[a]
Guinea pig	10	7.8	10	8.4
	2	4.8	2	6.1
Rat L	25	8.6	10	(9)
	5	4.8	2	(1–2)
Rat S	25	4.0	10	(9)
	5	0	2	(1–3)
Bovine	10	7.6	50	2.3
	2	6.2	10	0.5

[a]Disease index indicating disease severity based on a scale of 0–10.[19] Figures in parentheses are recalculated from a different disease severity scale of 0–4. [69]

TABLE 6. BIOASSAY OF MYELIN BASIC PROTEINS IN GUINEA PIGS

Myelin basic protein	Amount per animal[a] (μg)	No. clinically positive/No. tested	Disease index[b]
Guinea pig	5	9/10	6.2
	10	17/20	7.8
Human	5	4/5	5.3
	10	6/10	5.1
Bovine	5	4/5	5.4
	10	5/5	7.5
Rabbit	5	5/5	7.3
	10	10/10	7.4
Monkey	10	10/10	7.6

[a]Tested in groups of five animals.
[b]Disease severity, scale 0–10, based on combined histological and clinical ratings. [19]

TABLE 7. SUSCEPTIBILITY[a] OF GUINEA PIGS AND LEWIS RATS TO MAMMALIAN AND SUBMAMMALIAN MYELIN BASIC PROTEINS

Myelin basic protein	Guinea pigs	Lewis rats
Guinea pig	+ +	+ +
Rat L	+ +	+ +
Rat S	−(+ ?)	+ +
Bovine	+ +	+
Human	+ +	+
Monkey	+ +	+
Rabbit	+ +	+
Chicken	−	+
Turtle	−	+
Frog	−	+
Shark, carp	−	−

[a]+ +, active at 5–10 μg; +, active at 50 μg; −, inactive at 250 μg.

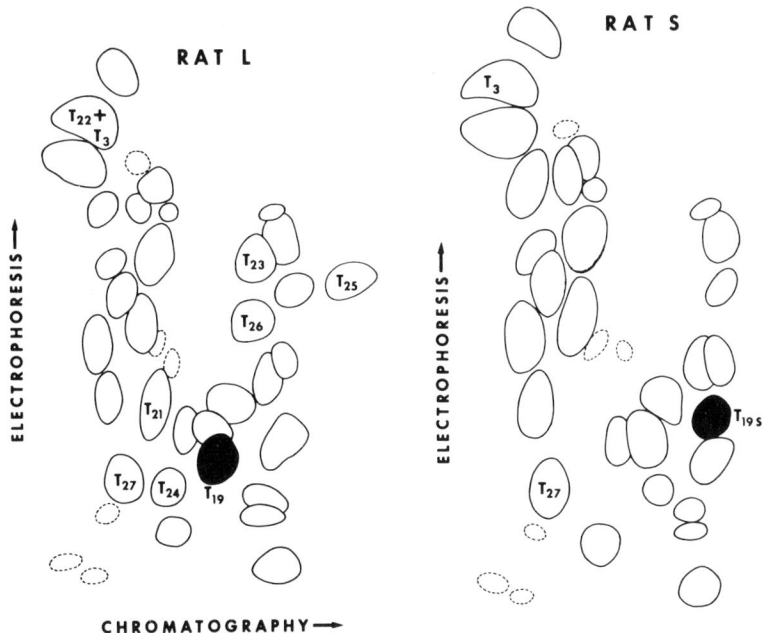

Fig. 8. Peptide maps of tryptic digests of rat L and rat S proteins.[74] Black areas were Ehrlich positive. T21 through T26 were not found in rat S but were present in rat L. Differences between T19 and T19S were confirmed by amino acid analysis. T19, (Phe, Ser, Trp, Gly$_2$, Ala, Gln) Lys; T19S, (Phe, Ser, Trp, Gly$_2$) Arg.; T27 (residues 160–162) was found in both proteins. This confirmed the location of the deletion as shown by brackets in Fig. 7.

Table 5). Since rat S and rat L are equally encephalitogenic in Lewis rats,[69] one must assume that the primary encephalitogenic determinant for the rat differs from that for the guinea pig.

This conclusion was also reached by Swanborg and Amesse,[75] who reported that modification of the tryptophan by 2-hydroxy-5-nitrobenzyl bromide did not change the encephalitogenic activity of myelin basic protein in the rat. McFarlin and Kibler have found that the 45-residue fragment isolated from guinea pig CNS tissue is as active in rats as the undegraded guinea pig myelin basic protein[143]. The 45-residue fragment from bovine CNS tissue, like the whole bovine myelin basic protein, is much less active in rats than the guinea pig fragment. Martenson et al.[144] have found no evidence for a second site outside the 44–89 residue fragment. Fragments 1–20, 21–116, and 117–170 were compared with protein on an equimolar basis. Only the 21–116 piece had significant activity in rats. The relative activity of bovine and guinea pig fragments (21–116) was the same as that of the parent proteins.

What do we know of the species specificity of the encephalitogenic determinants which have been defined? Two determinants of known sequence have been demonstrated in mammalian proteins; a third with mild activity has been reported to occur in the *N*-terminal region of two of these proteins (see Fig. 7). Sequence A has been reported to be the primary determinant in both human and bovine myelin basic proteins for induction of EAE in guinea pigs.[76,77] Although some evidence supports the concept of a single encephalitogenic site active in guinea pigs, the possibility that there are other sites which contribute to the activity of the whole molecule cannot be ignored. The bovine polypeptide which contains sequence B is mildly encephalitogenic in guinea pigs.[78] The first 20 residues (CB-1) isolated from the human protein after cyanogen bromide treatment have also been reported to have mild activity in guinea pigs.[60] The large amino-terminal piece obtained by splitting the molecule at the Trp-Gly bond by treatment with BNPS-skatole also appears to be mildly active in guinea pigs.[79]

The report by two groups[76,77] that sequence A is as active as either of the original proteins on a molar basis has not been confirmed by Alvord *et al.* (unpublished), who found that the encephalitogenic activity of the peptide is approximately equal to that of the protein on a weight basis. Even with this seventeenfold loss of activity, the Trp-peptide is much more active than fragment CB-1, the activity of which could be accounted for by contamination of the order of 1:1000.

Sequence A has also been reported to be active in rabbits,[80] but modification of the tryptophan by 2-nitrophenylsulfenyl chloride does not destroy the activity of the whole protein in rabbits,[81] suggesting that sequence A is not the primary determinant for the rabbit. Sequence B, which lies within the 45-residue polypeptide sequenced by Shapira *et al.*,[59] is also active in rabbits. In fact, this 45-residue polypeptide, which contains no tryptophan, is as active as the whole protein, suggesting that sequence B may be the primary encephalitogenic site for rabbits. Fragment CB-1 of the bovine protein has also been reported to have mild activity in rabbits.[78]

The encephalitogenic determinants recognized by the monkey have not been defined. Modification of the tryptophan in the intact protein by 2-hydroxy-5-nitrobenzyl bromide does not destroy its activity in monkeys.[82] Thus this species, like the rabbit and the rat, does not require the Trp region for encephalitogenicity. Kibler *et al.*[83] have found their 45-residue polypeptide to be encephalitogenic in both rhesus and squirrel monkeys, although Eylar *et al.*[82] reported that they were unable to induce EAE in four monkeys with 1 mg of a polypeptide which they claimed to be the same 45-residue fragment.

Recently, Behan and Martenson (unpublished) have induced EAE in rhesus monkeys with 0.5 or 1 mg of rat S myelin basic protein. The region which Kibler *et al.*[83] reported to be encephalitogenic in monkeys is located in the molecule on the amino-terminal side of Trp-116, all of which is present in the rat S molecule. The second region reported to be encephalitogenic in monkeys is located on the carboxyterminal end (residues 134–170).[82] Almost half of this peptide is missing from rat S which is distinguished from other mammalian myelin basic proteins by a deletion of a segment equivalent to residues 117–156 of the bovine or human proteins.[74] If there are two reactive sites in the bovine protein the ability of rat S to induce EAE in the monkey must be related entirely to the polypeptide isolated by Kibler et al.[83] We have tested fragments 1–116 and 117–170 in monkeys (Alvord, Kies and co-workers). Both fragments were highly active in amounts equivalent to 5 mg basic protein. From these data, one can assume that there are two regions in the molecule which are encephalitogenic in the monkey.

Since it is likely that different sites are encephalitogenic for the guinea pig, rabbit, and rat, and possibly monkey, it is impossible to predict which region of the molecule is encephalitogenic in the human. At the moment, it is also impossible to predict which species should be chosen for the experimental development of safe and effective modification of the proteins to be used for therapeutic trials in patients.

IV. PATHOGENESIS OF EAE

A. Delayed Hypersensitivity

1. Correlation of Disease with Delayed Skin Test to Myelin Basic Protein

Elucidation of the pathogenesis of EAE depended on an understanding of the immunological reactivity of the encephalitogen, and immunological studies were begun even before the protein was identified as a constituent of myelin. The observation by Shaw et al.[84] that EAE could be prevented or suppressed by repeated injections of encephalitogen (without mycobacteria) was made during an early study of the immunological behavior of encephalitogenic fractions.

Although the protein is antigenic (induces antibodies) its encephalitogenicity in guinea pigs is correlated with its ability to induce delayed hypersensitivity rather than with its antigenicity. In 1965, Shaw et al.[85] reported excellent correlation between skin reactivity to myelin basic protein and subsequent development of EAE. The skin reaction was of the delayed type, induced by homologous myelin basic protein in those animals which subsequently developed definite clinical signs of EAE (4–7 days after the development of a positive skin reaction).

Comparable skin reactivity was observed after sensitization with either heterologous or homologous myelin basic protein if homologous basic protein was used to elicit the skin reaction. It should be noted that if heterologous myelin basic protein is used for both sensitization and skin testing, the skin reactivity does not always correlate with disease. This has led several investigators to conclude erroneously that disease is not correlated with delayed hypersensitivity.[77,79,86–90] This was discussed by Shaw et al.,[85] who stated: "Following sensitization with heterologous preparations heterologous skin tests are larger than homologous and about one-fourth of the guinea pigs appear to be sensitized only to nonencephalitogenic components since they develop little or no EAE." In other words, a positive skin reaction unrelated to EAE can be detected if the heterologous myelin basic protein used for the skin test is the same one used for sensitization. Such positive skin reactions result from sensitization either to nonencephalitogenic regions of the molecule or to nonencephalitogenic contaminants in the protein preparation. In any event, such reactions will not interfere with the correlation between delayed hypersensitivity to the encephalitogenic component and disease induction if the tests are properly controlled. Skin tests with homologous encephalitogen are almost always positive unless the timing of the skin test is wrong—in mild disease, 10 days may be too early for the skin test to be positive, and in acute disease 10 days may be too late because of the anergic state which develops as the disease progresses.

Fragmentary data on the induction of delayed hypersensitivity with encephalitogenic peptides have been reported, but a complete analysis has not been published, to the author's knowledge. Lennon et al.[76] reported studies of delayed hypersensitivity in guinea pigs sensitized with encephalitogenic Trp-containing peptic peptides and human myelin basic protein. In a few guinea pigs sensitized with the protein, macrophage inhibition was elicited by the protein and by one of the two peptides. In guinea pigs immunized with peptides, positive skin tests and macrophage inhibition could be elicited with whole protein and with one of the encephalitogenic peptides. Eylar et al.[77] reported that synthetic encephalitogenic peptides failed to elicit delayed skin reactions in guinea pigs sensitized with bovine myelin basic protein. Burnett and Eylar[79] reported positive skin reactions in guinea pigs immunized with bovine myelin basic protein and skin tested with BNPS-skatole modified bovine protein and vice versa. Splitting of the molecule at the tryptophan yields two large peptides, both of which induce delayed hypersensitivity to the parent molecule but not to each other. This reactivity is obviously due to immunogenic groups unrelated to induction of EAE. The possibility that species-specific impurities or chemical modification of the fragments influenced the reaction was not considered.

In other fragmentary reports, Spitler et al.[145] attempted to demonstrate that EAE was independent of delayed hypersensitivity and Hashim and Schilling[146] reported that there are at least three nonencephalitogenic sequences in the basic protein molecule which induce delayed hypersensitivity. Unfortunately, the data reported by the two groups are not in agreement. The question of the structural specificity of cellular hypersensitivity in guinea pigs sensitized with homologous myelin basic protein is still unanswered.

2. Passive Transfer of EAE with Sensitized Cells

Detection of delayed hypersensitivity to the encephalitogen prior to onset of EAE is evidence in support of the hypothesis that the two phenomena are related but fails to prove that the same clone of sensitized cells is involved in both skin reaction and CNS lesion. That mononuclear cells participate in formation of the characteristic lesion in guinea pigs is well documented.[8,91–93] Levine's careful analysis of this phenomenon in rats also supports a determinant role for cellular invasion of the CNS as a primary factor in the disease.[22,94–96]

Further evidence supporting the role of sensitized lymphocytes in the development of EAE is their ability to induce the disease when transferred from sensitized donors to naive recipients. Paterson[97] succeeded in transferring histological but not clinical signs of disease in noninbred rats made tolerant to the donors' cells prior to transfer. Stone[98] was able to induce the complete syndrome in normal strain 13 guinea pigs by adoptive transfer of lymph node cells from sensitized histocompatible donors. He felt that transfers 5 days after donor sensitization were generally more successful than later transfers. However, the number of 5 day transfers far outweighed the number of attempted transfers on days 9, 10, 11, or 12 and possibly accounts for this bias. Our own results[99] do not substantiate his conclusion about optimal time of transfer. Successful transfers have been made repeatedly with cells obtained on the eleventh or twelfth day from sick guinea pigs immunized with 0.5 mg homologous myelin basic protein and 2.5 mg $H_{37}R_V$.

Adoptive transfer has been used as a means of investigating pathogenetic mechanisms in EAE. Falk et al.[100] showed that delayed hypersensitivity to homologous myelin basic protein as well as clinical and histological signs of EAE could be demonstrated in recipients of cells from donors sensitized with homologous white matter and mycobacteria. They also found that transfer of EAE was unsuccessful if either recipients or donors were pretreated with whole CNS tissue.[101] This schedule was similar to one shown by Alvord et al.[102] to be capable of preventing active induction of EAE. Protected donors apparently fail to produce specifically sensitized cells. Failure of protected recipients to mount a disease reaction with cells from a fully sensitized donor suggests either that the environment is hostile to the transferred cells or that auxiliary factors required for lesion formation are lacking in protected recipients. Experiments designed to establish participation of circulating antibody in this phenomenon were unsuccessful.

The studies reported by Falk et al.[100,101] were based mainly on cells obtained from donors sensitized with whole white matter and mycobacteria (the emulsion described by Stone[98]). Donor sensitization with 0.5 mg purified myelin basic protein plus 0.5 mg $H_{37}R_v$ was less consistently successful than donor sensitization with whole white matter.[100] Subsequent studies have shown that the critical component of the emulsion is not the encephalitogen but the absolute amount of mycobacteria.[99] Induction of sensitized cells capable of transferring EAE apparently requires a much more intensive sensitization than is required for active sensitization. Falk et al.[100] used 16 mg (dry weight) white matter plus 2.5 mg $H_{37}R_v$ to sensitize donors. Rauch and Griffin[103] reported that sensitization with 0.2 ml 50% whole tissue plus 5 mg $H_{37}R_a$ yielded cells fully capable of transferring EAE. The amounts of encephalitogen normally used for active induction—10–25 μg myelin basic protein—are inhibited by such large amounts of mycobacteria. Fortunately, mycobacterial inhibition is somehow related to an imbalance between the two antigens. If large amounts of both are incorporated in the emulsion used for donor sensitization, myelin basic protein is just as effective as whole CNS tissue for induction of cells capable of transferring EAE. Recipients of sensitized cells develop the same clinical and histological signs seen in guinea pigs actively sensitized in the same manner as the donors of the cells.

Levine has used the passive transfer technique in rats [104,105] as a means of studying the cytological character of the cells constituting the inflammatory infiltrate. Perivascular lesions characterized by small lymphocytes differed from those containing histiocytes and larger, activated lymphocytes only in duration; the small lymphocytes predominate in the lesions 1 day after transfer, whereas histiocytes and lymphocytes undergoing transformation predominate after a lapse of 3 days.[22,95] These observations are made possible by a special modification of the recipients' CNS (developed by Levine and Hoenig[106]) which insures production of the lesion at a predetermined location within 24 hr of the transfer of sensitized cells. Levine and Sowinski[107] have also reported that depletion of lymphocytes by immunosuppressant drug does not prevent lesion formation but changes the character of the cellular infiltrate.[96]

B. Role of Serum Antibody

1. In the Production of Lesions

Early attempts to define the role of humoral antibody in the induction of EAE yielded inconclusive information primarily because the studies were based on whole tissue sensitization. Nonencephalitogenic CNS fractions were frequently used to detect antibody, and any correlation between appearance of antibody and disease onset or concentration of antibody and disease severity was purely coincidental. After purified encephalitogen was available, the role of specific antibody in the pathogenesis of EAE was investigated by Kibler and Barnes.[108] They used I^{131}-labeled antigen to detect specific antibody by the

Farr technique in sera obtained from rabbits after a single encephalitogenic challenge. Although antibody to encephalitogen was induced in 50% of the rabbits, there was no significant correlation between humoral antibody production and EAE induction.

Lisak et al.[109] also observed little or no correlation between humoral antibody and EAE induction in guinea pigs. Antibody was detected in sera from only 12–13% of all guinea pigs immunized with pure homologous myelin basic protein. Although Lennon et al.[110] found antibody in sera from a much higher percentage of guinea pigs (85%), they reported no correlation between the presence or level of antibody and neurological disease. Their use of heterologous (human) myelin basic protein for both sensitization and radioimmunoassay probably accounts for the high percentage of guinea pigs producing antibody.

Presence of specific antibody prior to or coincident with disease onset does not prove that antibody to myelin basic protein plays a pathogenetic role, nor does apparent lack of specific antibody prove the converse. The fact that the many attempts to transfer disease passively with serum from immunized animals have been unsuccessful constitutes better evidence that antibody per se is not sufficient to induce lesions.[111] Even these experiments leave one with the uneasy feeling that incorrect timing of antiserum collection or transfer of inadequate amounts of specific antibody could account for the negative results. The latter is of particular concern because low levels of circulating antibody could be bound by the target organ and thus escape detection.

A recent report[112] that symptoms of EAE but not lesions were transferred by plasma of lethally irradiated, heavily sensitized donors has not been confirmed in our laboratory. Proponents of the antibody theory suggest that transfer of cells is effective only because of their ability to produce antibody in the recipients. However, Falk et al.[101] obtained no evidence that antibody was produced in significant amounts in recipients by cells from sensitized donors.

At the present time, there is no direct evidence that antibody specific to myelin basic protein is involved in the pathogenesis of EAE. It is difficult to understand how antibody to a nonencephalitogenic constituent of CNS could be involved. Therefore, the mere presence of CNS-specific antibody in sera of animals sensitized with whole CNS tissue cannot be accepted as proof that antibody is involved in pathogenesis of EAE.

2. In Prevention of EAE

Encephalitogen-induced prevention of EAE was first reported by Ferraro and Cazzullo,[113] who injected large amounts of whole CNS tissue in attempts to attract toxic elements of the encephalitogenic emulsion away from the natural target. The studies of Alvord and coworkers on encephalitogen-induced suppression came about as a result of their attempts to demonstrate delayed hypersensitivity in guinea pigs sensitized with encephalitogen plus adjuvant. In experiments designed to relate the size of the skin reaction to the amount of antigen used in the skin test, it was noted that increased amounts of encephalitogenic fractions injected intracutaneously decreased the severity of clinical signs.[84] Their subsequent analysis of this phenomenon in guinea pigs[114-117] has served as the prototype of many later studies on protection.[118-122]

Simultaneously with studies of immunologically induced protection against EAE, methods were developed for assay of specific antibody. Because guinea pigs normally do not produce precipitating antibody, even after hyperimmunization with myelin basic protein, other techniques were required for detection of specific antibody. Hemagglutination techniques were adapted by Hruby *et al.*[47] and passive cutaneous anaphylaxis by Falk *et al.*[123] to study the temporal relationship between antibody production and development of resistance to EAE. Lisak *et al.*[50] utilized a modified Farr technique to detect specific binding of I^{125}-labeled myelin basic protein to antibody in sera of protected animals.

There is no doubt that the production of antibody parallels the development of the EAE-resistant state.[124] An encephalitogenic emulsion which would normally induce EAE and delayed hypersensitivity to myelin basic protein in 100% of normal guinea pigs fails to induce any signs of disease (clinical or histological) in animals which have previously been injected nine or ten times (three times weekly) with 100 μg myelin basic protein in incomplete adjuvant. These guinea pigs also produce circulating antibody to myelin basic protein detectable by any of the three techniques mentioned above.

Despite the excellent correlation between antibody production and prevention of EAE, evidence was obtained which suggested that antibody may be as inconsequential in protection against EAE as it appears to be in the induction of EAE. Production of antibody can be blocked almost completely by methotrexate treatment during preimmunization with basic protein without altering the EAE-resistant state.[125]

This conclusion was based on the following observations:

a. Guinea pigs sensitized with myelin basic protein and Freund's complete adjuvant develop EAE and delayed hypersensitivity to myelin basic protein but produce no detectable antibody.

b. Daily treatment with methotrexate for 3 weeks *prior to sensitization* does not interfere with the development of either EAE or delayed hypersensitivity to myelin basic protein. (It had previously been reported[126] and was confirmed in this series of experiments that daily treatment with methotrexate starting *at the time of sensitization* and continuing for 2–3 weeks inhibits EAE induction.)

c. Injection with *myelin basic protein* in incomplete adjuvant three times weekly for 3 weeks *prior to sensitization* induces high levels of specific antibody and prevents development of EAE and delayed hypersensitivity to myelin basic protein.

d. Combined treatment with *methotrexate and myelin basic protein for 3 week prior to sensitization* diminishes the production of antibody to barely detectable amounts but does not interfere with the development of an EAE-resistant state. Thus antibody production seems not to be an important factor in protection resulting from repeated immunization with myelin basic protein.[125]

V. POSSIBLE RELATIONSHIP OF EAE TO HUMAN DEMYELINATIVE DISORDERS

Although many investigators believe that multiple sclerosis has an immunological basis,[8,127,128] the specific antigens involved and the mechanisms by which they promote the neurological damage are not known. EAE has offered the only experimental situation which can be used to study immunological phenomena associated with CNS tissue damage comparable to that found in multiple sclerosis. In fact, early studies on EAE were motivated almost entirely by the belief that the two diseases were fundamentally the same. Despite the wealth of information which has accumulated as a result of this belief, the understanding of the pathogenesis of multiple sclerosis has improved very little.

It the 26 years which have passed since Freund's adjuvant was found to be an essential part of the encephalitogenic emulsion,[4–7] the encephalitogen has been identified,[24] its encephalitogenic sequences have been defined[58,76–78] and many *in vitro* techniques have been developed[124,129–131] for studying the immunological reactions associated with EAE. Despite this, we still are unable to relate the experimental disease to human illness:

Why is Freund's adjuvant required for disease induction in some species but not in others? What replaces this component of the "encephalitogenic challenge" in patients with postinfectious allergic encephalitis?

What are the factors which predispose individuals to autoimmune CNS tissue damage?

Are exacerbations in human multiple sclerosis related in any way to fluctuations in immunological responsivity of the patient?

What is the reason for the encephalitogenicity of different regions of the myelin basic protein molecule in different species? Which amino acid sequence is encephalitogenic in humans?

These are only a few of the questions which remain unanswered. Hopefully, our attempts to answer them will yield information of help in formulating studies of patients. With respect to multiple sclerosis itself, perhaps we are asking the wrong questions: Is it possible that the sensitization is directed toward another constituent of CNS tissue? Or is the human reaction to myelin basic protein qualitatively different from the reactions studied in such great detail in various experimental animals?

There is little doubt that development of EAE is associated with development of delayed hypersensitivity to myelin basic protein,[85] yet no one has been able to demonstrate conclusively[132-137] that a state of delayed hypersensitivity to this protein is specifically related to multiple sclerosis. Is this because most of the sensitized cells are in the patient's lesions and not available to participate in skin reaction or other *in vitro* tests, or does the failure mean that different antigens are responsible for EAE and multiple sclerosis?

Is it possible that serum antibody plays a more important role in the production of demyelination in humans than in those experimental animals which have been studied most intensively? Bornstein and coworkers have demonstrated a globulin fraction in multiple sclerosis and "EAE" sera capable of inducing demyelination or preventing myelination under certain conditions in tissue culture.[138,139] Their results have been interpreted as indicating a relationship between EAE and multiple sclerosis. Although the globulins were found in sera from experimental animals as well as from patients with neurological disease, this does not prove that the same antigen is involved. Indeed, even in experimental animals the antigen which evokes the demyelinating antibody has not been identified. Globulin fractions which are toxic for neonatal glia cells in tissue culture have also been detected in sera of animals immunized with whole CNS tissue.[140] We believe that the demyelinating serum factor is completely unrelated to EAE, because sera from animals immunized with purified homologous myelin basic protein do not have this factor.[141] Likewise, the fact that similar factors have been demonstrated in sera from patients with motor system disease (amyotrophic lateral sclerosis) indicates lack of specificity of such antibodies to multiple sclerosis.

The detection and characterization of antibodies is complex, and perhaps our techniques are still inadequate. Lisak *et al.*[142] found no antibodies to myelin basic protein in sera of multiple sclerosis patients by a specific and (presumably) sensitive technique.[50] Perhaps we need to explore other methods for detection of CNS-specific antibodies in order to define the specificity of the serum fractions which induce demyelination or prevent myelination *in vitro*. Skepticism regarding the theory of a common mechanism for multiple sclerosis and EAE should not prevent us from searching for more concrete information regarding the specificity of the immune reaction in multiple sclerosis. There is a danger that important information may be overlooked if we limit ourselves to the concept of a common antigen (myelin basic protein) and fail to look for other constituents of CNS tissue as candidates for the dubious honor.

Levine[22] has suggested that "the relationship between multiple sclerosis and EAE will be clarified when the relationship between multiple sclerosis and acute disseminated encephalomyelitis has been established." Hopefully, while the neuropathologists are establishing the morphological (and pathogenetic?) interrelationships among human demyelinating diseases, the immunologists will succeed in elucidating the tangle of cellular interactions which are involved in an individual's response to antigenic stimulation.

ACKNOWLEDGMENTS

The author wishes to acknowledge the help of many present and previous coworkers, particularly of Dr. R. E. Martenson and Mrs. G. E. Deibler for their originality and ingenuity in their investigations of the chemistry and structure of myelin basic proteins; Mr. John Stream, Mrs. Elizabeth Brooks, and Mr. James Hodges for their assistance with bioassay; and Dr. E. C. Alvord, Jr., Dr. S. Levine, and Dr. R. F. Kibler for their advice and collaboration.

REFERENCES

1. R. Koritschoner and F. Schweinburg, Clinical and experimental observations on paralysis after injection of rabies vaccine, *Z. Immunitätsforsch.* **42**:217–283 (1925).
2. T. M. Rivers, D. H. Sprunt, and G. P. Berry, Observations on attempts to produce acute disseminated encephalomyelitis in monkeys, *J. Exptl. Med.* **58**:39–53 (1933).
3. T. M. Rivers and F. F. Schwentker, Encephalomyelitis accompanied by myelin destruction experimentally produced in monkeys, *J. Exptl. Med.* **61**:689–702 (1935).
4. E. A. Kabat, A. Wolf, and A. E. Bezer, Rapid production of acute disseminated encephalomyelitis in rhesus monkeys by injection of brain tissue with adjuvants, *Science* **104**: 362–363 (1946); E. A. Kabat, A. Wolf, and A. E. Bezer, The rapid production of acute disseminated encephalomyelitis in rhesus monkey by injection of heterologous and homologous brain tissue with adjuvants, *J. Exptl. Med.* **85**:117–130 (1947); E. A. Kabat, A. Wolf, and A. E. Bezer, Studies on the acute disseminated encephalomyelitis in rhesus monkeys, III, *J. Exptl. Med.* **88**:417–426 (1948).
5. I. M. Morgan, Allergic encephalomyelitis in monkeys in response to injection of normal monkey cord, *J. Bacteriol.* **51**:614–615 (1946); I. M. Morgan, Allergic encephalomyelitis in monkeys in response to injection of normal monkey nervous tissue, *J. Exptl. Med.* **85**: 131–194 (1947).
6. J. Freund, Some aspects of active immunization, *Ann. Rev. Microbiol.* **1**:291–308 (1947).
7. J. Freund, E. R. Stern, and T. M. Pisani, Isoallergic encephalomyelitis and radiculitis in guinea pigs after one injection of brain and mycobacteria in water-in-oil emulsion, *J. Immunol.* **57**:179–194 (1947).
8. E. C. Alvord, Jr., *in* "Handbook of Clinical Neurology" (P. J. Vinken and G. W. Bruyn, eds.) Vol. 9, pp. 500–571, North Holland Publishing Co., Amsterdam (1970).
9. S. Wright, The Effect of Inbreeding and Cross-Breeding on Guinea Pigs, U.S. Department of Agriculture Bulletin No. 1090 (1922).
10. S. Levine and E. J. Wenk, Induction of experimental allergic encephalomyelitis in rats without the aid of adjuvants, *Ann. N.Y. Acad. Sci.* **122**:209–224 (1965).
11. B. H. Waksman, *in* " 'Allergic' Encephalomyelitis" (M. W. Kies and E. C. Alvord, Jr., eds.) pp. 263–272, Charles C. Thomas, Springfield, Ill. (1959).

12. E. Witebsky, in " 'Allergic' Encephalomyelitis" (M. W. Kies and E. C. Alvord, Jr., eds.) pp. 321–334, Charles C. Thomas, Springfield, Ill. (1959).

13. B. Niedieck, E. Kuwert, O. Palacios, and O. Drees, Immunochemical and serological studies on the lipid hapten of myelin with relationship to experimental allergic encephalomyelitis (EAE), *Ann. N.Y. Acad. Sci.* **122:**266–276 (1965).

14. M. W. Kies and E. C. Alvord, Jr., in " 'Allergic' Encephalomyelitis" (M. W. Kies and E. C. Alvord, Jr., eds.) pp. 293–299, Charles C Thomas, Springfield, Ill. (1959).

15. J. Folch and M. B. Lees, Proteolipides, a new type of tissue lipoproteins, *J. Biol. Chem.* **191:** 807–817 (1951).

16. E. Roboz, N. Henderson, and M. W. Kies, A collagen-like compound isolated from bovine spinal cord—I, *J. Neurochem.* **2:**254–260 (1958).

17. M. W. Kies, E. C. Alvord, Jr., and E. Roboz, The allergic encephalomyelitic activity of a collagen-like compound isolated from bovine spinal cord—II, *J. Neurochem.* **2:**261–264 (1958).

18. M. W. Kies, J. B. Murphy, and E. C. Alvord, Jr., in "Chemical Pathology of the Nervous System" (J. Folch-Pi, ed.) pp. 197–204, Pergamon Press, London (1961).

19. E. C. Alvord, Jr., and M. W. Kies, Clinico-pathologic correlations in experimental allergic encephalomyelitis—II. Development of an index for quantitative assay of encephalitogenic activity of "antigens," *J. Neuropathol. Exptl. Neurol.* **18:**447–457 (1959).

20. B. H. Waksman, H. Porter, M. B. Lees, R. D. Adams, and J. Folch, A study of the chemical nature of components of bovine white matter effective in producing allergic encephalomyelitis in the rabbit, *J. Exptl. Med.* **100:**451–471 (1954).

21. R. F. Kibler, R. H. Fox, and R. Shapira, Isolation of a highly purified encephalitogenic protein from spinal cord, *Nature* **204:**1273–1275 (1964).

22. S. Levine, in "Immunological Disorders of the Nervous System" (L. P. Rowland, ed.) *Proc. Ass. Res. Nerv. Ment. Dis.* **49:**33–46 (1971).

23. M. W. Kies, S. Gordon, R. H. Laatsch, and E. C. Alvord, Jr., in "Fourth International Congress for Neuropathology," Vol. 1: "Histochemistry and Biochemistry," pp. 20–29, Georg Thieme Verlag, Stuttgart (1962).

24. R. H. Laatsch, M. W. Kies, S. Gordon, and E. C. Alvord, Jr., The encephalitogenic activity of myelin isolated by ultracentrifugation, *J. Exptl. Med.* **115:**777–788 (1962).

25. E. H. Eylar, J. Salk, G. C. Beveridge, and L. V. Brown, Experimental allergic encephalomyelitis. An encephalitogenic basic protein from bovine myelin, *Arch. Biochem. Biophys.* **132:**34–48 (1969).

26. M. W. Kies, J. B. Murphy, and E. C. Alvord, Jr., Fractionation of guinea pig brain proteins with encephalitogenic activity, *Fed. Proc.* **19:**207 (1960).

27. G. E. Deibler, R. E. Martenson, and M. W. Kies, Large scale preparation of myelin basic protein from central nervous tissue of several mammalian species, *Prep. Biochem.* **2(2):** 139–165 (1972).

28. M. W. Kies, E. B. Thompson, and E. C. Alvord, Jr., Studies on myelin protein–lipid complexes, *Abst. Vol. Sixth Internat. Congr. Biochem.* **8:**53 (1964).

29. L. F. Eng, F.-C. Chao, B. Gerstl, D. Pratt, and M. G. Tavaststjerna, The maturation of human white matter myelin. Fractionation of the myelin membrane protein, *Biochemistry* **7:**4455–4465 (1968).

30. L. Autilio, Fractionation of myelin proteins, *Fed. Proc.* **25:**764 (1966).

31. R. E. Martenson and F. N. LeBaron, Studies on the acid-extractable proteins of bovine brain white matter, *J. Neurochem.* **13:**1469–1479 (1966).

32. M. W. Kies, E. C. Alvord, Jr., R. E. Martenson, and F. N. LeBaron, Encephalitogenic activity of bovine basic proteins, *Science* **151:**821–822 (1966).

33. L. Ornstein, Disc electrophoresis—I. Background and theory, *Ann. N.Y. Acad. Sci.* **121(2):** 321–349 (1964); B. J. Davis, Disc electrophoresis—II. Method and application to human serum proteins, *Ann. N.Y. Acad. Sci.* **121(2):**404–427 (1964).

34. R. A. Reisfeld, U. J. Lewis, and D. E. Williams, Disk electrophoresis of basic proteins and peptides on polyacrylamide gels, *Nature (London)* **195**:281–283 (1962).

35. R. E. Martenson, G. E. Deibler, and M. W. Kies, Myelin basic proteins of the rat central nervous system. Purification, encephalitogenic properties, and amino acid compositions, *Biochim. Biophys. Acta* **200**:353–362 (1970).

36. E. W. Johns, The electrophoresis of histones in polyacrylamide gel and their quantitative determination, *Biochem. J.* **104**:78–82 (1967).

37. R. E. Martenson, G. E. Deibler, and M. W. Kies, Electrophoretic characterization of basic proteins in acid extracts of central nervous system tissue, *J. Neurochem.* **18**:2417–2426 (1971).

38. R. E. Martenson and M. K. Gaitonde, Electrophoretic analysis of the highly basic proteins of the rat brain fraction which induce experimental allergic encephalomyelitis, *J. Neurochem.* **16**:333–347 (1969); R. E. Martenson and M. K. Gaitonde, Comparative studies of highly basic proteins of ox brain and rat brain. Microheterogeneity of basic encephalitogenic (myelin) protein, *J. Neurochem.* **16**:889–898 (1969).

39. R. E. Martenson, G. E. Deibler, and M. W. Kies, *in* "Immunological Disorders of the Nervous System" (L. P. Rowland, ed.) *Proc. Ass. Res. Nerv. Ment. Dis.* **49**:76–93 (1971).

40. G. S. Baldwin and P. R. Carnegie, Isolation and partial characterization of methylated arginines from the encephalitogenic basic protein of myelin, *Biochem. J.* **123**:69–74 (1971).

41. G. E. Deibler and R. E. Martenson, Determination of methylated basic amino acids with the amino acid analyzer. Application to total acid hydrolyzates of myelin basic proteins, *J. Biol. Chem.* **248**:2387–2391 (1973); G. E. Deibler and R. E. Martenson, Chromatographic fractionation of myelin basic protein. Partial characterization and methylarginine contents of the multiple forms, *J. Biol. Chem.* **248**:2392–2396 (1973).

42. E. R. Einestein, D. M. Robertson, J. M. DiCaprio, and W. S. Moore, The isolation from bovine spinal cord of a homogeneous protein with encephalitogenic activity, *J. Neurochem.* **9**:353–361 (1962).

43. H. C. Rauch and S. Raffel, Immunofluorescent localization of encephalitogenic protein in myelin, *J. Immunol.* **92**:452–455 (1964).

44. A. Nakao, W. J. Davis, and E. R. Einstein, Basic proteins from the acidic extract of bovine spinal cord. I. Isolation and characterization; II. Encephalitogenic, immunologic and structural interrelationships, *Biochim. Biophys. Acta* **130**:163–170; 171–179 (1966).

45. H. C. Rauch and S. Raffel, Antigen uptake by specifically reactive cells in experimental allergic encephalomyelitis, *N.Y. Acad. Sci.* **122**:297–307 (1965).

46. H. C. Rauch and E. Roboz-Einstein, *in* "Pathogenesis and Etiology of Demyelinating Diseases" (K. Burdzy and P. Kallos, eds.) *Add. Internat. Arch. Allergy* **36**:376–386 (1969).

47. S. Hruby, E. C. Alvord, Jr., and C. M. Shaw, Relationships between antibodies and experimental allergic encephalomyelitis. I. Production of hemagglutinating and gel-precipitating antibodies in rabbits and guinea pigs, *Internat. Arch. Allergy* **36**:599–611 (1969).

48. M. W. Kies and E. A. Bump, A rapid qualitative test for detection of precipitating antibody to myelin basic protein, *Res. Commun. in Chem. Pathol. Pharmacol.* **4**:569–579 (1972).

49. R. S. Farr, A quantitative immunochemical measure of the primary interaction between I*BSA and antibody, *J. Infect. Dis.* **103**:239–262 (1958).

50. R. P. Lisak, R. G. Heinze, and M. W. Kies, Relationships between antibodies and experimental allergic encephalomyelitis. III. Coprecipitation and radioautography of ^{125}I-labeled antigen–antibody complexes for detection of antibodies to myelin basic protein, *Internat. Arch. Allergy* **37**:621–629 (1970).

51. M. W. Kies, Chemical studies on an encephalitogenic protein from guinea pig brain, *Ann. N.Y. Acad. Sci.* **122**:161–170 (1965).

52. M. W. Kies, Physico-chemical studies on the encephalitogenic protein from guinea pig brain, *Proc. Fifth Internat. Congr. Neuropathol., Zurich* (Internat. Congr. Series No. 100) p.p 257–258 (1965).

53. M. W. Kies and S. Schwimmer, Observations on proteinase in brain, *J. Biol. Chem.* **145:** 685–691 (1942).

54. R. F. Kibler and R. Shapira, Isolation and properties of an encephalitogenic protein from bovine, rabbit, and human central nervous system tissue, *J. Biol. Chem.* **243:**281– 286 (1968).

55. E. H. Eylar and G. A. Hashim, Allergic encephalomyelitis: The structure of the encephalitogenic determinant, *Proc. Natl. Acad. Sci.* **61:**644–650 (1968).

56. P. R. Carnegie, Digestion of an Arg-Pro bond by trypsin in the encephalitogenic basic protein of human myelin, *Nature* **223:**958–959 (1969).

57. E. H. Eylar and G. A. Hashim, The structure of the encephalitogenic protein of myelin, *Second Internat. Meeting Internat. Soc. Neurochem., Milan* (R. Paoletti, R. Famuagalli, and C. Galli, eds.) pp. 53–54 (1969).

58. R. F. Kibler, R. Shapira, S. McKneally, J. Jenkins, P. Selden, and F. Chou, Encephalitogenic protein: Structure, *Science* **164:**577–580 (1969).

59. R. Shapira, S. S. McKneally, F. Chou, and R. F. Kibler, Encephalitogenic fragment of myelin basic protein. Amino acid sequence of bovine, rabbit, guinea pig, monkey, and human fragments, *J. Biol. Chem.* **246:**4630–4640 (1971).

60. P. R. Carnegie, N-terminal sequence of an encephalitogenic protein from human myelin, *Biochem. J.* **111:**240–242 (1969).

61. G. A. Hashim and E. H. Eylar, The structure of the terminal regions of the encephalitogenic basic protein from bovine myelin, *Arch. Biochem. Biophys.* **135:**324–333 (1969).

62. E. H. Eylar and G. A. Hashim, Allergic encephalomyelitis: Cleavage of the C-tryptophyl bond in the encephalitogenic basic protein from bovine myelin, *Arch. Biochem. Biophys.* **131:**215–222 (1969).

63. G. A. Hashim and E. H. Eylar, The structure of the terminal regions of the encephalitogenic A1 protein, *Biochem. Biophys. Res. Commun.* **34:**770–776 (1969).

64. E. H. Eylar, Amino acid sequence of the basic protein of the myelin membrane, *Proc. Natl. Acad. Sci.* **67:**1425–1431 (1970).

65. M. W. Kies, E. B. Thompson, and E. C. Alvord, Jr., The relationship of myelin proteins to experimental allergic encephalomyelitis, *Ann. N.Y. Acad. Sci.* **122:**148–160 (1965).

66. M. W. Kies and E. C. Alvord, Jr., in "Pathogenesis and Etiology of Demyelinating Diseases" (K. Burdzy and P. Kallos, eds.) *Add. Internat. Arch. Allergy* **36:**182–202 (1969).

67. R. E. Martenson, G. E. Deibler, and M. W. Kies, Comparison of amino-acid sequences of hypothalamic peptide, brain-specific histone and myelin basic protein, *Nature New Biol.* **234:**87–89 (1971).

68. M. W. Kies, R. E. Martenson, and G. E. Deibler, in "Structural and Functional Proteins of the Nervous System" (A. N. Davison, I. G. Morgan, and P. Mandel, eds.) pp. 201– 214 Plenum Press, New York (1972).

69. R. E. Martenson, G. E. Deibler, M. W. Kies, S. Levine, and E. C. Alvord, Jr., Myelin basic proteins of mammalian and submammalian vertebrates: Encephalitogenic activities in guinea pigs and rats, *J. Immunol.* **109:**261–270 (1972).

70. H. C. Agrawall, N. L. Banik, A. H. Bone, M. L. Cuzner, A. N. Davison, and R. F. Mitchel, The chemical composition of dogfish myelin, *Biochem. J.* **124:**70P (1971).

71. C. W. Cotman and H. R. Mahler, Resolution of insoluble proteins in rat brain subcellular fractions, *Arch. Biochem. Biophys.* **120:**384–396 (1967).

72. E. Mehl, Comparison of the protein composition of myelin from different species, *Abst. First Internat. Congr. Neurochem., Strasbourg,* p. 154 (1967).

73. R. E. Martenson, G. E. Deibler, and M. W. Kies, Extraction of rat myelin basic protein free of other basic proteins of whole central nervous system tissue. An analysis of its electrophoretic heterogeneity, *J. Biol. Chem.* **244:**4268–4272 (1969).

74. R. E. Martenson, G. E. Deibler, M. W. Kies, S. S. McKneally, R. Shapira, and R. F. Kibler, Differences between the two myelin basic proteins of the rat central nervous system: A deletion in the smaller protein, *Biochim. Biophys. Acta* **263**:193–203 (1972).

75. R. H. Swanborg and L. S. Amesse, Experimental allergic encephalomyelitis: Species variability of the encephalitogenic determinant, *J. Immunol.* **107**:281–283 (1971).

76. V. A. Lennon, A. V. Wilks, and P. R. Carnegie, Immunologic properties of the main encephalitogenic peptide from the basic protein of human myelin, *J. Immunol.* **105**:1223–1229 (1970).

77. E. H. Eylar, J. Caccam, J. Jackson, F. C. Westall, and A. B. Robinson, Experimental allergic encephalomyelitis: Synthesis of disease-inducing site of the basic protein, *Science* **168**:1220–1223 (1970).

78. R. Shapira, F. C.-H. Chou, S. McKneally, E. Urban, and R. F. Kibler, Biological activity and synthesis of an encephalitogenic determinant, *Science* **173**:736–738 (1971).

79. P. R. Burnett and E. H. Eylar, Allergic encephalomyelitis. Oxidation and cleavage of the single tryptophan residue of the A1 protein from bovine and human myelin, *J. Biol. Chem.* **246**:3425–3430 (1971).

80. E. H. Eylar, F. C. Westall, and S. Brostoff, Allergic encephalomyelitis. An encephalitogenic peptide derived from the basic protein of myelin, *J. Biol. Chem.* **246**:3418–3424 (1971).

81. L.-P. Chao and E. R. Einstein, Localization of the active site through chemical modification of the encephalitogenic protein, *J. Biol. Chem.* **245**:6397–6403 (1970).

82. E. H. Eylar, S. Brostoff, J. Jackson, and H. Carter, Allergic encephalomyelitis in monkeys induced by a peptide from the A1 protein, *Proc. Natl. Acad. Sci.* **69**:617–619 (1972).

83. R. F. Kibler, P. K. Re', S. McKneally, R. Shapira, and M. F. Keeling, Biological activity of an encephalitogenic fragment in the monkey, *J. Biol. Chem.* **247**:969–972 (1972).

84. C.-M. Shaw, W. J. Fahlberg, M. W. Kies, and E. C. Alvord, Jr., Suppression of experimental "allergic" encephalomyelitis in guinea pigs by encephalitogenic proteins extracted from homologous brain, *J. Exptl. Med.* **3**:171–180 (1960).

85. C.-M. Shaw, E. C. Alvord, Jr., J. Kaku, and M. W. Kies, Correlation of experimental allergic encephalomyelitis with delayed-type skin sensitivity to specific homologous encephalitogen, *Ann. N.Y. Acad. Sci.* **122**:318–331 (1965).

86. D. Hughes and E. J. Field, Inhibition of macrophage migration *in vitro* by brain and encephalitogenic factor in allergic encephalomyelitis, *Internat. Arch. Allergy* **33**:45–58 (1968).

87. D. Hughes and S. E. Newman, Lymphocyte sensitivity to encephalitogenic factor in guinea pigs with experimental allergic encephalomyelitis as shown by *in vitro* inhibition of macrophage migration, *Internat. Arch. Allergy* **34**:237–256 (1968).

88. E. A. Caspary and E. J. Field, Encephalitogenic factor in experimental "allergic" encephalomyelitis, *Nature* **197**:1218 (1963).

89. R. H. Swanborg, Immunological response to altered encephalitogenic protein, *Fed. Proc.* **27**:620 (1968).

90. L. E. Spitler, E. H. Eylar, C. von Muller, and H. H. Fudenberg, Experimental allergic encephalitis: Cellular immunity to encephalitogenic protein without disease, *Fed. Proc.* **30**:305 (1971).

91. E. C. Alvord, Jr., and L. D. Stevenson, Experimental production of encephalomyelitis in guinea pigs, *Res. Publ. Ass. Nerv. Ment. Dis.* **28**:99–112 (1950).

92. B. H. Waksman and R. D. Adams, A histologic study of the early lesion in experimental allergic encephalomyelitis in the guinea pig and rabbit, *Am. J. Pathol.* **41**:135–162 (1962).

93. P. W. Lampert and M. W. Kies, Mechanisms of demyelination in allergic encephalomyelitis of guinea pigs. An electron microscopic study, *Exptl. Neurol.* **18**:210–223 (1967).

94. S. Levine and E. J. Wenk, Studies on the mechanism of altered susceptibility to experimental allergic encephalomyelitis, *Am. J. Pathol.* **39**(4):419–441 (1961).

95. S. Levine, Presidential address. Allergic encephalomyelitis: cellular transformation and vascular blockade, *J. Neuropathol. Exptl. Neurol.* **29**:6–20 (1970).

96. S. Levine and E. M. Hoenig, A new form of localized allergic encephalomyelitis featuring polymorphonuclear neutrophilic leukocytes, *Am. J. Pathol.* **64**:13–27 (1971).
97. P. Y. Paterson, *in* " 'Allergic' Encephalomyelitis" (M. W. Kies and E. C. Alvord, Jr., eds.) pp. 444–450, Charles C. Thomas, Springfield, Ill. (1959).
98. S. H. Stone, Transfer of allergic encephalomyelitis by lymph node cells in inbred guinea pigs, *Science* **134**:619–620 (1961).
99. M. W. Kies, A. R. Baig, and B. F. Driscoll, Unpublished.
100. G. A. Falk, M. W. Kies, and E. C. Alvord, Jr., Delayed hypersensitivity to myelin basic protein in the passive transfer of experimental allergic encephalomyelitis, *J. Immunol.* **101**:638–644 (1968).
101. G. A. Falk, M. W. Kies, and E. C. Alvord, Jr., Passive transfer of experimental allergic encephalomyelitis: Mechanisms of suppression, *J. Immunol.* **103**:1248–1253 (1969).
102. E. C. Alvord, Jr., C. -M. Shaw, S. Hruby, and M. W. Kies, Encephalitogen-induced inhibition of experimental allergic encephalomyelitis: Prevention, suppression and therapy, *Ann. N.Y. Acad. Sci.* **122**:333–345 (1965).
103. H. C. Rauch and J. Griffin, *in* "Pathogenesis and Etiology of Demyelinating Diseases" (K. Burdzy and P. Kallos, eds.) *Add. Internat. Arch. Allergy* **36**:387–400 (1969).
104. S. Levine and E. J. Wenk, Rapid passive transfer of allergic encephalomyelitis, *J. Immunol.* **99**:1277–1285 (1967).
105. S. Levine, E. J. Wenk, and E. M. Hoenig, Passive transfer of allergic encephalomyelitis between inbred rat strains: Correlation with transplantation antigens. *Transplantation* **5**:534–541 (1967).
106. S. Levine and E. M. Hoenig, Induced localization of allergic adrenalitis and encephalomyelitis at sites of thermal injury, *J. Immunol.* **100**:1310–1318 (1968).
107. S. Levine and R. Sowinski, Allergic encephalomyelitis: New form featuring polymorphonuclear leukocytes, *Science* **171**:498–499 (1971).
108. R. F. Kibler and A. E. Barnes, Antibody studies in rabbit encephalomyelitis induced by a water-soluble protein fraction of rabbit cord, *J. Exptl. Med.* **116**:807–825 (1962).
109. R. P. Lisak, R. G. Heinze, M. W. Kies, and E. C. Alvord, Jr., Antibodies to encephalitogenic basic protein in experimental allergic encephalomyelitis, *Proc. Soc. Exptl. Biol. Med.* **130**:814–818 (1969).
110. V. A. Lennon, S. Whittingham, P. R. Carnegie, T. A. McPherson, and I. R. MacKay, Detection of antibodies to the basic protein of human myelin by radioimmunoassay and immunofluorescence, *J. Immunol.* **107**:56–62 (1971).
111. M. W. Chase *in* " 'Allergic' Encephalomyelitis" (M. W. Kies and E. C. Alvord, Jr., eds.) pp. 348–374, Charles C. Thomas, Springfield, Ill. (1959).
112. H. Pabst and J.-M. Dupuy, Transfer of allergic encephalomyelitis with plasma factors, *Fed. Proc.* **29**:622 (1970).
113. A. Ferraro and C. L. Cazzullo, Prevention of experimental encephalomyelitis in guinea pigs, *J. Neuropathol. Exptl. Neurol.* **8**:61–68 (1949); A. Ferraro, L. Roizin, and C. L. Cazzullo, Experimental studies in allergic encephalomyelitis, *J. Neuropathol. Exptl. Neurol.* **9**:18–28 (1950).
114. M. W. Kies, C. -M. Shaw, W. J. Fahlberg, and E. C. Alvord, Jr., Factors affecting the suppression of allergic encephalomyelitis by homologous brain protein fractions, *Ann. Allergy* **18**:849–858 (1960).
115. C.-M. Shaw, E. C. Alvord, Jr., W. J. Fahlberg, and M. W. Kies, Specificity of encephalitogen-induced inhibition of experimental "allergic" encephalomyelitis in the guinea pig, *J. Immunol.* **89**:54–61 (1962).
116. E. C. Alvord, Jr., C.-M. Shaw, W. J. Fahlberg, and M. W. Kies, An analysis of various types of inhibition of experimental "allergic" encephalomyelitis in the guinea pig, *Z. Immunol. Forsch.* **126**:217–227 (1964).
117. E. C. Alvord, Jr, Pathogenesis of experimental allergic encephalomyelitis: Introductory remarks, *Ann. N.Y. Acad. Sci.* **122**:245–255 (1965).

118. E. R. Einstein, J. Csejtey, W. J. Davis, and H. C. Rauch, Protective action of the encephalitogen and other basic proteins in experimental allergic encephalomyelitis, *Immunochemistry* **5**:567–575 (1968).

119. G. A. Hashim and F. J. Schilling, Prevention of experimental allergic encephalomyelitis by nonencephalitogenic basic peptides, *Arch. Biochem. Biophys.* **156**:287–297 (1973).

120. R. H. Swanborg, Inhibition of experimental allergic encephalomyelitis (EAE) with modified encephalitogen, *Fed. Proc.* **30**:305 (1971).

121. E. H. Eylar, J. Jackson, B. Rothenberg, and S. W. Brostoff, Suppression of the immune response: Reversal of the disease state with antigen in allergic encephalomyelitis, *Nature* **236**:74–76 (1972).

122. D. Teitelbaum, A. Meshorer, T. Hirshfeld, R. Arnon, and M. Sela, Suppression of experimental allergic encephalomyelitis by a synthetic polypeptide, *Europ. J. Immunol.* **1**: 242–248 (1971).

123. G. A. Falk, R. G. Heinze, M. W. Kies, and E. C. Alvord, Jr., Skin-fixing antibody in experimental allergic encephalomyelitis, *J. Immunol.* **100**:321–328 (1968).

124. E.-C. Alvord, Jr., C.-M. Shaw, R. P. Lisak, G. A. Falk, and M. W. Kies, Relationships between antibodies and experimental allergic encephalomyelitis. V. Antibodies and delayed hypersensitivity in production and prevention of experimental allergic encephalomyelitis, *Internat. Arch. Allergy* **38**:403–412 (1970).

125. R. P. Lisak, G. A. Falk, R. G. Heinze, M. W. Kies, and E. C. Alvord, Jr., Dissociation of antibody production from disease suppression in the inhibition of allergic encephalomyelitis by myelin basic protein, *J. Immunol.* **104**:1435–1446 (1970).

126. M. W. Brandriss, J. W. Smith, and R. M. Friedman, Suppression of experimental allergic encephalomyelitis by antimetabolites, *Ann. N.Y. Acad. Sci.* **122**:356–368 (1965).

127. W. W. Tourtellotte, in "Immunological Disorders of the Nervous System" (L. P. Rowland, ed.), *Res. Publ. Ass. Nerv. Ment. Dis.* **49**:112–147 (1971).

128. C. E. Lumsden, The immunogenesis of the multiple sclerosis plaque, *Brain Res.* **28**: 365–390 (1971).

129. R. E. Rocklin, O. L. Meyers, and J. R. David, An *in vitro* assay for cellular hypersensitivity in man, *J. Immunol.* **104**:95–102 (1970).

130. J. R. David and P. Y. Paterson, *In vitro* demonstration of cellular sensitivity in allergic encephalomyelitis, *J. Exptl. Med.* **122**:1161–1171 (1965).

131. H. C. Rauch, R. W. Ferraresi, S. Raffel, and E. R. Einstein, Inhibition of *in vitro* cell migration in experimental allergic encephalomyelitis, *J. Immunol.* **102**:1431–1436 (1969).

132. R. E. Rocklin, W. A. Sheremata, R. G. Feldman, M. W. Kies, and J. R. David, The Guillain-Barré syndrome and multiple sclerosis, *New Engl. J. Med.* **284**:803–808 (1971).

133. R. F. Kibler, D. W. Paty, and V. Sherr, in "Immunological Disorders of the Nervous System" (L. P. Rowland, ed.) *Res. Publ. Ass. Nerv. Ment. Dis.* **49**:95–105 (1971).

134. D. Hughes, E. A. Caspary, and E. J. Field, Lymphocyte transformation induced by encephalitogenic factor in multiple sclerosis and other neurological diseases, *Lancet* **2**: 1205–1207 (1968).

135. P. C. Dau and R. D. A. Peterson, Transformation of lymphocytes from patients with multiple sclerosis. Use of encephalitogen of human origin, with a report of a trial of immunosuppressive therapy in multiple sclerosis, *Arch. Neurol.* **23**:32–40 (1970).

136. H. Bartfeldt and T. Atoynatan, *In vitro* delayed (cellular) hypersensitivity in multiple sclerosis to central nervous system antigens, *Internat. Arch. Allergy Appl. Immunol.* **39**: 361–367 (1970).

137. P. O. Behan, W. M. H. Behan, R. G. Feldman, and M. W. Kies, Cell-mediated hypersensitivity to neural antigens in humans and non-human primates with neurological diseases, *Arch. Neurol.* **27**:145–152 (1972).

138. M. B. Bornstein and H. Iwanami, Experimental allergic encephalomyelitis: Demyelinating activity of serum and sensitized lymph node cells on cultured nerve tissues, *J. Neuropathol. Exptl. Neurol.* **30**:240–248 (1971).

139. M. B. Bornstein, "A Tissue Culture Approach to Demyelinative Disorders," N. C. I. Monograph, No. 11, p. 197 (1963).
140. O. Berg and B. Kallen, An *in vitro* gliotoxic effect of serum from animals with experimental allergic encephalomyelitis, *Acta Pathol.* **54**:425–433 (1962).
141. F. J. Seil, G. A. Falk, M. W. Kies, and E. C. Alvord, Jr., The *in vitro* demyelinating activity of sera from guinea pigs sensitized with whole CNS and with purified encephalitogen, *Exptl. Neurol.* **22**:545–555 (1968); M. W. Kies, B. F. Driscoll, F. J. Seil, and E. C. Alvord, Jr., Myelination inhibition factor: dissociation from induction of experimental allergic encephalomyelitis, *Science* **179**:689–690 (1973).
142. R. P. Lisak, R. G. Heinze, G. A. Falk, and M. W. Kies, Search for anti-encephalitogen antibody in human demyelinative diseases, *Neurology* **18**:122–128 (1968).
143. D. E. McFarlin, S. E. Blank, R. F. Kibler, S. McKneally, and R. Shapira, Experimental allergic encephalomyelitis in the rat: response to encephalitogenic proteins and peptides, *Science* **179**:478–480 (1973).
144. R. E. Martenson, S. Levine, G. E. Deibler, and A. J. Kramer, Chemically derived fragments of guinea pig and bovine myelin basic proteins. Their encephalitogenic activity in Lewis rats, (submitted for publication).
145. L. E. Spitler, C. M. von Muller, H. H. Fudenberg, and E. H. Eylar, Experimental allergic encephalitis. Dissociation of cellular immunity to brain protein and disease production, *J. Exp. Med.* **136**:156–174 (1972).
146. G. A. Hashim and F. J. Schilling, Allergic encephalomyelitis: characterization of the determinants for delayed type hypersensitivity, *Biochem. Biophys. Res. Commun.* **50**:589–596 (1973); G. A. Hashim, F. Hwang, and F. J. Schilling, Experimental allergic encephalomyelitis: basic protein regions responsible for delayed hypersensitivity, *Arch. Biochem. Biophys.* **156**:298–309 (1973).

Chapter 6

DISORDERS OF FATTY ACIDS

John P. Blass

Departments of Biological Chemistry and Psychiatry and the Mental Retardation Center
University of California, Los Angeles
Los Angeles, California

and

Daniel Steinberg

Division of Metabolic Disease, Department of Medicine
University of California, San Diego
La Jolla, California

I. INTRODUCTION

Fatty acids are a major constituent of brain,[1-3] and furthermore they appear to have crucial roles in determining the properties of membranes. Consequently, any substantive derangement of neural tissues might be expected to have, as one of its effects, a change in fatty acid composition. One of the major problems in evaluating data on fatty acid composition is that of differentiating primary changes in fatty acid metabolism from changes secondary to other abnormalities.

The fact that fatty acids constitute such a major component of neural tissues makes it likely that primary disorders of fatty acid metabolism could lead to neurological disease. However, with the exception of the rare disorder phytanic acid storage disease, no well-defined disease states attributable to primary disorders of fatty acid metabolism have been clearly established. In the following discussion, we briefly review the fatty acid composition and metabolism of normal brain, discuss the available evidence implicating abnormalities of fatty acids in various disorders affecting the nervous system of man and other animals, and then discuss approaches which might be used to establish defects in brain fatty acid metabolism as primary pathogenetic events.

II. FATTY ACIDS IN NORMAL BRAIN AND NERVE

A. Distribution and Composition of Fatty Acids in Brain and Nerve

The fatty acids (Fig. 1) of neural tissues have been studied in great detail, particularly since the introduction of gas–liquid chromatography.[1-64] The fatty acid composition is known for specific brain lipids in a number of animal species[7,9,11,14-16,18-26,28-44,46,48,49,53-56,58-64] including man, [9-13,25,31-42,57-60] in individuals of different ages within the same species,[19,27,37,39,40,42,49,53,57,59] in different anatomical regions of the nervous system,[7,25,32,34,50,51] in different cell types,[46] and in different subcellular fractions.[8,17,23,25,29,30,47,50,52] Several reviews of the extensive literature are available.[1-5] In this section, we confine ourselves to seeking generalizations consistent with the available data.

In normal brain and nerve, fatty acids occur primarily as constituents of phospholipids and glycolipids. Free fatty acids are present only in very low concentrations,[62-64] and storage of fatty acids in triglycerides[3,47-49] or cholesterol esters[60] is also minimal. Indeed, the presence of free fatty acids and cholesterol esters may be a chemical indication of demyelination.[7,60,61,65-75]

The fatty acid composition of specific phospholipids and glycolipids in the brain varies with the membranes of which those lipids are constituents; myelin, in particular, has a unique fatty acid composition. When the fatty acid compositions of various specific lipid classes from different species of animals are compared, the differences from species to species are less striking than the differences from one lipid class to the next.[3,26,32-34,37,44,45,48,54] Fatty acid composition tends to be characteristic for a given lipid class (Fig. 2). The various subcellular fractions contain different types of membranes, and the fatty acid composition of a specific type of lipid may differ in different subcellular fractions.[8,17,47] The tendency of myelin to contain higher proportions of the longer-chain fatty acids is evident from the fatty acid composition of sphingomyelins isolated from various fractions of rat brain (Fig. 3). In fact, myelin has a highly characteristic fatty acid composition.[2-5,8,11,17,25,29-31,34,40,50] Sphingolipids in myelin contain longer-chain fatty acids as major constituents,[1-5,8,11,15,17,18,27,29-31,33-36,38-40,42,50,52,58] particularly 24:0 (lignoceric) and 24:1 (nervonic), and also traces of fatty acids of chain length greater than 30.[78] Cerebrosides and sulfatides, the glycosphingolipids found in greatest amounts

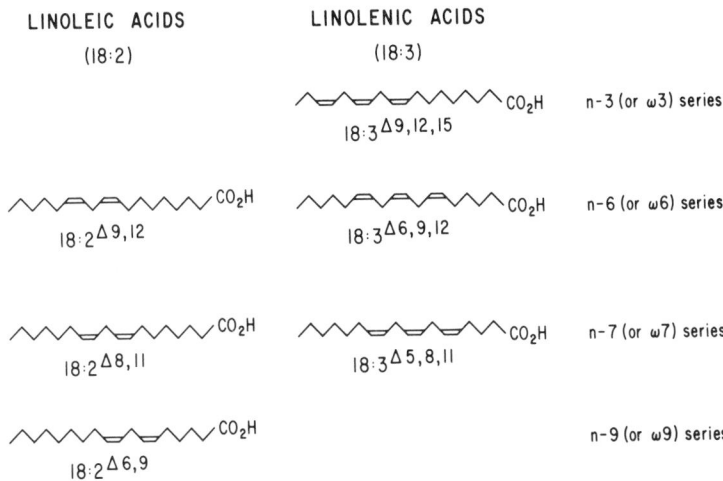

LINOLEIC ACIDS   LINOLENIC ACIDS
(18:2)   (18:3)

$18:3^{\Delta 9,12,15}$ n-3 (or ω3) series

$18:2^{\Delta 9,12}$ $18:3^{\Delta 6,9,12}$ n-6 (or ω6) series

$18:2^{\Delta 8,11}$ $18:3^{\Delta 5,8,11}$ n-7 (or ω7) series

$18:2^{\Delta 6,9}$ n-9 (or ω9) series

Fig. 1. Fatty acid nomenclature: structural relations among polyunsaturated fatty acids. The structures of several polyunsaturated fatty acids of biological interest are shown. Conventionally, the number preceding the colon refers to the number of carbon atoms in the chain and that after the colon to the number of double bonds.[76,77] The figures which follow the superscript Δ refer to the positions of the double bonds, counting the carboxyl carbon as 1. The position of the double bond closest to the terminal methyl group is given, relative to that methyl group, by the value of x in the notation $n - x$ or ωx. A series of fatty acids is known for each of the values of x. Examples using fatty acids of 18 carbon atoms are shown here. [The structures corresponding to 18:2 ($n - 3$) and 18:3 ($n - 9$) are not shown; they are not known to occur naturally.] All fatty acids with 18 carbon atoms and two double bonds (18:2) are called *linoleic acid* and all with three double bonds (18:3) *linolenic acid*. However, the small differences in the positions of the double bonds dramatically change the biological activities. The $n - 9$ series, derived from oleic acid ($18:1^{\Delta 9}$), and the $n - 7$ series, from palmitoleic acid ($16:1^{\Delta 7}$), do not cure the symptoms of essential fatty acid deficiency. The $n - 3$ or linolenic series cures the growth retardation, and the $n - 6$ or linolenic series cures all the symptoms. Since the double bond in the $n - 3$ or $n - 6$ position cannot be introduced by mammalian tissues, such fatty acids are essential components of a complete diet.

in myelin,[2,4,17,27,47,58,79] contain α-hydroxy fatty acids as well as unsubstituted fatty acids.[2,8,9,15,17,18,25,29-31,33,34,36,38-40,42,50,52,58,70] In myelin, longer-chain hydroxy fatty acids predominate, notably 24h:0 (cerebronic) and 24h:1 (hydroxynervonic). Glycerophosphatides from myelin contain lower proportions of polyunsaturated fatty acids and higher proportions of plasmalogens (fatty aldehydes) than do glycerophosphatides from other subcellular fractions, i.e., from other membranes.[17,23,29,30,32,34,41,43,44,47,53,58,59]

Variations in fatty acid composition of specific lipids in different parts of the brain and during development tend to reflect the membranes from which the lipids are derived. The fatty acid composition of lipids from white matter differs from that of lipids from gray matter.[9,10,12,25,31,32,34,51] Fewster and Mead[46]

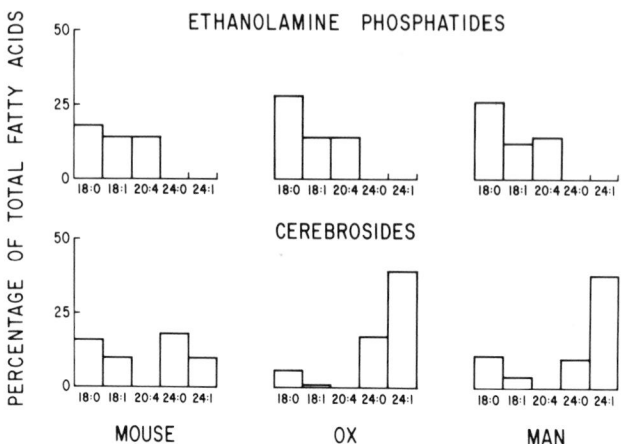

Fig. 2. Fatty acid compositions of selected lipids in brains of several species: two examples of the general similarity of fatty acid pattern in lipid classes from species to species. The bars represent, for each fatty acid shown, the percentage of total fatty acids in a specific lipid class for a particular species. Human ethanolamine glycerophosphatides were from gray matter, and mouse cerebrosides were from myelin. Data were selected from several studies.[8,15,32–34,44]

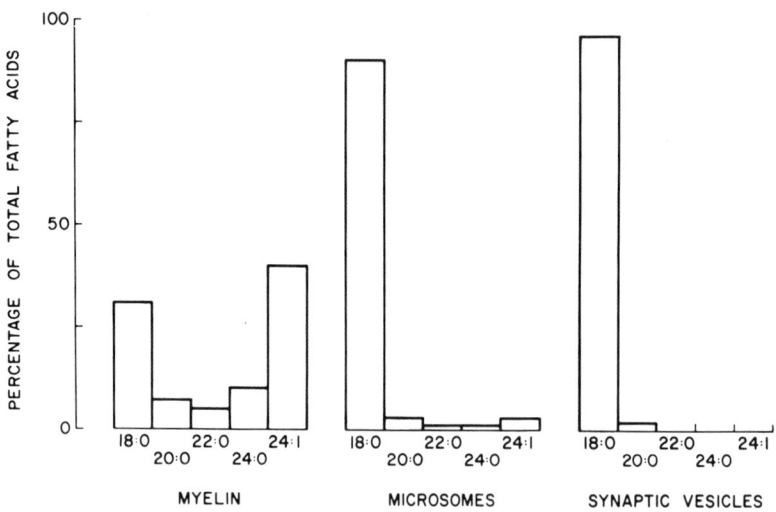

Fig. 3. Fatty acid composition of sphingomyelins from different subcellular fractions. The bars represent, for each fatty acid, the percentage of total fatty acids in sphingomyelins from different subcellular fractions of rat brain. Data are from Kishimoto et al.[17] Microsomes were "gray microsomes." Myelin and microsomes were prepared from whole brain, synaptic vesicles from cortex.

have demonstrated differences in fatty acid composition between lipids from neuronal-rich and glial-rich fractions. Myelin in the central nervous system is a derivative of the oligodendroglial cell membrane,[80,81] and the glial-rich fractions contain a higher proportion of the longer-chain fatty acids than do the neuronal-rich fractions. Recently, Norton and coworkers have extended such studies to include specific types of glial cells and have found relatively high proportions of longer-chain fatty acids in the oligodendroglial-rich fractions.[82] The fatty acid composition of brain lipids changes during development.[3,9,15,16, 19,27,37,38,42,49,52,53,57,59] Concomitant with the increase in myelin, the proportion of longer-chain fatty acids increases,[9,10,19,27,29,30,39,40,51,52] and, at least in human brain, the proportion of polyunsaturated fatty acids in glycerophosphatides falls and the proportion of 18:1 (oleic acid) and plasmalogens increases.[29,32,53,59]

It is not surprising to find that the fatty acid composition of the brain relates to the nature of the complex membrane systems of which these fatty acids are constituents. Numerous studies have indicated that fatty acids play an important role in determining the functional properties of the membranes into which they are incorporated. O'Brien[29,30] has reviewed these studies with particular reference to neural tissues. He has pointed out that short-range London–van der Waals interactions between fatty acid side chains can provide more energy of stabilization than do interactions between polar groups—approximately 18 vs. approximately 10 kcal per mole.[30] The very long fatty acid residues in myelin may allow interdigitation between adjacent lipid layers in membranes formed of biomolecular leaflets,[29,30,83–89] i.e., membranes of the type suggested by Danielli and Davson for myelin.[90] On the other hand, the folding of the carbon backbone in polyunsaturated fatty acids could interfere with the formation of membranes of that type.[29,30,84,91–93] Changes in neural membranes in disease could be expected either to cause or to result from changes in the fatty acid composition of the brain.

B. Origin of Brain Fatty Acids

Brain fatty acids appear to originate both by uptake of intact fatty acids transported via the blood from extraneural tissues and by *de novo* synthesis. Several lines of evidence indicate that fatty acids pass the blood–brain barrier and are incorporated intact into brain lipids. First, the simple presence of essential fatty acids in brain lipids proves the passage of some dietary fatty acids into the brain, since mammalian tissues are incapable of synthesizing such acids *de novo* or by desaturation.[94–100] Second, dietary alterations can lead to changes in the composition of polyunsaturated acids in both developing and mature brain.[1,3,101–106] For example, enriching the diet of young rats with corn oil leads to an increased proportion of $n-6$ polyunsaturated acids[101]; in adult rats, the $n-6$ fatty acid content of myelin is, in the long term, roughly proportional to the linoleate content of the diet.[102] Finally, several groups of workers have directly demonstrated uptake of radioactive fatty acids into neural tissues.[1,3,107–112] Dhopeschwarkar and Mead[107] found 0.02–0.03 % of an oral or intravenous dose of oleic acid-1-C^{14} to be incorporated into brain lipids; retention of up to 84 % of the radioactivity in the 1-position indicated incorporation of the intact acid. Evidence has been presented that incorporation is greater in younger animals.[108] Detailed studies of the cell types and subcellular fractions involved and of turnover rates of different lipid classes under specific physiological conditions would be of interest.

Synthesis of fatty acids by brain from lower molecular weight precursors is also well documented. With ^3H_2O as precursor, which presumably mixes readily with total body water, incorporation appears maximal during the stage of rapid myelination.[113] Incorporation of radioactivity from other precursors occurs also.[114–121] These data are somewhat more difficult to interpret because of the variety of metabolic pools in the brain for acetate[122–124] and for other substrates.[117,120,123,125–127] Brady and coworkers have described a cell-free preparation from brain that appears to synthesize fatty acids *de novo* not only from malonyl coenzyme A but also from crotonyl CoA or octanoyl CoA and from the coenzyme A derivatives of ketone bodies.[128,129] Aeberhard and Menkes[130] identified chain elongation mechanisms in cell-free preparations of brain. Bourre *et al.*[131] have presented kinetic evidence with cell-free systems from brain suggesting that there may be more than one system for chain elongation. Studies with carbon-labeled precursors are consistent with that interpretation but cannot yet be considered definitive because of the complexity of the metabolic pools involved.[114–127] Available evidence on the mechanisms of fatty acid desaturation in brain is consistent with the operation of mechanisms similar to those described in detail in other tissues[132]; studies with radioactive acetate in chick brain suggest that the systems for elongation of the $n-3$ and of the $n-6$ series of polyunsaturated fatty acids may not be identical.[121] The α-hydroxy acids in brain appear to arise by the α-oxidation pathway discussed below; odd-numbered fatty acids, which occur in increasing proportions in cerebrosides with age,[1–3,9,19,27,39,40,42] also appear to arise in part as products of α-oxidation.[118,119,133–136]

Oxidation of fatty acids by brain has been considered a quantitatively minor pathway,[137–140] glucose being the only significant caloric substrate. Two well-established observations supported this conclusion: first, perfusion studies showed the respiratory quotient (RQ) of the brain to be 1.0,[139,140] and, second, interruption of the supply of blood glucose to the brain led rapidly to convulsions and coma.[141] Owen *et al.*[142] recently reexamined brain metabolism in obese patients undergoing prolonged therapeutic starvation. By comparing the concentrations of metabolites in arterial and cerebral venous blood, they found significant utilization of ketone bodies but not of fatty acids. The RQ values were surprisingly low—0.52 to 0.72. The RQ for oxidation of acetoacetate is 1.0, for β-hydroxybutyrate 0.9, and for fatty acids about 0.7. Owen and coworkers suggested that CO_2 fixation in the brains of their patients might explain the anomalously low RQ values. Another explanation could be shifts in bicarbonate between blood and brain if the patients were not in a true steady state. Hawkins *et al.*[143] found that raising blood levels of ketones led to effective oxidation of these substrates even without starvation.[144] Thus starvation increases the oxidation of ketones, at least in part, by raising blood levels of these substances. Ketones may be particularly important as substrates in young animals[143-150] and in particular cell types in the nervous system.[150] Brain slices and mitochondria can oxidize ketone bodies effectively.[145,146] Rates of oxidation of fatty acids by brain slices or mitochondria from most species are low, although Beattie and Basford[152] reported that beef brain mitochondria oxidized fatty acids at one-third the rate of liver mitochondria.[1,146,151,154] Neural tissues appear to oxidize fatty acids by the mechanisms of β-oxidation described in detail in other tissues.[144,151-155] The physiological role of ketone or fatty acid metabolism in normal brain remains problematic.[144] As mentioned later in this chapter, two patients who failed to develop normal ketosis during fasting nevertheless developed no symptoms referable to the nervous system. The neurological effects of hypoglycemia are well established and make it difficult to accept a major quantitative role for fatty acids or ketone bodies as substrates for cerebral energy production in man under usual metabolic conditions.

Brain appears to be the only tissue containing significant activities of an α-oxidation system that converts particularly the long-chain fatty acids to α-hydroxy derivatives and then, by oxidative decarboxylation, shortens them by one carbon atom.[2,3,18,36,133,156-164] *In vivo* studies provide evidence for direct hydroxylation of lignocerate (24:0) to cerebronate (24h:0),[118,119,133,134,156,157] and the composition of the long-chain unsaturated hydroxy acids is again compatible with their origin by chain elongation (from palmitoleate and oleate) followed by hydroxylation.[18,31,33,36,39,42,50,52,160] The hydroxylation step presumed to initiate 1-carbon degradation has not been satisfactorily characterized in cell-free systems of brain,[2] but the further metabolism of α-hydroxy acids has been well studied. Levis and Mead[159] found that formation of the shortened acids from 18h:0 and from 23h:0 was enhanced by NAD and ATP but not CoA. The keto acid is believed to be an intermediate, based on the results of trapping experiments and the fact that the keto acid is decarboxylated by the same microsomal system even more rapidly than the hydroxy acid.[159,

[161-163] Decarboxylation of the keto acid is stimulated by ascorbate and by fumarate, but the mechanism remains uncertain. The a-oxidation system may be particularly important for degradation of longer-chain acids, since brain may lack the capacity to activate such long-chain acids to the coenzyme A derivatives.[113] The ω-oxidation, P450 system described in liver has not, to our knowledge, been described in brain.[165-168]

Fatty acids do turn over in brain. In some fractions, they turn over more rapidly and in others, such as myelin, more slowly.[1-5,113,169-173] Drugs,[174-176] hormones,[117,177] and dietary alterations[95-106,178,179] can influence turnover in complex ways. Interpretation of the effects of these agents is difficult, partly because results differ depending on such specific experimental parameters as the concentration of drug[174,175] or the nature of the radioactive precursor used.[117] Such variations fit with a role of fatty acids in neural tissues primarily as constituents of their extraordinarily specialized systems of membranes. Regional variations within cells, between cells, and among brain areas might account for complex and variable results.

Brain has the capacity for synthesis and degradation of fatty acids, including some specialized pathways. However, fatty acids have not been demonstrated to be a major substrate for energy metabolism in brain under ordinary conditions. This interpretation of the role of fatty acids in normal brain will be referred to in the following discussion of disorders affecting fatty acid metabolism.

III. PHYTANIC ACID STORAGE DISEASE (REFSUM'S DISEASE)

A. Clinical and Biochemical Abnormalities

During the early 1940s, Sigvald Refsum[180-182] recognized, as a distinct entity in the general category of heredofamilial ataxias, the syndrome he named *heredopathia atactica polyneuritiformis*. The cardinal and almost invariable signs are four: pigmentary degeneration of the retina; peripheral neuropathy, usually hypertrophic; cerebellar ataxia; and elevation of protein in the spinal fluid, without inflammatory cells.[181-183] Other common features, though less constant, include nerve deafness, ichthyosis, cataracts, nonspecific ECG changes, and minor skeletal abnormalities.[181-198] Patterns of inheritance suggest an autosomal recessive disease.[186-188] Onset is generally in the first two decades. The course is variable, usually slowly progressive, with occasional severe exacerbations often related to incidental infections or other stresses. The clinical findings in the approximately 40 patients described to date have been extensively reviewed.[181-198,360]

Postmortem studies indicated that the disorder was a lipidosis, with accumulation of fatty material in liver, spleen, and adventitial neural tissues.[199–202] In 1963, Klenk and Kahlke identified the accumulated material.[203–205] Using gas–liquid chromatography and mass spectroscopy, they showed that an unusual branched-chain acid, phytanic acid (3,7,11,15-tetramethylhexadecanoic acid) (Fig. 4), accounted for almost half of the total fatty acids in the liver. The presence of large quantities of this branched-chain acid, which is normally found in no more than trace quantities in human tissues,[206,207] has been almost universal in patients with the clinical picture of the disease.[25,182–195,197, 198,203–216] Phytanic acid accumulates in many tissues, including neural tissues[25,208,209] and specifically myelin.[25] However, two patients have been described with the clinical syndrome but without phytanic acid storage in their tissues.[198,200,361]

The biochemical characteristics of phytanic acid storage disease have been intensively studied, and the specific enzyme defect has been elucidated. The major normal pathway for synthesis and breakdown of phytanic acid in mammals, including man, is that outlined in Fig. 4.[217–224] Experiments with radioactive precursors demonstrated that phytanic acid is not formed at any significant rate *de novo* in mammalian tissues from low molecular weight precursors.[225–227] Long-term experiments with D_2O were conclusive in this re-

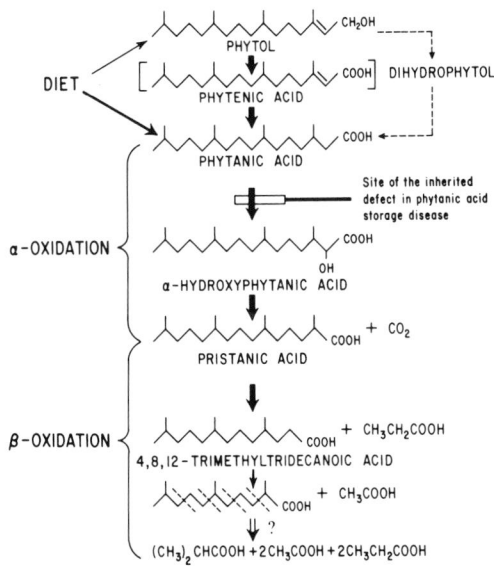

Fig. 4. Metabolic fate of phytol and phytanic acid. The studies establishing this pathway are discussed in detail elsewhere.[217–224] Available data suggest that phytenic acid is the main intermediate between phytol and phytanic acid.[216]

gard.[225,226] Phytanic acid in mammalian tissues apparently arises primarily from dietary phytanic acid and to a lesser extent from oxidation of dietary phytol.[225–236] Phytol is a constituent of chlorophyll and other plant products[235–236]; phytanic acid is found in significant concentrations in butterfat and in ruminant tissues.[183,228,229,237–246] The possibility that the diet includes still other precursors has not been excluded.

Oxidation of phytanic acid cannot proceed by the usual β-oxidation pathway because of the methyl group on the β-carbon. The major pathway appears to be via an initial hydroxylation at the α-position.[183,192,198,217–227] In the liver, this reaction is carried out primarily in heavy mitochondria.[220,221] It appears to be different in a number of ways from the brain α-oxidation system for long-chain acids discussed above. The phytanic acid hydroxylase in liver is mitochondrial, not microsomal; it is stimulated rather than inhibited by Fe^{3+} and inhibited rather than stimulated by Fe^{2+}, and, as discussed in more detail below, it appears to be coded on genetic loci distinct from the system for α-oxidation of straight-chain acids. The α-hydroxyphytanic acid is oxidatively decarboxylated to the next lower homologue, pristanic acid. The latter then undergoes successive β-oxidations, presumably yielding propionyl coenzyme A and acetyl coenzyme A alternately. However, direct evidence for the nature of the 2- and 3-carbon fragments has been limited. Recently, Hutton and Steinberg demonstrated the formation of propionate-C^{14} from phytanate-U-C^{14} using fibroblast cell lines from patients having a deficiency of propionic acid carboxylase.[362] Neither CO_2 fixation on the 3-methyl substituent followed by β-oxidation[247–249] nor ω-oxidation[250–253] appears to represent quantitatively significant pathways for the oxidation of phytanic acid.

The accumulation of phytanic acid in the tissues of patients appears to result from a specific inherited deficiency in the degradation of this compound—specifically in the α-hydroxylation system.[183,184,192,198,225,226,254–262] Both *in vivo* studies of patients[260] and *in vitro* studies of their cultured skin fibroblasts[254,256,261] showed severely deficient oxidation of radioactive phytanic acid but relatively normal metabolism of α-hydroxyphytanic acid, pristanic acid, and palmitic acid. The cultured cells were grown in media containing negligible amounts of phytol and phytanic acid; under these conditions, the cells from neither patients nor controls contained detectable phytanic acid. Thus dilution by endogenous nonradioactive phytanic acid or secondary metabolic effects of phytanic acid accumulation could not account for the low oxidation by the patients' cultured cells. Fibroblasts from 11 patients from eight families[261] oxidized phytanic acid at less than 3 % of the rate in control cells (Fig. 5). The demonstration of intermediate values for rates of oxidation in cells from the parents strongly supports the autosomal recessive pattern of inheritance inferred from genetic analyses. It also supports the assumption that the defect

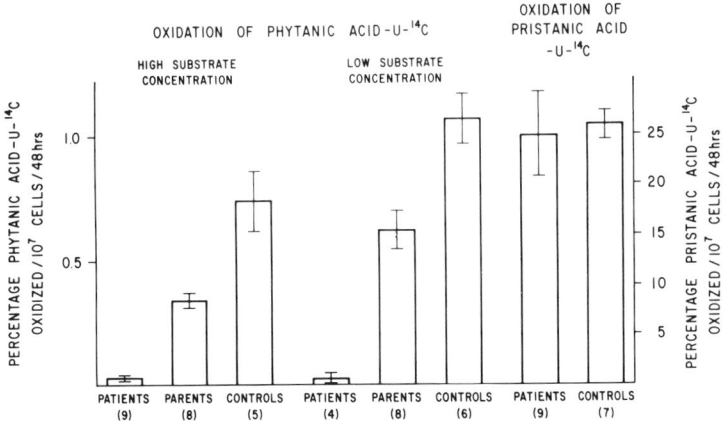

Fig. 5. Differentiation of homozygotes and heterozygotes in phytanic acid storage disease. Rates of oxidation of phytanic acid-U-C¹⁴ and pristanic acid-U-C¹⁴ by fibroblast cultures derived from skin biopsies of patients with phytanic acid storage disease, their parents, and controls are shown. Cultured skin fibroblasts from patients, their parents, and controls were incubated with phytanic acid-U-C¹⁴ or pristanic acid-U-C¹⁴, and the evolved C¹⁴O₂ was measured as described in detail by Herndon et al.[256,261]

in α-hydroxylation is a primary inherited defect. Whether the same mutation occurs in all patients with phytanic acid storage disease or whether different mutations affecting the same process occur remains unknown. Complementation studies and experiments with mixtures of cells have not been done.[263,264]

B. Pathophysiology of Phytanic Acid Storage Disease

Although the etiological aspects of this disorder have been largely elucidated, the pathogenesis remains unclear. The biochemical defect and the signs and symptoms of the disorder could be related in at least three ways: (1) The α-hydroxylation system for phytanic acid might be involved not only in phytanic acid oxidation but also in some key metabolic process in brain, such as the α-hydroxylation of very long chain fatty acids. In this case, the accumulation of phytanate might be coincidental and not of pathogenetic significance. (2) The enzyme might occur only in extraneural tissues, with phytanic acid as its primary substrate, and the phytanic acid accumulation might lead directly, or at least indirectly, to the signs and symptoms. (3) The defect might be linked genetically to some additional defect that is related by a direct pathophysiological mechanism to the clinical abnormalities.

Available evidence does not support the first possibility. A variety of studies both *in vitro* and *in vivo*, failed to demonstrate the presence of the phytanic acid oxidizing system in neural tissue.[248,265] For instance, rats during the stage of most rapid myelination were given injections of a mixture of lignoceric acid-H^3 (24:0) and phytanic acid-C^{14}.[265] Gas–liquid radiochromatography of the isolated fatty acids revealed the presence of H^3 in a whole series of straight-chain α-hydroxy fatty acids, confirming earlier studies of Hajra and Radin.[133, 134] Radioactive carbon was found in phytanic acid, but no radioactivity chromatographed with α-hydroxyphytanic acid. Small but significant amounts of C^{14} occurred in straight-chain acids, presumably representing reincorporation of small molecular fragments derived from phytanate degradation in the periphery. Furthermore, Stokke,[248] using 3,6-dimethyloctanoate as an analogue of phytanate, found no oxidation by brain slices or peripheral nerve from rats, guinea pigs, cats, or polecats. Analyses of postmortem specimens of neural tissues from patients with phytanic acid storage disease revealed no dificiencies in α-hydroxy fatty acids.[2,25,208] Nor were significant deficiencies in hydroxy acids demonstrable in biopsy specimens of skin, a tissue in which lipids may turn over more rapidly than in myelin.[265] While it remains impossible to prove the negative conclusion and while the relevant metabolic studies have not been done with human brain, available evidence suggests that normal brain does not have a significant capacity to form α-hydroxyphytanic acid. Thus defective activity of the enzyme in the brains of patients cannot be invoked to explain their neurological symptoms.

Another possibility is that the accumulation of phytanic acid in neural tissue (secondary to deficiency of the α-oxidation system in extraneural tissues) somehow leads to the clinical abnormalities. This would be analogous to the pattern in phenylketonuria, where deficiency of a hepatic enzyme leads to accumulation of phenylalanine and its metabolites, which in turn leads somehow to neurological disease. This hypothesis is strongly supported by evidence that patients on low-phytol, low–phytanic acid diets show clinical improvement accompanying reduction in their serum phytanic acid concentrations.[183,198, 226,228,229,252,262] The relationship between the clinical improvement and the reduction in serum phytanic acid levels is difficult to evaluate in a disorder characterized by spontaneous exacerbations and remissions. However, objective criteria (changes in nerve conduction velocity, changes in ECG, changes in quantitatively evaluated muscle strength) showed significant improvement,[228] and on return to a less restricted diet there was relapse in two carefully studied cases. On reinstatement of a rigidly controlled diet, there was a second remission which has been maintained in these two patients. A total of five patients have been followed on diet now for 3–5 years. Except for the relapse off diet just mentioned, all have shown some improvement in their peripheral neuropathy. Hearing and vision have not improved significantly.

If phytanic acid accumulation *per se* leads to neurological deterioration, it might be possible to develop an animal model. In animals fed a diet enriched in phytol or phytanic acid, significant concentrations of phytanic acid do indeed accumulate in many tissues, including brain.[198,223,224,227,230,232,233,248,266] However, these animals rapidly sicken and lose weight even relative to pair-fed controls if the high intake of phytol and phytanic acid is maintained.[230,233] Mice fed a 5% phytol diet die within 2 or 3 weeks. Despite extensive experimentation, feeding of a high-phytol diet by itself has not been associated with neurological abnormalities or pathological changes in the nervous systems of experimental animals. Whether nerve damage might develop if they lived long enough cannot be decided. Species examined included rodents,[198,223,227,230,232,233,248] monkeys,[198,230,266] and the Norwegian polecat,[248] a species chosen because of its low endogenous capacity to oxidize phytanic acid. The phytol-enriched diet has been fed through pregnancy, nursing, and growth and development of the offspring. The failure to induce neurological disease by extensive phytol feeding alone led us to consider the hypothesis that accumulation of phytanic acid might be *necessary* to predispose to the development of neurological disease without itself being *sufficient* to cause such disease.[267,268] This idea is consistent with the clinical observation that in phytanic acid storage disease, as in a number of other neurological disorders, clinical deterioration is often associated with intercurrent infection or other stress. Preliminary experiments,[267] in which the phytol-fed animals and pair-fed controls were injected with similar doses of antigen in adjuvant, suggested that the phytol-fed animals might be somewhat more susceptible to experimental allergic encephalomyelitis (Fig. 6). Although the relation of phytanic acid accumulation to neurological disease remains conjectural, phytanic acid accumulation itself appears damaging to animal tissues, at least under the stress of a challenge.

A number of mechanisms have been proposed by which phytanic acid accumulation could be deleterious. Incorporation of the "thorny" branched-chain phytanic acid molecule into biological membranes would be expected to distort the structure of membranes such as myelin, in which the fatty acids are closely packed. O'Brien[29,30] has discussed this possibility in detail, pointing out that the cross-sectional area of phytanic acid is half again as great as that of straight-chain fatty acids, and that the nonpolar interactions between side chains decrease with the fourth power of the distance between them. Both patients and animals with phytanic acid accumulation have decreased amounts of 18:1 (oleic acid) in their lipids, but aside from this the fatty acid composition is relatively normal. No specific role of 18:1 in biological membranes has been defined.[29,30] While the replacement of normally occurring fatty acids by phytanic acid may well account for the progressive changes in patients, no direct evidence for this hypothesis is yet available.

Other ways in which phytanate accumulation might lead to nerve changes

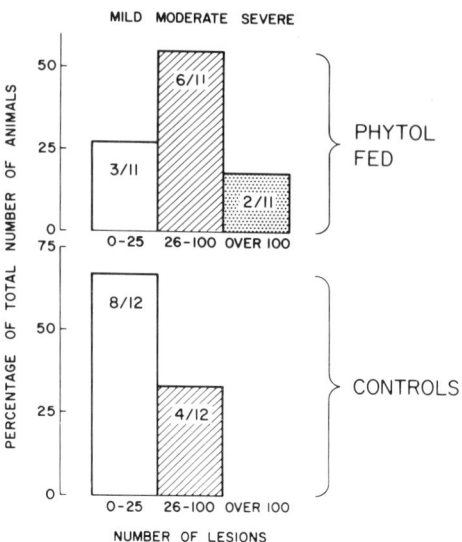

Fig. 6. Lesions of experimental allergic encephalomyelitis
in phytol-fed rats and in pair-fed controls. Three-month-
old rats fed a 2% w/v phytol diet for 16 days and their
pair-fed littermates[230] both received intraperitoneal injections of 1 ml of pertussis vaccine. The following day and
then another 3 days later, each animal received an injection into a hind footpad of 0.1 ml of a homogenate of two
whole guinea pig brains and spinal cords in 12 ml of complete Freund (Difco) adjuvant.[359] The diets were continued, and 14 days later animals were killed and the
brains and spinal cords examined histologically by a
double-blind technique. The bars represent the percentage
of animals in each group with mild disease (0–25 lesions),
moderate disease (26–100 lesions), or severe disease (>100
lesions). The frequency of the lesions did not appear to
correlate with the weights of the animals in each group.
Symptoms were not evaluated quantitatively but appeared more severe in the phytol-fed animals. There were
12 control and 11 phytol-fed animals, one phytol fed
animal having expired during the first week on the diet.
These results were originally reported by Blass *et
al.*[267]

have been proposed.[230] One is interference, competitively or noncompetitively,
with the function of essential compounds that also have a polyisoprenoid
structure, such as vitamin E, vitamin K, and coenzyme Q. Feeding of large
doses of fat-soluble vitamins did not ameliorate the systemic toxic effects of a
high phytanic acid diet.[230] Nevertheless, the phytanic acid might still be competing successfully, perhaps noncompetitively, at the site of vitamin *action*, in
which case even large doses of the vitamins might be ineffective. Phytanic acid

methyl groups are so spaced that they could interdigitate with the double bonds in polyunsaturated fatty acids and thus might interfere with their function. On the other hand, the characteristics of phytanic acid storage disease are not similar to those of essential fatty acid deficiency.[94,98-100] Furthermore, feeding large amounts of polyunsaturated essential fatty acids did not ameliorate the systemic toxic effects of a high phytanic acid diet in rats.[230] Finally, phytanic acid might interfere with specific enzyme systems; Try and Eldjarn[252] found that an apparent defect in ω-oxidation improved in patients after phytanic acid levels were decreased by dietary manipulation. Preliminary evidence suggests a possible defect in other pathways as well.[269] None of the proposed mechanisms discussed above are mutually exclusive, and certainly phytanic acid could exert more than one toxic effect.

Finally, it is possible that phytanic acid storage disease is characterized by more than a single genetic abnormality, a defect in some other systems being directly related to the clinical disorder. Although chromosomes appear normal in the few cases examined[198] and no other consistent chemical abnormalities have been described in these patients, these negative observations do not rule out still undiscovered associated inherited defects.

Patients with similar syndromes but with normal concentrations of phytanic acid have been found, consistent with the possibility that as yet undescribed defects can lead to the clinical abnormalities, perhaps by similar pathophysiological mechanisms. In phytanic acid storage disease, as in other lipidoses, the pathophysiological mechanisms relating the well-defined chemical abnormality to the phenotypic clinical changes remain to be clarified.

IV. DISORDERS WITH STORAGE OF CEROID

A number of disorders have been described that are characterized by progressive degenerative neurological disease, usually of childhood, with accumulation of fluorescent-staining material in the central nervous system.[35,270-276] Attempts have been made to classify these patients into subgroups on the basis of their clinical course and on the basis of the electron microscopic appearance of the stored materials,[270-273] which have recently been isolated although their chemical nature remains unknown.[274] Peroxidation of polyunsaturated lipids and cross-polymerization have been postulated to be important in their formation.[271]

In one disorder associated with ceroid storage (kinky-hair disease), O'Brien and Sampson[35] demonstrated a decrease in polyunsaturated fatty acids in brain phospholipids. Menkes et al.[275] originally described this condition in five patients from one family under the title "A sex-linked recessive disorder with retardation of growth, peculiar hair, and focal cerebral and cerebellar degeneration." Aguilar et al.[276] described six similar patients from another family. In the two of these latter patients who came to autopsy,[35] the concentration of 22:6 was low in both ethanolamine glycerophosphatides (7.7–18.6% vs. 20.8–24.9% in five controls) and serine glycerophosphatides (13.5–14.0% vs. 25.4–36.6% in five controls). Concentrations of other polyunsaturated acids [20:3 $(n-9)$, 20:3 $(n-6)$, 20:4, 22:5 $(n-6)$] were comparable to those of controls. In these studies, intensive effort was specifically directed to preventing oxidation of polyunsaturated fatty acids during isolation and analysis; concentrations of fatty aldehydes as well as of other polyunsaturated acids appeared normal.

Recent studies of a Swedish child with severe neurological disease and ceroid accumulation in the brain[13] revealed a striking diminution in certain unsaturated fatty acids, notably 22:6$(n-3)$ and 22:4$(n-6)$. The concentration of 20:4$(n-6)$ (arachidonic acid) was normal.

Rouser et al.[38] examined the fatty acid compositions of lecithins and sphingomyelins from the brain of one patient with cerebral degeneration of late adult life with accumulation of ceroid-like pigment (Alzheimer's disease) and from the brain of another patient with "senile cerebral cortical atrophy." They found no significant abnormalities but did not specifically analyze polyunsaturated fatty acids.

Low concentrations of specific polyunsaturated acids could reflect defects in biosynthesis of specific fatty acids (elongation and/or desaturation); abnormalities in incorporation of such compounds into complex lipids; increased breakdown, possibly due to selective peroxidation; or possibly defects in intestinal absorption of dietary polyunsaturates. Specific defects have not yet been demonstrated in the disorders associated with ceroid accumulation. Nevertheless, the possibility remains that specific inherited abnormalities of cerebral fatty acid metabolism may be involved in some of these conditions.

V. OTHER STORAGE DISORDERS

Several neurological disorders are characterized by the accumulation of specific materials in neural tissue and often in peripheral tissues. Two groups of these so-called storage disorders, namely, those characterized by the accumulation of mucopolysaccharides and of sphingolipids, are discussed in detail in Volumes 1 and 2 of this series. Both phytanic acid storage disease and disorders with accumulation of ceroid can also be considered to be storage disorders.

A number of abnormalities in fatty acid composition have been described in *disorders of sphingolipid metabolism*. The primary inherited defects in these disorders are known, and they do not directly involve metabolism of fatty acids.

Rouser *et al.*[38] found no fatty acids longer than 22 carbon atoms in sphingomyelins from the brain of a patient with Niemann–Pick disease and from the brain of a patient with metachromatic leukodystrophy. In a $3\frac{1}{2}$-year-old patient dying with GM_2-gangliosidosis (Tay–Sachs disease), proportions of these long-chain acids (about 12% of total sphingomyelin fatty acids) were similar to those found in normal infants and less than those found in normal adults (35% of total). Others have also found a decrease in longer-chain fatty acids in sphingomyelins of patients with Niemann–Pick disease and with metachromatic leukodystrophy.[39,40,277,278]

In *cerebrotendinous xanthomatosis,* a rare disorder characterized by the accumulation of cholestanol in brain and in peripheral xanthomas[279–281] and by progressive neurological disease, Menkes *et al.*[281,282] found no significant abnormalities in fatty acids.

In *Wolman's disease,* storage of cholesterol esters and of triglycerides accompanies severe, rapidly progressive neurological disease of infancy.[283–286] Histochemical evidence indicates that such patients are deficient in an esterase.[286] The stored cholesterol esters contain primarily oleic acid.[286] Examination of the fatty acid composition of other lipids from such patients and of the specificity of the deficient esterase for lipids containing various fatty acids would be of great interest.

In *mucopolysaccharidoses,* Phillipart[287,288] found no abnormalities in brain lipids, including fatty acids.

Other disorders in which the nature of the stored materials is more poorly understood have been described.[289] As the biochemical defects in such conditions become better defined, examination of the associated perturbations in cerebral fatty acid metabolism should help to elucidate the normal roles of specific fatty acids in the brain.

VI. DEMYELINATION OF KNOWN ETIOLOGY

Investigators in a number of laboratories have intensively studied fatty acids and other lipids in brains from patients with diseases characterized by loss of myelin. The most prominent diagnostic category in this group of disorders is multiple sclerosis. Loss of myelin would itself be expected to reduce the amounts of those fatty acids characteristic of myelin—notably the longer-chain unsubstituted and α-hydroxy acids. It therefore becomes critical to try to distinguish between changes which may *result from* loss of myelin and changes which may *cause* demyelination. One experimental approach to this problem would be to study the changes in fatty acids accompanying loss of myelin due to factors apparently unrelated to fatty acid metabolism and to compare them to changes in fatty acids in other demyelinating diseases.

A particularly clear example of changes in fatty acids due to demyelination is provided by the degeneration of nerve after section of an axon. Demyelination distal to a site of axonal damage (Wallerian degeneration) occurs in both the central and peripheral nervous systems, but because of the ready availability of the latter for surgical manipulation and because of its relative anatomical simplicity more detailed studies have been carried out in peripheral nerves.[7,61, 65-68,290-295] Berry et al.[7] described complex changes occurring after section of the sciatic nerves of cats. In general, the most striking changes occurred in cerebrosides and sphingomyelins, with decreases occurring in both longer-chain and shorter-chain fatty acids (e.g., at 8 days after section, 24:0, 24:1, 18:0, and 18:1 all decreased by 40–50%). The most severe changes occurred in the saturated and monounsaturated fatty acids, which have been postulated to play a role in stabilization of the myelin membrane.[29,30] The proportions of these acids in cholesterol esters and in free fatty acids increased. Polyunsaturated fatty acids also changed: at 8 days, 22:6 more than doubled while 22:5 decreased by a factor of 10. Thus changes in specific polyunsaturated fatty acids similar to those found in kinky-hair disease[35] can result from demyelination.

Other types of damage to peripheral nerves lead to different and distinct changes in fatty acids. Berry and Cevallos[61] found that organophosphorous poisoning of chickens led to decreases in both longer- and shorter-chain fatty acids in whole nerve; concentrations of 20:4 and 22:6 did not change. The alterations did not seem identical to those in sectioned cat nerve, but studies of chicken nerve were less extensive. Other workers have found that the patterns of demyelination appear to depend on the nature of the insult to the nerve.[296] Anatomical location was also important; changes occurred more rapidly in peripheral (tibial) than in central (optic) nerve.[68] Incorporation of radioactivity into fatty acids in peripheral nerve could be either increased or decreased by damage to the nerve, probably depending on whether damage, regeneration, or infiltration by other cell types predominated.[290,294,295] For instance, diphtheria toxin caused a fall in incorporation of radioactive acetate into peripheral nerve.[297]

Alterations in fatty acids can accompany disorders of myelin of known etiology in the central nervous system. For instance, lack of myelin is characteristic of untreated phenylketonuria.[298,299] Menkes[299,300] reported increased proportions of hydroxy fatty acids in cerebrosides from the brains of three such patients.[301] Eto and Suzuki[75] recently described a relationship between the fatty acids of lecithins and those of cholesterol esters in demyelination due to a number of causes. They suggested that transfer of the fatty acid in the β-position of lecithin to cholesterol esters is an important step in the mechanism of demyelination.

The changes in fatty acids that accompany demyelination in both the peripheral and central nervous systems indicate the difficulties in interpreting analytical studies in diseases of unknown etiology. As expected, the long-chain fatty acids characteristic of myelin decrease. Changes in other fatty acids also occur, however, and can appear surprisingly specific. For instance, the lack of 22:5 but not 22:6 in peripheral nerve during Wallerian degeneration might be misinterpreted as a primary factor in the loss of myelin. It is apparent that analytical results alone are insufficient to implicate a fatty acid abnormality as primary in a neurological disease.

VII. MULTIPLE SCLEROSIS

Fatty acids have been intensively studied in tissues from patients with multiple sclerosis. Indeed, this common and well-known disorder, characterized by intermittent and patchy demyelination in the central nervous system, has been extensively studied from a variety of viewpoints.[73,141,302,303] Abnormalities of fatty acids clearly occur in areas of frank demyelination.[11,304] Decreased amounts of polyunsaturated fatty acids have been found by some workers in areas grossly free of disease.[11,12,58,73] Deficiency of dietary "essential" fatty acids has been proposed as a cause of multiple sclerosis.[73,305,306]

A. Abnormalities in Fatty Acids in Multiple Sclerosis

The changes in fatty acids and other constituents of demyelinated plaques are consistent with loss of the typical myelin membrane. Reductions in total fatty acids, in α-hydroxy fatty acids, and in longer-chain fatty acids have been reported.[11,304] Polyunsaturated fatty acids appear to represent an increased proportion of total fatty acids[304] but to be decreased per unit wet weight.[11] Changes in other constituents include increases in water and ganglioside content,[304] decreases in cerebrosides,[70,74,307,308] and complex changes in phospholipids,[13,69,70–74,307–311] cholesterol esters,[71,74] and myelin proteins.[312] These changes *accompany* demyelination; the extent to which they are involved in *producing* demyelination remains to be elucidated.

Several groups have described slightly decreased proportions of unsaturated fatty acids in grossly normal tissues from patients with multiple sclerosis.[309, 314–319] Gerstl et al.[313] described concentrations at the lower end of the control ranges for trienes (6.3–14.1% *vs.* 7.6–21.7% in controls) and dienes (8.8–10.1% *vs.* 6.7–15.7% in controls). These are small differences, the significance of which is difficult to evaluate. The proportions of hexaenes, pentaenes, and tetraenes appeared entirely normal. In a later study,[12] these workers found decreased concentrations of polyunsaturated fatty acids in "grossly normal" areas in the brains of two patients with multiple sclerosis compared to the brains of seven controls, both in whole white matter (39.9–51.1 *vs.* 60.6–115.8 µmoles per gram dry weight) and in purified myelin (58.3–61.8 *vs.* 66.8–125.6 µmoles per gram dry weight). Decreases were noted both in monounsaturated and in polyunsaturated fatty acids.[11,12] Baker *et al.*[314] compared the composition of the major fatty acid components of a mixed lecithin–sphingomyelin fraction from white matter of six control brains and from "white matter which appeared normal on visual inspection" of the brains of nine patients with multiple sclerosis. In the material from the patients, the average proportion of total unsaturated fatty acids was 48.4% compared with 59.3% in controls. Most of this difference reflected the lower value for oleic acid (44.4% *vs.* 51.1%), but the arachidonic acid concentration was also lower (1.7% *vs.* 3.3%), as was that of palmitoleic acid (2.3% *vs.* 4.9%). Values for individual patients were not listed, and the significance of these rather small differences is difficult to evaluate. In later studies, these workers also found reduced concentrations of total linoleic acid and of cholesteryl linoleate[315–317] in serum of patients with "active" multiple sclerosis. The proportion of linoleic acid in platelets and red blood cells correlated with that in the plasma, as expected, and was also lower in patients with this disorder.[318] Clausen and Hansen[58] found 20:4 to be decreased in phosphatidylethanolamine in apparently normal white matter from patients with multiple sclerosis. Other workers have also found small differences in the content of polyunsaturated fatty acids in brains from patients with multiple sclerosis.[309,319]

It is notable that the specific changes reported by these various groups were not identical. Furthermore, several other groups of investigators failed to find significant abnormalities in fatty acids in the brains of patients with multiple sclerosis.[39,304,320] Mead *et al.*[320] took specific precautions to prevent oxidation of polyunsaturated fatty acids; in detailed studies they found no abnormalities in fatty acids or other lipids in purified myelin isolated from the brains of patients with multiple sclerosis. At the present time, a well-defined abnormality in polyunsaturated fatty acids has not been established in patients with multiple sclerosis.

B. Relation of the Analytical Findings to the Clinical Syndrome

An etiological role for a relative dietary deficiency of polyunsaturated fatty acids in multiple sclerosis has been proposed, based on several kinds of information.[305] The limited data suggesting that there may be a relative deficiency of polyunsaturated fatty acids in brain lipids of patients with multiple sclerosis have been summarized briefly above. As discussed previously in this chapter, the available amounts of polyunsaturated fatty acids in the diet could influence the amounts of polyunsaturated fatty acids found in brain lipids, particularly during early development when uptake of fatty acids is maximal.[108] Finally, the well-established epidemiological observation that multiple sclerosis is more common in certain latitudes[321] can be correlated with a lower dietary intake in those latitudes of marine oils and of certain vegetable oils such as soybean oil which are rich in polyunsaturated fatty acids. However, a number of problems remain concerning this proposal.

Interpretation of the reported decreases in polyunsaturated fatty acids is difficult for several reasons. As mentioned above, results have varied among different laboratories. For technical reasons, changes in polyunsaturated fatty acids are particularly hard to evaluate. Such acids oxidize easily. Even with the expert and meticulous technique necessary to measure these compounds accurately, secondary loss in damaged tissue is hard to rule out. Another problem is that alterations in the concentrations of a number of lipids, especially phospholipids, can occur in multiple sclerosis, even in unaffected areas of brain. Such changes have been described in some although not all[11,58,73,74,304,309,311] studies. Whether alterations in fatty acid composition are primary and predispose to the loss of certain classes of lipids or are secondary, simply reflecting such loss, remains unclear. Furthermore, alterations in apparently normal white matter may still be secondary to a demyelinating process. Damage to myelin presumably begins on a molecular level,[80] so that light microscopic or gross pathological changes may only delineate later stages of the process. In the central as in the peripheral nervous system,[61,65–68,75,172,290–301,322] demyelination from a variety of causes is associated with alterations in lipid patterns. The decision that a particular change, such as an alteration in fatty acid pattern, is a cause rather than an effect of demyelination is difficult. Experiments like those of Thompson and coworkers on extraneural tissues,[315–317] which may be less affected by the demyelinating process itself, may be particularly important in establishing specificity for particular disorders. Finally, a number of abnormalities other than those in lipids have been described in patients with multiple sclerosis.[69,73,302,323,324] These include, for example, decrease in pseudocholinesterase activity in plaques,[303] increase in concentrations of blood pyruvate after a glucose load,[323,324] immunological alterations and viral infections.[322] At this time, there seems to be no obvious *a priori* reason to consider one set of changes primary and all others secondary.

Moreover, dietary deficiencies of polyunsaturated fatty acid lead in both man and experimental animals to a disease characterized primarily by skin abnormalities but without clinical or pathological evidence of demyelination.[94–106,325–328] Mechanisms appear to exist in developing brain which allow selective uptake of required fatty acids.[1,3,5,101–106,179,329] In rats raised on a diet deficient in lipids, reinstitution of a normal diet allowed reaccumulation of 20:4 (arachidonic acid) and other $n-6$ polyunsaturated fatty acids.[179] In fact, the proportions of these acids in phosphoglycerides stabilized at levels higher than those in controls. It is hard to see why such mechanisms, if intact, would not protect against the effects of the sort of mild and probably intermittent dietary deficiencies which have been postulated in developed Northern countries.

Clausen and Møller[330] found that rats reared on a diet deficient in polyunsaturated fatty acids were more susceptible to experimental allergic encephalomyelitis than littermate controls. Symptoms developed in 17 of 19 deficient animals but in none of the 17 controls. Histological signs of inflammation and decreases in phosphatidylcholine and phosphatidylethanolamine were also somewhat more severe in the deficient animals. These results resemble those with animals poisoned with phytanic acid. The possibility that abnormalities in brain fatty acids may predispose to the development of demyelinating disease when combined with other insults to the brain remains an interesting hypothesis.

It is by no means certain that multiple sclerosis represents a single biological entity. Multiple sclerosis is classically defined as demyelinating disease in which lesions vary in time of onset and development and in location throughout the central nervous system.[141] Unpredictability is its hallmark. Any cause or interacting set of causes of patchy demyelination could by definition be the basis for the clinical syndrome we call multiple sclerosis. A large fraction of the group of individuals defined clinically as having multiple sclerosis may have some metabolic characteristics in common, but this disorder, like "pneumonia" or "jaundice," may well have more than one etiology.

VIII. OTHER HUMAN DISORDERS

Several other clinical disorders are peripherally relevant to a discussion of fatty acids in brain dysfunction.

Disorders of short-chain fatty acids are considered in detail elsewhere in this book. A number of such disorders are now known. It is of note that the injection of short-chain fatty acids can induce coma in experimental animals. These compounds do not appear to act by inhibition of energy production. The concentrations of ATP and creatine phosphate in brain actually increase. A primary effect on active transport of ions has been postulated.[331]

Myopathy with deficient production of ketones was recently reported by Engel and coworkers in a pair of twins.[332,333] These girls had an intermittent myopathy precipitated by starvation; they did not accumulate normal amounts of ketones in blood or urine when fasted or when on a high-fat diet but showed a brisk ketosis after ingesting medium-chain triglycerides. The primary biochemical defect is unknown. From the viewpoint of this chapter, the most interesting point about these patients is that they have no evidence of nervous system disease despite their lack of physiological ketosis. Thus the human brain may not need a large supply of ketones to withstand fasting.

Deficiencies of serum lipoproteins are associated with abnormalities in peripheral lipids and with neurological disease.[334-337] These disorders presumably result from inherited deficiencies of specific apoproteins[338,339] rather than abnormalities in lipid metabolism *per se*. Peripheral neuropathy has been reported in some patients with Tangier disease (deficiency of α-or high-density lipoprotein). In deficiency of β-lipoprotein (low-density lipoprotein), the neurological syndrome somewhat resembles that in phytanic acid storage disease.[336,340] Abnormalities in fatty acids occur in both the serum and the characteristic "thorny" red blood cells called *acanthocytes*.[340-343] In patients with the related syndrome of acanthocytosis and neurological disease but with normal serum β-lipoproteins,[344-346] serum and red cell lipids including fatty acids were normal. Lipid analysis of the central nervous systems of such patients would be of great interest, particularly if specific proteins are found to be deficient in the nervous system as well as in the peripheral tissues of such individuals.

IX. ANIMAL DISORDERS

Several animal disorders have been intensively studied as models to elucidate the pathophysiology of disorders affecting brain lipids.[347] In the *Quaking mouse*, an autosomal recessive mutation interferes with normal myelination. Baumann and coworkers studied the lipid and fatty acid compositon of brains of such mice in detail.[6,15] Reductions in total fatty acids of 24-carbon atoms (0.35% *vs.* normal 3.5%) accompanied reductions in galactolipids (7.7 *vs.* normal 12.5 mg/100 mg). The proportions of 24:0 and 24:1 were reduced in cerebrosides (11 *vs.* 47%), sulfatides (15 *vs.* 61%), and sphingomyelin; the proportions of 23h:0 and 24h:0 appeared normal.[15] Recently, these investigators found evidence for a deficiency in a fatty acid elongation system in cell-free preparations of the brains of such animals.[348] They have suggested that a defect in the biosynthesis of long-chain fatty acids might be a primary inherited abnormality in this disorder.[349] However, it should be noted that amounts of 24h:0, 23h:0, and 22:0 in glycolipids were normal. A variety of abnormalities in lipid metabolism exist in the brains of Quaking mice. These include not only quantitative differences in the composition of whole brain lipids and of isolated myelin [16,350-352] but also metabolic abnormalities,[352-354] including deficiency in the synthesis of galactosphingolipids in cell-free preparations.[353-355] The primary inherited abnormality appears to be transmitted as an autosomal recessive. Approximately two-thirds of phenotypically normal littermates of affected animals should be heterozygotes and one-third homozygous normals. Analysis of tissues from such littermates and from the parents, who are obligate heterozygotes, might clarify the nature of the primary inherited abnormality and facilitate further studies of the pathophysiology of this fascinating disorder.

The *Jimpy mouse* is another mutant with deficient myelination and autosomal recessive inheritance. Reductions in the proportions of longer-chain fatty acids in cerebrosides and sulfatides[16,356] and reductions in the activity of galactosyl-sphingosine transferase[353] have been found in the mutants.

In the *hereditary ataxic mouse*, lipids including fatty acids of whole forebrain and hindbrain, appeared normal.[45]

Mice with *hereditary muscular dystrophy* may have an abnormality affecting cerebral fatty acids. Brain homogenates from such mice incorporated twice as much radioactive acetate into fatty acid as did homogenates from normal littermates.[357] This finding is particularly provocative because brain is not believed to be primarily affected and is not histologically altered in this disorder.

X. CRITERIA FOR ESTABLISHING PRIMARY DISORDERS OF CEREBRAL FATTY ACID METABOLISM

The above discussion illustrates that, while there is suggestive evidence implicating fatty acid metabolism in a number of disease states, *primary* disorders of cerebral fatty acid metabolism have not as yet been firmly established.

Certain properties of brain suggest that such disorders may well exist: this tissue is rich in lipids; fatty acids appear to play an important structural role; certain specialized fatty acids (the longer-chain and α-hydroxy acids) exist primarily in brain, as does the enzyme system for oxidizing these compounds, the brain microsomal α-oxidation system. Mutations affecting key metabolic processes such as fatty acid oxidation may in some cases be lethal mutations. However, it seems reasonable to assume that at least partial defects in cerebral enzymes of fatty acid metabolism would be compatible with extra-uterine life and would be associated with neurological disorders.[268]

If the primary function of fatty acids in the brain is as components of membranes, patients with defects in fatty acid metabolism could be expected to have abnormalities in neural or glial membranes. A variety of patients with degeneration of myelin or other membranes are known, as discussed above. Abnormalities of fatty acids have been described in a number of such individuals, but in general it has not been clear whether the changes were causes or effects of the degeneration.

For definitive identification of specific inherited defects, several criteria must be met: (1) Analytical studies should be consistent with the proposed defect. (2) Studies of metabolic rates should indicate a defect in a particular pathway, while metabolism of related compounds should be normal. (3) The defect should be identified in cell-free systems, if possible by specific enzyme assay. (4) If the particular enzyme occurs in tissues not primarily affected by the disease process, demonstration of the defect in such tissues (e.g., neural or extraneural tissues structurally normal and functionally normal in other respects) would provide good evidence that it is not a *result* of the disease process. (5) Demonstration of the proposed enzyme defect in cultured cells would be especially convincing. (6) If the disease is inherited, particularly in a recessive pattern, demonstration in the tissues of heterozygotes of values intermediate between those for patients and those for normal subjects would suggest strongly that the inherited abnormality is primary for the specific enzyme examined. These criteria have in general been met in the study of the sphingolipidoses and a number of other genetic diseases involving the nervous system.[358]

The relationship between inherited or acquired disorders of fatty acid metabolism and neurological diseases may well prove to be complex. As noted above, preliminary studies in phytanic acid-poisoned rats[267] and essential fatty acid deficient rats[330] support the possibility that abnormalities in cerebral fatty acids can predispose to disease without being sufficient to cause symptoms by themselves. Neurological disorders secondary to a variety of causes, and in particular disorders affecting lipids, may be associated with abnormalities in brain fatty acids; the studies of Ito and Suzuki[75] suggest that fatty acids may be important in the mechanism of demyelination. The variety of defects found in the Quaking mouse illustrates the interactions which appear to exist between the metabolism of fatty acids and that of other compounds in the brain.

Basic knowledge of fatty acid metabolism in brain (and in other tissues) now appears to be adequate for definitive study of putative disorders of cerebral fatty acid metabolism. Hopefully, analysis of the neurological abnormalities associated with specific abnormalities of fatty acid metabolism will help to elucidate the role of these compounds not only in brain dysfunction but in normal cerebral metabolism as well.

REFERENCES

1. A. F. D'Adamo, *in* "Handbook of Neurochemistry" (A. Lajtha, ed.) Vol. 3, pp. 525–546, Plenum Press, New York (1970).
2. D. M. Bowen and N. S. Radin, Hydroxy fatty acid metabolism in brain, *Advan. Lipid Res.* **6:**255–272 (1968).
3. G. Rouser and A. Yamamoto, *in* "Handbook of Neurochemistry" (A. Lajtha, ed.) Vol. 1, pp. 121–170, Plenum Press, New York (1969).

4. L. C. Mokrasch, *in* "Handbook of Neurochemistry" (A. Lajtha, ed.) Vol. 1, pp. 171–193, Plenum Press, New York (1969).
5. G. Schettler, ed., "Lipids and Lipidoses," Springer-Verlag, New York (1967).
6. N. A. Baumann, C. M. Jacque, S. A. Pollet, and M. L. Harpin, Fatty acid and lipid composition of the brain of a myelin deficient mutant, the Quaking mouse, *Europ. J. Biochem.* **4:**340–344 (1968).
7. J. F. Berry, R. Cevallos, and R. R. Wade, Lipid class and fatty acid composition of intact peripheral nerve and during Wallerian degeneration, *J. Am. Oil Chem. Soc.* **42:** 492–500 (1965).
8. J. P. Blass, Fatty acid composition of cerebrosides in microsomes and myelin of mouse brain, *J. Neurochem.* **17:**545–549 (1970).
9. L. F. Eng, B. Gerstl, R. B. Hayman, Y. L. Lee, R. W. Tietsort, and J. K. Smith, The 2-hydroxy fatty acids in white matter of infant and adult brains, *J. Lipid Res.* **6:**135–139 (1965).
10. B. Gerstl, R. B. Hayman, M. G. Tavastjerna, and J. K. Smith, Fatty acids of white matter of human brain, *Experientia* **18:**131–133 (1962).
11. B. Gerstl, M. G. Tavastjerna, R. B. Hayman, L. F. Eng, and J. K. Smith, Alterations in myelin fatty acids and plasmalogens in multiple sclerosis, *Ann. N.Y. Acad. Sci.* **122:** 405–416 (1965).
12. B. Gerstl, M. G. Tavastjerna, R. B. Hayman, J. K. Smith, and L. F. Eng, Lipid studies of white matter and thalamus of human brains, *J. Neurochem.* **10:**889–902 (1963).
13. B. Hagberg, P. Sourander, and L. Svennerholm, Late infantile progressive encephalopathy with disturbed polyunsaturated fat metabolism, *Acta Paediat. Scand.* **57:**495–499 (1968).
14. B. J. Holub, A. Kuksis, and W. Thompson, Molecular species of mono, di, and triphosphoinositides of bovine brain, *J. Lipid Res.* **11:**558–564 (1970).
15. C. M. Jacque, M. L. Harpin, and N. A. Baumann, Brain lipid analysis of a myelin deficient mutant, the Quaking mouse, *Europ. J. Biochem.* **11:**218–224 (1969).
16. K. Joseph and E. Hogan, Sphingolipid fatty acid composition in murine genetic leukodystrophies, *Trans. Am. Soc. Neurochem.* **2:**85 (1971).
17. Y. Kishimoto, B. W. Agranoff, N. S. Radin, and R. M. Burton, Comparison of the fatty acids of lipids of subcellular brain fractions, *J. Neurochem.* **16:**397–404 (1969).
18. Y. Kishimoto and N. S. Radin, Isolation and determination methods for brain cerebrosides, hydroxy fatty acids, and unsaturated and saturated fatty acids, *J. Lipid Res.* **1:** 72–78 (1959).
19. Y. Kishimoto and N. S. Radin, Composition of cerebroside acids as a function of age, *J. Lipid Res.* **1:**79–82 (1959).
20. Y. Kishimoto and N. S. Radin, Structures of ester linked mono and diunsaturated fatty acids of pig brain, *J. Lipid Res.* **5:**98–108 (1965).
21. Y. Kishimoto and N. S. Radin, Determinations of brain gangliosides by determination of ganglioside stearic acid, *J. Lipid Res.* **7:**141–145 (1966).
22. Y. Kishimoto, M. Wajda, and N. S. Radin, 6-Acylgalactosyl ceramides of pig brain: Structure and fatty acid composition, *J. Lipid Res.* **9:**27–33 (1968).
23. F. A. Manzoli, S. Stefoni, L. Manzolli-Guidotti, and M. Barbieri, Fatty acids of myelin phospholipids, *FEBS Letters* **10:**317–320 (1970).
24. E. Mårtensson, A. Percy, and L. Svennerholm, Kidney glycolipids in late infantile metachromatic leukodystrophy, *Acta Paediat. Scand.* **55:**109 (1966).
25. M. C. McBrinn and J. S. O'Brien, Lipid composition of the nervous system in Refsum's disease, *J. Lipid Res.* **9:**552–561 (1968).
26. G. F. McMullin, S. C. Smith, and P. A. Wright, Tissue fatty acid composition in four diverse vertebrate species, *Comp. Biochem. Physiol.* **26:**211–221 (1968).
27. J. H. Menkes, M. Philippart, and M. C. Concone, Concentration and fatty acid composition of cerebrosides and sulfatides in mature and immature brain, *J. Lipid Res.* **7:**479–486 (1966).

28. K. Nishimura and T. Yamakawa, Isolation of cerebroside containing glucose and its possible significance in ganglioside metabolism, *Lipids* **3**:262–266 (1968).

29. J. S. O'Brien, Cell membranes—composition, structure, function, *J. Theoret. Biol.* **15**: 307–324 (1967).

30. J. S. O'Brien, Stability of the myelin membrane, *Science* **147**:1099–1107 (1965).

31. J. S. O'Brien, D. L. Fillerup, and J. F. Mead, Quantification and fatty acid composition of cerebroside sulfate in human cerebral gray and white matter, *J. Lipid Res.* **5**:109–116 (1964).

32. J. S. O'Brien, D. L. Fillerup, and J. F. Mead, Quantification and fatty acid and fatty aldehyde composition of ethanolamine, choline, and serine glycerophosphatides in human cerebral grey and white matter, *J. Lipid Res.* **5**:329–338 (1964).

33. J. S. O'Brien and G. Rouser, The fatty acid composition of brain sphingolipids: Sphingo-myelin, ceramide, cerebroside, cerebroside sulfate, *J. Lipid Res.* **5**:339–342 (1964).

34. J. S. O'Brien and E. L. Sampson, Fatty acid and fatty aldehyde composition of the major brain lipids in normal human grey matter, white matter, and myelin, *J. Lipid Res.* **6**: 545–551 (1965).

35. J. S. O'Brien and E. L. Sampson, Kinky hair disease. II. Biochemical studies, *J. Neuro-pathol. Exptl. Neurol.* **25**:523–530 (1966).

36. N. S. Radin and Y. Akahori, Fatty acids of human brain cerebrosides, *J. Lipid Res.* **2**: 335–341 (1961).

37. A. Rosenberg and N. Stern, Changes in sphingosine and fatty acid components of the gangliosides in developing rat and human brain, *J. Lipid Res.* **7**:122–131 (1966).

38. G. Rouser, G. Feldman, and C. Galli, Fatty acid compositions of human brain lecithin and sphingomyelin in normal individuals, senile cerebral cortical atrophy, Alzheimer's disease, metachromatic leukodystrophy, Tay–Sachs and Niemann–Pick diseases, *J. Am. Oil Chem. Soc.* **42**:411–412 (1965).

39. S. Stallberg-Stenhagen and L. Svennerholm, Fatty acid composition of human brain sphingomyelins: Normal variation with age and changes during myelin disorders, *J. Lipid Res.* **6**:146–155 (1965).

40. L. Svennerholm, *in* "Brain Lipids and Lipoproteins and the Leucodystrophies" (J. Folch-Pi and H. Bauer, eds.) pp. 104–119, Elsevier, New York (1963).

41. L. Svennerholm, Distribution and fatty acid composition of phosphoglycerides in normal human brain, *J. Lipid Res.* **9**:570–579 (1968).

42. L. Svennerholm and S. Stallberg-Stenhagen, Changes in the fatty acid composition of cerebrosides and sulfatides of human nervous tissue with age, *J. Lipid Res.* **9**:215–225 (1968).

43. H. Yabuuchi and J. S. O'Brien, Brain lipids. VI. Brain cardiolipin: Isolation and fatty acid positions, *J. Neurochem.* **15**:1383–1390 (1968).

44. H. Yabuuchi and J. S. O'Brien, Positional distribution of fatty acids in glycerophospha-tides of bovine grey matter, *J. Lipid Res.* **9**:65–67 (1968).

45. M. B. Weber, A study of brain lipids, nucleic acids and proteins in the ataxic (*axj*) mouse, *Neurology* **18**:243–249 (1968).

46. M. Fewster and J. F. Mead, Fatty acid and fatty aldehyde composition of glial cell lipids isolated from bovine white matter, *J. Neurochem.* **15**:1303–1312 (1968).

47. E. G. Lapetina, E. F. Soto, and E. De Robertis, Lipids and proteolipids in isolated subcellular membranes of rat brain cortex, *J. Neurochem.* **15**:437–445 (1968).

48. P. Lesch and K. Bernhard, Composition of the neutral lipids and their fatty acids in whale brain. II. Brains of finnwhales (*Balaenoptera physalus*), *Helv. Chim. Acta* **51**:652–660 (1968).

49. M. A. Wells and J. C. Dittmer, A comprehensive study of the postnatal changes in the concentration of lipids of developing rat brain, *Biochemistry* **6**:3169–3175 (1967).

50. J. S. O'Brien, E. L. Sampson, and M. B. Stern, Lipid composition of myelin from the peripheral nervous system, *J. Neurochem.* **14**:357–365 (1967).

51. J. Clausen, *in* "Handbook of Neurochemistry" (A. Lajtha, ed.) Vol. 1, pp. 273–300, Plenum Press, New York (1969).
52. L. F. Eng and E. P. Noble, The maturation of rat brain myelin, *Lipids* **3:**157–162 (1968).
53. E. Marshall, R. Fumagalli, R. Niemiro, and R. Paoletti, The change in fatty acid composition of rat brain phospholipids during development, *J. Neurochem.* **13:**857–862 (1966).
54. N. F. Avrova and S. A. Zabelinskii, Fatty acids and long chain bases of vertebrate brain gangliosides, *J. Neurochem.* **18:**675–681 (1971).
55. P. Stoffyn and J. Folch-Pi, Type of linkage binding fatty acids present in brain white matter proteolipid apoprotein, *Biochem. Biophys. Res. Commun.* **44:**157–161 (1971).
56. K. Samuelsson, Separation and identification of cerebrosides in cerebrospinal fluid by gas-liquid chromatography–mass spectrometry, *Scand. J. Lab. Clin. Invest.* **27:**381–391 (1971).
57. D. R. Illingworth and J. Glover, The composition of lipids in cerebrospinal fluid of children and adults, *J. Neurochem.* **18:**769–776 (1971).
58. J. Clausen and J. B. Hansen, Myelin constituents of the human central nervous system. Studies of phospholipid, glycolipid, and fatty acid pattern in normal and multiple sclerosis brains, *Acta Neurol. Scand.* **46:**1–17 (1970).
59. H. B. White, C. Galli, and R. Paoletti, Ethanolamine phosphoglyceride fatty acids in aging human brains, *J. Neurochem.* **18:**1337–1339 (1971).
60. C. Alling and L. Svennerholm, Concentration and fatty acid composition of cholesteryl esters of normal human brain, *J. Neurochem.* **16:**751–759 (1969).
61. J. F. Berry and W. H. Cevallos, Lipid class and fatty acid compositions of peripheral nerve from normal and organophosphorous poisoned chickens, *J. Neurochem.* **13:**117–124 (1966).
62. G. G. Lunt and C. E. Rowe, Production of unesterified fatty acid in the brain, *Biochim. Biophys. Acta* **152:**681–693 (1968).
63. C. E. Rowe, The occurrence and metabolism *in vitro* of unesterified fatty acid in mouse brain, *Biochim. Biophys. Acta* **84:**424–434 (1964).
64. N. G. Bazan, H. E.P. de Bazan, W. G. Kennedy, and C. D. Joel, Regional distribution and rate of production of free fatty acids in rat brain, *J. Neurochem.* **18:**1387–1393 (1971).
65. N. S. Burt, A. R. McNabb, and R. J. Rossiter, Chemical studies of peripheral nerve during Wallerian degeneration. II. Lipids after nerve crush, *Biochem. J.* **47:**318–323 (1950).
66. A. C. Johnson, A. R. McNabb, and R. J. Rossiter, Chemical studies of peripheral nerve during Wallerian degeneration, *Biochem. J.* **45:**500–507, (1949).
67. W. A. Mannell, Wallerian degeneration in the rat, a chemical study, *Can. J. Med. Sci.* **30:**173–179 (1952).
68. R. E. McCaman and E. Robins, Quantitative biochemical studies of Wallerian degeneration in the peripheral and central nervous systems, *J. Neurochem.* **5:**18–31 (1959).
69. H. Bauer and R. Heitmann, Chemical and serological investigations in multiple sclerosis, *Deutsch. Z. Nervenheilk.* **178:**47–77 (1958).
70. J. N. Cumings, Lipid chemistry of the brain in demyelinating diseases, *Brain* **78:**554–563 (1965).
71. C. Honneger, On thin layer chromatography of lipids—Investigations of brain samples from patients with multiple sclerosis and normals, *Helv. Chim. Acta* **45:**281–289 (1962).
72. D. W. Clarke and B. Gittens, The effect of serum from multiple sclerosis patients on the free fatty acid output of rat brain slices, *Can. J. Physiol. Pharmacol.* **46:**507–509 (1968).
73. R. H. S. Thompson, The biochemistry of multiple sclerosis, *in* "Scientific Basis of Medicine, Annual Reviews," pp. 283–301, Athlone Press, London (1961).
74. A. N. Davison and M. Wajda, Cerebral lipids in multiple sclerosis, *J. Neurochem.* **9:**427–432 (1962).
75. Y. Eto and K. Suzuki, Fatty acid composition of cholesterol esters in brains of patients with Schilder's disease, Gm_1-gangliosidosis, and Tay–Sachs disease, and its possible relation to the β-position fatty acids of lecithin, *J. Neurochem.* **18:**1007–1016 (1971).
76. "The Nomenclature of Lipids," *Biochemistry* **6:**3287–3292 (1967).

77. "The Nomenclature of Lipids," *Arch. Biochem. Biophys.* **123**:409–415 (1968).
78. S. G. Pakkala, D. L. Fillerup, and J. F. Mead, The very long chain fatty acids of human brain sphingolipids, *Lipids* **1**:449–450 (1966).
79. A. N. Davison and N. A. Gregson, The physiologic roles of cerebron sulphuric acid in the brain, *Biochem. J.* **85**:558–568 (1962).
80. R. P. Bunge, Glial cells and the central myelin sheath, *Physiol. Rev.* **48**:197–251 (1968).
81. S. Korey, "The biology of Myelin," Hoeber-Harper, New York (1959).
82. W. E. Norton and S. E. Poduslo, The bulk isolation and properties of bovine oligodendroglia, *Proc. Third Intern. Meeting Intern. Soc. Neurochem. Budapest*, p. 34 (1971).
83. F. A. Vandenheuvel, Study of the biologic structure at the molecular level with stereochemical projections. II. The structure of myelin in relation to other membrane systems, *J. Am. Oil Chem. Soc.* **42**:481–492 (1965).
84. F. A. Vandenheuvel, Structural studies of biologic membranes: The structure of myelin, *Ann. N.Y. Acad. Sci.* **122**:57–76 (1965).
85. J. B. Finean, The nature and stability of brain lipids, *Circulation* **26**:1151–1162 (1962).
86. J. B. Finean and R. E. Burge, The determination of the Fourier transform of the myelin layer from a study of swelling phenomena, *J. Mol. Biol.* **7**:672–682 (1963).
87. J. B. Finean, The nature and stability of lipid–protein–polysaccharide association in nerve myelin, *in* "Brain Lipids and Lipoproteins and the Leucodystrophies" (J. Folch-Pi and H. Bauer, eds.) pp. 57–63, Elsevier, New York (1963).
88. B. D. Ladbrooke, T. J. Jenkinson, V. B. Kamat, and D. Chapman, Physical studies of myelin: Thermal analysis, *Biochim. Biophys. Acta* **164**:101–109 (1968).
89. S. G. Kayser and S. Patton, Function of very long chain fatty acids in membrane structure: Evidence from milk cerebrosides, *Biochim. Biophys. Res. Commun.* **41**:1572–1578 (1970).
90. J. F. Danielli and H. A. Davson, A contribution to the theory of permeability of thin films, *J. Cell. Comp. Physiol.* **5**:495–508 (1935).
91. D. E. Green and S. Fleischer, The role of lipids in mitochondrial electron transfer and oxidative phosphorylation, *Biochim. Biophys. Acta* **70**:554–582 (1963).
92. A. A. Benson, On the orientation of lipids in chloroplast and cell membranes, *J. Am. Oil Chem. Soc.* **43**:265–270 (1966).
93. R. B. Park and T. Biggens, Quantasome: Size and composition, *Science* **144**:1009–1011 (1964).
94. G. O. Burr and M. M. Burr, A new deficiency disease produced by the rigid exclusion of fat from the diet, *J. Biol. Chem.* **82**:345–367 (1929).
95. J. F. Mead, The metabolism of the essential fatty acids. VI. Distribution of unsaturated fatty acids in rats on fat-free and supplemented diets, *J. Biol. Chem.* **227**:1025–1034 (1957).
96. J. F. Mead and D. R. Howton, Metabolism of essential fatty acids. VII. Conversion of γ-linolenic acid to arachidonic acid, *J. Biol. Chem.* **229**:575–582 (1957).
97. M. Guarnier and R. M. Johnson, The essential fatty acids, *Advan. Lipid Res.* **8**:115–174 (1970).
98. R. B. Alfin-Slater and L. Aftergood, Essential fatty acids reinvestigated, *Physiol. Rev.* **48**:758–784 (1968).
99. H. M. Sinclair, ed., "Essential Fatty Acids," Academic Press, New York (1958).
100. A. T. James and J. E. Lovelock, Essential fatty acids and human disease, *Brit. Med. Bull.* **14**:262–266 (1958).
101. L. A. Biran, W. Bartley, C. W. Carter, and A. Renshaw, Studies on essential fatty acid deficiency. Effect of the deficiency on the lipids in various rat tissues and the influence of dietary supplementation with essential fatty acids on deficient rats, *Biochem. J.* **93**:492–498 (1964).
102. L. Rathbone, The effect of diet on the fatty acid compositions of serum, brain, brain mitochondria and myelin in the rat, *Biochem. J.* **97**:620–628 (1965).
103. W. J. Culley and E. T. Mertz, Effect of restricted food intake on growth and composition of pre-weaning rat brain, *Proc. Soc. Exptl. Biol. Med.* **118**:233–235 (1965).

104. L. J. Machlin, G. J. Marco, and R. S. Gordon, Effect of diet and encephalomalacia on the fatty acid composition of the brain of young and old chickens, *J. Am. Oil Chem. Soc.* **39:** 229–232 (1962).

105. B. L. Walker, Maternal diet and brain fatty acids in young rats, *Lipids* **2:**497–500 (1967).

106. H. Mohrhauer and R. T. Holman, Alteration of the fatty acid composition of brain lipids by varying levels of dietary essential fatty acids, *J. Neurochem.* **10:**523–530 (1963).

107. G. A. Dhopeschwarkar and J. F. Mead, Fatty acid uptake by the brain. III. Incorporation of 1[^{14}C] oleic acid into the adult rat brain, *Biochim. Biophys. Acta* **210:**250–256 (1970).

108. E. T. Pritchard, The formation of phospholipids from ^{14}C-labelled precursors in developing rat brain *in vivo, J. Neurochem.* **10:**495–502 (1963).

109. H. Keen and C. Chlouverakis, Metabolism of isolated rat retina. The role of nonesterified fatty acid, *Biochem. J.* **94:**488–493 (1965).

110. C. H. Tator, J. R. Evans, and J. Olszewski, Tracers for detection of brain tumors. Evaluation of radioiodinated human serum albumin and radioiodinated fatty acid, *Neurology* **16:** 650–661 (1966).

111. G. R. Webster, The incorporation of long-chain fatty acids into phospholipids of respiring slices of rat cerebrum, *Biochem. J.* **102:**373–380 (1967).

112. J. Bernsohn, L. M. Stephanides, and H. Norgello, Incorporation of [1-^{14}C]linoleic acid into the central nervous system of the adult cat after intracisternal administration, *Brain Res.* **28:**327–337 (1971).

113. S. Gatt, Metabolism of 1[^{14}C]lignoceric acid in the rat, *Biochim. Biophys. Acta* **70:**370–380 (1963).

114. A. F. Adamo, L. I. Gidez, and F. M. Yatsu, Acetyl transport mechanisms. Involvement of *N*-acetyl aspartic acid in *de novo* fatty acid biosynthesis in the developing rat brain, *Exptl. Brain Res.* **5:**267–273 (1968).

115. K. Bernhard and W. Pedersen, Fatty acid synthesis in rat brain, *Helv. Chim. Acta* **46:** 2363–2368 (1967).

116. A. Etzrodt and H. Debuch, Incorporation of 1[^{14}C]acetate into the fatty acids and aldehydes of the ethanolamine-containing phospholipids in the brain of young rats, *Z. Physiol. Chem.* **351:**603–612 (1970).

117. E. Grossi, P. Paoletti, and M. Poggi, The effect of insulin on brain cholesterol and fatty acid biosynthesis, *World Neurol.* **3:**209–215 (1965).

118. A. K. Hajra and N. S. Radin, Isotopic studies of the biosynthesis of the cerebroside fatty acids in rats, *J. Lipid Res.* **4:**270–278 (1963).

119. A. J. Fulco and J. F. Mead, The biosynthesis of lignoceric, cerebronic and nervonic acids, *J. Biol. Chem.* **236:**2416–2420 (1967).

120. P. A. Srere and A. Bhaduri, Incorporation of radioactive citrate into fatty acids, *Biochim. Biophys. Acta* **59:**487–489 (1962).

121. K. Miyamoto, L. M. Stephanides, and J. Bernsohn, Acetate-1-^{14}C incorporation into polyunsaturated fatty acids of phospholipids of developing chick brain, *J. Lipid Res.* **8:**191–195 (1967).

122. S. Tucek, The use of choline acetyltransferase for measuring the synthesis of acetyl CoA and its release from brain mitochondria, *Biochem. J.* **104:**749–756 (1967).

123. D. D. Clarke, W. J. Nicklas, and S. Berl, Tricarboxylic acid cycle metabolism in brain. Effect of fluoroacetate and fluorocitrate on the labelling of glutamate, aspartate, glutamine and γ-amino butyrate, *Biochem. J.* **120:**345–351 (1970).

124. A. F. D'Adamo and A. P. D'Adamo, Acetyl transport mechanisms in the nervous system. The oxoglutarate shunt and fatty acid synthesis in the developing rat brain, *J. Neurochem.* **15:**315–323 (1968).

125. S. Berl and D. D. Clarke, *in* "Handbook of Neurochemistry" (A. Lajtha, ed.) Vol. 2, pp. 447–471, Plenum Press, New York (1969).

126. S. Berl, A. Lajtha, and H. Waelsch, Amino acid and protein metabolism. VI. Cerebral compartments of glutamic acid metabolism, *J. Neurochem.* **7:**186–197 (1961).

127. H. Waelsch, S. Berl, C. A. Rossi, D. D. Clarke, and D. P. Purpura, Quantitative aspects of CO_2 fixation in mammalian brain *in vivo, J. Neurochem.* **11**:717–728 (1964).

128. R. O. Brady, Biosynthesis of fatty acids. II. Studies with enzymes obtained from brain, *J. Biol. Chem.* **235**:3099–3103 (1960).

129. J. D. Robinson, R. M. Bradley, and R. O. Brady, Biosynthesis of fatty acids. III. Utilization of substituted acetyl coenzyme A derivatives as intermediates, *J. Biol. Chem.* **238**: 528–532 (1963).

130. E. Aeberhard and J. H. Menkes, Biosynthesis of long-chain fatty acids by subcellular particles of mature brain, *J. Biol. Chem.* **243**:3834–3840 (1968).

131. J. M. Bourre, S. Pollet, G. Dubois, and N. A. Baumann, Biosynthesis of long-chain fatty acids in mouse brain microsomes, *Compt. Rend. Acad. Sci., Ser. D* **271**:1221–1223 (1970).

132. K. Miyamoto, L. M. Stephanides, and J. Bernsohn, Incorporation of 1[^{14}C] linoleate and linolenate into polyunsaturated fatty acids of phospholipids of the embryonic chick brain, *J. Neurochem.* **14**:227–237 (1967).

133. A. K. Hajra and N. S. Radin, *In vivo* conversion of labelled fatty acid to the sphingolipid fatty acid in rat brain, *J. Lipid Res.* **4**:448–453 (1967).

134. A. K. Hajra and N. S. Radin, Biosynthesis of cerebroside odd-numbered fatty acids, *J. Lipid Res.* **3**:327–332 (1963).

135. W. Pedersen, L. Hausheer, and K. Bernhard, Further studies in neurochemistry: The incorporation of 1[^{14}C]propionate in the fatty acids of brain cerebrosides, *Helv. Chim. Acta* **46**:675–677 (1963).

136. J. D. Robinson, R. O. Brady, and R. M. Bradley, Biosynthesis of fatty acids. IV. Studies with inhibitors, *J. Lipid Res.* **4**:144–150 (1963).

137. H. McIlwain, "Biochemistry and the Central Nervous System," J. and A. Churchill, London (1959).

138. H. F. Bradford, Carbohydrate and energy metabolism, *in* "Applied Neurochemistry," Part II: "Metabolic Pathways" (A. N. Davison and John Dobbing, eds.) pp. 222–250, Blackwell Press, Oxford (1968).

139. L. Sokoloff, Metabolism of the central nervous system *in vivo, in* "Handbook of Physiology," Section I: "Neurophysiology" (H. W. Magoun, ed.) Vol. 3, pp. 1843–1864, Waverly Press, Baltimore (1960).

140. S. S. Kety, The general metabolism of the brain *in vivo, in* "The Metabolism of the Nervous System" (D. Richter, ed.) pp. 221–266, Pergamon Press, London (1957).

141. H. H. Merritt, "Textbook of Neurology" Lea and Febiger, Philadelphia (1963).

142. O. E. Owen, A. P. Morgan, H. G. Kemp, J. M. Sullivan, R. G. Herrera, and G. F. Cahill, Brain metabolism during fasting, *J. Clin. Invest.* **46**:1589–1595 (1967).

143. R. A. Hawkins, D. H. Williamson, and H. A. Krebs, Ketone body utilization by adult and suckling rat brain *in vivo, Biochem. J.* **122**:13–18 (1971).

144. U. Gottstein, W. Mueller, W. Berchoff, H. Gaertner, and K. Held, Utilization of non-esterified fatty acids and ketone bodies in the human brain, *Klin. Wschr.* **49**:406–411 (1971).

145. F. S. Rolleston and E. A. Newsholm, Effects of fatty acids, ketone bodies, lactate and pyruvate on glucose utilization by guinea-pig cerebral cortex slices, *Biochem. J.* **104**:519–523 (1967).

146. R. W. Von Korff, Personal communication.

147. M. A. Page, H. A. Krebs, and D. H. Williamson, Activities of the enzymes of ketone-body utilization in brain and other tissues of suckling rats, *Biochem. J.* **121**:49–53 (1971).

148. A. L. Smith, H. S. Satterthwaite, and L. Sokoloff, Induction of brain D(−)-β-hydroxy-butyrate dehydrogenase activity by fasting, *Science* **163**:79–81 (1969).

149. I. Pull and H. McIlwain, 3-Hydroxybutyrate dehydrogenase of rat brain on dietary change and during maturation, *J. Neurochem.* **18**:1163–1165 (1971).

150. R. Martinez and A. Toledano, Histochemical study of the metabolism of the ketone bodies in the nervous system. IV. Localization of the D(−)-β-hydroxybutyric dehydrogenase in the cerebellum and medulla oblongata, *Acta Histochem.* **38**:218–260 (1970).

151. P. M. Vignais, G. H. Gallagher, and I. Zabin, Activation and oxidation of long chain fatty acids by rat brain, *J. Neurochem.* **2:**283–287 (1958).
152. D. S. Beattie and R. E. Basford, Brain mitochondria. III. Fatty acid oxidation by bovine brain mitochondria, *J. Neurochem.* **12:**103–111 (1965).
153. K. G. Raju, Metabolism of acetate, propionate, butyrate, and glucose by bovine cerebral cortex slices, *Am. J. Physiol.* **219:**1739–1741 (1970).
154. M. E. Volk, R. H. Millington, and S. Weinhouse, Oxidation of endogenous fatty acids of rat tissues *in vitro, J. Biol. Chem.* **195:**493–501 (1952).
155. C. Allweis, T. Landau, M. Abeles, and J. Magnes, The oxidation of uniformly labelled albumin-bound palmitic acid to CO_2 by the perfused cat brain, *J. Neurochem.* **13:**795–804 (1966).
156. J. F. Mead and G. M. Levis, A one carbon degradation of the long chain fatty acids of brain sphingolipids, *J. Biol. Chem.* **238:**1634–1636 (1963).
157. G. M. Levis, The possible role of ascorbic acid in the α-hydroxyacid decarboxylase of brain microsomes, *Biochim. Biophys. Acta* **99:**194 (1965).
158. Y. Kishimoto and N. S. Radin, Occurrence of 2-hydroxy fatty acids in animal tissues, *J. Lipid Res.* **4:**139–143 (1963).
159. G. M. Levis and J. F. Mead, A 2-hydroxy acid decarboxylase in brain microsomes, *J. Biol. Chem.* **239:**77–80 (1964).
160. Y. Kishimoto and N. S. Radin, Structures of the 2-hydroxy unsaturated fatty acids of pig brain sphingolipids, *J. Lipid Res.* **5:**94–97 (1964).
161. W. E. Davis, A. K. Hajra, S. S. Parmer, N. S. Radin, and J. F. Mead, Decarboxylation of 2-keto fatty acids by brain, *J. Lipid Res.* **7:**270–276 (1966).
162. R. C. MacDonald and J. F. Mead, The alpha oxidation system of brain microsomes. Cofactors for alpha-hydroxyacid decarboxylation, *Lipids* **3:**275–283 (1968).
163. K. Lippel and J. F. Mead, Alpha-oxidation of 2-hydroxystearic acid *in vitro, Biochim. Biophys. Acta* **152:**669–680 (1968).
164. S. Hammarstrom, Configuration of 2-hydroxy fatty acids from brain cerebroside determined by gas chromatography, *FEBS Letters* **5:**192–195 (1969).
165. B. Preiss and K. Bloch, Omega-oxidation of long chain fatty acids in rat liver, *J. Biol. Chem.* **239:**85–88 (1964).
166. K. Wakabayashi and N. Shimazono, Studies on omega-oxidation of fatty acids *in vitro.* I. Overall reaction and intermediates, *Biochim. Biophys. Acta* **70:**132–142 (1963).
167. S. Bergström, B. Borgström, B. N. Tryding, and G. Westöö, Intestinal absorption and metabolism of 2,2-dimethylstearic acid in the rat, *Biochem. J.* **58:**604–608 (1954).
168. G. J. Antony and B. R. Landau, Relative contributions of alpha, beta and omega-oxidative pathways to *in vitro* fatty acid oxidation in rat liver, *J. Lipid Res.* **9:**267–269 (1968).
169. Y. Kishimoto, W. E. Davies, and N. S. Radin, Turnover of the fatty acids of rat brain gangliosides, glycerophosphatides, cerebrosides, and sulfatides as a function of age, *J. Lipid Res.* **6:**525–531 (1965).
170. Y. Kishimoto and N. S. Radin, Metabolism of brain glycolipid fatty acids, *Lipids* **1:**47–61 (1966).
171. L. N. Irwin and F. E. Samson, Content and turnover of gangliosides in rat brain following behavioral stimulation, *J. Neurochem.* **18:**203–211 (1971).
172. M. E. Smith, The metabolism of myelin lipids, *Advan. Lipid Res.* **5:**241–278 (1968).
173. F. N. LeBaron, Metabolism of myelin constituents, *in* "Handbook of Neurochemistry" (A. Lajtha, ed.) Vol. 3, pp. 561–573, Plenum Press, New York (1970).
174. E. Grossi, P. Paoletti, and R. Paoletti, The *in vitro* and *in vivo* effects of chlorpromazine on brain lipid synthesis, *J. Neurochem.* **6:**73–78 (1960).
175. R. Fumagalli, E. Grossi, and P. Paoletti, The effect of imipramine and desmethylimipramine on lipid biosynthesis in brain and liver, *J. Neurochem.* **10:**213–217 (1963).
176. W. L. Holmes, Drugs affecting lipid synthesis, *in* "Lipid Pharmacology" (R. Paoletti, ed.) pp. 131–184, Academic Press, New York (1964).

177. S. G. Eliasson, Lipid synthesis in peripheral nerve from alloxan diabetic rats, *Lipids* **1:** 237–240 (1966).
178. M. E. Smith, The effect of fasting on lipid metabolism of the central nervous system of the rat, *J. Neurochem.* **10:**531–536 (1963).
179. H. B. White, G. Galli, and R. Paoletti, Brain recovery from essential fatty acid deficiency in developing rats, *J. Neurochem.* **18:**869–882 (1971).
180. S. Refsum, Heredoataxia hemerolopica polyneuritiformis—A previously undescribed familial syndrome. A preliminary communication, *Nord. Med.* **28:**2682–2685 (1945).
181. S. Refsum, Heredopathia atactica polyneuritiformis, *Acta Psychiat. Neurol. Suppl.* **38:** 9–303 (1946).
182. S. Refsum, Heredopathia atactica polyneuritiformis reconsidered, *World Neurol.* **1:**334–337 (1960).
183. D. Steinberg, F. Q. Vroom, W. K. Engel, J. Cammermeyer, C. E. Mize, and J. Avigan, Refsum's disease: A recently characterized lipidosis involving the nervous system, *Ann. Int. Med.* **66:**365–395 (1967).
184. S. Refsum and L. Eldjarn, Heredopathia atactica polyneuritiformis—An inborn error in the metabolism of branched-chain fatty acids, in "Zukunft der Neurologie" (H. G. Bammer, ed.) pp. 36–44, Georg Thieme, Stuttgart (1967).
185. S. Refsum, L. Salomonsen, and M. Skatvedt, Heredopathia atactica polyneuritiformis in children, *J. Pediat.* **35:**335–343 (1949).
186. R. Richterich, S. Rosin, and E. Rossi, Refsum's disease (heredopathia atactica polyneuritiformis). An inborn error of lipid metabolism with storage of 3,7,11,15-tetramethyl hexadecanoic acid. Formal genetics, *Humangenetik* **1:**333–336 (1965).
187. S. Refsum, Heredopathia atactica polyneuritiformis, *Acta Genet. Stat. Med.* **7:**334–347 (1957).
188. R. Richterich, P. van Mechelen, and E. Rossi, Refsum's disease (heredopathia atactica polyneuritiformis): An inborn error of lipid metabolism with storage of 3,7,11,15-tetramethylhexadecanoic acid, *Am. J. Med.* **39:**230–236 (1965).
189. N. C. Nevin, J. N. Cumings, and F. McKeown, Refsum's syndrome, heredopathia atactica polyneuritiformis, *Brain* **90:**419–428 (1967).
190. K. Try, Heredopathia atactica polyneuritiformis (Refsum's disease). The diagnostic value of phytanic acid determination in serum lipids, *Europ. Neurol.* **2:**296–314 (1969).
191. M. D. Toussaint, C. Coers, and M. N. Toppet, Heredopathia atactica polyneuritiformis (Refsum's syndrome). Clinical inquiry and biopsy, *Bull. Soc. Belg. Ophthalmol.* **122:**383 (1959).
192. W. Kahlke, Refsum's syndrome, in "Lipids and Lipidoses" (A. Schettler, ed.) pp. 352–383, Springer-Verlag, New York (1967).
193. S. Refsum, Heredopathia atactica polyneuritiformis, a metabolic disease of nervous tissue, *Nord. Med.* **73:**570 (1965).
194. J. Dereux and S. E. Gruner, Refsum's disease—Biopsy study of a case using the electron microscope, *Rev. Neurol. (Paris)* **109:**564 (1963).
195. S. Refsum, Diagnosis and differential diagnosis of herodopathia atactica polyneuritiformis, *Deutsch. Z. Nervenheilk.* **195:**257–262 (1969).
196. E. M. Ashenhurst, J. H. D. Millar, and T. G. Milliken, Refsum's syndrome affecting a brother and two sisters, *Brit. Med. J.* **2:**415–417 (1958).
197. K. Try, O. Stokke, and L. Eldjarn, Two new cases of heredopathia atactica polyneuritiformis (Refum's disease) with demonstrated phytanic acid accumulation, *Scand. J. Clin. Lab. Invest.* **17:**(Suppl. 86), 195 (1965).
198. D. Steinberg and J. H. Herndon, Jr., Refsum's disease: Phytanic acid storage disease, in "Shy's The Cellular and Molecular Basis of Neurologic Disease" (G. M. Shy, E. S. Goldensohn, and S. H. Appel, eds.) Lea and Febiger, Philadelphia (in press).
199. J. Cammermeyer, Neuropathological changes in hereditary neuropathies: Manifestation of the syndrome heredopathia atactica polyneuritiformis in the presence of interstitial hypertrophic polyneuropathy, *J. Neuropathol. Exptl. Neurol.* **15:**340–367 (1956).

200. E. H. Kolodny, W. K. Hass, B. Lane, and W. D. Drucker, Refsum's syndrome: Report of a case including electron microscopic studies of the liver, *Arch. Neurol.* **12**:583–596 (1965).

201. L. van Bogaart, P. van Mechelen, J. J. Martin, and G. C. Guazzi, On the neuropathology of Refsum–Thiebaut disease, *Rev. Neurol.* **116**:229–240 (1967).

202. M. Fardeau and W. K. Engel, Ultrastructural study of a peripheral nerve biopsy in Refsum's disease, *J. Neuropathol. Exptl. Neurol.* **28**:278–294 (1969).

203. E. Klenk and W. Kahlke, On the existence of 3,7,11,15-tetramethylhexadecanoic acid (phytanic acid) in the cholestero esters and other lipid fractions of organs from an illness of unknown genesis (thought to be heredopathia atactica polyneuritiformis, Refsum's syndrome), *Z. Physiol. Chem.* **33**:133–139 (1963).

204. W. Kahlke, Refsum's syndrome—Lipidchemical investigations in 9 cases, *Klin. Wschr.* **42**:1011–1016 (1964).

205. W. Kahlke and R. Richterich, Refsum's disease (heredopathia atactica polyneuritiformis), an inborn error of lipid metabolism with storage of 3,7,11,15-tetramethylhexadecanoic acid. II. Isolation and identification of the storage product, *Am. J. Med.* **39**:237–241 (1965).

206. G. J. Kremer, On the existence of 3,7,11,15-tetramethylhexadecanoic acid in the lipids of normal sera, *Klin. Wschr.* **43**:517–518 (1965).

207. J. Avigan, The presence of phytanic acid in normal human and animal plasma, *Biochim. Biophys. Acta* **116**:391–394 (1966).

208. R. P. Hansen, 3,7,11,15-Tetramethylhexadecanoic acid: Its occurrence in the tissues afflicted with Refsum's syndrome, *Biochim. Biophys. Acta* **106**:304–310 (1965).

209. J. Dereux, A. Lowenthal, Y. Mardens, and D. Korcher, Phytanic acid levels in serum and central nervous system in Refsum's disease, *Pathol. Europ.* **3**:468–473 (1968).

210. S.-C. Tsai, Oxidation of phytanic acid as related to Refsum's disease, *N. Y. State J. Med.* **69**:3149–3152 (1969).

211. K. A. Karlsson, A. Norrby, and B. Samuelsson, Use of thin-layer chromatography for the preliminary diagnosis of Refsum's disease (heredopathia atactica polyneuritiformis), *Biochim. Biophys. Acta* **144**:162–164 (1967).

212. W. S. Alexander, Phytanic acid in Refsum's syndrome, *J. Neurol. Neurosurg. Psychiat.* **29**:412–416 (1966).

213. M. Bonduelle, P. Gouygnes, G. Lormeau, G. Deloux, P. Laudat, and L. M. Wolf, Refsum's disease; studies of lipids of serum and urine, *Rev. Neurol.* **115**:933–942 (1966).

214. D. Steinberg, Remarks on the biochemical basis of Refsum's disease, *Nord. Med.* **73**:571–572 (1965).

215. L. Eldjarn, Biochemical points of view on the origin of phytanic acid, *Nord. Med.* **73**:571 (1965).

216. J. H. Baxter and G. W. A. Milne, Phytenic acid: Identification of five isomers in chemical and biological products of phytol, *Biochim. Biophys. Acta* **176**:265–277 (1969).

217. J. Avigan, D. Steinberg, A. Gutman, C. E. Mize, and G. W. A. Milne, Alpha-decarboxylation, an important pathway for degradation of phytanic acid in animals, *Biochem. Biophys. Res. Commun.* **24**:838–844 (1966).

218. C. E. Mize, D. Steinberg, J. Avigan, and H. M. Fales, A pathway for oxidative degradation of phytanic acid in mammals, *Biochem. Biophys. Res. Commun.* **25**:359–365 (1966).

219. C. E. Mize, J. Avigan, D. Steinberg, R. C. Pittman, H. M. Fales, and G. W. A. Milne, A major pathway for the mammalian oxidative degradation of phytanic acid, *Biochim. Biophys. Acta* **176**:720–739 (1969).

220. S.-C. Tsai, J. H. Herndon, Jr., B. W. Uhlendorf, H. M. Fales, and C. E. Mize, The formation of alpha-hydroxyphytanic acid from phytanic acid in mammalian tissues, *Biochem. Biophys. Res. Commun.* **28**:571–577 (1967).

221. S.-C. Tsai, J. Avigan, and D. Steinberg, Studies on the alpha-oxidation of phytanic acid by rat liver mitochondria, *J. Biol. Chem.* **244**:2682–2692 (1969).

222. E. Klenk and G. J. Kremer, Investigations of the metabolism of phytol, dihydrophytol and of phytanic acid, *Z. Physiol. Chem.* **343**:39–51 (1965).

223. R. P. Hansen, F. B. Shorland, and I. A. M. Prior, The fate of phytanic acid when administered to rats, *Biochim. Biophys. Acta* **116**:178–180 (1966).

224. R. P. Hansen, F. B. Shorland, and I. A. M. Prior, The occurrence of 4,8,12-trimethyltridecanoic acid in the tissues of rats fed high levels of phytanic acid, *Biochim. Biophys. Acta* **152**:642–644 (1968).

225. D. Steinberg, J. Avigan, C. Mize, L. Eldjarn, K. Try, and S. Refsum, Conversion of U-C^{14}-phytol to phytanic acid and its oxidation in heredopathia atactica polyneuritiformis, *Biochem. Biophys. Res. Commun.* **19**:783–789 (1965).

226. D. Steinberg, C. E. Mize, J. Avigan, H. M. Fales, L. Eldjarn, K. Try, O. Stokke, and S. Refsum, Studies on the metabolic error in Refsum's disease, *J. Clin. Invest.* **46**:313–322 (1967).

227. C. E. Mize, J. Avigan, J. H. Baxter, H. M. Fales, and D. Steinberg, Metabolism of phytol-U-^{14}C and phytanic acid-U-^{14}C in the rat, *J. Lipid Res.* **7**:692–697 (1966).

228. R. A. P. Kark, W. K. Engel, J. P. Blass, D. Steinberg, and G. O. Walsh, Heredopathia atactica polyneuritiformis (Refsum's disease)—A second trial of dietary therapy in two patients, *Birth Defects: Orig. Art. Ser. (Nerv. Syst.)* **7**:53–55 (1971).

229. L. Eldjarn, K. Try, O. Stokke, A. W. Munthe-Kaas, S. Refsum, D. Steinberg, J. Avigan, and C. Mize, Dietary effects on serum phytanic acid levels and on clinical manifestations in heredopathia atactica polyneuritiformis, *Lancet* **1**:691–693 (1966).

230. D. Steinberg, J. Avigan, C. E. Mize, J. H. Baxter, J. Cammermeyer, H. M. Fales, and P. F. Highet, Effects of dietary phytol and phytanic acid in animals, *J. Lipid Res.* **7**:684–691 (1966).

231. W. Stoffel and W. Kahlke, The transformation of phytol into 3,7,11,15-tetramethylhexadecanoic (phytanic) acid in heredopathia atactica polyneuritiformis (Refsum's syndrome), *Biochem. Biophys. Res. Commun.* **19**:33–36 (1965).

232. J. H. Baxter, D. Steinberg, C. E. Mize, and J. Avigan, Absorption and metabolism of uniformly ^{14}C-labeled phytol and phytanic acid by the intestine of the rat studied with thoracic duct cannulation, *Biochim. Biophys. Acta* **137**:277–290 (1967).

233. D. Steinberg, J. Avigan, C. Mize, and J. Baxter, Phytanic acid formation and accumulation in phytol-fed rats, *Biochem. Biophys. Res. Commun.* **19**:412–416 (1965).

234 F. B. Shorland, R. P. Hansen, and I. A. M. Prior, The effect of phytanic acid on the fatty acid composition of the lipids of the rat with further observations on its metabolism, *Proc. Seventh Intern. Congr. Nutr.* **5**:339 (1966).

235. J. H. Baxter and D. Steinberg, Absorption of phytol from dietary chlorophyll in the rat, *J. Lipid Res.* **8**:615–620 (1967).

236. J. H. Baxter, Absorption of chlorophyll phytol in normal man and in patients with Refsum's disease, *J. Lipid Res.* **9**:636–641 (1968).

237. R. P. Hansen and J. D. Morrison, The isolation and identification of 2,6,10,14-tetramethylpentadecanoic acid from butter fat, *Biochem. J.* **93**:225–228 (1964).

238. R. P. Hansen, Occurrence of 2,6,10,14-tetramethylpentadecanoic acid in sheep fat, *Chem. Ind.* **28**:1258–1259 (1965).

239. R. P. Hansen, 4,8,12-Trimethyltridecanoic acid: Its isolation and identification from sheep perinephric fat, *Biochim. Biophys. Acta* **164**:550–557 (1968).

240. R. P. Hansen, The isolation and identification of 4,8,12-trimethyltridecanoic acid from butter fat, *J. Dairy Res.* **36**:77–85 (1969).

241. W. R. H. Duncan and G. A. Garton, Blood lipids. III. Plasma lipids of the cow during pregnancy and lactation, *Biochem. J.* **89**:414–419 (1963).

242. A. K. Lough, Blood lipids. 4. The isolation of 3,7,11,15-tetramethylhexadecanoic acid (phytanic acid) from ox plasma lipids, *Biochem. J.* **91**:584–588 (1964).

243. R. P. Hansen, 3,7,11,15-Tetramethylhexadecanoic acid: Its occurrence in sheep fat, *New Zealand J. Sci.* **8**:158–160 (1965).

244. R. P. Hansen, Occurrence of 3,7,11,15-tetramethylhexadecanoic acid in ox perinephric fat, *Chem. Ind.* **7**:303–304 (1965).

245. S. Patton and A. A. Benson, Phytol metabolism in the bovine, *Biochim. Biophys. Acta* **125:** 22–32 (1966).
246. W. Sonneveld, P. H. Begemann, G. J. van Beers, R. Keuning, and J. C. M. Schogt, 3,7,11, 15-Tetramethylhexadecanoic acid, a constituent of butter fat, *J. Lipid Res.* **3:**351–355 (1962).
247. L. Eldjarn, K. Try, and O. Stokke, The existence of an alternative pathway for the degradation of branched-chain fatty acids, and its failure in heredopathia atactica polyneuritiformis (Refsum's disease), *Biochim. Biophys. Acta* **116:**395–397 (1966).
248. O. Stokke, Alpha-oxidation of fatty acids in various mammals, and a phytanic acid feeding experiment in an animal with a low alpha-oxidation capacity, *Scand. J. Clin. Lab. Invest.* **20:**305–312 (1967).
249. O. Stokke, Evidence against a CO_2-fixation mechanism in the degradation of a beta-methyl-substituted fatty acid in mammals, *Biochim. Biophys. Acta* **176:**230–236 (1969).
250. L. Eldjarn, Heredopathia atactica polyneuritiformis (Refsum's disease)—A defect in the omega-oxidation mechanism of fatty acids, *Scand. J. Clin. Lab. Invest.* **17:**178–181 (1965).
251. L. Eldjarn, K. Try, and O. Stokke, The ability of patients with heredopathia atactica polyneuritiformis to omega-oxidize and degrade several isoprenoid branch-chained fatty structures, *Scand. J. Clin. Lab. Invest.* **18:**141–150 (1966).
252. K. Try and L. Eldjarn, Normalization of the tricaprin test for omega-oxidation in Refsum's disease upon lowering of serum phytanic acid, *Scand. J. Clin. Lab. Invest.* **20:**294–296 (1967).
253. K. Try, The *in vitro* omega-oxidation of phytanic acid and other branched-chain fatty acids by mammalian liver, *Scand. J. Clin. Lab. Invest.* **22:**224–230 (1968).
254. D. Steinberg, J. H. Herndon, Jr., B. W. Uhlendorf, C. E. Mize, J. Avigan, and G. W. A. Milne, Refsum's disease: Nature of the enzyme defect, *Science* **156:**1740–1742 (1967).
255. D. Steinberg, J. Avigan, C. E. Mize, J. H. Herndon, Jr., H. M. Fales, and G. W. A. Milne, The nature of the metabolic defect in Refsum's disease, *Pathol. Europ.* **3:**450–458 (1968).
256. J. H. Herndon, Jr., D. Steinberg, B. W. Uhlendorf, and H. M. Fales, Refsum's disease: Characterization of the enzyme defect in cell culture, *J. Clin. Invest.* **48:**1017–1032 (1969).
257. L. Eldjarn, O. Stokke, and K. Try, Alpha-oxidation of branched-chain fatty acids in man and its failure in patients with Refsum's disease showing phytanic acid accumulation, *Scand. J. Clin. Lab. Invest.* **18:**694–695 (1966).
258. O. Stokke, K. Try, and L. Eldjarn, Alpha-oxidation as an alternative pathway for the degradation of branched-chain fatty acids in man, and its failure in patients with Refsum's disease, *Biochim. Biophys. Acta* **144:**271–284 (1967).
259. K. Try, Indications of only a partial defect in the alpha-oxidation mechanism in Refsum's disease, *Scand. J. Clin. Lab. Invest.* **20:**255–262 (1967).
260. C. E. Mize, J. H. Herndon, Jr., J. P. Blass, G. W. A. Milne, C. Follansbee, P. Laudat, and D. Steinberg, Localization of the oxidative defect in phytanic acid degradation in patients with Refsum's disease, *J. Clin. Invest.* **48:**1033–1040 (1969).
261. J. H. Herndon, Jr., D. Steinberg, and B. W. Uhlendorf, Refsum's disease: Defective oxidation of phytanic acid in tissue cultures derived from homozygotes and heterozygotes, *New Engl. J. Med.* **281:**1034–1038 (1969).
262. D. Steinberg, C. E. Mize, H. J. Herndon, Jr., H. M. Fales, W. K. Engel, and F. Q. Vroom, Phytanic acid in patients with Refsum's syndrome and response to dietary treatment, *Arch. Int. Med.* **125:**75–87 (1970).
263. J. C. Fratantoni, C. W. Hall, and E. F. Neufeld, Hurler and Hunter syndromes—Mutual correction of the defect in cultured fibroblasts, *Science* **162:**570–572 (1968).
264. B. S. Danes and A. G. Bearn, Correction of cellular metachromasia in cultured fibroblasts in several inherited mucopolysaccharidoses, *Proc. Natl. Acad. Sci.* **67:**357–364 (1970).
265. J. P. Blass, J. Avigan, and D. Steinberg, α-Hydroxy fatty acids in hereditary ataxic polyneuritis (Refsum's disease), *Biochim. Biophys. Acta* **187:**36–41 (1969).
266. J. Cammermeyer, Personal communication.

267. J. P. Blass, J. Avigan, and R. G. Clark, Effects of phytol feeding and experimental allergic encephalomyelitis on myelin synthesis, *Fed. Proc.* **28**:838 (1969).
268. J. P. Blass, R. A. P. Kark, and W. K. Engel, Clinical studies of a patient with pyruvate decarboxylase deficiency, *Arch. Neurol.* **25**:449–460 (1971).
269. J. P. Blass, Unpublished observations.
270. W. Zeman and P. Dyken, Neuronal ceroid-lipofuchsinosis (Batten's disease)—Relationship to amaurotic familial idiocy, *Pediatrics* **44**:570–583 (1969).
271. W. Zeman, The neuronal ceroid-lipofuchsinoses–Batten's syndrome, *Trans. Am. Soc. Neurochem.* **2**:51 (1971).
272. P. E. Duffy, M. Kornfeld, and K. Suzuki, Neurovisceral storage disease with curvilinear bodies, *J. Neuropathol. Exptl. Neurol.* **27**:351–370 (1968).
273. M. E. Richardson and J. H. Bornhofen, Early childhood cerebral lipidosis with prominent myoclonus, *Arch. Neurol.* **18**:34–43 (1968).
274. W. S. Harcourt and E. A. Porta, Ceroid, *Am. J. Med. Sci.* **250**:324–345 (1965).
275. J. H. Menkes, M. Alter, G. K. Steigleder, D. R. Weaky, and J. H. Sung, A sex-linked recessive disorder with retardation of growth, peculiar hair, and focal cerebral and cerebellar degeneration, *Pediatrics* **29**:764–779 (1962).
276. M. J. Aguilar, D. L. Chadwick, K. Okuyama, and S. Kamoshita, Kinky hair disease. I. Clinical and pathological features, *J. Neuropathol. Exptl. Neurol.* **25**:507–522 (1966).
277. H. Pilz and H. Jatzkewitz, Thin layer chromatographic determination of C_{18} and C_{24} sphingomyelin in normal and pathologic brain including a case of Niemann–Pick disease, *J. Neurochem.* **11**:603–611 (1964).
278. W. T. Norton and S. Poduslo, cited in W. Kahlke, Metachromatic leukodystrophy, *in* "Lipids and Lipidoses" (G. Schettler, ed.) pp. 310–331, Springer-Verlag, New York (1967).
279. L. van Bogaert, H. J. Schere, and E. Epstein, "A Cerebral Form of Generalized Cholesterosis," Masson et Cie, Paris (1937).
280. M. Philippart and L. van Bogaert, Cholestanolosis (cerebrotendinous xanthomatosis). A follow-up study on the original family, *Arch. Neurol.* **21**:603–610 (1969).
281. J. H. Menkes, J. R. Schimschock, and P. D. Swanson, Cerebrotendinous xanthomatosis—The storage of cholestanol within the nervous system, *Arch. Neurol.* **19**:47–53 (1968).
282. J. H. Menkes, Personal communication.
283. M. Wolman, Involvement of nervous tissue in primary familial xanthomatosis with adrenal calcification, *Pathol. Europ.* **3**:259–265 (1968).
284. G. C. Guazzi, J. J. Martin, M. Philippart, H. Roels, C. Hooff, H. van der Eecken, M. J. Delbeke, and L. Urints, Wolman's disease. Distribution and significance of the central nervous system lesions, *Pathol. Europ.* **3**:266–277 (1968).
285. A. C. Crocker, G. F. Vawter, E. B. D. Newhauser, and A. Rosowsky, Wolman's disease—Three new patients with a recently described lipidosis, *Pediatrics* **35**:627–640 (1965).
286. B. D. Lake and A. D. Patrick, Wolman's disease—Deficiency of E600 resistant acid esterase activity with storage of lipids in lysosomes, *J. Pediat.* **76**:262–266 (1970).
287. M. Philippart, Gargoylism, *Rein et Fois: Maladies Nutr.* **9**:245–249 (1966).
288. M. Philippart, Personal communication.
289. D. S. Fredrickson, Classification and features of lipidoses affecting the nervous system, *Pathol. Europ.* **3**:121–142 (1968).
290. G. Majno and M. L. Karnovsky, A biochemical and morphological study of myelination and demyelination. II. Lipogenesis *in vitro* by rat nerves following transection, *J. Exptl. Med.* **108**:197–213 (1958).
291. J. Domonkos and L. Heiner, Decomposition of phospholipids during Wallerian degeneration, *J. Neurochem.* **15**:87–91 (1968).
292. N. Miani, The relationship between axon and Schwann cell. Phospholipid metabolism of degenerating and regenerating peroneal–tibial nerves of the rabbit *in vitro*, *J. Neurochem.* **9**:525–536 (1952).

293. M. A. Stewart, J. V. Passonneau, and O. H. Lowry, Substrate changes in peripheral nerve during substrate ischaemia and Wallerian degeneration, *J. Neurochem.* **12**:719–727 (1965).

294. E. T. Pritchard and R. J. Rossiter, Chemical studies of peripheral nerve during Wallerian degeneration. X. *In vitro* incorporation of ^{14}C-labelled precursors into phosphatides, *J. Neurochem.* **3**:341–346 (1958).

295. D. Kline, W. L. Magee, E. T. Pritchard, and R. J. Rossiter, Chemical studies of peripheral nerve during Wallerian degeneration. VII. Labelling of phospholipid and cholesterol from carboxy-^{14}C acetate, *J. Neurochem.* **3**:52–58 (1958).

296. E. R. Peterson and M. R. Murray, Patterns of peripheral demyelination *in vitro, Ann. N.Y. Acad. Sci.* **122**:39–50 (1965).

297. G. Majno and M. L. Karnovsky, Experimental study of diptheritic polyneuritis in the rabbit and guinea pig. II. The effect of diptheria toxin on lipid biosynthesis by guinea pig nerve, *J. Neuropathol. Exptl. Neurol.* **19**:7–24 (1960).

298. N. Malamud, Neuropathology of phenylketonuria, *J. Neuropathol. Exptl. Neurol.* **25**: 254–268 (1966).

299. J. H. Menkes, The pathogenesis of mental retardation in phenylketonuria and other inborn errors of amino acid metabolism, *Pediatrics* **39**:297–308 (1967).

300. J. H. Menkes, Cerebral lipids in phenylketonuria, *Pediatrics* **37**:967–978 (1966).

301. J. L. Foote, R. J. Allen, and B. W. Agranoff, Fatty acids in esters and cerebrosides of human brain in phenylketonuria, *J. Lipid Res.* **6**:518–524 (1965).

302. L. C. Scheinberg and S. R. Korey, Multiple sclerosis, *Ann. Rev. Med.* **13**:411–430 (1962).

303. R. E. Caspary, Demyelinating diseases and allergic encephalomyelitis. A comparative review with special reference to multiple sclerosis, *in* "Biochemical Aspects of Neurological Disorders" (J. N. Cumings, ed.) pp. 44–61, Blackwell Press, Oxford (1968).

304. Y. Kishimoto, N. S. Radin, W. W. Turtellotte, J. A. Parker, and H. H. Itabashi, Gangliosides and glycerophospholipids in multiple sclerosis white matter, *Arch. Neurol.* **16**:44–54 (1967).

305. J. Bernsohn and L. M. Stephanides, Aetiology of multiple sclerosis, *Nature* **215**:821–823 (1967)

306. Editorial: Fatty acids and multiple sclerosis, *Lancet* **2**:708–709 (1967).

307. H. Jatzkewitz, The role of cerebroside sulfuric esters in leukodystrophy and a new method for the quantitative ultramicrodetermination of the brain sphingolipids, *in* "Brain Lipids and Lipoproteins and the Leucodystrophies" (J. Folch-Pi and H. Bauer, eds.) pp. 147–152, Elsevier, New York (1963).

308. H. Jatzkewitz, A new method for quantitative ultramicrodetermination of sphingolipids from brain, *Z. Physiol. Chem.* **336**:25–39 (1964).

309. C. M. Plum and S. E. Hansen, The cerebral lipids in multiple sclerosis, *Acta Psychiat. Neurol. Scand.* **141**:83–92 (1960).

310. J. N. Cumings and H. Goodwin, Sphingolipids and phospholipids of myelin in multiple sclerosis, *Lancet* **2**:664–665 (1968).

311. H. P. Schwarz, L. Dreisbach, M. Barrionevo, A. Kleschik, and I. Kostyk, Chromatography of sphingolipids in human brain, *J. Lipid Res.* **2**:208–214 (1961).

312. P. J. Riekkinen, J. Palo, A. U. Arstila, H. J. Savolainen, U. K. Rinne, E. K. Kaualo, and H. Frey, Protein composition of multiple sclerosis myelin, *Arch. Neurol.* **24**:545–549 (1971).

313. B. Gerstl, M. J. Kahnke, J. K. Smith, M. G. Tavastjerna, and R. B. Hayman, Brain lipids in multiple sclerosis and other diseases, *Brain* **84**:310–319 (1961).

314. R. W. R. Baker, R. H. S. Thompson, and K. J. Zilkha, Fatty-acid composition of brain lecithins in multiple sclerosis, *Lancet* **1**:26–27 (1963).

315. R. W. R. Baker, R. H. Thompson, and K. J. Zilkha, Serum fatty acids in multiple sclerosis, *J. Neurol. Neurosurg. Psychiat.* **27**:408–414 (1964).

316. R. W. R. Baker, R. H. S. Thompson, and K. J. Zilkha, Changes in the amounts of linoleic acid in the serum of patients with multiple sclerosis, *J. Neurol. Neurosurg. Psychiat.* **29**: 95–98 (1966).

317. R. W. R. Baker, H. Sanders, R. H. S. Thompson, and K. J. Zilkha, Serum cholesterol linoleate levels in multiple sclerosis, *J. Neurol. Neurosurg. Psychiat.* **28**:212–217 (1965).

318. S. Gul, A. D. Smith, R. H. S. Thompson, H. Payling-Wright, and K. J. Zilkha, Fatty acid composition of phospholipids from platelets and erythrocytes in multiple sclerosis, *J. Neurol. Neurosurg. Psychiat.* **33**:506–510 (1970).

319. J. N. Cumings, R. C. Shortman, and T. Skirbic, Lipid studies in blood and brain in multiple sclerosis and motor neurone disease, *J. Clin. Pathol.* **18**:611–644 (1965).

320. J. F. Mead, Personal communication.

321. M. G. McCall, T. L. G. Brereton, A. Dawson, K. Millingen, J. M. Sutherland, and E. D. Acheson, Frequency of multiple sclerosis in three Australian cities—Perth, Newcastle and Hobart. *J. Neurol. Neurosurg. Psychiat.* **31**:1–9 (1968).

322. D. S. P. Patterson, S. Terlecki, J. T. Done, D. Sweasey, and C. N. Herbert, Neurochemistry of the spinal cord in experimental border disease (hypomyelinogenesis congenita) of lambs, *J. Neurochem.* **18**: 883–894 (1971).

323. D. H. Henneman, M. D. Altschule, R. M. Gonce, and L. Alexander, Carbohydrate metabolism in brain disease—glucose metabolism in multiple sclerosis, *Arch. Neurol. Psychiat.* **72**: 688–695 (1954).

324. I. C. K. McKenzie and G. R. Webster, Studies on intermediate carbohydrate metabolism in multiple sclerosis, *J. Neurol. Neurosurg. Psychiat.* **23**: 127–132 (1960).

325. A. E. Hansen, R. A. Stewart, G. Hughes, and L. Soderhjelm, Relation of linoleic acid to infant feeding, *Acta Paediat.* **51**: (Suppl. 137), 5–41 (1962).

326. R. Caren and L. Carbo, Plasma fatty acids in pancreatic cystic fibrosis and liver disease, *J. Clin. Endocrinol.* **26**:470–477 (1966).

327. S. Futterman, J. L. Downer, and A. Hendrickson, Effect of essential fatty acid deficiency on the fatty acid composition, morphology and electroretinographic response of the retina, *Invest. Ophthalmol.* **10**:151–156 (1971).

328. D. N. Menton, The effects of essential fatty acid deficiency on the fine structure of mouse skin, *J. Morphol.* **132**:181–205 (1970).

329. C. Galli, H. B. White, and R. Paoletti, Brain lipid modifications induced by essential fatty acid deficiency in growing male and female rats, *J. Neurochem.* **17**:347–355 (1970).

330. J. Clausen and J. Møller, Allergic encephalomyelitis induced by brain antigen after deficiency in polyunsaturated fatty acids during myelination, *Intern. Arch. Allergy* **36**: 224–233 (1969).

331. C. O. Walker, D. W. McCandless, J. D. McGarry, and S. Schenker, Cerebral energy metabolism in short-chain fatty acid induced coma, *J. Lab. Clin. Med.* **76**:569–583 (1970).

332. W. K. Engel, N. A. Vick, C. J. Glueck, and R. I. Levy, A skeletal muscle disorder associated with intermittent symptoms and a possible defect of lipid metabolism, *New Engl. J. Med.* **282**:697–704 (1970).

333. R. Bressler, Carnitine and the twins, *New Engl. J. Med.* **282**:745–746 (1970).

334. D. S. Fredrickson, Familial high density lipoprotein deficiency: Tangier disease, *in* "The Metabolic Basis of Inherited Disease" (J. B. Stanbury, J. B. Wyngaarden, and D. S. Fredrickson, eds.) pp. 486–508, McGraw-Hill, New York (1966).

335. W. Kahlke, Tangier disease, *in* "Lipids and Lipidoses" (G. Schettler, ed) pp. 401–411, Springer-Verlag, New York (1967).

336. J. F. Schwarz, L. P. Rowland, H. Eder, P. A. Marks, E. F. Osserman, E. Hirschberg, and H. Anderson, Bassen–Kornzweig syndrome—Deficiency of serum β-lipoprotein, *Arch. Neurol.* **8**:438–454 (1963).

337. W. K. Engel, J. D. Dorman, R. I. Levy, and D. S. Fredrickson, Neuropathy in Tangier disease, *Arch. Neurol.* **17**:1–9 (1967).

338. A. M. Gotto, R. I. Levy, K. John, and D. S. Fredrickson, On the protein defect in a-β-lipoproteinemia, *New Engl. J. Med.* **289**:813–818 (1971).

339. D. S. Fredrickson, R. I. Levy, and F. T. Lindgren, A comparison of heritable abnormal lipoprotein patterns as defined by 2 different techniques, *J. Clin. Invest.* **47**:2446–2457 (1969).

340. H. Mars, L. A. Lewis, A. L. Robertson, A. Butkus, and G. H. Williams, Familial hypo-β-lipoproteinemia: A genetic disorder of lipid metabolism with nervous system involvement, *Am. J. Med.* **46**:886–900 (1969).

341. P. T. Kuo and D. R. Bassett, Blood and tissue lipids in a family with hypo-β-lipoproteinemia, *Circulation* **26**:660 (1962).

342. G. B. Phillips, Quantitative chromatographic studies of plasma and red blood cell lipids in patients with acanthocytosis, *J. Lab. Clin. Med.* **59**:357–363 (1962).

343. P. Ways, C. F. Reed, and D. J. Hanahan, Abnormalities of erythrocyte and plasma lipids in acanthocytosis, *J. Clin. Invest.* **42**:1248–1260 (1963).

344. J. W. Estes, T. J. Morley, I. M. Levine, and C. P. Emerson, A new hereditary acanthocytosis syndrome, *Am. J. Med.* **42**:868–881 (1967).

345. I. M. Levine, J. W. Estes, and J. M. Looney, Hereditary neurologic disease with acanthocytosis, *Arch. Neurol.* **19**:403–409 (1968).

346. E. M. R. Critchley, D. B. Clark, and A. Wikler, Acanthocytosis and neurologic disorder without a-β-lipoproteinemia, *Arch. Neurol.* **18**: 134–140 (1968).

347. R. L. Sidman, M. C. Green, and S. H. Appel, "Catalog of the Neurologic Mutants of the Mouse," Harvard University Press, Cambridge, Mass. (1965).

348. S. Pollet, J. M. Bourre, G. Dubois, and N. Baumann, Biosynthesis of long chain fatty acids in mouse brain microsomes, *Proc. Third Intern. Meeting Intern. Soc. Neurochem.*, *Budapest*, p. 250 (1971).

349. N. A. Baumann, M. L. Harpin, and J. M. Bourre, Long chain fatty acid formation—Key step in myelination studied in mutant mice, *Nature* **227**:960–961 (1970).

350. H. Singh, N. Spritz, and B. Geyer, Brain myelin in the Quaking mouse, *J. Lipid Res.* **12**:473–481 (1971).

351. R. M. C. Dawson and N. Clarke, Cerebral phospholipids in "Quaking" mice, *J. Neurochem.* **18**:1313–1316 (1971).

352. G. Hauser, J. Eichberg, and S. Jacobs, Polyphosphoinositide levels and biosynthesis in Quaking mouse brain, *Biochem. Biophys. Res. Commun.* **43**:1072–1080 (1971).

353. N. M. Neskovic, J. L. Nussbaum, and P. Mandel, Glycolipid metabolism in a myelination disorder of Jimpy and Quaking mice, *Brain Res.* **21**:39–53 (1970).

354. E. Constantino-Ceccarini and P. Morell, Quaking mice. *In vitro* studies of brain sphingolipid biosynthesis, *Brain Res.* **29**:75–84 (1971).

355. T. Kurihara, J. L. Nussbaum, and P. Mandel, 2′, 3′-cyclic nucleotide 3′-phosphohydrolase in the brain of the "Jimpy" mouse, a mutant with deficient myelination, *Brain Res.* **13**: 401–403 (1969).

356. J. L. Nussbaum, N. Neskovic, and P. Mandel, Fatty acid composition of phospholipids and glycolipids in Jimpy mouse brain, *J. Neurochem.* **18**:1529–1543 (1971).

357. J Rabinowitz, Enzymic studies on dystrophic mice and their litter-mates (lipogenesis and cholesterol-genesis), *Biochim. Biophys. Acta* **43**:337–338 (1960).

358. W. L. Nyhan, ed., "Amino Acid Metabolism and Genetic Variation," McGraw-Hill, New York (1968).

359. P. Lampert and S. Carpenter, Electron microscopic studies on the vascular permeability and the mechanism of demyelination in experimental allergic encephalitis, *J. Neuropathol. Exptl. Neurol.* **24**:11–24 (1965).

360. D. Steinberg, Phytanic acid storage disease (Refsum's syndrome), *in* "The Metabolic Basis of Inherited Disease" (J. B. Stanbury, J. B. Wyngaarden, and D. S. Fredrickson, eds), McGraw-Hill, New York (1972).

361. H. J. Kayden, T. J. Reagan, C. E. Mize, J. H. Herndon, Jr., and D. Steinberg, Diffuse cerebral sclerosis erroneously reported as Refsum's syndrome, *Arch, Neurol.* **28**:304 (1973).

362. D. Hutton and D. Steinberg, Identification of propionate as a degradation product of phytanic acid oxidation in rat and human tissues, *J. Biol. Chem.* (in press).

Chapter 7

BILIRUBIN ENCEPHALOPATHY

Marilyn Louise Cowger

*Albany Medical College of Union University
and
State University of New York at Albany
Albany, New York*

I. INTRODUCTION AND HISTORICAL PERSPECTIVE

So-called physiological jaundice is a mild, transient increase in concentration of serum bilirubin of the unconjugated variety seen during the first few days of life of the newborn infant. It occurs regularly and has been recognized for many years. In fact, the first mention of jaundice occurring in the newborn may go back as far as 1473. The very early history of jaundice in the newborn has been reviewed in a number of works, to which the interested reader is referred.[1-4]

When one superimposes upon this "physiological jaundice" additional mechanisms which further increase bilirubin production or decrease bilirubin excretion, then serum bilirubin concentrations may rise so high that the welfare of the infant is in jeopardy: unconjugated bilirubin, unbound to protein, is cytotoxic and can lead to severe brain damage or even death. Such mechanisms include increased red cell destruction regardless of cause (e.g., erythroblastosis fetalis, defined as sensitization of newborn red cells due to antigenic differences with the mother, congenital hemolytic anemias, and sepsis), delay of liver maturation, inborn errors of bilirubin metabolism, and the presence of factors inhibitory to the hepatic mechanism for detoxifying bilirubin, to mention a few.

It took many years before the relationship of bilirubin to the brain damage seen in some jaundiced infants was understood. The term *kernicterus,* the yellow staining of certain nuclei of the brain, was originally coined by Schmorl in 1903[3] (although the condition had been mentioned earlier by Orth in 1875[2,3]) to describe a gross anatomical finding. The brain damage associated with nuclear staining was not recognized as a complication of severe jaundice until the later 1940s.[5,6] However, the relationship was substantiated by the demonstration in 1950 that exchange transfusion prevented kernicterus[6,7] and, furthermore, controlled hyperbilirubinemia.[6] The study by Hsia et al.[8] demonstrated the quantitative relationship between the development of kernicterus and serum bilirubin concentrations above 20 mg %. The term *kernicterus* was changed in meaning to include, in addition to the anatomical finding, a clinical picture occurring in the newborn period characterized by muscular rigidity, opisthotonus, ocular signs, and disturbance of respiration with the possibility of death. Survivors manifested variable neurological sequelae ranging from mild motor disturbances to severe athetoid cerebral palsy, often with mental retardation, difficulties in vertical gaze, and deafness. This clinical complex was at first recognized to be associated primarily with erythroblastosis fetalis, but the work of Zuelzer and Mudgett[9] demonstrated that kernicterus could result from a variety of causes.

In 1953, Claireaux et al.[10] showed that kernicterus was secondary to indirect (unconjugated) bilirubin deposited in the brain by extracting the pigment from brain and comparing the spectral curves and the chromatographic characteristics of the extracted pigment with a standard bilirubin preparation. The pioneering work of Day[11] showing that bilirubin depressed the respiration of chopped rat brain soon followed. With the demonstration that kernicterus was secondary to the deposition of bilirubin in the brain, that bilirubin was cytotoxic, and that its distribution was not exclusively limited to brain nuclei, the term *bilirubin encephalopathy* became more appropriate, and this is the preferred term today.

The prevention of hyperbilirubinemia in the newborn infant and its clinical management have depended on the acquisition of the appropriate physiological and biochemical knowledge. After the recognition by Landsteiner and Wiener[12] in 1940 of the Rh antigen on infants' red cells, two events followed. In 1946, Wallerstein[13] performed one of the first exchange transfusions Such a procedure can accomplish two things: remove many of the sensitized red cells (if erythroblastosis fetalis is the clinical problem) and remove some of the accumulated bilirubin. In 1961, Finn et al.[14] demonstrated that Rh sensitization could be prevented by passively administering Rh antibody during gestation to the Rh-negative mother not yet immunized, thus eliminating one of the major causes of severe hyperbilirubinemia. In 1963, Liley[15] reported the first successful intrauterine transfusion, a procedure which decreases the number of stillbirths due to erythroblastosis fetalis.

Other developments were aimed at reducing serum bilirubin levels. In 1958, Cremer *et al.*[16] demonstrated that infants placed under blue fluorescent light had lower concentrations of serum bilirubin. Phototherapy for the treatment of neonatal jaundice soon became popular in Europe and South America but saw little use in the United States until the later 1960s.[17] At present, it is widely used, but controversy still exists because of concern over the lack of identification of the breakdown products and their possible toxicity as well as incomplete knowledge of the biological effects of continuous light on newborn infants.[18] In 1966, Yaffe *et al.*[19] showed that phenobarbital and related compounds reduced the serum bilirubin concentration of a hyperbilirubinemic infant. It was presumed that the glucuronide-conjugating ability of the liver had been enhanced. Many investigations followed in which the drug was administered to the mother before delivery and to the infant following delivery or just to the infant, and most studies demonstrated reduction of serum bilirubin concentrations. However, phenobarbital probably affects bilirubin metabolism in multiple ways, as well as affecting many other metabolic pathways in the liver. Currently it is not widely used in therapy. (See Behrman and Fisher[20] for review of phenobarbital studies and recommendations for use.) Still another approach to lowering serum concentrations of bilirubin in the neonate is to reduce the fraction derived from its enterohepatic circulation. This was attempted initially by the oral administration of charcoal[21] and cholestyramine[22] but more recently possibly achieved through the use of agar.[23]

II. CHEMISTRY AND METABOLISM OF BILIRUBIN WITH SPECIAL NOTE OF VARIATIONS SEEN IN THE NEWBORN PERIOD

Bilirubin is a linear tetrapyrrole most likely in the lactam configuration (Fig. 1). It is soluble in aqueous alkaline solutions and in chloroform, but its true solubility at physiological pH is very low. When dissolved in sodium hydroxide followed by neutralization, the true solubility of bilirubin is only about 0.1 μM at pH 7.4 and ionic strength 0.1 M. At this pH, a stable colloidal solution forms which makes the apparent solubility appear higher.[24] As the pH is raised, the true solubility increases; at pH 8.2, colloid formation is no longer seen. Bilirubin is known to have a high lipid solubility.

In the presence of light, bilirubin decomposes rapidly, biliverdin being an early intermediate followed by degradation to a number of water-soluble derivatives. Oxygen is necessary for the photodecay of bilirubin in alkaline solutions. The addition of albumin stabilizes solutions in the dark but under illumination may accelerate the early phase of photodecomposition to biliverdin.[25]

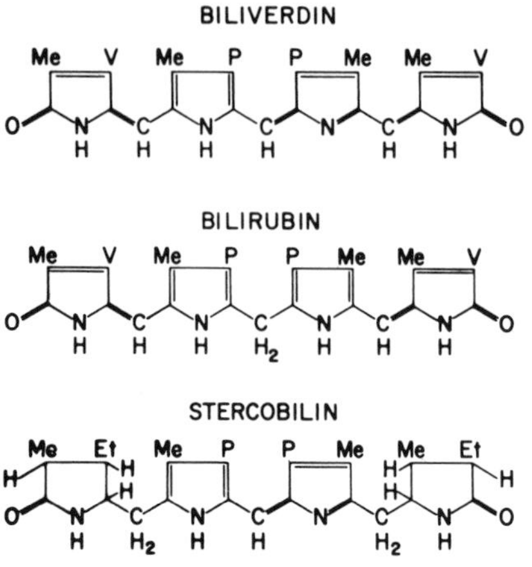

BILIVERDIN

BILIRUBIN

STERCOBILIN

Fig. 1. Molecular structure of the common bile pigments. Me, $-CH_3$; V, $-CH=CH_2$; P, $-CH_2-CH_2-COOH$; Et, $-CH_2-CH_3$.

When reacted with diazotized sulfanilic acid, bilirubin forms a violet color which can be quantified: the well-known van den Bergh reaction.[26] The addition of diazotized sulfanilic acid to serum results in a violet color which develops immediately (the direct reaction); following the addition of an accelerator such as alcohol, further color develops. The intensity of the total color developed is directly proportional to the total serum bilirubin concentration, and the difference between the total and the direct reaction is the so-called indirect-reacting bilirubin. In the past, the direct fraction has been correlated with the conjugated form of bilirubin or bilirubin diglucuronide, but current evidence suggests that the direct van den Bergh reaction overestimates the amount of bilirubin diglucuronide.[27] Normal serum has very little bilirubin of the conjugated variety.

Approximately 85% of bilirubin arises from the breakdown of the hemoglobin heme of senescent red cells; 15–20% may arise from other sources. The portion of bile pigment that arises much earlier than that resulting from the breakdown of red cells at the end of their 120 day normal life span (the late peak of bile pigment formation) has been extensively investigated.[28–35] This fraction has been referred to as the early-labeled peak of bile pigment. The early-labeled peak was initially defined by studies using N^{15}-labeled glycine as a precursor of heme followed by isolation of labeled stercobilin in stool.[28,29] In more recent years, the early-labeled fraction has been studied using glycine-C^{14} and δ-aminolevulinic acid-C^{14} (ALA-C^{14}), isolating bilirubin from plasma and bile.[30–34] The early-labeled fraction has at least two components: a component arising within 12–24 hr following administration of glycine-C^{14} or within 1–2 hr following ALA-C^{14} and thought to represent hepatic hemes (cytochromes, catalase, tryptophan pyrrolase) and a second, larger peak, probably erythropoietic in origin and arising 3–5 days following glycine-C^{14} administration. The erythropoietic component may represent developing red cells which are destroyed in the marrow and is strikingly increased in conditions with ineffective erythropoiesis. The nonerythropoietic component can be augmented by increasing the turnover of liver hemes such as by the administration of phenobarbital.[35] There is some suggestion that the early-labeled peak may be increased in newborn infants,[36] and currently this problem is being investigated by techniques measuring the production of carbon monoxide, which quantitatively reflects the breakdown of heme.

The production of bilirubin from heme has been ascribed to an enzymatic process involving a mixed-function oxygenase, microsomal heme oxygenase, catalyzing the oxidation of heme at the α-methene bridge with the liberation of carbon monoxide to biliverdin, coupled thence with soluble NADPH-dependent biliverdin reductase to form bilirubin.[37] It is interesting that most degradative pathways are oxidative in character, but the production of bile pigment is reductive (see Fig. 1 for structures of the three major pigments). The conversion of heme to bilirubin takes place in the reticuloendothelial system, and the insoluble bile pigment is then transported firmly bound to serum albumin as a soluble complex to the liver. The binding of bilirubin by serum albumin is extremely important, since unbound unconjugated bilirubin is highly toxic to some tissues.

Once at the liver, bilirubin crosses the liver cell membrane by a mechanism not yet fully understood. In recent years, two liver cytoplasmic protein fractions, Y and Z, which may be important to the transport of bilirubin into the liver cell, have been described; they bind bilirubin, sulfobromophthalein (BSP), and other organic anions.[38] Y is found only in the liver, has a high affinity for bilirubin, and increases in concentration following the administration of phenobarbital. Z has also been described in intestinal mucosa and binds bilirubin once Y is saturated. Several compounds (e.g., flavaspidic acid and bunamiodyl) which can produce an unconjugated hyperbilirubinemia compete with bilirubin for binding to Z protein. The intracellular concentration of Y is relatively deficient in newborn rhesus monkey liver, and the plasma clearance of BSP is impaired.[39] Likewise, in the human neonate BSP clearance is reduced.[40] This suggests that one factor in the production of "physiological jaundice" in the human newborn could be a relative deficiency of Y protein, which might limit the uptake of bilirubin from the plasma.

The next step in bilirubin metabolism in mammalian liver is the conjugation of one molecule of bilirubin with two molecules of glucuronic acid, converting a highly lipid-soluble molecule to a water-soluble ester glucuronide no longer capable of easily crossing many cell membranes, yet somehow in proper configuration for excretion into the bile. The enzyme responsible for this reaction is glucuronyl transferase (uridine diphosphate glucuronic acid trans-glucuronylase, or UDPGT) located in the microsomal fraction. Many other compounds are conjugated via this enzyme (steroids, drugs such as chloramphenicol, salicylates, sulfonamides, and menthol), and it is uncertain whether a separate enzyme exists specifically for the conjugation of bilirubin (see Fig. 2 for the metabolic sequence). For quite a few years, "physiological jaundice" of newborns was almost solely attributed to a relative deficiency of this hepatic microsomal enzyme in the neonate, but this view has been challenged in the last few years,[41–43] warranting the consideration of other mechanisms for this phenomenon. To further complicate this issue, recent work suggests that the conjugation of bilirubin is a far more complex process that involves a variety of mono- and disaccharides, in addition to glucuronic acid.[119]

Following excretion of the conjugated pigment into the bile and thence into the intestine, a reductive process once again occurs, brought about by the action of intestinal bacteria in the lower ileum and colon. A variety of pigments are produced, but normally the pigment excreted in the greatest amount is *l*-stercobilinogen. Alteration of the intestinal flora by antibiotics may result in the excretion of *d*- or *i*-urobilinogens. There is an enterohepatic circulation of these bile pigments with absorption through the intestinal wall and presentation of the pigment primarily to the liver for re-excretion in the bile, with a small fraction being excreted by the kidney into the urine. Figure 3 is a schematic version of the production and excretion of bilirubin.

Once again, the newborn infant has a significant variation in metabolism. Initially, his gastrointestinal tract is not colonized with bacteria, and the bile pigment excreted into the stool for varying periods of time is bilirubin. β-Glucuronidase, present in the gastrointestinal tract of infants, hydrolyzes the glucuronic acid moiety from the bilirubin molecule. Unconjugated bilirubin is

1. Glucose + ATP $\xrightarrow[\text{hexokinase}]{\text{Mg}^{2+}}$ glucose-6-P + ADP

2. Glucose-6-P $\xleftrightarrow[\text{phosphoglucomutase}]{\text{Mg}^{2+}}$ glucose-1-P

3. Glucose-1-P + UTP $\xleftrightarrow[\substack{\text{UDPG-}\\\text{pyrophosphorylase}}]{}$ UDPG + PP$_i$

4. UDPG + 2NAD$^+$ $\xrightarrow[\substack{\text{UDPG-}\\\text{dehydrogenase}}]{}$ UDPGA + 2NADH + 2H$^+$

5. UDPGA + bilirubin $\xrightarrow[\substack{\text{glucuronyl}\\\text{transferase}}]{}$ bilirubin diglucuronide + UDP

Fig. 2. Pathway for the formation of bilirubin diglucuronide. UDPGA, uridine diphosphate glucuronic acid; UDPG, uridine diphosphate glucose.

then available for an enterohepatic circulation and may contribute to newborn jaundice. Poland and Odell[23] demonstrated that by feeding newborns agar, which stabilizes bilirubin, no rise in the serum bilirubin concentration occurred after the thirteenth hour. These investigators feel that their study supports a proposal that the reabsorption of the bilirubin within the intestinal tract at the time of birth may be primarily responsible for "physiological jaundice."

It would thus appear that "physiological jaundice" in the neonate may well be the result of several factors. First, there may be accelerated bilirubin production. Newborn red cell survival times, particularly in the premature infant, may be slightly reduced.[44] Also, there may be an increase in the bile pigment production reflected in the early-labeled peak.[36] Second, a relative deficiency of Y protein may be decreasing the hepatic uptake of bilirubin from plasma. Third, there may be a relative UDPGT deficiency and in addition deficient uridine diphosphate glucuronic acid formation which would contribute to deficient activity of UDPGT. Fourth, there may be a significant enterohepatic circulation of bilirubin. (For a review of some of these features, see Arias.[45])

III. PROTEIN BINDING OF BILIRUBIN

The protein binding of bilirubin, well recognized since the early 1930s, was systematically investigated by Martin in 1949.[46] Dialysis studies with crystalline human serum albumin suggested that 2–3 moles of bilirubin was bound per mole of albumin between pH 7.4 and 7.9, that pH between 5 and 8 did not affect the absorption maxima of the bilirubin–albumin complex, and that under certain conditions globulins also bound bilirubin. Since these early studies, there have been many investigations with many controversies concerning the binding of bilirubin by protein. This is an important question, and the normal determinants of the albumin binding of bile pigments need to be defined. As long as unconjugated bilirubin is firmly attached to serum albumin, it does not seem to readily cross most cell membranes; it is thus rendered nontoxic. Over the past few years, attempts have been made with varying success to measure the number of binding sites on the serum albumin of newborns available to bind bilirubin. Such methods have included dye binding,[47,48] column chromatography,[49,50] and a method taking advantage of agents known to displace bilirubin from its primary binding site.[51] It was hoped that this kind of information would provide better criteria for the need for exchange transfusion. In the past, the main determinant has been the serum bilirubin, but in actuality this reflects only the vascular pool rather than giving any notion about the total body distribution of the pigment.

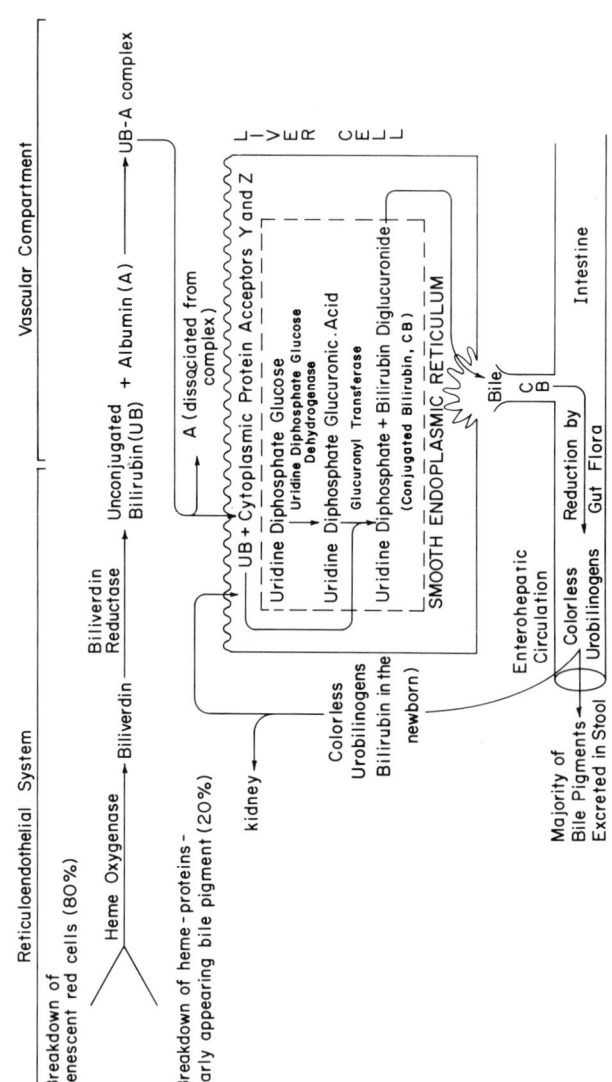

Fig. 3. Schematic version of the production and excretion of bilirubin.

A. Influence of Molar Ratios and pH on Binding of Bilirubin and Albumin

A difference exists in the ability of the albumin of various animal species to bind bilirubin.[52] A frequently quoted figure is that 1 mole of bovine serum albumin (BSA) binds 1 mole of bilirubin and that 1 mole of human serum albumin (HSA) can bind 2 moles of bilirubin.[52,53] Many studies of bilirubin binding have been done with BSA because of its greater definition. Probably no one will deny that HSA is different in its ability to bind bilirubin, it most likely having a stronger affinity than BSA. However, the confusion arises when one attempts to state that specific molar ratios of bilirubin to albumin at a given pH may or may not represent a hazard to the newborn. Schmid et $al.$[52,53] have stated that bilirubin is bound to human albumin in a 2:1 ratio, but Odell[54] has demonstrated displacement of bilirubin from albumin into mitochondria when a 1:1 ratio is exceeded. Furthermore, there is some controversy about the effect of pH on these ratios. Odell[55] noted increasing amounts of bilirubin in mitochondria as the hydrogen ion concentration of the medium increases.

Some of these controversies can be eliminated by looking carefully at the particular experimental designs, which have varied greatly from one investigator to another. With bilirubin and albumin complexes at ratios less than 1:1, BSA binds bilirubin tightly throughout the pH range of 5.0–8.5.[56] When this ratio is exceeded, secondary binding sites can be demonstrated up to 4:1 bilirubin: albumin at pH 8.5. Below pH 5.5, there is evidence of only one binding site. Between pH 5.5 and 8.5, there is evidence of two binding configurations. Human albumin has two principal binding sites. In contrast to BSA, the choice of the buffer greatly affects the spectrum of bilirubin bound to HSA, and the presence of phosphate augments the pH-dependent spectral transitions. The binding constant of the first site of human albumin is reported to be approximately 10^8, at both pH 7.5 and 7.0.[57] However, when one adds another phase (cholestyramine, Sephadex, tissue culture cells, mitochondria, etc.), the equilibria are significantly altered. It must be remembered that in the intact organism other cell surfaces may be interacting with the bilirubin–albumin complex. Under conditions of high molar ratios where primary binding sites have been exceeded or under conditions of low pH, perhaps other tissue surfaces become more competitive with albumin for the binding of bilirubin. These factors, rather than absolute molar ratios or pH, may be the primary factors determining body distribution of bilirubin.

B. Other Factors Which Affect Binding of Bilirubin by Albumin

In 1955, a nursery "epidemic" of bilirubin encephalopathy was reported leading to the recognition that certain drugs could displace bilirubin from its albumin binding sites and lead to the production of bilirubin encephalopathy at lower concentrations of serum bilirubin.[58] Some of the sulfonamides (sulfamethoxypyridazine, sulfadiazine, sulfanilic acid, and sulfasoxazole), sodium salicylate, nonesterified fatty acids, caffeine sodium benzoate, diazepam, and perhaps other compounds have this potential. Recent evidence shows that it is the sodium benzoate in the last two drugs that is actually the displacing agent.[59] *In vivo*, these compounds have been associated with increased numbers of cases of bilirubin encephalopathy in Gunn rats (a strain of rats lacking liver glucuronyl transferase so that the homozygous animal develops kernicterus).[60] *In vitro*, spectral shifts can be observed when these compounds are added to bilirubin-albumin complexes.[61] It has been said that these compounds compete with bilirubin for the same binding sites on serum albumin.[61] However, the binding constant of the primary site on BSA for bilirubin is in the range of 10^6[62] and has been estimated to be 10^8 for HSA.[57] In our experience, it has taken about 300 times more drug, mole for mole, to see any displacement of bilirubin from BSA (1:1 ratio) or HSA (1.8:1 bilirubin:albumin) into tissue culture cells.[63] The binding constants for some of the abovementioned drugs are in the range of 10^3 to 10^5.[64] Some thoughts to consider are (1) there is no displacement of bilirubin from primary sites and only displacement from the weaker secondary sites; (2) the binding sites for bilirubin and the competitive drugs are not the same; the drugs may bind at an adjacent site and perhaps change the conformation of the protein so that bilirubin is "squeezed off" of the surface of the protein; and (3) albumin of newborns is different and is much less efficient in its ability to bind bilirubin. The evidence for the last point is controversial.[50,65]

Other factors affecting the solubility of bilirubin and the binding of bilirubin and albumin are the salt concentrations and the type of salt present. Burnstine and Schmid[66] studied the solubility of bilirubin at various salt concentrations and noted a "salting in" and "salting out" effect. In the pH range of 7.2–7.9, decreasing hydrogen ion concentration of the buffer resulted in an abrupt increase in solubility. The pH at which the change occurred was dependent on the molarity of the buffer. At the ionic strength of extracellular fluid (0.150) and pH 7.4, it was estimated that 5 mg % was in solution. Odell[54] noted an increase in the amount of bilirubin displaced from albumin into mitochondria when a salt was substituted for sucrose as the suspending medium. Blauer and King[62] could demonstrate striking differences in the optical rotary dispersion (ORD) complex of bilirubin and BSA which depended on the order of addition of reagents as well as the presence of salts which decreased the molar amplitude of the complex. Wennberg and Rasmussen[57] reported that the choice of the buffer affected the spectral characteristics of bilirubin–HSA complexes.

IV. PROBLEMS HAMPERING RESEARCH

Much effort has been spent in attempting to understand the basic chemistry of bilirubin, its interrelationship with proteins, and its role as a cytotoxic agent. However, dispite the effort there is much controversy and there are many unanswered questions. Part of the problem has to do with the nature of the compound itself (its insolubility, photolability, and rapid degradation in alkaline solutions).

The purification of bilirubin is also difficult in that the point is reached where degradation is occurring faster than purification. There can be significant variations in the quality of preparations. This hampers basic research and in the past has also produced errors in the clinical determination of serum bilirubin. This is a serious matter, since serum bilirubin concentrations are still the most frequently used criterion for exchange transfusion. This procedure has a significant mortality in the newborn infant and therefore should be undertaken prudently.

V. PATHOLOGICAL LESION OF BILIRUBIN ENCEPHALOPATHY

A. Gross Observations

Since the advent of exchange transfusion, brains of human infants with the clinical syndrome of bilirubin encephalopathy have been rarely available for study. Recent studies have utilized Gunn rat brains, which have a similar pathological lesion. In infants dying in the newborn period (acute state), the gross appearance, except for icterus, is unremarkable.[67] Some brains appear swollen, and occasionally small hemorrhages are found in the dura, subdural or subarachnoid space, brain substance, or walls of the lateral ventricles. Deeper yellow staining in certain areas is demonstrable on coronal section of brain. The pattern of staining varies among affected brains.

The areas most commonly pigmented in 55 cases analyzed by Zuelzer and Mudgett[9] were, in order of frequency, hippocampus, basal ganglia, nuclei of pons and medulla, mesial portions of the thalamus, flocculus and dentate nucleus of the cerebellum, and hypothalamic area. Claireaux[68,69] reported the commonest areas of staining in 35 cases as follows: basal ganglia, 32; cerebellum, 25; hippocampus, 24; medulla oblongata, 24; subthalamic nuclei (Luys), 19; thalamus, 18; corpus striatum, 18; fourth ventricle, 18; dentate nucleus, 17. The distribution appears to follow regional blood flow.[70]

On the other hand, in infants dying after the neonatal period (chronic stage) the brain is normal in external appearance. On section, pigmentation has usually vanished.[68]

B. Observations with Light Microscopy

The degree of microscopic cellular damage seen depends on the duration of life, according to Vaughan *et al.*[6] In infants dying within the first 72 hr, no *specific* changes of bilirubin encephalopathy were observed, even though gross pigmentation was severe. In infants dying after 72 hr, *characteristic* changes could be observed in some of the stained areas. In infants who had lived long enough so that pathological changes could be observed, Haymaker *et al.*[67] noted in 87 cases that some of the grossly icteric areas were characterized by pathological changes but others were not. The globus pallidus and the red zone of the substantia nigra always had neuronal necrosis. Necrosis of the subthalamic nuclei occurred in 60% of his cases, and 50% had necrosis of the hippocampus. There was severe damage to the mammillary bodies of the hypothalamus in all cases examined, to the third and fourth cranial nerve nuclei in 60% of cases, to the interstitial nuclei in all cases examined, and to the lateral cuneate nucleus in 40% of cases. In the pons and medulla oblongata, the central pontine nuclei and papilioform nuclei were involved in 40% of cases. In the cerebellum, the flocculi and vermis were usually free of damage, but the dentate and roof nuclei were damaged in 50% of cases. The nuclei of the floor of the fourth ventricle usually showed only small numbers of damaged cells. On the other hand, in contrast with its yellow color grossly, the thalamus was only occasionally damaged. Schutta and Johnson[70] point out the striking uptake of pigment by the Gasserian ganglion with lack of necrosis, whereas the Purkinje cells of the cerebellum take up relatively little pigment but are severely damaged.

According to Haymaker *et al.*,[67] the earliest manifestation of neuronal necrosis was profound alteration of the cytoplasm with loss of the Nissl substance. The nuclear chromatin became swollen and distorted and later hyperchromatic. Studies by Blanc and Johnson[71] of brain cells from the icteric areas of Gunn rat brain revealed a sequence of changes. Early alterations included the rounding up of cells and the appearance of granules in the cytoplasm. The nuclei lost their normal contour and became eccentric and eventually pyknotic. In severely damaged cells, the cytoplasmic granules became coarse, and the cells shrank and eventually lysed. Recently, Schutta and Johnson[70] have carried out detailed studies on the cerebellum of Gunn rats. Swelling and vacuolization of the Purkinje cell cytoplasm were recognizable on the third day after birth in the homozygous animal. Animals with severe clinical signs had more abnormal Purkinje cells than less affected ones.

Observations in infants dying in the chronic stage (after the third week of life) are quite different. Areas of necrosis with severe loss of neurons and reparative gliosis are seen. The subthalamic nuclei and dentate nuclei may be severely affected, as is the corpus striatum. In two patients of Claireaux's surviving beyond the neonatal period, the only changes seen were in the hippocampus.[68] Some detailed observations on human cases may be found in the report by Haymaker et al.[67] Blanc and Johnson[71] reported ganglial cell loss and focal gliosis in the basal nuclei of some Gunn rats 4 weeks or more of age and with a clinical history of kernicterus.

C. Observations with Electron Microscopy

In recent years, electron microscopy of human brain, rat brain, and brain tissue culture cells has been utilized. Fine structural studies of the cerebellum of Gunn rats and myelinating cerebellar tissue culture cells have been carried out by Schutta and Johnson.[70,72] In the early stages of Purkinje cell damage, membranous whorls were seen in the cytoplasm (Fig. 4A). These changes were nonspecific, being found in some normal cells and in cells exposed to various chemicals and infectious agents. In severely affected animals, many Purkinje cells showed these complex intracytoplasmic membranous bodies, which were the most striking abnormality seen. Cells in which the cytoplasm was virtually replaced with these bodies seldom survived. In some cells, membranous bodies were converted to lipofuscin-like granules. At a more advanced stage of damage, the cell membrane became disrupted and then disappeared. Mitochondrial abnormalities were also seen and will be briefly described below.[73]

VI. EXTRANEURAL PATHOLOGICAL LESIONS INDUCED BY BILIRUBIN

Although the brain damage appears to be the most significant both for the immediate mortality and for the long-range effects (mental retardation, deafness, and cerebral palsy), damage has been observed in other tissues. There have been frequent comments concerning jaundice of other areas of the body in infants dying of bilirubin encephalopathy. Bilirubin crystals have been reported in a number of sites, in particular the kidney, and necrosis of the renal medulla has been specifically mentioned.[9] Bernstein and Landing[74] carried out a systematic study of extraneural lesions and compared infants with hyperbilirubinemia with and without kernicterus. They observed necrosis of multiple organs in infants with kernicterus, particularly in the gastrointestinal tract, spleen, renal medulla, adrenal, testis, respiratory tract, and bone marrow. The distribution was similar to the distribution of bilirubin crystals, and therefore they felt justified in attributing the lesions to a direct toxic effect of bilirubin.

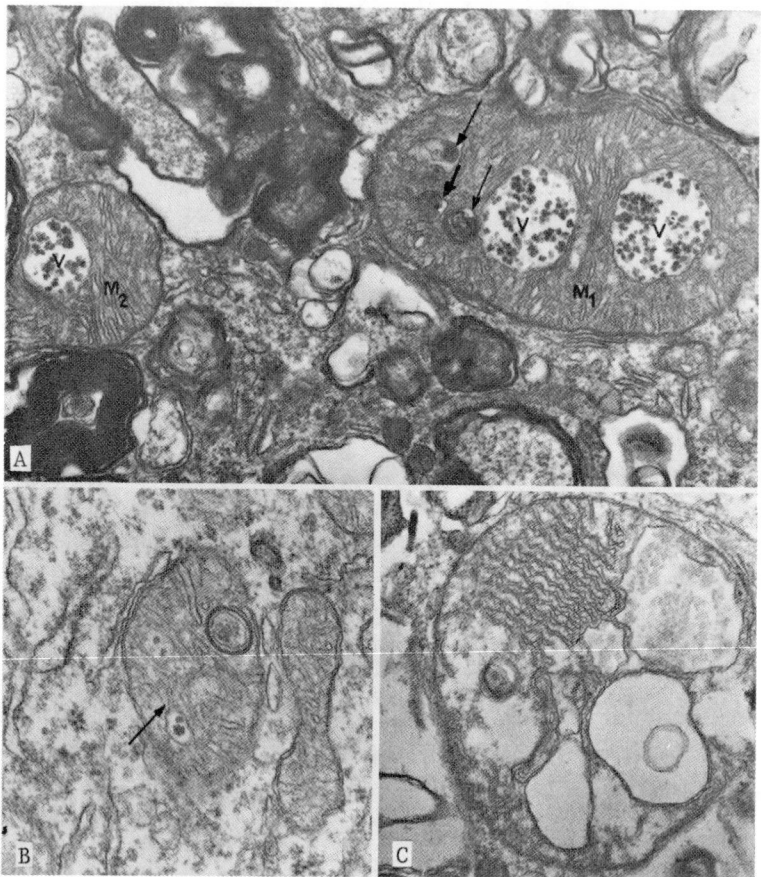

Fig. 4. A, Enlarged mitochondria in a Purkinje cell profile which also shows numerous cytoplasmic membranous inclusions. The mitochondrion marked M_1 shows two vesicles filled with glycogen-like granules (V) and several myelin bodies (arrows). The mitochondrion marked M_2 also contains a vesicle with granules. $\times 29,000$. B, Mitochondrion with a small vesicle containing several glycogen-like granules. Arrow indicates continuity with a crista. This mitochondrion also contains a lamellar inclusion. $\times 42,000$. C, Enlarged mitochondrion with angular cristae and vesicles, some of which are filled with glycogen-like granules. $\times 29,000$ (reproduced at 60% of original size). [From Schutta *et al.*,[73] reproduced by permission.]

VII. METABOLIC LESIONS INDUCED BY BILIRUBIN

A. Effects of Bilirubin on Tissue Respiration

Since the pioneering observations of Day[11] that bilirubin depressed the respiration of chopped rat brain, a number of laboratories have investigated the effects of this bile pigment on cellular oxidation and energy transfer. Early workers found that bilirubin inhibited respiration of a variety of fresh tissues.[75-79] Zetterström and Ernster[80] in 1956 reported that bilirubin at a concentration of 3×10^{-4} M almost completely uncoupled phosphorylation and partially depressed respiration in rat liver mitochondria. Ernster[81] later presented evidence that bilirubin behaved as a detergent type of uncoupler. Quastel and Bickis[79] also noted an uncoupling effect of this pigment in ascites tumor cells. More recent studies of the respiratory lesion have dealt with effects of bilirubin on whole cells in tissue culture, isolated mitochondria and submitochondrial particles, and purified preparations of respiratory enzymes.

1. Respiratory Effects of Bilirubin on Tissue Culture Cells

In a strain of mouse epithelial cells adapted to grow in protein-free media,[82,83] bilirubin was found to enhance glycolysis, as seen from increased glucose utilization and lactate production; this could be observed at bilirubin concentrations as low as 2.5 μM. Oxygen uptake was increased at low concentrations of bilirubin (5–10 μM) but began to decrease above 10 μM and was abolished at 50 μM bilirubin. Presumably, an uncoupling of oxidative phosphorylation occurred in these cells as a primary mechanism, but respiratory inhibition became the predominant effect at the higher bilirubin concentrations (greater than 10 μM). Intracellular ATP content was severely decreased; this effect was also directly related to concentration.

2. Effects of Bilirubin on Purified Respiratory Enzymes

Bilirubin has been shown to inhibit reduced nicotinamide adenine dinucleotide (NADH) oxidase beginning at 5 μM, with 50% inhibition occurring somewhere between 10 and 15 μM.[82] Succinate oxidase was less sensitive to bilirubin; no inhibition occurred until the concentration of the bile pigment exceeded 20 μM. NADH–cytochrome c reductase activity was also inhibited, but cytochrome c oxidase activity was not affected. These studies suggested that the locus of bilirubin inhibition might be similar to that of Amytal, a classical electron transport inhibitor.

3. Effects of Bilirubin on Mitochondria

Recent studies of bilirubin effects on mitochondrial reactions have been reported. Diamond and Schmid[84] did not observe uncoupling in mitochondrial suspensions prepared from whole brains of normal newborn guinea pigs until bilirubin had been added to attain a concentration of 6.7 μM. Yet mitochondria from whole brains or cerebella of newborn guinea pigs with symptoms of bilirubin encephalopathy were not uncoupled, suggesting that uncoupling of oxidative phosphorylation was not the primary cause of toxicity. Menken and

Weinbach,[85] employing polarographic methods, could not demonstrate any difference in the P:O ratios or respiratory control indices of rat brain mitochondria from neonatal kernicteric and nonjaundiced Gunn rats. However, it should be mentioned that Schenker *et al.*[86] found cerebellar ATP concentrations of kernicteric Gunn rats to be lower than those of control animals. In their studies, oxygen consumption of kernicteric cerebellum was approximately 20% lower than that of the controls. Vogt and Basford,[87] using a much higher concentration of bilirubin (0.4 mM), demonstrated an uncoupling in brain mitochondria from mature and weanling rats. They found that bilirubin inhibited both endogenous and DNP-activated ATPase activities but had no effect on Mg^{2+}-stimulated ATPase.

Recently, Mustafa *et al.*[88] conducted a detailed study of bilirubin effects on mitochondrial reactions. A four- to fivefold stimulation of respiration at low concentrations of bilirubin (15–20 μM) but an inhibition at high concentrations (greater than 50 μM) was observed in hepatic as well as cardiac mitochondria. However, respiratory stimulation was not observed in brain mitochondria but rather only inhibition. Respiratory control was found to be more sensitive to bilirubin effects than phosphorylation (Fig. 5A,B). The K_m value (50% of maximum as defined in the original article) for bilirubin was about 4 μM in rat liver mitochondria and 2.5 μM in brain mitochondria. A complete abolishment

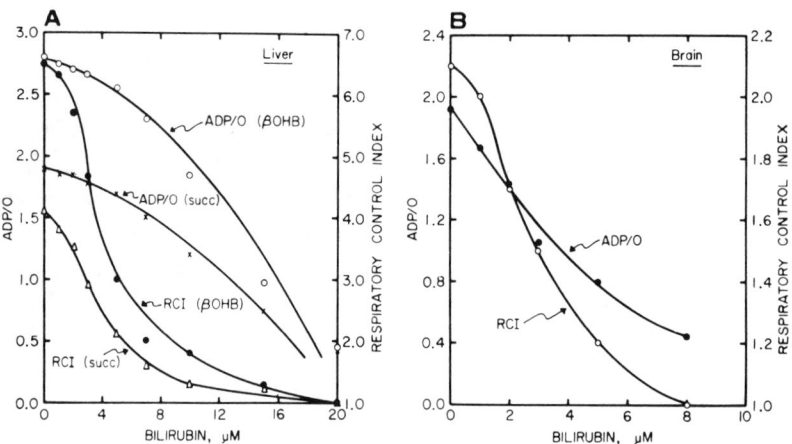

Fig. 5. Effect of bilirubin on phosphorylation. A, Rat liver mitochondria. Basal medium contained 180 mM sucrose, 18 mM mannitol, 8 mM $MgCl_2$, 5 mM potassium phosphate, 5 mM tris–chloride, and mitochondria (3.6 mg of mitochondrial protein per milliliter). Other additions, where made, were 10 mM β-hydroxybutyrate (βOHB), 10 mM succinate (succ), 150 μM ADP, and bilirubin as indicated. B, Rabbit brain mitochondria. Conditions were as in A, except that 1.9 mg protein per milliliter was used. RCI, respiratory control index. [From Mustafa *et al.*,[88] reproduced by permission.]

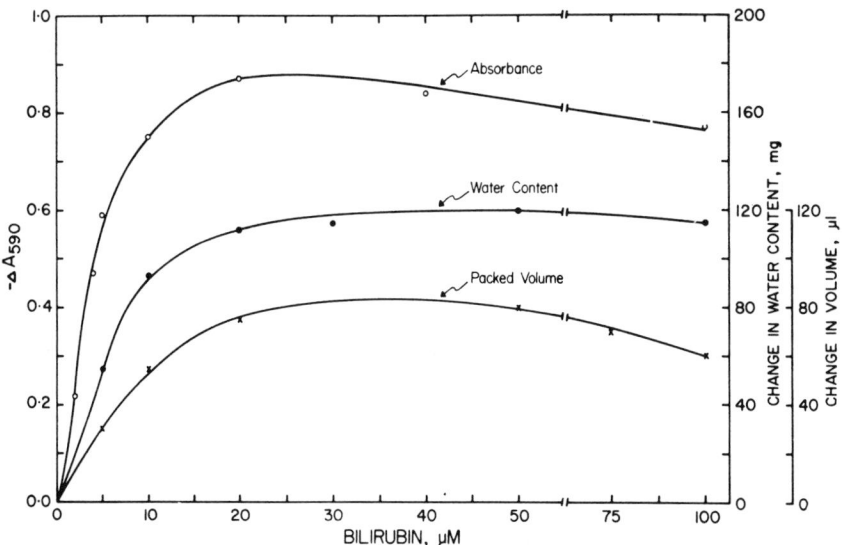

Fig. 6. Effect of bilirubin concentration on swelling of mitochondria measured by changes in absorbance, water content, and packed volume. Reaction mixture contained 20 mM sucrose, 50 mM mannitol, 50 mM tris–chloride, 7 mM $MgCl_2$, 10 mM potassium phosphate, and 10 mM succinate, pH 7.5. The amounts of mitochondria used, in terms of protein, were, for measurements by absorbance, 3 mg/3 ml; water content, 15 mg/12 ml; and packed volume, 17 mg/12 ml. [From Mustafa et al.,[88] reproduced by permission.]

of respiratory control required 8 and 18 μM bilirubin for mitochondria of brain and liver, respectively. Oxidative phosphorylation of brain mitochondria was somewhat more sensitive than that of liver or heart mitochondria. Bilirubin at 8 μM caused almost a complete uncoupling in brain mitochondria but less than 50% inhibition in liver mitochondria.

A major effect of bilirubin was the induction of an energy-dependent, large-amplitude, irreversible swelling of mitochondria,[88] as demonstrated by a drop in optical density at 590 nm and an increase in the packed volume and water content of these subcellular organelles (Fig. 6). Observations of swelling were similar for cerebral mitochondria and liver and heart. The K_m value of bilirubin for overall swelling was about 2.5 μM in liver mitochondria and 2 μM in brain mitochondria. However, the rate and extent of swelling in liver mitochondria were always higher than those observed in brain mitochondria; heart mitochondria occupied an intermediate position. Bilirubin thus appears to be a transport-inducing agent similar in some regards to gramicidin, valinomycin, and the nonactins. However, unlike these agents, bilirubin appears to initiate a unidirectional movement of ions and water into mitochondria so that irreversible swelling occurs. Thus, in contrast to some of the previously reported results, certain mitochondrial reactions are exquisitely sensitive to bilirubin so that the bilirubin effect on respiration should be considered as one of the important mechanisms of its cytotoxicity.

Electron micrographs of cerebellar cells in tissue culture exposed to 5 mg % bilirubin demonstrated that mitochondria increased in size and accumulated glycogen-like granules within 12 hr[73] (Fig. 4A,B,C). These changes were similar to those seen in the cerebella of kernicteric Gunn rats[70] but in the *in vitro* system occurred over a much shorter time interval. Whether the enlargement is at all related to the swelling phenomenon observed in mitochondrial suspensions is unknown.

These reaction parameters (respiratory control, respiration, oxidative phosphorylation, and swelling) are essentially the manifestations of mitochondrial membranes. Further studies have suggested that the alteration of these processes occurs because of binding of bilirubin to the lipid of mitochondrial membranes.

B. Effects of Bilirubin on Lipids

The lipophilic and lipid-soluble characteristics of bilirubin have been known for many years. Kahán et al.[89] studied the lipid extracts from rat brain incubated with bilirubin as well as brains from rats made kernicteric *in vivo* and from kernicteric human infants. They suggested that bilirubin was possibly bound to cerebral gangliosides. Recently, Mustafa and King[90] have reported a more direct demonstration of bilirubin binding to various lipids and phospholipids. Free bilirubin in solution showed an absorption maximum of 440 mμ with a millimolar extinction coefficient of about 52. The addition of mitochondria caused a shift with a broad peak appearing at 450 mμ and a shoulder around 492 mμ. Upon addition of bovine serum albumin (BSA) to the solution, the characteristic spectrum of the bilirubin–albumin complex emerged (Fig. 7). When mitochondrial lipid was substituted for mitochondria, the same spectral shift occurred and, once again, BSA restored the characteristic bilirubin–albumin complex. Mitochondria from which lipid was exhaustively extracted did not cause this spectral shift. In the presence of exogenously added lipid, the uncoupling action of bilirubin on mitochondria was prevented, although respiratory control was still somewhat affected. The rate of mitochondrial swelling caused by bilirubin was significantly diminished in the presence of external lipid, although the overall swelling was not affected unless the lipid concentration was very high. Bilirubin was found to bind to many different lipids including various cells and cellular components containing lipid, e.g., ascites tumor cells, ascites cell mitochondria, tissue culture cells, yeast cell fragments, and erythrocyte membrane. Furthermore, bilirubin bound to ascites

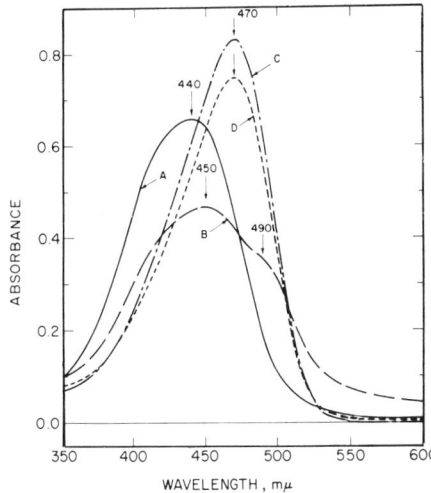

Fig. 7. Effect of mitochondria on the absorption spectrum of bilirubin. Basal medium contained 50 mM potassium phosphate and 50 mM tris–chloride, pH 7.5. In addition, curve A contained 13.2 μM bilirubin; curve B, 13.2 μM bilirubin and 20 μg (protein) of rat liver mitochondria per milliliter; curve C, 13.2 μM bilirubin and 6 mg of bovine serum albumin per milliliter, curve D, same as curve B but also with 6 mg of bovine serum albumin per milliliter. Identical spectra are obtained by substituting 15 μg of mitochondrial lipid per milliliter for the mitochondria. [From Mustafa and King,[90] reproduced by permission.]

cell lipid, asolectin, triolein, and commercial vegetable oils. The spectra obtained were all virtually the same, and in all cases BSA displaced the lipid from lipid-bound bilirubin to form a bilirubin–albumin complex.

C. Effects of Bilirubin on Cell Membrane Systems

In a very early study, Stenhagen and Rideal[91] reported the ability of a variety of tetrapyrroles including bilirubin to penetrate artificial lipid and protein membranes. Bilirubin was found to penetrate lipid monolayers, and this was more effective at pH 7.2 than at 7.8 or 8.2. In view of the lipophilia of bilirubin, it would be reasonable to assume that it can penetrate biological membranes. The form of bilirubin distributed to body tissues has been quite clearly shown to be unconjugated bilirubin. Conjugated bilirubin, a more water-soluble compound, does not appear to cross biological membranes. Unconjugated bilirubin bound to sufficient serum albumin is nontoxic and in this complex form does not have access to most body tissues. However, despite the strong interaction between bilirubin and albumin, bilirubin does cross certain biological membranes such as those of liver and placenta.[92]

1. The Plasma Membrane

Small amounts of bilirubin are taken up by the red cell membrane.[93–95] Cheung et al.[96] reported a disruption of electrolyte transport across the red blood cell membrane. However, inordinately high concentrations of bilirubin were used before effects were seen. This same group also reported bilirubin effects on the plasma membrane of the platelet.[97] Odell et al.[98] have demonstrated an effect of bilirubin on the kidney membrane resulting in altered renal transport. In a special strain of tissue culture cells grown in protein-free media, several effects of bilirubin have been shown on the plasma membrane.[99] Following a lag phase (1–2 hr), the length of which was directly dependent on the bilirubin concentration (2.5–25 μM), large dye molecules penetrated the cells and protein molecules were extruded from it. Earlier membrane changes in the handling of small molecules by these same cells were reflected by a change of K^+ transport within the first hour of exposure to the bile pigment (both in the energy-related K^+ influx and the passive process of K^+ efflux).

2. The Lysosomal Membrane

In this same line of tissue culture cells and in rat liver preparations,[83,99] it was demonstrated that bilirubin also increased the permeability of lysosomal membranes. At 500 μM concentrations, the availability of the lysosomal enzyme acid phosphatase (a measure of membrane permeability) increased. This concentration is much higher than those needed to affect plasma or mitochondrial membranes. However, it is quite possible that lysosomes can concentrate bilirubin. Novikoff[100] studied the kidney cells of rats during development of bile nephrosis and stained sections for bilirubin and acid phosphatase. Both stains were localized in the same droplets, which eventually lost their acid phosphatase stainability and were thought to fuse into large bilirubin-containing masses. Allison et al.[101] treated several strains of tissue culture cells with "photosensitizers" bearing striking structural resemblance to bilirubin and found that they localized microscopically in lysosomes. Finally, when L929 tissue culture cells were treated with bilirubin, the pigment appeared to localize in granules.[83] A pure speculation is that perhaps the nonspecific patchy cytoplasmic damage seen in electron microscopic studies of bilirubin-damaged cells might be the result of bilirubin-induced increased permeability of the lysosome.

3. The Mitochondrial Membrane

The effects of bilirubin on mitochondrial membranes have been discussed in previous sections. All of the changes in mitochondrial reactions induced by bilirubin are mitochondrial membrane functions. Electron micrographs prepared from bilirubin-treated mitochondria[102] demonstrated swelling as manifested by an increase in size and disappearance of cristae structure (Fig. 8A, B). In the case of extensive swelling, a loss of the outer membrane and disruption of the inner membrane occurred. However, when purified outer and inner membranes were treated with bilirubin, no gross lesions were seen. It was

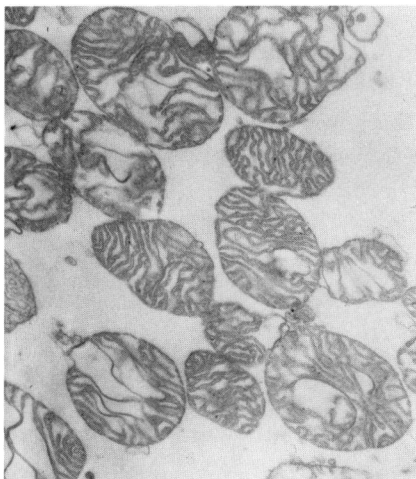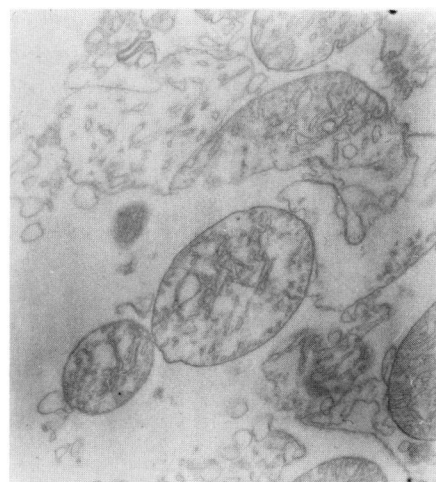

Fig. 8. A, Electron micrograph of control beef heart mitochondria before swelling. ×40,128. B, Beef heart mitochondria treated with 10 μM bilirubin. The dense cristae have disappeared. Lightly stained vesicles have appeared. Some mitochondria have lost the outer membrane, and in others the inner membrane appears fragmented. ×34,884. [M. G. Mustafa and M. L. Cowger, unpublished data.]

suggested that bilirubin complexes with the mitochondrial lipid and subtly alters the membrane, leading to ion transport accompanied by an influx of water and eventual osmotic disruption of this subcellular organelle.

D. Other Metabolic Reactions

Other metabolic reactions that bilirubin may be involved in have been described. In tissue culture cells and ascites tumor cells, bilirubin inhibited the uptake of glycine and its incorporation into protein.[79,82] In red blood cells, it has been suggested that bilirubin may interfere with the hexose monophosphate shunt.[103] Labbe et al.[104] demonstrated that heme biosynthesis was inhibited by bilirubin. Thaler[105] showed that bilirubin stimulated the synthesis of hepatic and cerebral lipids in Gunn rats. In the same animals, DNA synthesis was suppressed by bilirubin.

In icteric brain[106] and in tissue culture cells stained with bilirubin,[83] the yellow color disappeared after a few hours. This was independent of any effect of light. As a possible explanation, a numbr of enzymes have been demonstrated that can induce the oxidation of bilirubin. Sumner and Nyman[107] demonstrated that peroxidase catalyzed the conversion of bilirubin to biliverdin and suggested that this system operated in liver to oxidize bilirubin. Brodersen and Bartels[106] reported that a number of enzyme proteins (hemoglobin, cytochrome c, peroxidase, and xanthine oxidase) can oxidize bilirubin. These authors have also isolated an oxidase from guinea pig brain which catalyzes the oxidation of bilirubin to a diazo-negative product. It may thus be possible that tissues other than the liver have mechanisms for detoxifying bilirubin, although these are nonspecific.

VIII. WHY IS BRAIN DAMAGE THE MAIN PATHOLOGICAL PHENOMENON INDUCED BY BILIRUBIN?

For many years, it was felt that kernicterus was a phenomenon only of the neonate. However, this is no longer felt to be true, since several cases in older children and one in a teenager have now been reported.[108,109] Furthermore, in animal experiments it has been possible to induce kernicterus in adult animals.[65] Formerly, it was explained that the newborn had a more permeable blood–brain barrier so that bilirubin was offered easy access to brain tissue. The early evidence for a more permeable barrier was based on the observation that trypan blue introduced into the bloodstream was not excluded from the newborn brain and its entry was restricted by the mature brain.[110] Later, evidence with P^{32} suggested a concept of a "developing blood–brain barrier."[111] However, many disagree that the blood–brain barrier is more permeable in the neonate.[110,112] Dobbing[110] has recently suggested that the impression of a more permeable blood–brain barrier in the very young may represent a reflection of enhanced activity associated with the "growth spurt" of brain in early life. The problems of bilirubin and the blood–brain barrier have been recently reviewed.[113] From an intuitional point of view, it is difficult to see why a highly lipid-soluble molecule which has been shown to form a lipid complex and to cross biological membranes easily would have difficulties penetrating the lipid boundaries of the so-called blood–brain barrier. It is more likely that the complexing of bilirubin with serum albumin is the factor that prevents the entry of bilirubin into the brain.

Why then is bilirubin encephalopathy more common in the newborn infant? A number of factors exist coincidentally at the time of birth and are not likely to be repeated at any one time in later life. Newborn infants and in particular the premature (who tend to get bilirubin encephalopathy at lower concentrations of serum bilirubin than full-term infants) have lower concentrations of serum albumin than older infants, children, and adults. The chances for situations resulting in the development of anoxia and acidosis are greater in the newborn period. These two factors probably enhance the ability of bilirubin to penetrate the brain. Another factor leading to the increased vulnerability of the newborn is that at multiple points throughout the metabolic pathway of bilirubin detoxification and excretion the newborn demonstrates deficiencies (see Section II) allowing the accumulation of unconjugated bilirubin. At the same time, the newborn and especially the premature may have a slightly shortened red cell life span.[44] Thus bilirubin production may be increased. Production may also be enhanced via the shunt pathway, as reflected in an increased early labeled fraction.[36]

Let us take it for granted that no significant problem exists in the passage into the brain of bilirubin which is unconjugated and not bound to protein and that for bilirubin there is no age difference in the function of the blood–brain barrier. Some very fundamental and as yet unanswered questions concerning the effects of bilirubin on the brain and other tissues are left. What accounts for the pattern of distribution of bilirubin in the brain? Why are some tissues more vulnerable to the effects of bilirubin?

Several explanations have been invoked to explain the pattern of distribution of bilirubin staining in the brain. There are those who feel that there must be preexisting damage (in particular anoxic damage) of the brain[114] and that it is damaged tissue that selectively picks up bilirubin when hyperbilirubinemia exists. The adherents of this idea point out that damaged areas of the adult central nervous system readily take up circulating bilirubin[67] and that similarities exist between the regional distribution of damage in anoxia and kernicterus. In some animal systems, prior anoxia does seem to be a prerequisite to the development of kernicterus; in others, it does not.[115] Diamond and Schmid[65] noted that there was no difference in pigment distribution in newborn icteric guinea pigs from one area of brain to another, whereas bilirubin loads given to adult Gunn rats with preexisting brain damage resulted in the pigment being highest in the brain stem and cerebellum. These same two authors made yet a second suggestion to explain pigment distribution, namely, that certain areas of the brain are more vulnerable to the direct toxic effects of bilirubin. Once damaged, these areas then tend to retain the pigment in greater amounts. Chen et al.[116] attributed the access of bilirubin to cells to be a result of the increased permeability of capillary and cell membranes resulting from anoxia. Others deny the similarities in the distribution of anoxic lesions and kernicterus. The argument that in adults primarily only predamaged areas become bilirubin stained may be meaningless as far as the newborn is concerned. It has been pointed out that in adults there may be no boundary in the necrotic areas even to the bilirubin–albumin complex. (For a review of this whole question, see Bakay[113] and Schutta and Johnson.[117])

The prior damage argument cannot explain the point mentioned in an earlier section, namely, that some areas which are heavily stained with bile pigment show very little damage, whereas other areas show minimal staining but are profoundly damaged. It is true that acidosis and anoxia do seem to potentiate the development of kernicterus at lower blood bilirubin concentrations. Perhaps this reflects a situation in which primary albumin binding sites are saturated and the weaker secondary sites are unable to compete, with changes in extravascular tissues induced by acidosis and anoxia bestowing on them a greater avidity for bilirubin. There is a much simpler explanation offered for the pattern of bilirubin staining in the newborn brain. There is good suggestive evidence that the distribution pattern in humans follows the patterns of blood flow in the brain.[117, 118] Thus, once the albumin binding of bilirubin (which may be influenced by many previously discussed factors) is exceeded, bilirubin can permeate brain tissue following such a distribution pattern.

Even with this simpler explanation for the brain distribution of bilirubin, we are still left with the problem of why certain tissues display such striking sensitivity to bilirubin and others seem relatively resistant. If the postulate that bilirubin damages cells because it damages a variety of cell membrane systems holds, then it becomes more difficult to understand why there are such differences in vulnerability from one tissue to another. Several possibilities come to mind: (1) Certain tissues depend more on the metabolic processes damaged by bilirubin (oxidative metabolism, for example). (2) The mitochondria or lysosomes or cell membranes are different from one tissue to another. (3) There are metabolic differences that are unique from one cell type to another, and it is these metabolic processes that bilirubin is primarily affecting in some cell types. (4) Perhaps tissues differ in their ability to oxidize bilirubin nonspecifically and thus in their ability to protect themselves from bilirubin toxicity. However, the latter would not explain why some heavily stained areas show very little damage.

Thus, despite many years of diligent investigation, there are still many problems to solve. We have done much on the clinical level to reduce the gross brain damage and diminish the number of deaths from bilirubin encephalopathy. However, until there is better understanding of the basic mechanisms involved, we cannot hope to bring the problem of bilirubin encephalopathy into the file of medical problems solved.

REFERENCES

1. L. K. Diamond, in "Bilirubin Metabolism in the Newborn" (D. Bergsma, D. Hsia, and C. Jackson, eds.) Birth Defects: Orig. Art. Ser. 6:3–6 (1970).
2. F. H. Allen and L. K. Diamond, "Erythroblastosis Fetalis," pp. 6–8, Little, Brown, Boston (1957).
3. A. Claireaux, Hemolytic disease of the newborn, Part 1: A clinical–pathological study of 157 cases, Arch. Dis. Child. 25:61–80 (1950).

4. L. K. Diamond, K. D. Blackfan, and J. M. Baty, Erythroblastosis fetalis and its association with universal edema of the fetus, icterus gravis neonatorum and anemia of the newborn, *J. Pediat.* **1**:269–309 (1932).

5. P. L. Mollison and M. Cutbush, Haemolytic disease of the newborn: Criteria of severity, *Brit. Med. J* **1**:123–130 (1949).

6. V. C. Vaughan, F. H. Allen, and L. K. Diamond, Erythroblastosis fetalis. IV. Further observations on kernicterus, *Pediatrics* **6**:706–716 (1950).

7. F. H. Allen, L. K. Diamond, and V. C. Vaughan, Erythroblastosis fetalis. VI. Prevention of kernicterus, *Am. J. Dis. Child.* **80**:779–791 (1950).

8. D. Y. Y. Hsia, F. H. Allen, S. S. Gellis, and L. K. Diamond, Erythroblastosis fetalis. VIII. Studies of serum bilirubin in relation to kernicterus, *New Engl. J. Med.* **247**:668–671 (1952).

9. W. W. Zuelzer and R. T. Mudgett, Kernicterus, etiologic study based on an analysis of 55 cases, *Pediatrics* **6**:452–474 (1950).

10. A. E. Claireaux, P. G. Cole, and G. H. Lathe, Icterus of the brain in the newborn, *Lancet* **2**:1226–1230 (1953).

11. R. L. Day, Inhibition of brain respiration *in vitro* by bilirubin. Reversal of inhibition by various means, *Proc. Soc. Exptl. Biol. Med.* **85**:261–264 (1954).

12. K. Landsteiner and A. S. Wiener, An agglutinable factor in human blood recognized by immune sera for rhesus blood, *Proc. Soc. Exptl. Biol. Med.* **43**:223 (1940).

13. H. Wallerstein, Treatment of severe erythroblastosis by simultaneous removal and replacement of the blood of the newborn infant, *Science* **103**:583–584 (1946).

14. R. Finn, C. A. Clarke, W. T. A. Donohoe, R. B. McConnell, P. M. Sheppard, D. Lehane, and W. Kulke, Experimental studies on the prevention of Rh haemolytic disease, *Brit. Med. J.* **1**:1486–1490 (1961).

15. A. W. Liley, Intrauterine transfusion of foetus in haemolytic disease, *Brit. Med. J.* **2**: 1107–1109 (1963).

16. R. J. Cremer, P. W. Perryman, and D. H. Richards, Influence of light on the hyperbilirubinemia of infants, *Lancet* **1**:1094–1097 (1958).

17. J. Lucey, M. Ferreiro, and J. Hewitt, Prevention of hyperbilirubinemia of prematurity by phototherapy, *Pediatrics* **41**:1047–1054 (1968).

18. R. E. Behrman and D. Y. Y. Hsia, Summary of a symposium on phototherapy for hyperbilirubinemia, *J. Pediat.* **75**:718–726 (1969).

19. S. J. Yaffe, G. Levy, T. Matsuzawa, and T. Baliah, Enhancement of glucuronide-conjugating capacity in a hyperbilirubinemic infant due to apparent enzyme induction by phenobarbital, *New Engl. J. Med.* **275**:1461–1466 (1966).

20. R. E. Behrman and D. E. Fisher, Phenobarbital for neonatal jaundice, *J. Pediat.* **76**: 945–948 (1970).

21. R. A. Ulstrom and E. Eisenklam, The enterohepatic shunting of bilirubin in the newborn infant. 1. Use of an oral activated charcoal to reduce normal serum bilirubin values, *J Pediat.* **65**:27–37 (1964).

22. R. Lester, L. Hammaker, and R. Schmid, A new therapeutic approach to unconjugated hyperbilirubinemia, *Lancet* **2**:1257 (1962).

23. R. L. Poland and G. B. Odell, Physiologic jaundice, the enterohepatic circulation of bilirubin, *New Engl. J. Med.* **284**:1–6 (1971).

24. R. Brodersen and J. Theilgaard, Bilirubin colloid formation in neutral aqueous solution, *Scand. J. Clin. Lab. Invest.* **24**:395–398 (1969).

25. J. D. Ostrow and R. V. Branham, *in* "Bilirubin Metabolism of the Newborn" (D. Bergsma, D. Hsia, and C. Jackson, eds.) *Birth Defects: Orig. Art. Ser.* **6**:93–99 (1970).

26. T. K. With, "Bile Pigments, Chemical, Biological, and Clinical Aspects," pp. 295–338, Academic Press, New York (1968).

27. R. Brodersen, Bilirubin diglucuronide in normal human blood serum, *Scand. J. Clin. Lab. Invest.* **18**:361–379 (1966).

28. C. H. Gray, A. Neuberger, and P. H. A. Sneath, Studies in congenital porphyria. 2. Incorporation of ^{15}N in the stercobilin in the normal and in the porphyric, *Biochem. J.* **47:** 87–92 (1950).

29. I. M. London, R. West, D. Shemin, and D. Rittenberg, On the origin of bile pigment in normal man, *J. Biol. Chem.* **184:**351–358 (1950).

30. L. G. Israels, J. Skanderbeg, H. Guyda, W. Zingg, and A. Zipursky, A study of the early-labelled fraction of bile pigment: The effect of altering erythropoiesis on the incorporation of [2-^{14}C] glycine into haem and bilirubin, *Brit. J. Haematol.* **9:**50–62 (1963).

31. T. Yamamoto, J. Skanderbeg, A. Zipursky, and L. G. Israels, Early appearing bilirubin: Evidence for two components, *J. Clin. Invest.* **44:**31–41 (1965).

32. S. H. Robinson, M. Tsong, B. W. Brown, and R. Schmid, The sources of bile pigment in rat: Studies of "early-labeled" fraction, *J. Clin. Invest.* **45:**1569–1586 (1966).

33. G. W. Ibrahim, S. Schwartz, and C. J. Watson, Early labeling of bilirubin from glycine and δ-aminolevulinic acid in bile fistula dogs with special reference to stimulated erythropoiesis, *Metabolism* **15:**1129–1139 (1966).

34. S. H. Robinson, The origins of bilirubin, *New Engl. J. Med.* **279:**143–149 (1968).

35. R. Schmid, H. S. Marver, and L. Hammaker, Enhanced formation of rapidly labelled bilirubin by phenobarbital: Hepatic microsomal cytochromes as a possible source, *Biochem. Biophys. Res. Commun.* **24:**319–328 (1966).

36. M. F. Vest, *in* "Bilirubin Metabolism" (I. A. D. Bouchier and B. H. Billing, eds.) pp. 47–53, Blackwell Scientific Publications, Oxford (1967).

37. R. Tenhunen, H. S. Marver, and R. Schmid, Microsomal heme oxygenase. Characterization of the enzyme, *J. Biol. Chem.* **244:**6388–6394 (1969).

38. A. J. Levi, Z. Gatmaitan, and I. M. Arias, Two hepatic cytoplasmic protein fractions, Y and Z, and their possible role in the hepatic uptake of bilirubin, sulfobromophthalein, and other anions, *J. Clin. Invest.* **48:**2156–2167 (1969).

39. A. J. Levi, Z. Gatmaitan, and I. M. Arias, Deficiency of hepatic organic anion-binding protein, impaired organic anion uptake by liver, and "physiologic" jaundice in newborn monkeys, *New Engl. J. Med.* **283:**1136–1139 (1970).

40. S. Sussman, J. V. Carbone, G. Grodsky, V. Hjelte, and P. Miller, Sulfobromophthalein sodium metabolism in newborn infants, *Pediatrics* **29:**899–906 (1962).

41. G. J. Dutton, D. E. Langelaan, and P. E. Ross, High glucuronide synthesis in newborn liver: Choice of species and substrates, *Biochem J.* **93:**4p–5p (1964).

42. R. Brodersen, J. Jacobsen, H. Hertz, H. Rebbe, and B. Sørensen, Bilirubin conjugation in the human fetus, *Scand. J. Clin. Lab. Invest.* **20:**41–48 (1967).

43. L. Strebel and G. B. Odell, UDP glucuronyl transferase in rat liver: Genetic variation and maturation, *Pediat. Res.* **3:**351–352 (1969).

44. M. F. Vest and H. Grieder, Erythrocyte survival in the newborn infant, as measured by chromium[51] and its relation to the postnatal serum bilirubin level, *J. Pediat.* **59:**194–199 (1961).

45. I. M. Arias, *in* "Bilirubin Metabolism of the Newborn" (D. Bergsma, D. Hsia, and C. Jackson, eds.) *Birth Defects: Orig. Art. Ser.* **6:**55–59 (1970).

46. N. H. Martin, Preparation and properties of serum and plasma proteins. XXI. Interactions with bilirubin, *J. Am. Chem. Soc.* **71:**1230–1232 (1949).

47. W. J. Waters, The reserve albumin binding capacity as a criterion for exchange transfusion, *J. Pediat.* **70:**185–192 (1967).

48. E. G. Porter and W. J. Waters, A rapid micromethod for measuring the reserve albumin binding capacity in serum from newborn infants with hyperbilirubinemia, *J. Lab. Clin. Med.* **67:**660–668 (1966).

49. N. A. Kaufmann, J. Kapitulnik, and S. H. Blondheim, The absorption of bilirubin by Sephadex and its relationship to the criteria for exchange transfusion, *Pediatrics* **44:** 543–548 (1969).

50. W J. Keenan, J. E. Arnold, and J. M. Sutherland, Serum bilirubin binding determined by Sephadex column chromatography, *J. Pediat.* **74:**813 (1969).

51. G. B. Odell, S. N. Cohen, and P. C. Kelly, Studies in kernicterus. II. The determination of the saturation of serum albumin with bilirubin, *J. Pediat.* **74:**214–230 (1969).
52. R. Schmid, I. Diamond, L. Hammaker, and C. B. Gundersen, Interaction of bilirubin with albumin, *Nature* **206:**1041–1043 (1965).
53. J. D. Ostrow and R. Schmid, The protein-binding of C^{14}-bilirubin in human and murine serum, *J. Clin. Invest.* **42:**1286–1299 (1963).
54. G. B. Odell, The distribution of bilirubin between albumin and mitochondria, *J. Pediat.* **68:**164–180 (1966).
55. G. B. Odell, Influence of *p*H on distribution of bilirubin between albumin and mitochondria, *Proc. Soc. Exptl. Biol. Med.* **120:**352–354 (1965).
56. R. P. Wennberg and M. L. Cowger, Spectral identification of albumin–bilirubin complexes, *Pediat. Res.* **3:**376–377 (1969).
57. R. P. Wennberg and L. Fraser Rasmussen, The influence of albumin on bilirubin distribution, *Abst. Forty-first Ann. Meeting Soc. Pediat. Res.,* p. 182 (1971).
58. W. A. Silverman, D. H. Andersen, W. A. Blanc, and D. N. Crozier, A difference in mortality rate and incidence of kernicterus among premature infants allotted to two prophylactic antibacterial regimens, *Pediatrics* **18:**614–625 (1956).
59. D. Schiff, G. Chan, and L. Stern, Fixed drug combinations and the displacement of bilirubin from albumin, *Pediatrics* **48:**139–141 (1971).
60. L. Johnson, F. Sarmiento, W. A. Blanc, and R. Day, Kernicterus in rats with an inherited deficiency of glucuronyl transferase, *Am. J. Dis. Child.* **97:**591–608 (1959).
61. G. B. Odell, The dissociation of bilirubin from albumin and its clinical implications, *J. Pediat.* **55:**268–279 (1959).
62. G. Blauer and T. E. King, Interaction of bilirubin with bovine serum albumin in aqueous solution, *J. Biol. Chem.* **245:**372–381 (1970).
63. M. L. Cowger, Unpublished data.
64. I. Moriguchi, S. Wada, and T. Nishizawa, Protein bindings. III. Binding of sulfonamides to bovine serum albumin, *Chem. Pharm. Bull.* **16:**601–605 (1968).
65. I. Diamond and R. Schmid, Experimental bilirubin encephalopathy. The mode of entry of bilirubin-^{14}C into the central nervous system, *J. Clin. Invest.* **45:**678–689 (1966).
66. R. C. Burnstine and R. Schmid, Solubility of bilirubin in aqueous solutions, *Proc. Soc. Exptl. Biol. Med.* **109:**356–358 (1962).
67. W. Haymaker, C. Margoles, A. Pentschew, H. Jacob, R. Lindenberg, L. S. Arroyo, O. Stochdorph, and D. Stowens, *in* "Kernicterus and Its Importance in Cerebral Palsy," pp. 21–228, Charles C Thomas, Springfield, Ill. (1961).
68. A. E. Claireaux, *in* "Kernicterus" (Andrew Sass-Kortsák, ed.) pp. 140–149, University of Toronto Press, Toronto (1961).
69. A. E. Claireaux, Hemolytic disease of the newborn, Part II. Nuclear jaundice (kernicterus), *Arch. Dis. Childh.* **25:**71–80 (1950).
70. H. S. Schutta and L. Johnson, Bilirubin encephalopathy in the Gunn rat: A fine structure study of the cerebellar cortex, *J. Neuropathol. Exptl. Neurol.* **26:**377–396 (1967).
71. W. A. Blanc and L. Johnson, Studies on kernicterus. Relationship with sulfonomide intoxication, report on kernicterus in rats with glucuronyl transferase deficiency, and review of pathogenesis, *J. Neuropathol. Exptl. Neurol.* **18:**165–189 (1959).
72. D. H. Silberberg and H. S. Schutta, The effects of unconjugated bilirubin and related pigments on cultures of rat cerebellum, *J. Neuropathol. Exptl. Neurol.* **26:**572–583 (1967).
73. H. S. Schutta, L. Johnson, and H. S. Neville, Mitochondrial abnormalities in bilirubin encephalopathy, *J. Neuropathol. Exptl. Neurol.* **29:**296–305 (1970).
74. J. Bernstein and B. H. Landing, Extraneural lesions associated with neonatal hyperbilirubinemia and kernicterus, *Am. J Pathol.* **40:**371–384 (1962).
75. R. L. Day, Inhibition of brain respiration *in vitro* by bilirubin: Reversal of inhibition by various means, *Am. J. Dis. Child.* **88:**504–506 (1954).
76. W. J. Waters and W. R. Bowen, Bilirubin encephalopathy: Studies relating to cellular respiration, *Am. J. Dis. Child.* **90:**603 (1955).

77. R. Day, Kernicterus, further observations on the toxicity of heme pigments, *Pediatrics* **17**:925–928 (1956).
78. W. R. Bowen and W. J. Waters, Bilirubin encephalopathy: Studies related to the site of inhibitory action of bilirubin on brain metabolism, *Am. J. Dis. Child.* **93**:21–22 (1957).
79. J. H. Quastel and I. J. Bickis, Metabolism of normal tissues and neoplasms *in vitro, Nature (London)* **183**:281–286 (1959).
80. R. Zetterström and L. Ernster, Bilirubin, an uncoupler of oxidative phosphorylation in isolated mitochondria, *Nature (London)* **178**:1335–1337 (1956).
81. L. Ernster, *in* "Kernicterus" (A. Sass-Kortsák, ed.) pp. 174–192, University of Toronto Press, Toronto (1961).
82. M. L. Cowger, R. P. Igo, and R. F. Labbe, The mechanism of bilirubin toxicity studied with purified respiratory enzyme and tissue culture systems, *Biochemistry* **4**:2763–2770 (1965).
83. M. L. Cowger, Mechanism of bilirubin toxicity on tissue culture cells: Factors that affect toxicity, reversibility by albumin and comparison with other respiratory poisons and surfactants, *Biochem. Med.* **5**:1–16 (1971).
84. I. Diamond and R. Schmid, Oxidative phosphorylation in experimental bilirubin encephalopathy, *Science* **155**:1288–1289 (1966).
85. M. Menken and E. C. Weinbach, Oxidative phosphorylation and respiratory control of brain mitochondria isolated from kernicteric rats, *J. Neurochem.* **14**:189–193 (1967).
86. S. Schenker, D. W. McCandless, and P. E. Zollman, Studies of cellular toxicity of unconjugated bilirubin in kernicteric brain, *J. Clin. Invest.* **45**:1213–1220 (1966).
87. M. T. Vogt and R. E. Basford, The effect of bilirubin on the energy metabolism of brain mitochondria, *J. Neurochem.* **15**:1313–1320 (1968).
88. M. G. Mustafa, M. L. Cowger, and T. E. King, Effects of bilirubin on mitochondrial reactions, *J. Biol. Chem.* **244**:6403–6414 (1969).
89. I. L. Kahán, M. Timár, and M. Földi, Bilirubin-binding cerebral lipid, *Acta Paediat. Acad. Sci. Hung.* **9**:121–131 (1968).
90. M. G. Mustafa and T. E. King, Binding of bilirubin with lipid, a possible mechanism of its toxic reaction in mitochondria, *J. Biol. Chem.* **245**:1084–1089 (1970).
91. E. Stenhagen and E. K. Rideal, The interaction between porphyrin, lipoid, and protein monolayers, *Biochem. J.* **33**:1591–1598 (1939).
92. R. Lester, R. E. Behrman, and J. F. Lucey, Transfer of bilirubin-C[14] across monkey placenta, *Pediatrics* **32**:416–419 (1963).
93. D. Watson, The absorption of bilirubin by erythrocytes, *Clin. Chim. Acta* **7**:733–734 (1962).
94. F. A. Oski and J. L. Naiman, Red cell binding of bilirubin, *J. Pediat.* **63**:1034–1037 (1963).
95. N. A. Kaufmann, A. J. Simcha, and S. H. Blondheim, The uptake of bilirubin by blood cells from plasma and its relationship to the criteria for exchange transfusion, *Clin. Sci.* **33**:201–208 (1967).
96. W. H. Cheung, A. Sawitsky, and H. D. Isenberg, The effect of bilirubin on the mammalian erythrocyte, *Transfusion* **6**:475–486 (1966).
97. U. Suvansri, W. H. Cheung, and A. Sawitsky, The effect of bilirubin on the human platelet, *J. Pediat.* **74**:240–246 (1969).
98. G. B. Odell, J. C. Natzschka, and G. N. B. Storey, Bilirubin nephropathy in the Gunn rat, *Am. J. Physiol.* **212**:931–938 (1967).
99. M. L. Cowger and M. G. Mustafa, Some membrane effects of bilirubin, *Pediat. Res.* **5**:419–420 (1971).
100. A. B. Novikoff, *in* "Lysosomes" (A. V. S. de Reuck and M. P. Cameron, eds.) Ciba Foundation Symposium, p. 36, Little, Brown, Boston (1963).
101. A. C. Allison, I. A. Magnus, and M. R. Young, Role of lysosomes and of cell membranes in photosensitization, *Nature* **209**:874–878 (1966).
102. M. G. Mustafa and M. L. Cowger, Unpublished data.

103. I. Matsuda, M. Tashimo, and A. Takase, Effects of bilirubin on glucose oxidation in red cells, *Experientia* **25:**865–866 (1969).
104. R. F. Labbe, M. R. Zaske, and R. A. Aldrich, Bilirubin inhibition of heme biosynthesis, *Science* **129:**1741–1742 (1959).
105. M. M. Thaler, *in* "Bilirubin Metabolism of the Newborn" (D. Bergsma, D. Hsia, and C. Jackson, eds.) *Birth Defects: Orig. Art. Ser.* **6:**128–130 (1970).
106. R. Brodersen and P. Bartels, Enzymatic oxidation of bilirubin, *Europ. J. Biochem.* **10:** 468–473 (1969).
107. J. B. Sumner and M. Nyman, The oxidation of bilirubin by peroxidase, *Science* **102:** 209 (1945).
108. I. M. Rosenthal, H. J. Zimmerman, and N. Hardy, Congenital nonhemolytic jaundice with disease of the central nervous system, *Pediatrics* **18:**378–386 (1956).
109. W. A. Gardner and B. W. Konigsmark, Familial nonhemolytic jaundice: Bilirubinosis and encephalopathy, *Pediatrics* **43:**365–376 (1969).
110. J. Dobbing, *in* "Progress in Brain Research" (A. Lajtha and D. H. Ford, eds.) Vol. 29, pp. 417–427, Elsevier, New York (1968).
111. L. Bakay, Studies on blood–brain barrier with radioactive phosphorus. III. Embryonic development of the barrier, *Arch. Neurol. Psychiat.* **70:**30–39 (1953).
112. I. Diamond, Kernicterus: Revised concepts of pathogenesis and management, *Pediatrics* **38:**539–542 (1966).
113. L. Bakay, *in* "Progress in Brain Research" (A. Lajtha and D. H. Ford, eds.) Vol. 29, pp. 315–319, Elsevier, New York (1968).
114. H. Chen, C.-S. Lin, and I.-N. Lien, Ultrastructural studies in experimental kernicterus, *Am. J. Pathol.* **48:**683–711 (1966).
115. B. Rozdilsky, *in* "Kernicterus" (A. Sass-Kortsák, ed.) pp. 161–166, University of Toronto Press, Toronto (1961).
116. H. Chen, C.-S. Lin, and I.-N. Lien, Vascular permeability in experimental kernicterus, an electron-microscopic study of the blood–brain barrier, *Am. J. Pathol.* **51:**69–99 (1967).
117. H. S. Schutta and L. Johnson, Clinical signs and morphologic abnormalities in Gunn rats treated with sulfadimethoxine, *J. Pediat.* **75:**1070–1079 (1969).
118. M. Reivich, G. Isaacs, E. Evarts, and S. Kety, The effect of slow wave sleep and REM sleep on regional cerebral blood flow in cats, *J. Neurochem.* **15:**301–306 (1968).
119. B. Billing and F. Jansen, Enigma of bilirubin conjugation, *Gastroenterology* **61:**258–260 (1971).

Chapter 8

THE ACTION OF THYROID HORMONES AND THEIR INFLUENCE ON BRAIN DEVELOPMENT AND FUNCTION

Louis Sokoloff

National Institute of Mental Health
Bethesda, Maryland

and

Charles Kennedy

Georgetown University Hospital
Washington, D.C.

I. INTRODUCTION

The actions of hormones are generally restricted to a limited number of specific target tissues. The thyroid hormones differ from most in the wide diversity of tissues and organs in which they exert their effects. Indeed, in mammals it appears that only the tissues of the reticuloendothelial system, the testis, and the mature brain are immune to their actions,[1] and even in these tissues, as clearly evidenced in brain,[2] it is likely that the hormones are active during some period in the life of the organism. The diversity of the thyroid hormones' target tissues is matched by the multiplicity of their biochemical effects. Hardly an area of metabolism remains unaffected by their actions. The diffuseness of their biochemical effects could reflect different and multiple mechanisms of actions. It seems more parsimonious, however, to assume a single mechanism of action on a process so fundamental that it either influences directly a wide variety of biochemical reactions in many areas of metabolism or triggers a succession of events most of which no longer reflect the direct influence of the hormone but more remote consequences of its primary action.

The manifestations of the actions of the thyroid hormones vary with the tissues and their degree of maturation and development. In the fully grown and mature mammal, the effects of thyroid dysfunction are most prominently manifested by changes in energy exchange, but brain is one of the organs which do not share in this effect. In the young, developing animal, the effects of the thyroid hormones on growth, development, and maturation predominate, and the brain is one of the organs most dramatically affected. Cretinism leads, therefore, not only to physical dwarfism but also to mental deficiency resulting from a morphologically and functionally underdeveloped brain.

The effects of the thyroid hormones in brain are manifested quite differently from those in other tissues. Most tissues, no matter how greatly altered functionally and biochemically, undergo relatively little anatomical change under the influence of these hormones. In contrast, there are major morphogenetic components to their actions in brain. The striking morphological reorganization, indeed, differentiation, which characterizes the postnatal development of the mammalian brain is largely dependent on the thyroid hormones. The sensitivity of the brain to the thyroid hormones is limited only to its period of development and maturation; once fully developed it no longer exhibits any effects clearly attributable to a direct action of the hormones. It is probably because their actions are exerted at a time when rapid changes are in progress that the morphogenetic effects of the thyroid hormones predominate in brain. The basic biochemical mechanism of action is likely to be the same in brain as in other tissues.

A. Chemical Nature of Thyroid Hormones

The term "thyroid hormones" applies to two closely related amino acids, L-thyroxine and L-triiodothyronine, which are known to be synthesized in the thyroid gland, secreted into the circulation, and responsible for all the effects of the thyroid hormones on bodily functions. Thyroxine appears to be the predominant species, alone accounting for more than 95% of the circulating hormonal iodine.[3] Other thyroactive compounds with related chemical structures, e.g., the tetraiodo- and triiodo- acetic and propionic analogues, have been identified in peripheral tissues, but these are probably metabolites of the hormones which have retained some biological activity.

Both hormones are iodinated diphenyl ether derivatives of L-alanine with a phenolic group in the 4'-position (Fig. 1). Although their chemical structures are quite simple, the exact structural requirements for thyromimetic activity

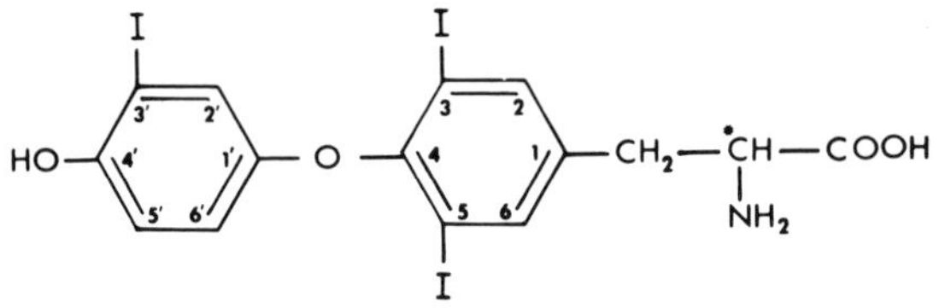

3, 5, 3'-Triiodo-L-Thyronine

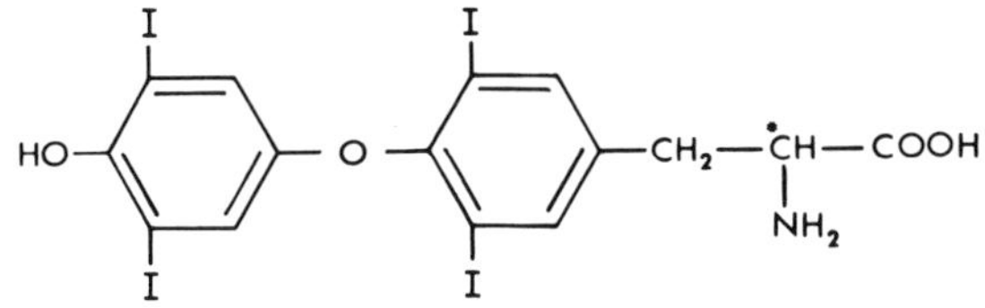

L-Thyroxine

*Asymmetric carbon

Fig. 1. Chemical structure of the thyroid hormones.

have not been precisely defined.[4,5] The alanine side-chain is not absolutely essential; analogues with other side chains, such as acetic, propionic, or pyruvic acid, retain activity, although relative potency is altered.[4,5] The D-alanine analogue, D-thyroxine, has only a fraction of the biological activity of L-thyroxine *in vivo*, but this probably reflects its more rapid metabolism and excretion[6]; in almost all *in vitro* actions of thyroxine in cell-free enzyme systems, the D- and L-isomers are essentially equally effective. The side-chain, therefore, probably plays little role in the specific biochemical mechanism of action of the hormones but may influence biological activity *in vivo* by affecting the hormones' binding and storage on plasma proteins, distribution, metabolic degradation, or excretion.

The iodine atoms *per se* are not essential; bromo-, fluoro-, chloro-, and nitro-analogues exhibit activity, although with decreasing effectiveness in that order.[4,5] Some type of substitution at the 3- and 5-position is essential, and iodine substituents are associated with the greatest amount of activity. Alkyl substitution in these positions, for example, results in a striking reduction, but still some residual activity; however, thyronine and 3',5'-diiodothyronine are devoid of activity.[5] Substitution at the 3'- and 5'-position is not an absolute requirement for activity, only for maximal activity. For example, 3,5-diiodothyronine has definite although slight activity, and alkyl substitution at the 3'- and/or 5'-position results in enhanced activity, even greater than that of the analogous iodinated molecules.[4,7] Triiodothyronine is reputed to be more active than thyroxine, but the difference may reflect extracellular mechanisms rather than structure–function relationships in the specific mode of action of the hormone intracellularly. Triiodothyronine is less firmly bound to plasma proteins than thyroxine,[8–10] and a given dose then leads to a higher free hormone concentration in the blood and more rapid distribution to the tissues.

The phenolic structure of the thyroid hormones has been suspected to be of fundamental importance to their functional activity. Niemann and Mead[11] studied a variety of isomers of thyroxine with the phenolic hydroxyl group displaced to various positions of the outer ring. Only those isomers capable of oxidation to a quinoid structure were active, and they postulated that thyromimetic activity is correlated with the oxidation–reduction potential of the system as a whole. More recently, however, O-alkyl derivatives, in which the phenolic structure is converted to an ether structure, have been found to be active,[4,5,12] and Jorgensen et al.[13] have observed considerable thyromimetic activity with 3,5-diiodo-4(2′,3′-dimethylphenoxy)-DL-phenylalanine, an analogue with no substituent at the 4′-position and incapable of being metabolically converted to a phenolic structure. Therefore, the functional role of the phenolic structure, if any, remains obscure.

The ether structure is essential, although a sulfur can be substituted for the oxygen bridge with retention of some activity.[4,14] Whether it need be a diphenyl ether is uncertain, but this appears unlikely, because benzyloxy derivatives of diiodotyrosine exhibit slight degrees of thyromimetic activity.[4]

B. General Biological and Biochemical Effects of Thyroid Hormones

The thyroid hormones have broad and diffuse effects which are manifested in all organ systems and have ramifications in almost every aspect of bodily function and metabolism. Extensive studies of these effects and their underlying biochemical mechanisms have led to a voluminous literature which can only briefly and selectively be considered in this chapter. There are a number of excellent reviews available for more detailed and comprehensive treatment of the subject.[8,15–22] This section is concerned only with the general effects of thyroid hormones, regardless of the tissue or organism in which observed. It is understood, however, that many of the effects in other tissues apply to and form the basis of some of the effects observed in the nervous system. The effects of thyroid hormones in the nervous system and, as far as possible, the mechanisms of these effects will be treated in subsequent sections.

1. Growth and Metamorphosis

In mature mammals, the most prominent effects of thyroid hormones are on metabolic rates. In immature mammals, thyroid hormones also profoundly influence growth and development. Thyroid deficiency in early postnatal life leads to stunted growth and incomplete maturation. Growth and development can be reinstituted, however, by adequate thyroid replacement therapy, provided treatment is initiated early enough. Mild juvenile hyperthyroidism may, on the other hand, lead to accelerated growth and the ultimate achievement of a greater than projected body stature. Severe hyperthyroidism generally retards the growth of the young, possibly because of a relative nutritional deficiency imposed by the increased demands of the accelerated metabolic rate.[23]

The most dramatic example of the effects of thyroid hormones on developmental processes is their action in amphibian metamorphosis.[24] The metamorphosis of the tadpole into an adult frog is absolutely dependent on thyroid hormones. It is a period of phenomenal structural, functional, and biochemical reorganization of the entire organism into essentially a new phenotype. The process obviously involves major revision of gene expression, but the mechanism by which thyroid hormones initiate it is unknown. Protein and RNA synthesis are both increased or at least altered in many organs, and the tissue levels of various enzymes are drastically changed.[25,26] Once the mature adult state is achieved, however, thyroid hormones have no apparent further effects. The actions of thyroid hormones in brain bear a striking similarity to their role in metamorphosis. Thyroid hormones promote and are essential for the postnatal maturation of the brain, but once mature the brain appears to become insensitive to their actions.

2. Protein and Nitrogen Metabolism

The thyroid hormones produce prominent and, in some ways, paradoxical effects on nitrogen metabolism. Protein synthesis is stimulated in all tissues that also respond with increased oxygen consumption.[27] In the growth-retarded, hypothyroid, immature mammal, this effect may be anabolic and lead to positive nitrogen balance and stimulation of growth,[8] but excessive doses become catabolic and retard growth despite the continued stimulation of protein synthesis.[8,28] Thyroid hormones also stimulate protein synthesis in most tissues of fully grown animals,[27] but urinary nitrogen excretion is usually strikingly increased.[8,15,29,30] With large doses, negative nitrogen balance and weight loss ensue. The apparent paradox may again be the result of a relative nutritional insufficiency associated with the increased metabolic rate. Adequate caloric intake restores normal nitrogen balance,[30] and magnesium and vitamin supplementation has been reported to reverse the catabolic effects of excessive thyroid hormone administration.[23] Hyperthyroidism is also characterized by a creatinuria which probably reflects decreased creatine uptake by muscle tissue.[31] Creatine phosphate deficiency may cause contracting muscle to be more directly dependent on ATP generation from oxidative phosphorylation and might account for the increased oxygen consumption per given amount of work in hyperthyroid muscle.[30]

3. Energy Exchange

Perhaps the earliest discovery concerning the functions of the thyroid was in regard to its effects on oxygen consumption and heat production.[32] In mammals, thyroid deficiency is associated with a severe decrease in oxygen consumption; complete thyroidectomy can lead eventually to as much as a 50% decrease in basal metabolic rate. The hypometabolism is reversed by the administration of thyroid hormones, which, if administered in excess, can drive the basal metabolic rate (BMR) above normal. The stimulation of oxygen consumption occurs after a latent period of several hours following the administration of a dose of hormone; the latent period with triiodothyronine is shorter than with thyroxine, probably because of its lesser binding to plasma proteins and greater availability for diffusion into the cells. All tissues except mature brain, testis, and spleen[1] participate in the augmentation of oxidative metabolism, and all classes of foodstuffs, carbohydrate, protein, and lipid, are the substrates for the increased oxygen utilization.[30] Thyroid hormones cause rapid depletion of liver glycogen stores,[33-35] partly because of increased carbohydrate utilization but also because of the potentiation of the glycogenolytic effects of epinephrine.[36] Synergism with epinephrine may also contribute to the mobilization of fatty acids from adipose tissue and the elevated plasma free fatty levels in hyperthyroidism.[37,38] Serum cholesterol levels are inversely related to thyroid activity even though its synthesis is stimulated by thyroid hormones.[39]

The increased oxidative metabolism does not appear to lead to comparable increases in available energy for physical work. This is most evident in muscle, which becomes noticeably inefficient in the hyperthyroid state. The increment in total dody oxygen consumption associated with a given load of muscular effort is greatly increased in hyperthyroidism,[30] the additional energy apparently bissipated as heat. It has been suggested that the role of the thyroid is primarily in heat production and thermoregulation by inducing non-energy-conserving, heat-producing, calorigenic shunts in electron transport, e.g., NADPH–cytochrome reductase.[40,41] The effects of thyroid hormones on oxidative metabolism are, in fact, limited to warm-blooded animals and are not seen in poikilothermic species. Increased thyroid activity is unquestionably an important component of the mechanism of acclimation to cold,[42] but it is doubtful whether so limited a function is consistent with the numerous other peripheral actions of thyroid hormones.

4. Mitochondrial Structure and Function

Because of their prominent effects on oxygen consumption and energy exchange, numerous studies have been directed at the action of thyroid hormones on the structure and function of mitochondria, the cellular organelles which carry out the bulk of the cell's energy-yielding oxidative processes.

Initial efforts were focused on the *respiratory electron transport chain.* Thyroxine itself has been considered a possible electron carrier which could undergo oxidation–reduction in its phenolic ring to and from a quinoid structure.[11] It has even been proposed as a component of an additional site for coupling of electron transport to phosphorylation in the electron transport chain.[43] Experimental evidence in support of this hypothesis is lacking, and in fact it appears unlikely in view of the demonstrated biological activity of analogues which lack the 4'-phenolic hydroxyl group and cannot therefore form quinoid structures.[4,5,12] Components of the mitochondrial respiratory chain, such as cytochrome c[44,45] and ubiquinone,[46] and the microsomal NADPH–cytochrome c reductase system[40] are increased in hyperthyroidism and reduced in hypothyroidism, but these changes appear to be secondary adaptations to the changes in metabolic rate rather than the cause.[45,46] Thyroid hormones *in vivo* induce mitochondrial L-α-glycerophosphate dehydrogenase,[47] a flavin-linked dehydrogenase. Recent studies suggest, however, that this effect may be indirect. Thyroid hormones stimulate flavokinase and the conversion of riboflavin to the flavin coenzymes, FMN and FAD,[48,49] which raises the possibility that the increase in α-glycerophosphate dehydrogenase may represent not so much increased enzyme synthesis as stabilization and protection of the enzyme from degradation by converting the apoenzyme to its flavin-holoenzyme form. This is an intriguing finding which serves to explain, a number of hitherto poorly understood actions of thyroid hormones, including their interactions with riboflavin metabolism,[49,50] but its significance for the fundamental action of the hormones remains to be clarified.

The oxidation of tricarboxylic acid cycle substrates in the mitochondria provides electrons which are transmitted along the various components of the respiratory electron transport chain, terminating eventually in the reduction of oxygen to water. The energy released by this oxidation is partially conserved in high-energy phosphate bonds, most notably ATP. The mechanism by which phosphate bond synthesis is coupled to electron transport constitutes one of the major unsolved problems in biochemistry. Most of the ATP of the cell is synthesized by this process of *oxidative phosphorylation.* Since ATP is the essential energy donor for most of the energy-consuming reactions in the cell, it is clear that oxidative phosphorylation is of fundamental importance to all aspects of cellular metabolism and function. The possibility that thyroid hormones exert their biological effects by acting on this process has long been, and continues to be, under serious consideration.

Dinitrophenol, a substituted phenol like the thyroid hormones, is the classical uncoupler of oxidative phosphorylation. It dissociates electron transport from the phosphorylation process so that electron transport no longer drives ATP synthesis.[51] The uptake of inorganic phosphate into ATP is inhibited, and in fact the opposite reaction, ATP hydrolysis by mitochondria, is stimulated. The P/O ratio, the number of high-energy phosphate bonds formed per atom of oxygen consumed, is a measure of the efficiency of the energy conservation, and this value is depressed by dinitrophenol. The thyroid hormones have similar although not identical effects on oxidative phosphorylation.[19] Mitochondria isolated from thyroxine-treated animals or preincubated with thyroid hormones *in vitro* exhibit depressed P/O ratios.[19,52-55] The oxygen consumption, however, is not increased by the thyroid hormones *in vitro;* it may, in fact, be reduced. Mitochondrial ATPase is stimulated but only weakly compared to the effect of dinitrophenol. Furthermore, relatively high doses or concentrations are required to produce these effects, and, in contrast to dinitrophenol, thyroid hormones depress oxidative phosphorylation only in intact mitochondria and not in actively phosphorylating submitochondrial particles.[56]

The earliest effects elicited by thyroid hormones *in vivo* are those observed on mitochondrial energy-conserving functions. Hoch[57] has observed almost immediate effects of minute doses of thyroid hormones on rat liver mitochondrial respiratory control. In normal "tightly coupled" mitochondria, respiration is regulated by the levels of phosphate acceptors and is restricted by low levels of ADP. Thyroid hormones loosen the coupling and reduce the effectiveness of this control; in effect, oxygen consumption appears to be stimulated by removing the brake imposed by limiting amounts of phosphate acceptor, i.e., ADP. Human skeletal muscle mitochondria from hyperthyroid patients show the same loosening of coupling.[58] Such an effect suggests some interaction of thyroid hormones with the coupling reactions between oxidation and phosphorylation, but it does not necessarily mean uncoupling. Uncoupling implies energy wastage which may be compatible with calorigenesis or the increased metabolic rate caused by thyroid hormones, but it cannot be reconciled with their stimulation of energy-requiring processes, such as protein and RNA synthesis, growth, and maturation. A change in respiratory control, however, is not inconsistent with these effects. It does not by itself explain them, but it may be a reflection of some mitochondrial–hormone interaction, other than uncoupling, which leads to profound changes in cellular functions. Uncoupling may, however, be of significance in extreme thyrotoxic states when the concentrations of the thyroid hormones become relatively enormous.[59]

Closely related to and possibly responsible for the effects on oxidative phosphorylation are the actions of thyroid hormones on *mitochondrial structure*. Low concentrations of hormone, much lower than those required to depress the P/O ratio, cause mitochondria to swell.[60-63] Mitochondria normally exhibit swelling–contraction cycles,[62] but hypothyroidism reduces and hyperthyroidism enhances their fragility and tendency to swell.[61,63] Mitochondria from tissues which are physiologically thyroxine-insensitive, including adult brain, do not swell in response to thyroxine.[64] The morphological change is because of water accumulation and reflects functional and structural alteration of the mitochondrial membrane.[65] Uncoupling of oxidative phosphorylation is not the cause; dinitrophenol, the classical uncoupler, not only does not cause swelling but also reverses thyroxine-induced swelling.[61,62] Indeed, it may be that the swelling may be responsible for the apparent uncoupling.[62] Thyroactive compounds and a variety of swelling agents, which include Ca^{2+}, inorganic phosphate, and glutathione, promote the release from the mitochondrial membrane of a protein-bound, fatty acid derivative, designated "U factor," which causes mitochondrial swelling, uncouples oxidative phosphorylation, and stimulates mitochondrial ATPase activity.[62] ATP, which reverses the swelling effect, causes the incorporation of U factor into the cardiolipin and phosphatidic acid fractions of the mitochondrial membranes.[62]

It is clear that thyroid hormones produce profound alterations in the cytostructural organization and function of the mitochondria. These effects are real and relevant, but the mechanism of each of them remains obscure. It is probable that none of them represents the basic mechanism of action of the hormone and that all of them are consequences of a primary mode of action. They do provide strong evidence, however, that the primary locus of the hormone action resides within the mitochondria. Indeed, recent studies have provided evidence of tight and specific binding of thyroxine to the inner mitochondrial membrane.[66]

5. Ion Transport

The enhancement of energy exchange by thyroid hormones has led to an examination of their effects on energy-consuming processes. A number of effects on such processes, including stimulation of RNA and protein synthesis, which will be discussed below, have been observed. Recent studies have directed attention to ion transport. Ions are generally transported across cell membranes by active transport which utilizes ATP as the energy source, and increased ion transport could be expected to lead to accelerated ATP hydrolysis and inevitably increased oxygen consumption. It has long been known that altered thyroid state results in disorders in electrolyte balance. Recent results have suggested that thyroid hormones may stimulate Mg^{2+}, Ca^{2+}, Na^+, and K^+ transport into or out of cells.[67-69] Because ouabain, which inhibits Na^+,K^+-

dependent ATPase and the transport of these ions, has been found to block the increased oxygen consumption of liver slices from hyperthyroid animals, it has been proposed that the stimulation of Na^+ and K^+ transport is the mechanism by which thyroid hormones stimulate metabolic rate.[68,69] Thyroid hormones have also been found to promote the more rapid development of the Na^+,K^+-ATPase in developing rat brain.[70]

6. Catecholamine Metabolism

The thyroid hormones potentiate the action of the adrenergic amines. Cardiovascular and metabolic responses to a given dose of epinephrine are more intense and prolonged in hyperthyroidism and reduced in hypothyroidism.[36, 38,71–74] Epinephrine has a calorigenic action qualitatively like that of the thyroid hormones, and many of the physiological and behavioral manifestations of hyperthyroidism resemble those of increased sympathetic nervous activity and can be ameliorated by adrenolytic drugs. There have been reports that thyroid hormones lower catechol-O-methyl transferase[75] and monoamine oxidase[76] activities, but there has been controversy concerning the significance of these effects in the overall metabolic degradation and excretion of the catecholamines.[77,78] Catecholamines are, however, largely inactivated by uptake and binding in sympathetic nerve endings within the tissues; thyroid hormones have been reported to diminish this process.[79] Brewster et al.[80] have argued that the actions of the thyroid hormones are mediated entirely by the sympathetic amines. It is likely that these amines initiate or exaggerate some of the effects attributed to the thyroid hormones, but sympathetic blockade by means of dibenzyline or guanethidine does not eliminate or even alter their most fundamental metabolic actions.[81,82] The sympathetic nervous system and the adrenergic amines may modulate but do not mediate the effects of the thyroid hormones.

7. Protein and Nucleic Acid Synthesis

Numerous studies have established that thyroid hormones stimulate protein synthesis.[2,22,27,28,59,83–88] Since structural and enzymatic proteins are so fundamentally involved in all aspects of cellular morphology, metabolic activity, and function, an effect on protein synthesis could readily ramify into all the diverse and multiple actions ascribed to the thyroid hormones. The administration of thyroid hormones to normal animals stimulates the incorporation of amino acids into protein in tissues, such as liver, kidney, and heart, which also respond with increased metabolic rates; no effects on protein synthesis are observed in mature brain, testis, or spleen, all tissues which retain normal rates of oxygen utilization in hyperthyroidism[1,2,27] (Fig. 2). The increased metabolic

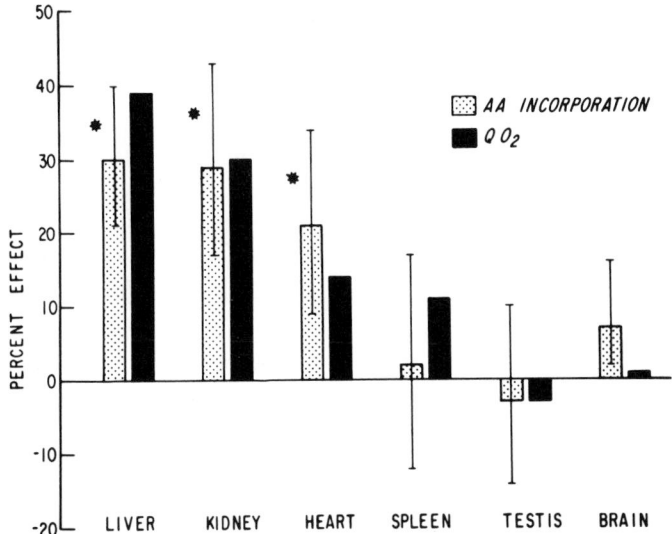

Fig. 2. Distribution of effects of hyperthyroidism on protein synthesis *in vivo* (AA) and on tissue oxygen consumption (QO₂) in various organs. The values are the means plus or minus standard errors. Asterisk indicates statistically significant effect ($p < 0.05$). [Data on protein synthesis from Michels *et al.*[27]; data on tissue oxygen consumption from Gordon and Heming.[1]]

rate appears to be secondary to the effect on protein synthesis; it follows the stimulation of protein synthesis by a number of hours[84] and is prevented or abruptly reversed by inhibitors of protein synthesis.[89,90] Contrary to most effects of thyroid hormones *in vivo*, the stimulation of protein synthesis occurs with essentially no latent period.[88] Hypothyroidism leads to reduced rates of protein synthesis[59,84] which are restored to normal by thyroid hormone replacement therapy.[84]

The effects on protein synthesis are not only observed *in vivo* and in cell-free tissue preparations from treated animals but can also be reproduced in cell-free preparations from normal animals by thyroid hormones added *in vitro*.[59] Stimulations occur with thyroxine concentrations as low as 10^{-7} M and increase with increasing hormone concentration until an optimum is reached at about 4×10^{-4} M; above this concentration, the effect abruptly reverses from stimulation to inhibition.[59] The reversal may reflect the binding of Mg^{2+} by high concentrations of thyroxine and may be of significance in severe thyrotoxic states.[59] The mechanism of the stimulation of protein synthesis has been extensively studied in the *in vitro* system and has been found to be a translational effect at the level of the transfer of tRNA-bound amino acid to ribosomal protein.[83,85] It is exerted primarily on the elongation or completion of the nascent polypeptide chain; initiation of new protein chains appears to be secondarily stimulated, perhaps because accelerated completion and release of existing polypeptide chains free the ribosomes and template RNA more rapidly for recycling.[87] The effect is not limited to a single specific species of protein but is observed in a large variety of tissues synthesizing different types of protein[27, 59,87,91–93]; in fact, it can be elicited in the synthesis of artificial polypeptides directed by synthetic messenger polyribonucleotides.[86]

The thyroid hormones do not, however, stimulate protein synthesis directly. They first interact with mitochondria in an energy-dependent reaction leading to some product or consequence which is actually responsible for the enhanced ribosomal protein-synthesizing activity.[59,83,87,88,94] The identity of the products of the thyroxine–mitochondrial reaction has not yet been established, nor is there any convincing evidence that it is a single unique compound rather than a readjustment of concentrations of existing substances, but it is known that the protein synthesis-stimulating activity is heat-stable, dialyzable, and acid-labile.[94] Mitochondria from thyroxine-insensitive tissue, such as mature brain, are incapable of reacting with thyroid hormones to produce this stimulating activity,[2,94,95] a deficiency which may be the basis of the lack of responsiveness of these tissues to the hormones.

When administered *in vivo*, thyroid hormones also stimulate nuclear RNA polymerase activity and the synthesis of ribosomal and, perhaps, also messenger RNA.[88,96] These effects result ultimately in increased cellular contents of functional ribosomes and therefore also result in a second basis for increased rates of protein synthesis.[28,84,88,96] The effects on nucleic acid synthesis cannot, however, be reproduced by thyroid hormones *in vitro*, are completely prevented *in vivo* by mild fasting of the animal, and appear several hours after the initial mitochondria-dependent cytoplasmic stimulation of protein synthesis.[88] They probably represent a secondary, positive feedback, cellular adaptation to the earlier cytoplasmic stimulation of protein synthesis. An initial mitochondria-dependent cytoplasmic stimulation of the existing cellular protein-synthesizing machinery is followed by a nuclear-mediated cellular adaptive increase in the amount of protein-synthesizing machinery.[88,94] Although secondary, the nuclear-mediated response may represent the mechanism for the selective transcription of genetic information which is so obviously displayed in thyroid hormone–induced amphibian metamorphosis.[25,26] The actions of thyroid hormones on protein and nucleic acid synthesis offer thus far the most comprehensive explanation of their multiple physiological and biochemical effects.

8. Effects on Specific Enzyme Activities

Thyroid hormones exert effects on a number of specific enzymes and the chemical reactions which they catalyze. Wolff and Wolff[97] have listed almost 70 enzymatic activities which are influenced by thyroid hormones *in vivo* and more than 30 which are affected by the hormones *in vitro*. Most of those altered only *in vivo* probably reflect secondary changes and may bear no special relationship to the specific molecular mechanism of hormonal action. Some of the actions *in vitro,* however, are on isolated, purified enzymes. Unfortunately, the classes of enzymes represented are so diverse that no common denominator on which to base a single mechanism of action has been recognized. The nearest approach to this goal has been achieved by Wolff and Wolff,[98] who found that thyroxine inhibits a large number of zinc-containing, NAD^+-linked dehydrogenases. Among these are the dehydrogenases which catalyze the oxidation of glutamate, malate, triosephosphate, lactate, and alcohol. The effect on glutamate dehydrogenase has been found to be accompanied by a dissociation of the enzyme protein into smaller subunits.[99,100] Thyroxine forms complexes with a number of metal ions, but its action on the dehydrogenases does not appear to be related to complex formation with the Zn^{2+} contained in them.[97,98] The effect on alcohol dehydrogenase has been attributed to the noncompetitive inhibition by thyroid hormones of the binding of the nicotinamide portion of the NAD^+ cofactor to the enzyme.[101] The action of thyroid hormones on this class of dehydrogenases may eventually prove to be a model for the specific chemical action of the thyroid hormones, but it is at present inadequate to explain the multiplicity of their physiological and biochemical effects.

The search for a single mechanism of action is confounded by the effects on other classes of enzymes. Creatine phosphokinase is strongly inhibited by thyroid hormones.[102,103] The mechanism is unknown, but this effect may have important implications for the accumulation and utilization of energy stores in muscle. Thyroxine inhibits a number of pyridoxal phosphate–dependent enzymes, but the effect appears to be due to pyridoxal phosphate dificiency.[104] Since pyridoxal phosphate is the cofactor in the various transamination and amino acid decarboxylation reactions, this action could be of importance in the nervous system by virtue of its effects on γ-aminobutyric acid metabolism and the synthesis of the biogenic amines. Thyroxine forms a highly insoluble complex with Mg^{2+}[105] and inhibits a number of Mg^{2+}-dependent enzymatic activities through this action.[97] Indeed, thyroxine administration causes a rapid fall in serum Mg^{2+} levels, and magnesium deficiency causes catabolic effects similar to those observed in hyperthyroidism.[106] Mg^{2+} deficiency also causes uncoupling of oxidative phosphorylation, and magnesium reverses the uncoupling action of thyroxine both *in vivo* and *in vitro*.[107,108] Some of the debilitating effects of thyrotoxicosis may, in fact, be secondary to absolute or relative magnesium deficiency and can be reversed by magnesium administration.[23,106,107]

One of the most dramatic effects of thyroid hormones is the induction of mitochondrial L-α-glycerophosphate dehydrogenase activity.[47] Greater than tenfold increases in enzymatic activity can be achieved in liver and kidney following a single dose of hormone *in vivo*. There are no direct effects on the enzyme *in vitro*. Only the flavin-linked mitochondrial enzyme is induced; the soluble NAD^+-linked α-glycerophosphate dehydrogenase is unaffected. The two enzymes and α-glycerophosphate and its oxidation product, dihydroxyacetone phosphate, constitute a shuttle mechanism by which the hydrogen of extramitochondrial NADH is transported into the mitochondria and made available to the respiratory electron transport chain. It is a major pathway for the reoxidation of NADH produced extramitochondrially during aerobic glycolysis. The mechanism of the enhancement of enzymatic activity is still unknown, but recent studies indicate that the effect may not be an induction of enzyme synthesis but rather a preservation of the enzyme from degradation. Thyroid hormones have been found to stimulate flavokinase, the enzyme which phosphorylates riboflavin to FMN, and thus to accelerate the synthesis of both FMN and FAD.[48,49] FAD is the coenzyme for α-glycerophosphate dehydrogenase, and it is not uncommon for saturation of an enzyme with its coenzyme to stabilize it and protect it from degradation. The significance of this effect to the overall action of thyroid hormones remains, however, to be determined.

9. Mechanism of Action

The relative simplicity of the chemical structure of the thyroid hormones and the conspicuousness of their biological effects have encouraged intense efforts to elucidate their biochemical effects, all capable of explaining some but none consistent with all of the biological actions of the hormones. In no case has a biochemical effect been precisely and unequivocally defined at the molecular level. The multiplicity of effects has confounded rather than facilitated attempts to establish the primary mode of action. Numerous ostensibly unrelated biochemical processes and pathways are accelerated or slowed; a variety of enzymatic activities are activiated, induced, or repressed; and even *in vitro* a number of purified enzymes, which catalyze distinctly different classes of reactions and operate by obviously different mechanisms, are either stimulated or inhibited by thyroid hormones. Many of these effects, particularly those observed following *in vivo* administration, reflect secondary or even more remote consequences of earlier more specific biochemical actions or intracellular conditions arising from earlier effects.

Regardless of the mechanism, however, a number of these effects operate in the nervous system and play an important role in the specific actions of the thyroid hormones in the brain.

II. GENERAL ROLE OF THYROID HORMONES IN THE STRUCTURAL AND FUNCTIONAL MATURATION OF THE NERVOUS SYSTEM

A. Morphology

Just as thyroid hormones influence the gross morphology of somatic tissues during development so also do they play a major role in changing the morphology of the brain. When amphibian metamorphosis is prevented by prior removal of the thyroid gland, the larval appearance of the brain of the persisting tadpole is in sharp contrast to that of the frog.[109] The cerebral hemispheres are smaller, and the ventricles are larger and have thin walls. The optic lobes fail to broaden and attain the pyramidal configuration seen in the frog, and the diencephalon remains narrow and elongated. In mammals also, thyroid hormones clearly affect both the ultimate size and shape of the brain. In cretinous man, brain weight is lower than normal by as much as 40%.[110,111] Cerebral gyri are narrow, and the thickness of the cortex is less than in the normal brain. The underlying white matter is also diminished in bulk. In other mammalian species, the lesser growth of the brain of thyroid-deficient animals has been consistently seen.[112,113]

The histological substrate of impaired cerebral growth in hypothyroidism has been most completely studied in the rat. A number of considerations make this species particularly suitable for observing thyroid-dependent changes over a large fraction of the maturational period. Its nervous system is functionally very immature at birth; thyroidectomy can be readily accomplished immediately after birth, either chemically or surgically, and daily replacement therapy can be readily instituted. Litters are large, permitting control animals to be littermates.

Normally in the first 2 postnatal weeks several changes take place in the rat cerebral cortex, the most prominent of which are (1) an enlargement of perikarya, (2) a reduction in cell density as a result of growth of axons and dendrites,[112] and (3) an ingrowth of axonal processes from the thalamus and elsewhere which contributes to the wider spacing of cell bodies.[114] Thyroidectomy at birth curtails the rate of these changes, resulting in a cortex in which cells are hypoplastic and are more closely aggregated. A lesser dilution of cell bodies by intervening neuropil is seen which is most notable in layer 4 because of a lesser growth of axons originating in the thalamus.[115] Basal dendrites are restricted in their growth and branching, with a resulting impoverishment of axodendritic connectivity. Synaptogenesis is severely reduced.[116]

It is perhaps paradoxical that hyperthyroidism imposed during development should also result in an impoverishment in synaptogenesis. Nicholson and Altman explain the reduction in synapse formation in the divergent states as resulting from two different mechanisms. In thyroid deficiency it is the result of a generalized retardation in differentiation of neurons, while in thyroid excess there is an inhibition of mitotic activity in the germinal matrix reducing the number of stem cells from which neurons are formed. Excess thyroid hormone appears to induce premature differentiation, resulting in a transient excess in relative number of synapses, but an ultimate deficiency. However, one cannot conclude from these observations that differentiation at the expense of cell division is a generalized action of thyroid hormones. In amphibians, for instance, thyroid hormones have been shown to have a marked stimulatory effect on mitosis in certain regions of the larval brain.[117,118] One may infer, therefore, that the particular biological action of thyroid hormones is determined by the cell itself, depending on its locus, its origin, and its stage of development. Seemingly it speeds the process already in the ascendancy.

Further illustrating a cell-determined action of thyroid hormones is their role in myelin formation. This activity, which is confined to the postnatal period in the rat, is delayed in animals thyroidectomized at birth.[119] In tissue culture of the newborn rat cerebellum, the addition of thyroxine to the medium advances the time of appearance of myelin by several days.[120] The likelihood that there is a similar action on myelin formation in the peripheral nervous system is suggested by the abnormally slow motor nerve conduction velocity found in hypothyroid infants[121] and the high values found in a newborn infant of a thyrotoxic mother.[122]

Also sensitive to the presence or absence of thyroid hormones during cerebral maturation is the microvasculature. In cretinous animals, capillary density in the cortex is much lower than in euthyroid controls.[123] This reduction in normal capillary density has the effects both of reducing the area of capillary surface available for the exchange of substrates and products of metabolism and of increasing the mean capillary distance for their diffusion or transport from each cell. Such undervascularization could be seen as a possible mechanism whereby cerebral growth and function might be impaired by local undernutrition. However, in view of the known biological action of thyroid hormones to increase the metabolic rate of developing brain *in vitro*,[124,125] where the vasculature plays no role, it seems more likely that the lower capillary density of the cretinous brain is secondary to the lower metabolic rate. There is a considerable body of evidence that local metabolic rate, by its effect on local oxygen tension, is a primary determinant of vascular growth, at least during the developmental period.[126–130]

B. Electrical Activity

The role of thyroid hormones is also seen in the changing electrical activity of the cortex during development. Electroencephalographic tracings have been made in the awake, alert rat in the euthyroid and hypothyroid states at various times of postnatal life.[131] In the adult animal which has been thyroidectomized at birth, there is a much lower amplitude than in normal controls. Auditory arousal and photic stimulation fail to influence the tracing as they do in the normal animal. If thyroidectomy is deferred until the day 24 of postnatal life, by which time much of adult behavior has been attained, the EEG abnormalities described above are not found. A more detailed analysis of the effect of thyroid hormones on the development of electrocortical activity has been made in a study of the evoked response elicited by stimulation of the medial thalamus.[132] Thyroidectomy at any age results in an increase in both the latency and the duration of the response wave. The amplitude, however, which is related to age, is reduced only if thyroidectomy has been carried out at birth. Hormone replacement therapy in neonatally thyroidectomized animals restores the latency and duration of the response to normal, but the depressed amplitude remains unchanged. It is concluded that there is a bimodal effect of the hormones on electrocortical activity, one of which is maturational and probably related to the underdeveloped neuropil and one which is metabolic and readily correctable by replacement therapy.

C. Behavior and Learning in Experimental Animals

The behavior of the cretinous animal stands in sharp contrast to that of the animal with normal thyroid function. In the first several days of life in the newborn rat, there emerges a series of innately organized responses such as a startle response to sound, a placing response (the forepaws are placed on a horizontal bar when the chin is touched to it), and the ability to land on four feet when released in the air in the inverted position. These responses are all delayed in their appearance in the neonatally thyroidectomized animal, but they can be advanced in their day of first appearance by giving excess thyroid hormone.[133, 134] Similar observations have been made in the mouse.[125] The fact that these reflexes are influenced only in the time of their appearance and are not modified in quality indicates that the action of thyroid is on the rate of emergence of a predetermined pattern of function.

This is in contrast to the situation involving the role of thyroid hormones on the development of a capacity for learning, a function which is not a stereotyped response to a uniform stimulus but rather one requiring the storage and recall of information derived from sequential stimuli. Techniques for assessing the learning capacity in the rat have included a T-maze with a variety of external cues, conditioned avoidance tests, and closed-field tests.[134] The scoring systems used in these tests weigh differentially such factors as alertness, speed of performance, and ability to retain a series of cues. Hypothyroidism induced at any age was shown to decrease both alertness and speed of performance: therefore, a test situation and scoring system had to be devised in which the effects of these aspects of behavior were excluded or minimized. Detailed studies proved that the Hebb–Williams closed-field test, which had proved useful in detecting small cortical lesions, was most sensitive in assessing learning capacity in the rat.[135] In this test, rats thyroidectomized at birth performed far more poorly than did littermate controls. Their performance was found to be improved if thyroidectomy was delayed, the longer the delay the better the performance. If the delay was as long as 24 days, no impairment of learning capacity could be measured. A similar graded improvement in scores was found in animals thyroidectomized at birth and then given hormone replacement by regular injection. The younger the age at which thyroid hormone replacement therapy was begun, the less the deficit in learning capacity. If treatment was begun as early as 10 days of age, there was virtually complete protection against learning impairment.

The ability of thyroid hormones, if given in excess, to hasten the appearance of innately organized responses of early life suggested the possibility that an excess given over a longer period of development might actually enhance learning capacity. The hypothesis was tested in experiments involving separate evaluations of three thyroid hormone analogues: triiodothyronine, L-thyroxine, and triiodothyropropionic acid.[136] Doses were adjusted so as not to impair body growth. Contrary to the speculation, all three agents resulted in an ultimate impairment in learning ability. This observation may be the functional counterpart of the finding that synaptogenesis is curtailed in the animal made hyperthyroid during development.[116]

D. Maturation of Brain Function in Man

Thyroid hormones act on the functional development of the nervous system in man in a manner closely resembling their action in experimental animals. However, additional variables are present. Whereas experimental work has largely been carried out under conditions which are all or none with respect to the level of thyroid function, clinical observations, aided by laboratory measurements, have been made in variable degrees of deficit or excess. This variability cannot be monitored during gestation, when there is a changing relative contribution of thyroid hormones from maternal and fetal circulation. Additionally, man's nervous system is more complex and matures over a much longer period than that of the experimental animal, thereby broadening the spectrum of potential functional disability. These variables, however, have not obscured what has been well established in other species, namely, that thyroid hormones are essential to normal maturation of the nervous system and that their deficiency early in life for more than a brief period results in irreversible diminution of ultimate functional capacity.

1. Sporadic Cretinism

Most of the detailed studies on the developing nervous system in man during thyroid hormone deficit have been carried out in sporadic cretinism. This is the commonest form of cretinism in the United States[137] and is usually the result of failure of development of the thyroid gland during embryogenesis. While the cause of such failure is unknown, there is reason to believe that in some instances at least it is genetically determined.[138] In a few cases, it may be due to an isolated deficiency of thyrotropin.[139] Some sporadic cretins have no thyroid tissue, while others have ectopic rests which are functional and partially attenuate the severity of signs.[140,141] Among sporadic cretins are also a number who are goitrous. In most of these, it has been possible to establish evidence for a defect in thyroid hormone biosynthesis at one of the following steps: (a) the transport of iodide into the thyroid gland, (b) the oxidation of iodide and the iodination of tyrosine residues, (c) the coupling of iodotyrosines to form iodo-thyronines, (d) the deiodination of iodotyrosine resulting in a loss of precursors in the urine, and (e) an impairment in the synthesis of thyroglobulin. In rare cases, there may be a widespread defect in somatic cells which prevents their responding to the circulating hormone. These disorders are discussed in detail elsewhere.[142]

At birth, the sporadic cretin may appear normal or may have slight changes in appearance or behavior which become more prominent in subsequent weeks or months.[143] In order of frequency of occurrence these changes include um-bilical hernia, lethargy, retarded growth, respiratory difficulty, and dry skin. The tongue may be large and protrude. In only a few is a goiter present. Less obvious earlier signs, but hardly peculiar to the cretin, are slowness in feeding, infre-quency and an altered character of the cry, and constipation.[144] The develop-mental milestones of infancy are delayed. During early childhood, there is a striking inertia and indifference to the environment which later merges into stubbornness and antagonistic behavior.[145] Speech development is delayed, seemingly for reasons of difficulty both with articulation and with the mental processes essential to language. A certain number have overt abnormalities in motor function, usually in the form of incoordination, mild ataxia, or tre-mor.[146] Although hypotonia and hyporeflexia may be prominent in the neonate with hypothyroidism, in the untreated cretin in later childhood an increase in muscle tone may be present which is accompanied by hyperactive deep tendon reflexes.[146] These are altered in quality, there being prolongation of the time for both contraction and relaxation of the muscle following the stretch stimulus.[147,148] Occasionally the cretinous child is seizure-prone; grand mal, akinetic, and psychomotor attacks have been described.[146] The EEG may have paroxysmal discharges, but in most cretins the record is characterized by slow waves, often of low amplitude; sometimes stereotyped wave forms are seen.[149,152] Although myxedema is present in varying degrees, several of its manifestations reported in adults have not been noted in children. The paranoid suspiciousness, auditory hallucinations, and depression of myxedema madness[153,154] have no counterpart in childhood. Likewise, coma with hypothermia seems not to occur below the age of 30.[155-157] A number of other neurological signs commonly found in long-standing myxedema such as polyneuritis, optic neuritis, anosmia, dysgeusia, and paresthesias also are rarely seen in the young. [158,159] Cerebellar ataxia, while sometimes seen in childhood, is usually mild, whereas in adult life it may be severe and be the primary neurological manifestation. It may be that the absence of these many signs in childhood only reflects the fact that a long period of thyroid deficit is a precondition for their appearance. Deafness in sporadic cretinism is uncommon, but its familial occurrence in association with goiter and mild congenital hypothyroidism has long been known.[160,161] Study of patients with this disorder indicates that they are among those who fail to organify iodine, but they are distinct from others of this type in that the defective organification is only partial. The cause for the deafness is obscure. It has been thought to be the result of some toxic metabolic product arising from the failure in iodine organification[161] or possibly a maturational defect not directly related to thyroid hormone deficiency but rather one linked to an unrelated genetic trait.[162] Congenital deafness has also been reported in a few patients with evidence of impaired peripheral response to thyroid hormone.[163] In summary, the sporadic cretin is characterized primarily by slow and ultimately defective mental developmental accompanied by occasional minor neurological signs and, rarely, deafness.

Early efforts to treat the sporadic cretin with thyroid extract indicated the potential for reversal of their symptoms and signs and for improvement in their ultimate mental capacity. However, the inconsistency of outcome in many early reports[164-169] led to the thesis that many cretins had an unrelated hereditary form of mental deficiency. As large numbers came under observation for prolonged periods and criteria for adequacy of treatment became established, this concept lost support and the spectrum of ultimate outcome is now seen solely in the light of thyroid deficiency. This is determined by a number of variables: (a) the age, including fetal age, of onset of the deficiency, (b) the adequacy of dosage and the constancy with which euthyroidism is maintained, and (c) the duration of the deficiency. The earlier the start of replacement therapy, the greater the likelihood of ultimately attaining a normal mental capacity.[146,170] The crucial nature of even a few months' delay in affecting the outcome has been clearly shown.[144] Instances of failure can be explained by there having been a thyroid deficiency *in utero* which resulted in irreversible injury to maturation potential before treatment was started. As had been found in experimental studies, there is an age when thyroid deficiency is no longer capable of having a permanently damaging effect on brain development. This appears to be about 2 years.[146]

The behavioral effect of replacement therapy in the sporadic cretin is two-fold in nature: (a) one which is prompt and characterized by an increase in alertness and responsiveness[171] and (b) one which is long term and paralleled by structural changes accompanying the thyroid-induced acceleration of brain growth and maturation. This may persist over several years and constitutes a recovery period from the previous abnormal growth patterns.[145] Because of this lag, a satisfactory evaluation in terms of maximal mental performance must be deferred for a considerable period after attainment of the euthyroid state. In some cretins who have gone untreated and therefore have suffered irretrievable loss of brain growth potential, the giving of thyroid may lead to a persisting pattern of deviant behavior marked by restlessness, hyperactivity, and aggressiveness.[145]

2. Endemic Cretinism

In certain mountainous regions of the world, iodine is deficient in the diet. The scarcity of this element results in underproduction of thyroid hormones and marked hypertrophy of the gland (goiter). Moderate degrees of myxedema in the female population appear not to prevent pregnancy, but the offspring are severely cretinous at birth. They differ from most sporadic cretins in having a slower course of motor and mental development and additionally have more widespread neurological changes.[172–176] Defective development of upper motor neurons in the pyramidal system has been said to result in severe weakness and hypertonia of both arms and legs.[172,174] Walking is achieved late in childhood and is characterized by symmetrical partial flexion of the hips and knees, and a tendency to internal rotation of the hips (spastic diplegia). In some patients, rigidity and dystonia in the limbs, neck, and trunk are superimposed. Deafness is severe so that speech does not develop, and severe internal strabismus impairs normal vision. Growth failure and sometimes myxedema accompany the neurological signs.

While this disorder was once considered a distinct entity ("nervous cretinism"),[172] many variations in the symptom–sign complex were found when an entire population was studied and the state of thyroid function assayed by measurements of iodine excretion, serum levels of protein-bound iodine, and I^{131} uptake.[174] The degree of growth failure, the presence or absence of myxedema, and the extent of neurological involvement were found to be particularly variable. Occasionally mental retardation was present with neither deafness nor any motor abnormality. Deaf-mutism was seen in some individuals who were not mentally retarded as evidenced by the ability to learn and communicate by sign language. Nevertheless, spastic diplegia, deaf-mutism, and mental deficiency were far more commonly seen as a triad than as isolated signs. The authors concluded that the neurological syndrome is the result of thyroid hormone deficiency throughout gestation. The variability of ultimate deficit appeared to be explained by differences in the degree and timing of iodine deficiency in both pregnant mother and postnatal infant.

That the triad of disability had its origin in the early part of gestation is suggested by the study of Pharoah in the Western Highlands of New Guinea.[177] Here endemic cretinism is prevalent. A controlled evaluation was made on the prevention of cretinism by the use of intramuscular injections of iodized oil in a population of 8000. It was found that after 4 years there had been a fivefold reduction in the incidence of the cretinous syndrome in the offspring of mothers who had received the oil compared to that in those whose mothers were untreated. In all but one of the mothers who had received the oil yet had a cretinous baby, the injection had been given after the onset of the pregnancy. While it is remotely possible that elemental iodine is essential to embryogenesis in a manner not involving thyroid hormones[178] the tentative inference is that the severe neurological picture is the result of thyroid deficiency. The lesser neurological impairment of the sporadic cretin is because it has the continuous benefit of maternal supplies and because its deficiency occurs only late in pregnancy when these supplies are not sufficiently augmented by hormones derived from the fetal thyroid gland. There is some suggestion that the motor disorder and the hearing loss of the endemic cretin may improve if thyroid hormone is given in childhood.[172] Because of the well-known improvement which takes place spontaneously in many stationary neurological syndromes during development, further study will be required to resolve this question.

3. Hyperthyroidism

Hyperthyroidism is a disorder of unknown cause, but with familial tendencies, which results in excessive secretion of thyroid hormones during development.[179,180] The signs and symptoms are similar to those of thyrotoxicosis in adults, but the onset may be insidious in the young. Restlessness and emotional instability with hyperkinesis may direct attention to psychological factors and the occurrence of frank choreiform involvements may be mistakenly considered to be due to Sydenham's chorea.[181] Seizures and upper motor neuron syndromes, occasionally seen in adults with thyrotoxicosis,[182] are rare in children, but tremor, involving cranial musculature as well as that of the extremities, is common to all ages. Myopathic states are also seen. The neurological manifestations are usually accompanied by a growth spurt with advancement of bone age, and in late childhood by precocious puberty. While spontaneous hyperthyroidism occasionally occurs in early infancy,[183,184] the condition, when found in the neonatal period, is secondary to overactivity of the maternal gland during pregnancy.[185,186] In neonatal thyrotoxicosis, a transient advancement in neuromuscular development has been noted,[187,188] but no instance of impaired ultimate mental capacity has been reported as might be expected from observations of experimental animals.[136] The absence of such mental deficiency in man may simply reflect the fact that the thyrotoxic state, when recognized in infancy, is successfully controlled early in its course. This contrasts with the conditions in animal experiments in which the abnormal metabolic state was sustained throughout the entire period of maturation.

E. Conclusions Regarding the Role of Thyroid Hormones in the Functional Development of the Nervous System

It is now appreciated that normal structural and functional maturation of the nervous system is exquisitely sensitive to the presence of thyroid hormones. Deficiency early in life not only results in abnormal function at the time but leads to aberrant and irreversible structural changes which are accompanied by impaired capacity to learn, respond, and perform. In the rat, a period of total deficiency after birth may last as long as 10 days without causing ultimate irreversible deficit. In the case of man, it has not been possible to determine the length of such a benign period. The problem is complex in human life because of a variability in the degree of thyroid deficiency in the naturally occurring disorders and the fact that the end point of mental capacity, measured many years after the hormone deficit, is affected by many other factors. Some of these factors, such as parental mental capacity and educational opportunity, can be reasonably well assessed, but the contribution of nutritional status, intercurrent disease, head trauma, and emotional factors is more difficult to establish. Adding further complexity to the problem in human life is the difficulty in determining whether or not thyroid function is adequate in the neonatal period. Signs by which this has been judged in the past, such as bone age, must necessarily reflect a preexisting deficit of uncertain duration. This also applies to many clinical stigmata.[144,146] Those aspects of behavior which promptly follow the onset of deficiency such as feeding difficulty, constipation, and altered cry[144] are highly nonspecific.

It is therefore essential to define the limits of normal thyroid function by physiological criteria and to be able to apply these repeatedly in any given infant. The permissible limits of normality of thyroid function tests may differ from those established for late infancy of childhood. The high indices of thyroid activity found in the neonate[189,190] may well reflect a mandatory augmentation of gland function which is essential to meet the demands of the extremely rapid brain and body growth at the time. The finding of relatively depressed values for butanol-extractable iodine in the sera of babies of short gestational age[191] who have no clinical signs of cretinism may have particular significance in the light of the high incidence of suboptimal brain development known to occur in such infants.

III. BIOCHEMICAL AND METABOLIC EFFECTS OF THYROID HORMONES IN NERVOUS TISSUES

In most mammalian tissues, the manifestations of the actions of thyroid hormones are confined more or less to quantitative changes in biochemical and metabolic processes which proceed whether or not the hormones are present. Only the rates and the balance among these metabolic activities are altered, and these tissues generally exhibit no gross qualitative changes in structure or function specifically identified only with the actions of the thyroid hormones.

The situation is different in brain, where the hormones produce a variety of changes ranging from subtle, barely detectable, quantitative alterations in the rates of chemical reactions or functional activities to major reorganization of structure at both the microscopic and macroscopic levels and, not surprisingly, dramatic modification of behavioral functions. The thyroid hormones promote the major structural, biochemical, and functional reorganization which characterizes the postnatal maturation of the brain but seemingly have little or no effect once the brain is mature. Their action in the nervous system, therefore, resembles their role in amphibian metamorphosis; the hormones are essential to initiate the sequence of changes which lead to the transformation of a tadpole into a frog but once the adult stage is reached are without further apparent effect.

It is unlikely that the different pattern of their effects in brain, as compared to other tissues, reflects a different biochemical mechanism of the hormones' action. It is probably related to the timing of the action. In most mammalian tissues, thyroid hormones are active all through life. In the brain, responsiveness to the hormones is confined only to the period of development and maturation, a time when the plasticity of the tissue is great and profound modifications of structure and function are occurring. It is to be expected that a fundamental biochemical effect would have more diverse consequences in a rapidly changing tissue than in one which is more or less stable and merely maintaining or renewing itself.

Many of the biochemical and metabolic effects of thyroid hormones observed in other tissues and described above are operative in and relevant to brain, at least during the duration of its thyroxine-sensitivity. The brain, however, enjoys certain unique biochemical properties, and these serve to modify some of the general effects of the hormones or to lead to specific effects not seen in other tissues. These will now be considered here.

A. Energy Exchange

The characteristic action of thyroid hormones on metabolic rate also occurs in brain, but it is clearly age-dependent. Fazekas et al.[124] studied the effects of thyroxine administration on the metabolic rate of the cerebral cortex of the rat from birth to early adulthood. Cerebral cortical oxygen consumption in the rat is normally low at birth and remains so for approximately the first 10 days of life. It then rises along a sigmoid curve until it reaches the adult level at approximately 45 days of age (Fig. 3). Thyroxine administration to newborn rats shifts this curve to the left; it causes an earlier and steeper rise of the curve and an

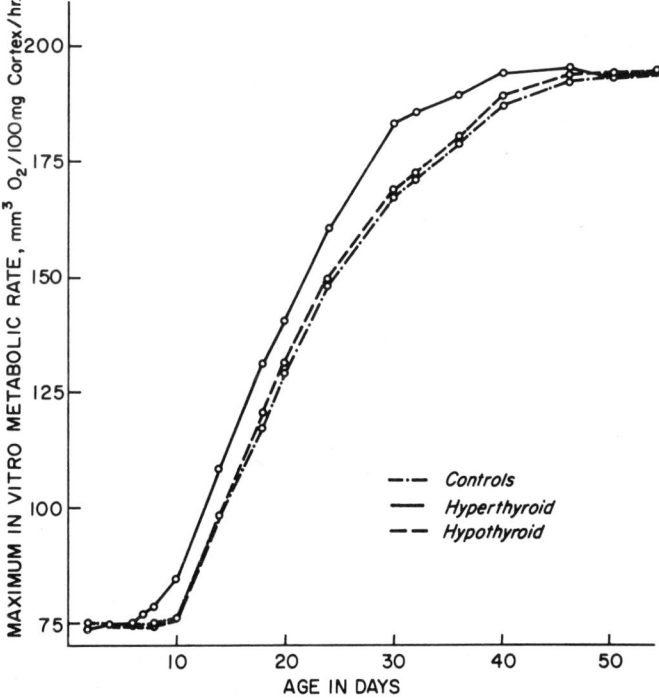

Fig. 3. The normal postnatal development of cerebral cortical oxygen consumption in the rat and the effects of altered thyroid state on this development. [From Fazekas et al.[124]]

advancement of the time at which the normal adult metabolic rate is achieved. Once the adult level is reached, however, further administration of thyroxine is without effect.

An excess of thyroid hormone causes, therefore, an earlier maturation of the energy exchange of the brain and stimulates cerebral oxygen consumption only during the brain's growth and development. Thyroid hormones appear to have no effects on cerebral metabolic rate in the fully mature brain. Cerebral O_2 consumption is clearly normal in human adult hyperthyroidism,[192–194] but in human adult hypothyroidism one study reported it to be depressed[195] and another found it to be normal.[194] The apparent discrepancy is probably due to differences in the severity and duration of the hypothyroidism. It is likely that simple deficiency of thyroid hormone in the adult does not in itself lower cerebral metabolic rate, but when allowed to progress to myxedema, secondary effects, possibly originating in other tissues, may intervene to cause a decrease of the energy exchange of the brain. Hypothyroidism present at birth or before a critical time in the course of brain development, if untreated, almost certainly leads to low cerebral metabolic rates in adulthood in view of the impaired morphological and functional maturation of the brain[196] as well as the delayed development of certain oxidative enzymes, such as succinic dehydrogenase, in such circumstances.[197,198]

The shape of the developmental curve for oxygen consumption more or less parallels and probably reflects maturational changes in the biochemical machinery for energy exchange as well as the development of functional activities which expend energy. A number of mitochondrial enzymes of oxidative metabolism show a similar time course of development in the brain. Succinic dehydrogenase,[197,198] cytochrome oxidase,[197,199] and D-β-hydroxybutyrate dehydrogenase,[199] for example, show similar developmental curves in the brain, although the last differs somewhat in that instead of leveling off at a constant concentration in mature brain, it gradually declines following weaning back to low concentrations again.[199] Some of these enzymes appear to be under the influence if not the control of the thyroid hormones. Neonatal thyroidectomy leads to retardation and diminution of the normal rise of succinic dehydrogenase activity in brain which can be restored to normal by hormone replacement therapy begun in the rat before the tenth day of age[197]; if delayed beyond that age the enzyme activity never achieves the normal adult level. Surprisingly, the development of cytochrome oxidase activity does not appear to be affected by neonatal thyroidectomy,[197] although in most thyroxine-sensitive tissues other components of the electron transport chain have been found to change with the level of thyroid activity.[44–46] Effects of hyperthyroidism on cytochrome oxidase activity in brain have not been reported, but postnatal hyperthyroidism has been found to advance the maturation of D-β-hydroxybutyric dehydrogenase activity.[200] The effects of thyroid deficiency on the development of the latter enzyme activity remain to be studied.

Several glycolytic enzymes show similar developmental patterns which are dependent on thyroid hormones. Hexokinase, phosphofructokinase, and pyruvic kinase increase in activity in brain during postnatal development.[201] The increases in all three are diminished by neonatal thyroidectomy and restored by replacement hormone therapy if begun sufficiently early in life.[201] Hyperthyroidism induced and maintained from birth by hormone administration does not, however, accelerate or enhance this development.[201]

Metabolic rates are determined not only by the levels of the enzymes which carry out the reactions. The enzymes are usually present in excess in the tissues, and other factors play key roles in the regulation of the reaction velocities. Thus oxygen consumption and the rates of other reactions of energy metabolism are largely controlled by the level of ATP and the ATP/ADP ratio. Through this homeostatic mechanism, the rate of ATP generation is adjusted to ATP utilization. The rise of oxygen consumption in the brain with development reflects not only changes in the levels of the enzymes of energy exchange but also increasing rates of energy utilization associated with the development of cerebral functional activities. Since nervous tissue functions largely through electrical activity, ion transport, particularly of Na^+ and K^+, is important in establishing and maintaining the membrane potentials and neuronal excitability. Na^+- and K^+-dependent ATPase is a membrane-bound enzyme which is believed to be involved in the mechanism of the K^+,Na^+ pump and couples the energy derived from ATP hydrolysis to the active transport of these ions. Na^+,K^+-ATPase activity is low in the newborn brain and increases in activity with development.[70,198] This increase is severely diminished by neonatal thyroidectomy.[70,198] Thyroid hormones may therefore also influence cerebral energy exchange by their effects on an important ATP-utilizing process, specifically Na^+,K^+-ATPase. Indeed, it has recently been proposed that the calorigenic effects of thyroid hormones are mediated primarily by their stimulation of the Na^+,K^+-pump.[68,69] Mg^{2+}-stimulated ATPase exhibits a similar developmental pattern and impaired development in hypothyroidism in brain.[70,198]

B. Lipid Synthesis

The thyroid hormones promote the process of myelination,[120,196] and it is therefore not surprising that they play a role in the synthesis of a number of lipids which are components of myelin. Neonatal thyroidectomy impedes myelination, and although phospholipid deposition appears to be unaffected,[202] the formation of cholesterol, cerebrosides, sulfatides, and saturated fatty acids is severely reduced.[203,204] Recent studies indicate that thyroid hormones stimulate and thyroid deficiency impairs the brain microsomal fatty acid–elongation system involved in the synthesis of saturated fatty acids.[204] The effects of thyroid hormones on myelination are not confined, however, to myelin lipids; hyperthyroidism in early postnatal life has been found to accelerate the deposition of proteolipid protein in brain.[200]

C. Miscellaneous Enzymes

Thyroid status affects a number of specific enzymes.[205] Those associated with oxidative metabolism, glycolysis, and ATPase have been discussed above. In addition, however, glutamic decarboxylase,[113,198] GABA transaminase,[198] and acetylcholinesterase[206] are all decreased by neonatal thyroidectomy. All these enzymes are involved in some aspect of neurotransmitter functions, and these effects are consistent with the deficient nerve-ending proliferation and synaptogenesis observed in postnatal hypothyroidism.[113] Aspartate transaminase[207] is also lowered by neonatal thyroidectomy. Not all enzymes, however, are altered by thyroid function. Glutamic dehydrogenase,[113] alanine transaminase,[113] aldolase,[197] and lactic dehydrogenase[113] do not appear to be altered by thyroid deficiency in early life. Clearly, the actions of thyroid hormones on the levels of enzymes are broad but not entirely indiscriminate. The basis for the selectivity is obscure.

D. Nucleic Acid and Protein Synthesis

Normal maturation is associated with changes in DNA, RNA, protein, and specific enzyme concentrations and quantities in the brain. There may be some increase in total DNA content of the brain, probably mainly because of glial proliferation but also because of neuronal multiplication in the cerebellum, which matures somewhat later than the cerebrum. Brain DNA concentration, however, falls as a result of the tremendous growth of perikaryonal cytoplasm, proliferation of axonal and dendritic processes, and myelin deposition, all of which lead to dilution of the DNA confined to the nuclei.[196] RNA and protein contents and concentrations rise, however, because of their rapid synthesis during the developmental period of the brain. All of these changes are delayed and reduced in magnitude by thyroid deficiency at or shortly after birth.[113, 207,208]

Many of these effects are consequences of the role played by thyroid hormones in biosynthetic processes. It has become increasingly clear that an action on protein synthesis may underlie many of the effects of thyroid hormones in brain as well as in other tissues.[209] The rate of protein synthesis is normally three or four times greater in immature than in mature brain,[92,95] a difference undoubtedly reflecting the perikaryonal growth, axonal and dendritic proliferation, and myelination which are occurring during the process of maturation. Like their effects on cerebral O_2 consumption, thyroid hormones stimulate protein synthesis in the immature but not the mature brain.[92,95] These effects on protein synthesis are observed *in vivo*[27,210,211] and also *in vitro*.[92,95]

The *in vitro* effect has made possible studies of the mechanism of the effect. It will be recalled from previous discussion that mitochondria are involved in the mechanism of the stimulation of protein synthesis by thyroid hormones in liver and other thyroxine-sensitive tissues. A preliminary interaction between mitochondria and the hormone yields a product or consequence which is responsible for the enhancement of ribosomal protein biosynthetic activity.[94] The same mechanism applies to immature brain. By mixing various subcellular organelles from mature brain and immature brain or liver *in vitro*, it has been possible to demonstrate that the loss of thyroxine-sensitivity in mature brain is the result of some change in its mitochondria.[2,95] Immature brain mitochondria share with the mitochondria from other thyroxine-sensitive tissues the capacity to participate with thyroid hormones in the necessary prerequisite interaction for the stimulation of protein synthesis. The mitochondria from mature brain have apparently lost this ability, and this may be the basis for the restriction of thyroid hormone activity in brain to early postnatal life.

Thyroid hormones also stimulate RNA synthesis.[96] This effect occurs later than the initial translational effect on protein synthesis,[88] but it serves to amplify and prolong the effects of thyroid hormones on protein synthesis.[88] The mechanism of the effect on RNA synthesis is unknown, but it is clearly secondary either to the earlier translational effect on protein synthesis or to some other intracellular change induced by the hormones' actions. Even though secondary, however, this effect is part of the constellation of the overall pattern of thyroid hormone effects and may provide the basis for the selectivity of the hormones in controlling the activities of individual enzyme systems and processes.

The actions of thyroid hormones on protein and RNA synthesis go far to explain many of their major effects in the central nervous system. As has been found in other tissues[84] and in the body as a whole,[89,90] the effects of thyroid hormones on metabolic rate are secondary to those on protein synthesis. The influence of thyroid hormones on the morphological and functional development of the brain and on the time course of the appearance of various enzymatic activities is fully consistent with the effects on protein synthesis. Finally, the change in thyroxine sensitivity with maturation of the brain and its implications in regard to the critical periods of development can at present be explained only in relation to the role of mitochondria in the mechanism of the thyroxine stimulation of protein synthesis.[2,94]

REFERENCES

1. E. S. Gordon and A. E. Heming, The effect of thyroid treatment on the respiration of various rat tissues, *Endocrinology* **34**:353–360 (1944).
2. L. Sokoloff, The mechanism of action of thyroid hormones on protein synthesis and its relationship to the differences in sensitivities of mature and immature brain, in "Protein Metabolism of the Nervous System" (A. Lajtha, ed.) pp. 367–382, Plenum Press, New York (1970).
3. J. Wynn, Organic iodine constituents in human serum, *Arch. Biochem. Biophys.* **87**:120–124 (1960).
4. H. A. Selenkow and S. P. Asper, Jr., Biological activity of compounds structurally related to thyroxine, *Physiol. Rev.* **35**:426–474 (1955).
5. T. C. Bruice, N. Kharasch, and R. J. Winzler, A correlation of thyroxine-like activity and chemical structure, *Arch. Biochem. Biophys.* **62**:306–317 (1956).
6. D. F. Tapley, F. F. Davidoff, W. B. Hatfield, and J. E. Ross, Physiological disposition of D- and L-thyroxine in the rat, *Am. J. Physiol.* **197**:1021–1027 ((1959).
7. C. M. Greenberg, B. Blank, F. R. Pfeiffer, and J. F. Pauls, Relative activities of several 3'- and 3':5'-aryl thyromimetic agents, *Am. J. Physiol.* **205**:821–826 (1963).
8. R. Pitt-Rivers and J. R. Tata, "The Thyroid Hormones," Pergamon Press, London (1960).
9. J. R. Tata and C. J. Shellabarger, An explanation for the difference between the responses of mammals and birds to thyroxine and triiodothyronine, *Biochem. J.* **72**:608–613 (1959).
10. K. Sterling and M. Tabachnick, Determination of the binding constants for the interaction of thyroxine and its analogues with human serum albumin, *J. Biol. Chem.* **236**:2241–2243 (1961).
11. C. Niemann and J. F. Mead, The synthesis of DL-3,5-diiodo-4-(3',5'-diiodo-2'-hydroxy-phenoxy)-phenylalanine, a physiologically active isomer of thyroxine, *J. Am. Chem. Soc.* **63**:2685–2687 (1941).
12. K. Tomita, H. A. Lardy, D. Johnson, and A. Kent, Synthesis and biological activity of O-methyl derivatives of thyroid hormones, *J. Biol. Chem.* **236**:2981–2986 (1961).
13. E. C. Jorgensen, N. Zenker, and C. Greenberg, Thyroxine analogues. III. Antigoitrogenic and calorigenic activity of some alkyl substituted analogues of thyroxine, *J. Biol. Chem.* **235**:1732–1737 (1960).
14. C. R. Harington, Synthesis of a sulfur-containing analogue of thyroxine, *Biochem. J.* **43**:434–437 (1948).
15. D. Marine, The physiology and principal interrelations of the thyroid, in "Glandular Physiology and Therapy," pp. 315–333, American Medical Association, Chicago (1935).
16. S. B. Barker, Mechanism of action of thyroid hormone, *Physiol. Rev.* **31**:205–243 (1951).
17. J. Wolff and R. C. Goldberg, Disorders of iodine metabolism, in "Biochemical Disorders in Human Disease" (R. H. S. Thompson and E. J. King, eds.) pp. 289–351, Academic Press, New York (1957).
18. S. B. Barker, Peripheral actions of thyroid hormones, *Fed. Proc.* **21**:635–641 (1962).
19. F. L. Hoch, Biochemical actions of thyroid hormones, *Physiol. Rev.* **42**:605–673 (1962).
20. J. H. Means, L. J. DeGroot, and J. B. Stanbury, "The Thyroid and Its Diseases," 3rd ed., McGraw-Hill, New York (1963).
21. R. Pitt-Rivers and W. R. Trotter, "The Thyroid Gland" (R. Pitt-Rivers and W. R. Trotter, eds.) Vol. 1, Butterworths, London (1964).
22. L. Sokoloff, Action of thyroid hormones, in "Handbook of Neurochemistry" (A. Lajtha, ed.) Vol. 5, Part B, pp. 525–549, Plenum Press, New York (1971).
23. S. N. Gershoff, J. J. Vitale, I. Antonowicz, M. Nakamura, and E. E. Hellerstein, Studies of interrelationships of thyroxine, magnesium, and vitamin B_{12}, *J. Biol. Chem.* **231**:849–854 (1958).
24. J. F. Gudernatsch, Feeding experiments on tadpoles. II. A further contribution to the knowledge of organs with internal secretion, *Am. J. Anat.* **15**:431–480 (1914).

25. P. P. Cohen, Biochemical aspects of metamorphosis: Transition from ammontelism to ureotelism, *Harvey Lectures* **60**:119–154 (1965).
26. E. Frieden, Thyroid hormones and the biochemistry of amphibian metamorphosis, *Rec. Progr. Hormone Res.* **23**:139–194 (1967).
27. R. Michels, J. Cason, and L. Sokoloff, Thyroxine: Effects on amino acid incorporation into protein *in vivo, Science* **140**:1417–1418 (1963).
28. J. R. Tata, Biological action of thyroid hormones at the cellular and molecular levels, *in* "Actions of Hormones on Molecular Processes" (G. Litwack and D. Kritchevsky, eds.) pp. 58–131, Wiley, New York (1964).
29. H. F. Müller, Beitrage zur Kenntniss der Basedowischen Krankheit, *Deutsch. Arch. Klin. Med.* **51**:335–412 (1893).
30. W. M. Boothby and I. Sandiford, The total and the nitrogenous metabolism in exophthalmic goiter, *J.A.M.A.* **81**:795–800 (1923).
31. C. D. Fitch, R. Coker, and J. S. Dinning, Metabolism of creatine-1-C^{14} by vitamin E-deficient and hyperthyroid rats, *Am. J. Physiol.* **198**:1232–1234 (1960).
32. A. Magnus-Levy, Über den respiratorischen Gewechsel unter dem Einfluss der Thyroidea sowie unter verschiedenen pathologischen Zustanden, *Berl. Klin. Wschr.* **32**:650–652 (1895).
33. I. A. Mirsky and R. H. Broh-Kahn, The effect of experimental hyperthyroidism on carbohydrate metabolism, *Am. J. Physiol.* **117**:6–12 (1936).
34. R. Sternheimer, The effect of a single injection of thyroxine on carbohydrates, protein, and growth in the rat liver, *Endocrinology* **25**:899–908 (1939).
35. S. D. Burton, E. Robbins, and S. O. Byers, Effect of hyperthyroidism on glycogen content of the isolated rat liver, *Am. J. Physiol.* **188**:509–513 (1957).
36. K. R. Hornbrook, P. V. Quinn, J. H. Siegel, and T. M. Brody, Thyroid hormone regulation of cardiac glycogen metabolism, *Biochem. Pharmacol.* **14**:925–936 (1965).
37. C. Rich, E. L. Bierman, and I. L. Schwartz, Plasma nonesterified fatty acids in hyperthyroid states, *J. Clin. Invest.* **38**:275–278 (1959).
38. M. Vaughan, An *in vitro* effect of triiodothyronine on rat adipose tissue, *J. Clin. Invest.* **46**:1482–1491 (1967).
39. K. Fletcher and N. B. Myant, Influence of the thyroid on the synthesis of cholesterol by liver and skin *in vitro, J. Physiol* **144**:361–372 (1958).
40. A. H. Philips and R. H. Langdon, The influence of thyroxine and other hormones on hepatic TPN-cytochrome reductase activity, *Biochim. Biophys. Acta* **19**:380–382 (1956).
41. V. R. Potter, Possible biochemical mechanisms underlying adaptation to cold, *Fed. Proc.* **17**:1060–1063 (1958).
42. J. S. Hart, Metabolic alterations during chronic exposure to cold, *Fed. Proc.* **17**:1045–1054 (1958).
43. R. D. Dallam and R. B. Howard, Thyroxine-enhanced oxidative phosphorylation of rat liver mitochondria, *Biochim. Biophys. Acta* **37**:188–189 (1960).
44. D. L. Drabkin, Cytochrome C metabolism and liver regeneration. Influence of thyroid gland and thyroxine, *J. Biol. Chem.* **182**:335–349 (1950).
45. H. M. Klitgard, Effect of thyroidectomy on cytochrome C concentration of selected rat tissues, *Endocrinology* **78**:642–644 (1966).
46. S. Pedersen, J. R. Tata, and L. Ernster, Ubiquinone (coenzyme Q) and the regulation of basal metabolic rate by thyroid hormones, *Biochim. Biophys. Acta* **69**:407–409 (1963).
47. Y. P. Lee, A. E. Takemori, and H. Lardy, Enhanced oxidation of α-glycerophosphate by mitochondria of thyroid-fed rats, *J. Biol. Chem.* **234**:3051–3054 (1959).
48. R. S. Rivlin and R. G. Langdon, Effects of thyroxine upon biosynthesis of flavin mononucleotide and flavin adenine nucleotide, *Endocrinology* **84**:584–588 (1969).
49. R. S. Rivlin, Regulation of flavoprotein enzymes in hyperthyroidism, *Advan. Enzyme Regulation* **8**:239–250 (1970).
50. R. S. Rivlin, Medical progress: Riboflavin metabolism, *New Engl. J. Med.* **283**:463–472 (1970).

51. W. F. Loomis and F. Lipmann, Reversible inhibition of the coupling between phosphorylation and oxidation, *J. Biol. Chem.* **173**:807–808 (1948).
52. G. F. Maley and H. A. Lardy, Metabolic effects of thyroid hormones *in vitro*. II. Influence of thyroxine and triiodothyronine on oxidative phosphorylation, *J. Biol. Chem.* **204**:435–444 (1953).
53. F. L. Hoch and F. Lipmann, The uncoupling of respiration and phosphorylation by thyroid hormones, *Proc. Natl. Acad. Sci.* **40**:909–921 (1954).
54. G. F. Maley and H. A. Lardy, Efficiency of phosphorylation in selected oxidations by mitochondria from normal and thyrotoxic rat livers, *J. Biol. Chem.* **215**:377–388 (1955).
55. H. G. Klemperer, The uncoupling of oxidative phosphorylation in rat liver mitochondria by thyroxine, triiodothyronine, and related substances, *Biochem. J.* **60**:122–135 (1955).
56. C. Cooper and A. L. Lehninger, Oxidative phosphorylation by an enzyme complex from extracts of mitochondria. I. Span β-hydroxybutyrate to oxygen, *J. Biol. Chem.* **219**:489–505 (1956).
57. F. L. Hoch, Rapid effects of a subcalorigenic dose of L-thyroxine on mitochondria, *J. Biol. Chem.* **241**:524–525 (1966).
58. L. Ernster, D. Ikkos, and R. Luft, Enzymic activities of human skeletal muscle mitochondria: A tool in clinical metabolic research, *Nature* **184**:1851–1854 (1959).
59. L. Sokoloff and S. Kaufman, Thyroxine stimulation of amino acid incorporation into protein, *J. Biol. Chem.* **236**:795–803 (1961).
60. D. F. Tapley, C. Cooper, and A. L. Lehninger, The action of thyroxine on mitochondria and oxidative phosphorylation, *Biochim. Biophys. Acta* **18**:597–598 (1955).
61. D. F. Tapley, The effect of thyroxine and other substances on the swelling of isolated rat liver mitochondria, *J. Biol. Chem.* **222**:325–339 (1956).
62. A. L. Lehninger, Water uptake and extrusion by mitochondria in relation to oxidative phosphorylation, *Physiol. Rev.* **42**:467–517 (1962).
63. D. F. Tapley, Mode and site of action of thyroxine, *Proc. Mayo Clin.* **39**:626–636 (1964).
64. D. F. Tapley and C. Cooper, Effect of thyroxine on the swelling of mitochondria isolated from various tissues of the rat, *Nature* **178**:119 (1956).
65. G. E. Paget and J. M. Thorp, An effect of thyroxine on the fine structure of the rat liver cell, *Nature* **199**:1307–1308 (1963).
66. M. J. Dimino and F. L. Hoch, Localization of endogenous and exogenous thyroid hormone in rat liver mitochondria, *Fed. Proc.* **31**:213 (1972) (abst.).
67. S. Wallach, J. V. Bellavia, P. J. Gamponia, and P. Bristrim, Thyroxine-induced stimulation of hepatic cell transport of calcium and magnesium, *J. Clin. Invest.* **51**:1572–1577 (1972).
68. Ismail-Beigi and I. S. Edelman, Mechanism of thyroid calorigenesis: Role of active sodium transport, *Proc. Natl. Acad. Sci.* **67**:1071–1078 (1970).
69. Ismail-Beigi and I. S. Edelman, The mechanism of the calorigenic action of thyroid hormone, *J. Gen. Physiol.* **57**:710–722 (1971).
70. T. Valcana and P. S. Timiras, Effect of hypothyroidism on ionic metabolism and Na-K activated ATP phosphohydrolase activity in the developing rat brain. *J. Neurochem.* **16**:935–943 (1969).
71. E. Goetsch, Newer methods in the diagnosis of thyroid disorders: Pathological and clinical, *N.Y. J. Med.* **18**:259–267 (1918).
72. O. Thibault, Action renforcatrice de la thyroxine sur l'effect inhibiteur de l'adrénaline sur l'intestin de lapin isolé, *Compt. Rend. Soc. Biol. Paris* **142**:499–504 (1948).
73. H. E. Swanson, Interrelationships between thyroxine and adrenalin in the regulation of oxygen consumption in the albino rat, *Endocrinology* **59**:217–225 (1956).
74. T. S. Danowski, A. C. Heineman, Jr., J. V. Bonessi, and C. Moses, Hydrocortisone and/or desiccated thyroid in physiologic dosage. XIV. Effects of thyroid hormone excesses on pressor activity and epinephrine responses, *Metabolism* **13**:747–752 (1964).
75. A. D'Iorio and J. Leduc, The influence of thyroxine on the *O*-methylation of catechols, *Arch. Biochem. Biophys.* **87**:224–227 (1960).

76. M. H. Zile, Effect of thyroxine and related compounds on monamine oxidase activity, *Endocrinology* **66**:311–312 (1960).

77. T. S. Harrison, Adrenal medullary and thyroid relationships, *Physiol. Rev.* **44**:161–185 (1964).

78. R. P. Zimon, E. V. Flock, G. M. Tyce, S. G. Sheps, and C. A. Owen, Jr., Effect of thyroid hormones on metabolism of DL-norepinephrine by isolated rat liver, *Endocrinology* **80**: 808–814 (1967).

79. R. J. Wurtman, I. J. Kopin, and J. Axelrod, Thyroid function and the cardiac disposition of catecholamines, *Endocrinology* **73**:63–74 (1963).

80. W. R. Brewster, Jr., J. P. Isaacs, P. F. Osgood, and T. L. King, The hemodynamic and matabolic interrelationships in the activity of epinephrine, norepinephrine, and the thyroid hormones, *Circulation* **13**:1–20 (1956).

81. A. Surtshin, J. K. Cordonnier, and S. Lang, Lack of influence of the sympathetic nervous system on the calorigenic response to thyroxine, *Am. J. Physiol.* **188**:503–506 (1957).

82. W. Y. Lee, D. Bronsky, and S. S. Waldstein, Studies of thyroid and sympathetic nervous system interrelationships. II. Effects of guanethidine on manifestations of hyperthyroidism, *J. Clin. Endocrinol. Metab.* **22**:879–885 (1962).

83. L. Sokoloff, S. Kaufman, P. L. Campbell, C. M. Francis, and H. V. Gelboin, Thyroxine stimulation of amino acid incorporation into protein. Localization of stimulated step, *J. Biol. Chem.* **238**:1432–1437 (1963).

84. J. R. Tata, L. Ernster, O. Lindberg, E. Arrhenius, S. Pedersen, and R. Hedman, The action of thyroid hormones at the cell level, *Biochem. J.* **86**:408–428 (1963).

85. L. Sokoloff, P. L. Campbell, C. M. Francis, and C. B. Klee, Thyroxine stimulation of amino acid incorporation into ribosomal protein, *Biochim. Biophys. Acta* **76**:329–332 (1963).

86. L. Sokoloff, C. M. Francis, and P. L. Campbell, Thyroxine stimulation of amino acid incorporation into protein independent of any action on messenger RNA synthesis, *Proc. Natl. Acad. Sci.* **52**:728–736 (1964).

87. R. L. Krause and L. Sokoloff, Effects of thyroxine on initiation and completion of protein chains of hemoglobin *in vitro*, *J. Biol. Chem.* **242**:1431–1438 (1967).

88. L. Sokoloff, P. A. Roberts, M. M. Januska, and J. E. Kline, Mechanisms of stimulation of protein synthesis by thyroid hormones *in vivo*, *Proc. Natl. Acad. Sci.* **60**:652–659 (1968).

89. J. R. Tata, Inhibition of the biological action of thyroid hormones by actinomycin D and puromycin, *Nature* **197**:1167–1168 (1963).

90. W. P. Weiss and L. Sokoloff, Reversal of thyroxine-induced hypermetabolism by puromycin, *Science* **140**:1324–1326 (1963).

91. T. F. Necheles, Peptide synthesis in bone marrow: Insulin and thyroxin effects, *Am. J. Physiol.* **203**:693–696 (1962).

92. S. Gelber, P. L. Campbell, G. E. Deibler, and L. Sokoloff, Effects of L-thyroxine on amino acid incorporation into protein in mature and immature rat brain, *J. Neurochem.* **11**:221–229 (1964).

93. D. M. Brown, Thyroxine stimulation of amino acid incorporation into protein of skeletal muscle *in vitro*, *Endocrinology* **78**:1252–1254 (1966).

94. L. Sokoloff, Role of mitochondria in the stimulation of protein synthesis by thyroid hormones, *in* "Some Regulatory Mechanisms for Protein Synthesis in Mammalian Cells" (A. Pietro, M. R. Lamborg, and F. T. Kenney, eds.) pp. 345–367, Proceedings of the Third Kettering Symposium, 1968, Academic Press, New York (1968).

95. C. B. Klee and L. Sokoloff, Mitochondrial differences in mature and immature brain. Influence on rate of amino acid incorporation into protein and responses to thyroxine, *J. Neurochem.* **11**:709–716 (1964).

96. J. R. Tata and C. C. Widnell, Ribonucleic acid synthesis during the early action of thyroid hormones, *Biochem. J.* **98**:604–620 (1966).

97. E. C. Wolff and J. Wolff, The mechanism of action of the thyroid hormones, *in* "The Thyroid Gland" (R. Pitt-Rivers and W. R. Trotter, eds.) Vol. 1, pp. 237–281, Butterworths, London (1964).

98. J. Wolff and E. C. Wolff, The effect of thyroxine on isolated dehydrogenases, *Biochim. Biophys. Acta* **26**:387–396 (1957).

99. J. Wolff, The effect of thyroxine on isolated dehydrogenases. II. Sedimentation changes in glutamic dehydrogenase, *J. Biol. Chem.* **237**:230–235 (1962).

100. J. Wolff, The effect of thyroxine on isolated dehydrogenases. III. The site of action of thyroxine on glutamic dehydrogenase, the function of adenine and guanine nucleotides, and the relation of kinetic to sedimentation changes, *J. Biol. Chem.* **237**:236–242 (1962).

101. K. McCarthy, W. Lovenberg, and A. Sjoerdsma, The mechanism of the inhibition of horse liver alcohol dehydrogenase by thyroxine and related compounds, *J. Biol. Chem.* **243**:2754–2760 (1968).

102. B. A. Askonas, Effect of thyroxine on creatine phosphokinase activity, *Nature* **167**:933–934 (1951).

103. S. A. Kuby, L. Noda, and H. A. Lardy, Adenosine triphosphate–creatine transphosphorylase. III. Kinetic studies, *J. Biol. Chem.* **210**:65–82 (1954).

104. A. Horvath, Inhibition by thyroxine of enzymes requiring pyridoxal-5-phosphate, *Nature* **179**:968 (1957).

105. H. A. Lardy, Effect of thyroid hormones on enzyme systems, *in* "The Thyroid," pp. 90–101, Brookhaven Symposium in Biology, No. 7, 1954, Brookhaven National Laboratory, Upton, New York (1955).

106. J. J. Vitale, D. M. Hegsted, M. Nakamura, and P. Connors, The effect of thyroxine on magnesium requirement, *J. Biol. Chem.* **226**:597–601 (1957).

107. J. J. Vitale, M. Nakamura, and D. M. Hegsted, The effect of magnesium deficiency on oxidative phosphorylation, *J. Biol. Chem.* **228**:573–576 (1957).

108. S. H. Mudd, J. H. Park, and F. Lipmann, Magnesium antagonism of the uncoupling of oxidative phosphorylation by iodothyronines, *Proc. Natl. Acad. Sci.* **41**:571–576 (1955).

109. B. M. Allen, Brain development in anuran larvae after thyroid or pituitary gland removal, *Endocrinology* **8**:639–651 (1924).

110. R. D. Adams and N. P. Rosman, Hypothyroidism: Neuromuscular system, *in* "The Thyroid" (S. C. Werner and S. H. Ingbar, eds.) 3rd ed. pp. 771–780, Harper and Row, New York (1971).

111. P. Marie, C. Trétiakoff, and E. Stumfer, Étude anatomopathologique des centres nerveux dans un cas de myxoedème congénital avec crétinisme, *L'Encephale* **15**:601–608 (1920).

112. J. T. Eayrs and S. H. Taylor, The effect of thyroid deficiency induced by methyl thiouracil on the maturation of the central nervous system, *J. Anat. (London.)* **85**:350–358 (1951).

113. R. Balázs, S. Kovács, P. Teichgräber, W. A. Cocks, and J. T. Eayrs, Biochemical effects of thyroid deficiency on the developing brain, *J. Neurochem.* **15**:1335–1349 (1968).

114. J. T. Eayrs and B. Goodhead, Postnatal development of the cerebral cortex in the rat, *J. Anat. (London.)* **93**:385–402 (1959).

115. J. T. Eayrs, The cerebral cortex of normal and hypothyroid rats, *Acta Anat.* **25**:160–183 (1955).

116. J. L. Nicholson and J. Altman, Synaptogenesis in the rat cerebellum: Effects of early hypo- and hyper-thyroidism, *Science* **176**:530–531 (1972).

117. P. Weiss and F. Rosetti, Growth responses of opposite sign among different neuron types exposed to thyroid hormones, *Proc. Natl. Acad Sci.* **37**:540–556 (1951).

118. T. Ferguson, Thyroxine effects upon the mitotic activity of medulla oblongata after unilateral excision in embryos of the frog, *Gen. Comp. Endocrinol.* **7**:74–79 (1966).

119. N. P. Rosman, M. J. Malone, M. Helferstein, and E. Kraft, The effect of thyroid deficiency on myelination of the brain, *Neurology* **22**:99–106 (1972).

120. M. Hamburgh, Evidence for a direct effect of temperature and thyroid hormone on myelinogenesis *in vitro, Develop. Biol.* **13**:15–30 (1966).

121. A. Moosa and V. Dubowitz, Slow nerve conduction velocity in cretins, *Arch. Dis. Child.* **46**:852–854 (1971).

122. F. J. Schulte, G. Albert, and R. Michaelis, Gestationsalter und Nervenleitgeschwindigkeit bei normalen und abnormalen Neugeborenen, *Deutsch. Med. Wschr.* **94**:599–601 (1969).

123. J. T. Eayrs, The vascularity of the cerebral cortex in normal and cretinous rats, *J. Anat.* **88**:164–173 (1954).

124. J. F. Fazekas, F. B. Graves, and R. W. Alman, The influence of the thyroid on cerebral metabolism, *Endocrinology* **48**:169–174 (1951).

125. M. Hamburgh and E. Vicari, Effect of thyroid hormone on nervous system maturation, *Anat. Rec.* **127**:302 (1957).

126. E. H. Craigie, Vascular patterns of the developing nervous system, *in* "Biochemistry of the Developing Nervous System" (H. Waelsch, ed.) pp. 28–51, Academic Press, New York (1955).

127. E. Scharrer, The blood vessels of the nervous tissue, *Quart. Rev. Biol.* **19**:308–318 (1944).

128. R. L. Friede, "Histochemical Atlas of Tissue Oxidation in the Brain Stem of the Cat," Karger, Basel (1961).

129. M. M. Brand and A. Bignami, The effects of chronic hypoxia on the neonatal and infantile brain, *Brain* **92**:233–254 (1969).

130. C. Kennedy, G. D. Grave, J. W. Jehle, and L. Sokoloff, Changes in blood flow in the component structures of the dog brain during postnatal maturation, *J. Neurochem.* **19**:2423–2433 (1972).

131. P. B. Bradley, J. T. Eayrs, and K. Schmaboch, The electroencephalogram of normal and hypothyroid rats, *Electroencephalog. Clin. Neurophysiol.* **12**:467–477 (1960).

132. P. B. Bradley, J. T. Eayrs, A. Glass, and W. Heath, The maturational and metabolic consequences of neonatal thyroidectomy upon the recruiting response in the rat, *Electroencephalog. Clin. Neurophysiol.* **13**:577–586 (1961).

133. J. T. Eayrs and W. A. Lishman, The maturation of behavior in hypothyroidism and starvation, *Brit. J. Anim. Behav.* **3**:17–24 (1955).

134. J. T. Eayrs, Thyroid and central nervous development, *in* "The Scientific Basis of Medicine Annual Reviews," pp. 317–339, Athlone, London (1966).

135. J. T. Eayrs, Age as a factor determining the severity and reversibility of the effects of thyroid deprivation in the rat, *J. Endocrinol.* **22**:409–419 (1961).

136. F. Khamsi and J. T. Eayrs, A study of the effects of thyroid hormones on growth and development, *Growth* **30**:143–156 (1966).

137. E. A. Carr, W. H. Beierwaltes, J. V. Neel, R. Davidson, G. H. Lowrey, V. N. Dodson, and J. H. Tanton, The various types of thyroid malfunction in cretinism and their relative frequency, *Pediatrics* **28**:1–16 (1961).

138. T. H. Shepard, Phenylthiocarbamide non-tasting among congenital athyrotic cretins: Further studies in an attempt to explain the increased incidence, *J. Clin. Invest.* **40**:1751–1757 (1961).

139. K. Miyai, M. Azukizawa, and Y. Kumahara, Familial isolated thyrotropin deficiency with cretinism, *New Engl. J. Med.* **285**:1043–1048 (1971).

140. G. Little, C. K. Meador, R. Cunningham, and J. A. Pittman, "Cryptothyroidism," the major cause of sporadic "athyreotic" cretinism, *J. Clin. Endocrinol. Metab.* **25**:1529 (1965).

141. E. M. McGirr and J. H. Hutchison, Dysgenesis of the thyroid gland as a cause of cretinism and juvenile myxedema, *J. Clin. Endocrinol. Metab.* **15**:668–679 (1955).

142. J. B. Stanbury, Familial goiter, *in* "The Metabolic Basis of Inherited Disease" (J. B. Stanbury, J. B. Wyngaarden, and D. S. Fredrickson, eds.) 3rd ed. Chapter 10, pp. 223–265, McGraw-Hill, New York (1972).

143. G. H. Lowrey, R. H. Aster, E. A. Carr, G. Ramon, W. H. Beierwaltes, and N. R. Spafford, Early diagnostic criteria of congenital hypothyroidism, *A.M.A. J. Dis. Child.* **96**:131–143 (1958).

144. S. Raiti and G. H. Newns, Cretinism: Early diagnosis and its relation to mental prognosis, *Arch. Dis. Childh.* **46**:692–694 (1971).

145. J. Money, Psychologic studies in hypothyroidism, *Arch. Neurol. Psychiat.* **76**:296–309 (1956).
146. D. W. Smith, R. M. Blizzard, and L. Wilkins, The mental prognosis in hypothyroidism of infancy and childhood, *Pediatrics* **19**:1011–1022 (1957).
147. R. L. Wall, H. J. Umlauf, and L. J. Geppert, Muscle reflex patterns in infancy and childhood, *Pediatrics* **64**:701–710 (1964).
148. M. Goldberg and F. C. Larson, The Achilles reflex: A diagnostic test of thyroid dysfunction, *Lancet* **1**:243–245 (1963).
149. R. Harris, M. Della Rovere, and P. F. Prior, Electroencephalographic studies in infants and children with hypothyroidism, *Arch. Dis. Child.* **40**:612–617 (1965).
150. E. A. Nieman, The electroencephalogram in congenital hypothyroidism: A study of 10 cases, *J. Neurol. Neurosurg. Psychiat.* **24**:50–57 (1961).
151. Y. Taher, M. Gabr, O. Shahin, and M. K. A. Khalek, Electroencephalographic changes in cretinism, *Clin. Electroencephalog.* **1**:6–12 (1970).
152. A. Topper, Mental achievement of congenitally hypothyroid children, *A.M.A. J. Dis. Child.* **81**:233–249 (1951).
153. R. Asher, Myxoedemetous madness, *Brit. Med. J.* **2**:555–562 (1949).
154. W. M. Easson, Myxedema with psychosis, *Arch. Gen. Psychiat.* **14**:277–283 (1966).
155. M. Malden, Hypothermic coma in myxedema, *Brit. Med. J.* **2**:764–766 (1955).
156. V. K. Summers, Myxedema coma, *Brit. Med. J.* **2**:366–368 (1953).
157. E. H. Jellinek, Fits, faints, coma and dementia in myxoedema, *Lancet* **2**:1010–1012 (1962).
158. S. N. Nickel and B. Frame, Neurologic manifestations of myxedema, *Neurology* **8**:511–516 (1958).
159. G. M. Cremer, N. P. Goldstein, and J. Paris, Myxedema and ataxia, *Neurology* **19**:37–46 (1969).
160. V. Pendred, Deaf-mutism and goitre, *Lancet* **2**:532 (1896).
161. F. R. Fraser, Association of congenital deafness with goitre (Pendred's syndrome), *Ann. Hum. Genet.* **28**:201–249 (1965).
162. W. R. Trotter, Deafness and thyroid dysfunction, *Brit. Med. Bull.* **16**:92–98 (1960).
163. S. Refetoff, L. T. DeWind, and L. J. DeGroot, Familial syndrome combining deaf-mutism, stippled epiphyses, goiter and abnormally high PBI: Possible target organ refractoriness to thyroid hormone, *J. Clin. Endocrinol. Metab.* **27**:279–294 (1967).
164. A. Gesell, C. S. Amatruda, and C. S. Culotta, Effect of thyroid therapy on the mental and physical growth of cretinous infants, *Am. J. Dis. Child.* **52**:1117–1138 (1936).
165. A. Lewis, A study of cretinism in London with especial reference to mental development and problems of growth, *Lancet* **1**:1505–1509 (1937).
166. L. Wilkins, The rates of growth, osseous development and mental development in cretins as a guide to thyroid treatment, *J. Pediat.* **12**:429 (1938).
167. R. P. Goodkind and H. L. Higgins, Hypothyroidism in infants and children, *New Engl. J. Med.* **224**:722–726 (1941).
168. H. Bruch and D. J. McCune, Mental development of congenitally hypothyroid children: Its relationship to physical development and adequacy of treatment, *Am. J. Dis. Child.* **67**:205–224 (1944).
169. L. S. Radwin, J. P. Michelson, B. Kramer, and A. B. Berman, End results in treatment of congenital hypothyroidism, *Am. J. Dis. Child.* **78**:821–843 (1949).
170. E. B. Man, A. C. Mermann, and R. E. Cooke, The development of children with congenital hypothyroidism, *J. Pediat.* **63**:926–941 (1963).
171. W. H. Gantt and W. Fleischmann, Effect of thyroid therapy on the conditioned reflex function in hypothyroidism, *Am. J. Psychiat.* **104**:673–681 (1948).
172. R. McCarrison, Observations on endemic cretinism in the Chitral and Gilgit valleys, *Proc. Roy. Soc. Med.* **2**:1–36 (1909).

173. G. Raman and W. H. Beierwaltes, Correlation of goiter, deaf-mutism and mental retardation with serum thyroid hormone levels in non-cretinous inhabitants of a severe endemic goiter area in India, *J. Clin. Endocrinol. Metab.* **19**:228–233 (1959).

174. J. C. Choufoer, M. Van Rhijn, and A. Querido, Endemic goiter in western New Guinea. II. Clinical picture, incidence and pathogenesis of endemic cretinism, *J. Clin. Endocrinol. Metab.* **25**:385–402 (1965).

175. L. C. G. Lobo, F. Pompeu, and D. Rosenthal, Endemic cretinism in Goiaz, Brazil, *J. Clin. Endocrinol. Metab.* **23**:407–412 (1963).

176. R. Fierro-Benitez, W. Penafiel, L. J. DeGroot, and I. Ramirez, Endemic goiter and endemic cretinism in the Andrean region, *New Engl. J. Med.* **280**:296–302 (1969).

177. P. O. D. Pharoah, I. H. Buttfield, and B. S. Hetzel, Neurological damage to the fetus resulting from severe iodine deficiency during pregnancy, *Lancet* **1**:308–310 (1971).

178. Editorial "New light on endemic cretinism," *Lancet* **2**:365–366 (1972).

179. K. M. Saxena, J. D. Crawford, and N. B. Talbot, Childhood thyrotoxicosis: A long-term perspective, *Brit. Med. J.* **2**:1153–1158 (1964).

180. T. McKendrick and G. H. Newns, Thyrotoxicosis in children: A follow-up study, *Arch. Dis. Childh.* **40**:71–76 (1965).

181. J. C. Syner, P. S. Fancher, and J. W. Kemble, Chorea associated with hyperthyroidism, *U.S. Armed Forces Med. J.* **5**:61–67 (1954).

182. J. Logothetis, Neurologic and muscular manifestations of hyperthyroidism, *Arch. Neuro* **5**:533–544 (1961).

183. L. L. Levitsky, E. Trias, and M. S. Grossman, Spontaneous thyrotoxicosis in infancy: Report of a case, *Pediatrics* **46**:627–629 (1970).

184. H. E. Leszynsky, E. Gross-Kieselstein, and A. Abrahamov, Hyperthyroidism in a 3-month-old baby, *Pediatrics* **47**:1069–1073 (1971).

185. J. M. McKenzie, Neonatal Grave's disease, *J. Clin. Endocrinol. Metab.* **24**:660–668 (1964).

186. M. J. Hoffman, B. S. Hetzel, and J. Manson, Neonatal thyrotoxicosis, *Aust. Ann. Med.* **15**:262–265 (1966).

187. C. Farrehi, Accelerated maturity in fetal thyrotoxicosis, *Clin. Pediat.* **7**:134–137 (1968).

188. C. Farrehi, M. Mitchell, and D. M. Fawcett, Heart failure in congenital thyrotoxicosis, *Pediatrics* **37**:460–466 (1966).

189. D A. Fisher, Pediatric aspects, *in* "The Thyroid" (S. C. Werner and S. H. Ingbar, eds.) 3rd ed. pp. 665–681, Harper and Row, New York (1971).

190. D. A. Fisher, W. D. Odell, C. J. Hobel, and R. Garza, Thyroid function in the term fetus, *Pediatrics* **44**:526–535 (1969).

191. A. N. Marks and E. B. Man, Serum butanol-extractable iodine concentrations in prematures, *Pediatrics* **35**:753–758 (1965).

192. P. Scheinberg, Cerebral circulation and metabolism in hyperthyroidism, *J. Clin. Invest.* **29**:1010–1013 (1950).

193. L. Sokoloff, R. L. Wechsler, R. Mangold, K. Balls, and S. S. Kety, Cerebral blood flow and oxygen consumption in hyperthyroidism before and after treatment, *J. Clin. Invest.* **32**:202–208 (1953).

194. S. Sensenbach, L. Madison, S. Eisenberg, and L. Ochs, The cerebral circulation and metabolism in hyperthyroidism and myxedema, *J. Clin. Invest.* **33**:1434–1440 (1954).

195. P. Scheinberg, E. A. Stead, Jr., E. S. Brannon, and J. V. Warren, Correlative observations on cerebral metabolism and cardiac output in myxedema, *J. Clin. Invest.* **29**:1139–1146 (1950).

196. J. T. Eayrs, Endocrine influence on cerebral development, *Arch. Biol. Liege* **75**:529–565 (1964).

197. M. Hamburgh and L. B. Flexner, Biochemical and physiological differentiation during morphogenesis. XXI. Effect of hypothyroidism and hormone therapy on enzyme activities of the developing cerebral cortex of the rat, *J. Neurochem.* **1**:279–288 (1957).

198. C. A. Garcia Argiz, J. M. Pasquini, B. Kaplun, and C. J. Gomez, Hormonal regulation of brain development. II. Effect of neonatal thyroidectomy on succinate dehydrogenase and other enzymes in developing cerebral cortex and cerebellum of the rat, *Brain Res.* **6**;635–646 (1967).

199. C. B. Klee and L. Sokoloff, Changes in D(—)-β-hydroxybutyrate dehydrogenase activity during brain maturation in the rat, *J. Biol. Chem.* **242**:3880–3883 (1967).

200. G. D. Grave, S. Satterthwaite, C. Kennedy, and L. Sokoloff, Accelerated postnatal development of D(—)-β-hydroxybutyrate dehydrogenase activity in the brain in hyperthyroidism, *J. Neurochem.* **20**:495–501 (1973).

201. W. S. Schwark, R. L. Singhal, and G. M. Ling, Metabolic control mechanisms in mammalian systems. Regulation of key glycolytic enzymes in developing brain during experimental cretinism, *J. Neurochem.* **19**:1171–1182 (1972).

202. A. Cuaron, J. Gamble, N. B. Myant, and C. Osorio, The effect of thyroid deficiency on the growth of the brain and on the deposition of brain phospholipids in foetal and newborn rabbits, *J. Physiol. (Lond.)* **168**:613–630 (1963).

203. P. Walravens and H. P. Chase, Influence of thyroid on formation of myelin lipids, *J. Neurochem.* **16**:1477–1484 (1969).

204. J. Grippo and J. H. Menkes, Effect of thyroid on fatty acid biosynthesis in brain, *Pediat. Res.* **5**:466–471 (1971).

205. N. B. Myant, The role of the endocrine glands in mammalian brain development, *in* "Advances in Experimental Medicine and Biology" (Vol. 13 of *Chemistry and Brain Development,* R. Paoletti and A. N. Davison, eds.), pp. 227–237, Plenum Press, New York (1971).

206. S. E. Geel and P. S. Timiras, Influence of neonatal hypothyroidism and of thyroxine on the acetylcholinesterase and cholinesterase activities in the developing central nervous system of the rat, *Endocrinology* **80**:1069–1074 (1967).

207. J. M. Pasquini, B. Kaplun, C. A. Garcia Argiz, and C. J. Gomez, Hormonal regulation of brain development. I. The effect of neonatal thyroidectomy upon nucleic acids, protein, and two enzymes in developing cerebral cortex and cerebellum of the rat, *Brain Res.* **6**:621–634 (1967).

208. S. E. Geel and P. S. Timiras, The influence of neonatal hypothyroidism and thyroxine on the ribonucleic acid and deoxyribonucleic acid concentrations of rat cerebral cortex, *Brain Res.* **4**:135–142 (1967).

209. L. Sokoloff, Action of thyroid hormones and cerebral development, *Am. J. Dis. Child.* **114**:498–506 (1967).

210. L. Schneck, D. H. Ford, and R. Rhines, The uptake of S[35]-L-methionine into the brain of euthyroid and hyperthyroid neonatal rats, *Acta Neurol. Scand.* **40**:285–290 (1965).

211. S. Geel, T. Valcana, and P. S. Timiras, Effect of neonatal hypothyroidism and of thyroxine on L-[14C] leucine incorporation into protein *in vivo* and the relationship to ionic levels in the developing brain of the rat, *Brain Res.* **4**:143–150 (1967).

Chapter 9

BIOLOGY OF THE STRIATUM[1]

André Barbeau

Department of Neurobiology
Clinical Research Institute of Montreal
Montreal, Quebec, Canada

I. INTRODUCTION

In the last 20 years, there has been a significant advance in our understanding of the pathophysiology and biochemistry of disorders of the basal ganglia. The first steps were taken with the demonstration of the role of copper and ceruloplasmin in hepatolenticular degeneration, or Wilson's disease. A few years later, it was found that dopamine is highly concentrated in the basal ganglia of the brain and that it is deficient in Parkinson's disease. This led to the use of the precursor L-dopa in the therapy of Parkinson's disease, a development which proved to be important both for the patients and for the attention paid by neurobiologists to this part of the brain.

It is not my purpose to review these aspects of the problem, for I have done so on previous occasions.[1,2] I shall rather focus on recent developments which indicate that there are a number of previously unsuspected neurotransmitter systems in the brain, all interacting in some way with the striatum. From such recent knowledge, one can take a fresh look into the role played by the

[1]The studies from the author's laboratory quoted in this chapter were supported in part by grants from the Medical Research Council of Canada, the Committee to Combat Huntington's Disease (Los Angeles Chapter), the United Parkinson Foundation, and the W. Garfield Weston Foundation.

striatum in the physiology of the brain. It is opportune to stop and analyze our present understanding of the striatum, because new therapeutic approaches for Parkinson's disease and Huntington's chorea are being introduced. If we are to derive the maximum of knowledge from our future clinical observations with these drugs and substances, we must have a framework, however temporary, with which to make comparisons. Biological details of each individual disease of the extrapyramidal system have been avoided purposely because these have been presented recently in many reviews and monographs.[1,2,7,78-80] The present overview is meant only as background for the understanding of these illnesses.

The knowledge acquired recently about the anatomy of new pathways within the brain and the biochemistry of neurotransmitters and modulators[3] has permitted a clearer understanding of the functions of the striatum. In recent papers, I have gradually evolved the framework which I shall detail herein.[4-7] Over the years, many functions have been assigned to the striatum, some of which have been neglected or forgotten; others have received more important emphasis. A few aspects of the clinical physiology of this part of the brain are summarized in Table 1 and discussed in the following sections.

II. ROLE OF THE STRIATUM IN MOTILITY

A. "Filter" Mechanism in Tone Control

The study of Buchwald et al.[8] and those of many others[9] indicate that the striatum holds a key place in sensorimotor integration. Input from proprioceptive and visual pathways, among others, is received from brain stem and thalamic relays. The main function of the striatum in these circumstances appears to be inhibition of the effector mechanism in the pallidum.[10] It is of interest that early stimulation studies were essentially negative.[11] Indeed, only when the striatum was stimulated with the experimental animal moving was the "arrest reaction" described.[12] The output end of this mechanism involves pathways leaving the basal ganglia toward the thalamus and the cortex and toward many brain stem nuclei (subthalamic nucleus, red nucleus, substantia nigra and tegmental nuclei of the reticular formation).

TABLE 1. FUNCTIONS OF THE STRIATUM

1. Feedback "filter" function in postural tone control mechanisms
2. "Set" function in the state of preparedness for movement
3. "Trigger" function in the initiation of movements
4. "Strategy" programming in some mental functions
5. Feedback integration and modulation of autonomic and neurohumoral homeostasis

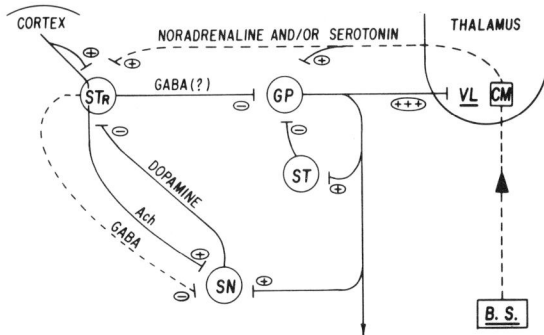

Fig. 1. Neurotransmitter pathways in the basal ganglia. GABA: gamma amino butyric acid, GP: globus pallidus, ST: subthalamus, STr: striation, Ach: acetyl choline, SN: substantia nigra, VL: ventrolateral nucleus of the thalamus, CM: center median, B.S.: brain stem, and DA: dopamine.

Recent evidence indicates that GABA-ergic intrabasal-ganglia fibers may be the modulator of this powerful striatal inhibition on the globus pallidus (Fig. 1). If the globus pallidus is viewed as the "effector" mechanism of the extrapyramidal system, and the center most responsible for controlling "*tone*" in lower levels in conjunction with the cerebellum, then it is easy to understand that the neostriatum (caudate and putamen) can play a "*filter*" role for all the feedback impulses containing mainly proprioceptive and visual information from the periphery. Variations in the relative importance of this feedback information, compared with the motor and postural program previously elaborated in higher centers and transmitted from the frontal lobe to the striatum, will permit the fine modulation of the GABA-ergic pathway, and thus the amount of inhibition on the effector center of the pallidum.

The intricate connections within the basal ganglia, modeled on what has been described for the cerebellum, are illustrated in Fig. 2.

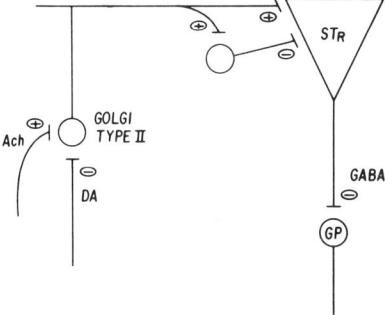

Fig. 2. Proposed model for intrastriatal connections.

In this respect, the elegant studies of Albe-Fessard *et al.*[9] have clearly demonstrated that the same striatal neuron can receive inputs from both proprioceptive and visual afferents and respond with an algebraic summation of these stimuli. It is also important to remember that the brain appears to function mainly through inhibition of inhibition when it wants to produce excitation, rather than through direct overactivation of a given "effector." The many inhibitory loops converging and acting on the striatum are compatible with this concept of a "filter" mechanism for the integration of divergent proprioceptive and visual feedback information into a coherent modulation of tone control.

It is indeed evident that the smooth regulation of muscle tone is important for the maintenance of posture and for the rhythmic alternation of agonist and antagonist muscle contractions necessary in walking. However, no actual movement can take place unless the organism is properly prepared for this movement, triggered adequately, and driven with sufficient energy. We will see that the striatum plays a role in each of these successive mechanisms.

B. "Set" Function of the Striatum

"*Set*" is the elaboration of the state of preparedness for movement which includes modifications in attention, in vigilance, and in the tone of the muscles to be used. It also involves the organization of a multitude of secondary automatisms and attitudes designed to facilitate execution of the movement. Every movement requires a certain degree of muscle preparedness and of postural adjustment before it can be carried out. This "set" permits the immediate responsiveness of the limbs to any motor drive. If it is impaired, there will be an increase in the reaction time, and specifically in the initiation time, before the patterned order is properly interpreted and executed. There will also be sluggish and unbalanced postural adjustments. Such a defect in "set" can be seen in the akinesia of animals treated with reserpine and of patients with Parkinson's disease. In both cases, the concentration of dopamine in the striatum is decreased.[13,14] Studies in Parkinson's disease and with the use of the precursor L-dopa have permitted a clearer delineation of the symptom "akinesia" and of its biochemistry.[6,15]

C. Definition of Akinesia

1. Clinical

In clinical terms, akinesia as seen in Parkinson's disease is a symptom complex manifested by a number of phenomena just recently better identified. Although various degrees of involvement are known and the presence of other associated symptoms is the most common situation, it is possible in a few patients to observe an almost pure akinetic syndrome without rigidity and tremor. We recently have studied 50 such patients and isolated the following components of akinesia.[6]

a. Defect in Motor Initiative. A defect in motor initiative is mainly recognized during evaluation of the reaction time. Brumlik and Boshes[17] have pursued this approach with refined techniques and report that the delay is more important at the level of the "central integrative" mechanism than at the periphery. Every new movement is a major effort which is easier abandoned. A restriction of proprioceptive and visual afferences, or their central integration, is probably involved in the mechanism, as demonstrated by the fact that akinesia is more severe when the patient attempts to enter a small doorway or tunnel or when he wants to cross a wide black band painted on the floor. During these attempts, "freezing," a sense of heaviness, anxiety, and even postural instability may be manifest.

Akinesia is not only a slow initiation but also includes a *decreased impulsion to move*, as defined long ago by Kleist[18] and as called *adynamie volitionnelle* by Ajuriaguerra.[16] We prefer to call this defect a lack in *kinetic motivation* because it probably includes a factor of vigilance and a strong, but nonspecific, emotional content.

The third aspect of this lack of motor initiative is the loss of associated movements, particularly the large number of acquired automatisms with which we accompany willful gestures. The clinical expression of this defect has long been recognized through the decreases in facial mimicry, rhythmic or postural limb movements, speech volume, and writing ability. Although an important component of the clinical picture in akinetic patients, and one which contributes most to the diagnosis of this entity, we feel that the decrease in associated movements is a secondary, albeit immediate, effect of the primary defect in motor initiative and particularly "set." It is as if this constellation of minimal defects exists to preserve kinetic energy for the more vital mass movements which will be set in motion by the trigger mechanism.

b. Defect in the Kinetic Melody. The kinetic "melody" is the ability to change rapidly from one motor pattern to the next in a smooth flow of movements as dictated by circumstances or willful decisions. A patient with Parkinson's disease walking fairly easily in a straight line will be unable, when ordered, to shift rapidly and smoothly to another direction without stopping, turning slowly, and reinitiating the slow and difficult creation of a new repetitive motor pattern. He acts as if at any one time he is capable of conceptualizing, on a motor plane, only a single plan or program. Any change requires the reformulation of both strategy and tactics, with the inherent time lag being considerably increased. In physiological terms, the akinetic patient must redefine his motor attitude for each consecutive postural pattern. Every movement having to be "thought out" in detail, the result is a considerable decrease in total activity.

    c. Defect in the Strategy of Learning. A defect in the strategy of learning is manifested by bradypsychia and a constructional defect akin to apraxia. This defect will be described in more detail below.

    d. Rapid Fatigability. The akinetic Parkinsonism patient tires very easily, as clearly demonstrated by Schwab and Zieper[19] with a bulb ergograph. This finding has been confirmed in all our own patients.

2. Biochemical

    In biochemical terms, akinesia has now been better defined and correlated with a specific deficiency in dopamine. In animals, the best model for the reproduction of this symptom is obtained after administration of reserpine. Unfortunately, this substance equally depletes the brain of dopamine, serotonin, noradrenaline, and even acetylcholine, and it is therefore very difficult to identify which neurotransmitter is most involved. Repletion of certain amines with L-dopa, but not with 5-hydroxytryptophan (5-HTP), can correct akinesia,[13] but again it is unclear whether dopamine or noradrenaline is responsible. Moreover, as we will see later, it is important to avoid equating akinesia with decreased motility.

    In patients with Parkinson's disease, our early studies[20,21] had demonstrated a correlation between the decrease in urinary excretion of dopamine and the presence of clinical akinesia. The excretion of homovanillic acid (HVA), the principal metabolite of dopamine, is decreased only in severe, long-standing akinesia.[22] However, it is much more exact to evaluate the state of cerebral dopamine metabolism through the concentration of HVA in cerebrospinal fluid (CSF), especially after a probenecid test. In these conditions, concentrations of HVA are lowest in the most severely akinetic patients.[23,24] Recently, Hornykiewicz and collaborators (unpublished observations, 1972) have confirmed that the striatal concentration of dopamine is lowest in the patients with Parkinson's disease who had the most severe akinesia clinically.

    Thus akinesia appears well correlated with a defect in metabolism of dopamine. This is further supported by the observation that akinesia is also the symptom best corrected in patients given L-dopa.[2] However, it is not easy to separate the respective roles of dopamine and noradrenaline in the increased motility induced by L-dopa. Furthermore, Birkmayer and Hornykiewicz[25] failed to modify motility with threodihydroxyphenylserine (threo-DOPS), a substance which is transformed directly into noradrenaline without the dopamine step.

    More specific stimulation of dopamine receptors in the brain can be obtained with apomorphine or pyribedil. In animals, both substances when given in overdose do not produce hypermotility, but rather a variety of stereotyped behaviors.[26,27] In man, the use of these dopamine agonists, and particularly apomorphine, has resulted in a clinical paradox attested to by a number of recent publications.[28-32] Indeed, in all these studies carried out in patients with Parkinson's disease, the symptom which was first corrected was not akinesia but tremor, and secondarily rigidity. The same observations have recently been made with pyribedil.[2] Thus the specific stimulation of dopamine receptors, when used alone, will only correct akinesia (in dopamine deficiency states) but will not produce an increase in motility. Once the "set" has been corrected and the body is ready to respond to command, there still must be a "trigger" mechanism for movement to occur. In fact, because of the state of hyperresponsiveness of the receptors induced by the drug (reserpine) or disease-related dopamine deficiency, the slightest overstimulation of these same receptors results in stereotyped movements and not in hypermotility.[33]

Snyder et al.[34,35] have used the divergent properties of D- and L-amphetamine to differentiate between the behavioral effects of dopamine and noradrenaline. Their studies, along with my own clinical observations,[2] indicate that dopamine is the most likely mediator of "set" and that noradrenaline modulates "drive." Except in exceptional circumstances, only a small rise in noradrenaline will be necessary to increase motor drive once "set" has been overcome.

Thus L-dopa first corrects "set" by stimulating quiescent but hypersensitive dopamine receptors probably in the striatum and then increases motility by activating the noradrenergic "drive" mechanisms in the brain stem and hypothalamus. This two-step action of L-dopa best corresponds to available clinical experience as we see it today. Both clinically and experimentally, overstimulation of these same dopamine receptors results in abnormal involuntary movements.[36]

D. "Trigger" Function of the Striatum

There is still considerable controversy as to the role of the striatum in the trigger (démarrage) mechanism of movement. Actually, it is more probable that willfull commands, as they are carried from the frontal cortex to the substantia nigra, do not relay directly into the striatum but only send collaterals into it. The transmitter involved in this process is still unknown, but it could be acetylcholine, as demonstrated by the studies of Olivier et al.[37] on the striatonigral pathway. Jung and Hassler[12] feel that this system acts on the substantia nigra, which then starts movements through a combination of ascending and descending influences. There is indeed some evidence that the implantation of eserine

[2]Presentations by Charpentier, et al., Fieschi et al., and Calne et al., at the Symposium International Trivastal, Monastir, November 1972.

directly into the substantia nigra can release dopamine within the striatum and presumably modify tone through inhibition of the powerful striatopallidal inhibitory pathway. In this way, a sudden "trigger" impulse would stimulate dopamine release in the striatum, modify "set" and its muscle tone component, and, at the same time, through descending pathways to the spinal cord, modify the more important preparedness mechanism situated in the γ system.

E. The "Drive" Mechanism

As indicated above, a decrease in motility is not synonymous with akinesia. Indeed, once the "set" defect has been corrected (e.g., with L-dopa, apomorphine, or pyribedil), there still is need for stimulation of other systems before movement can occur. The first indication along these lines came from the observation that inhibitors of monoamine oxidase can improve motility but that, while serotonin and noradrenaline concentrations are increased by this drug, there is very little change in the brain dopamine concentration. Studies by Anden et al.[38,39] clearly indicate that while dopamine is involved in the correction of akinesia in man and animals, noradrenaline is much more important in the escape reactions which underly motility and drive. To overcome the roadblock of akinesia induced by a deficiency of dopamine, a very strong stimulus such as fear would be required unless the dopamine deficit is first corrected. This occurs in "paradoxical kinesia." Thus in our view the clinical effect of L-dopa on motility is the composite result of these two successive actions on dopaminoceptive and then noradrenoceptive receptors, first in the striatum and then in the brain stem and hypothalamus.

The role of noradrenaline in drive mechanisms is further demonstrated in animals when noradrenergic receptors are specifically stimulated by threo-DOPS or clonidine. In this case, one observes hypermotility, aggressiveness, and hyperexcitability. Clonidine produces a marked decrease in brain stem noradrenaline concentration, but at the same time this hyperstimulation of noradrenaline receptors is accompanied by a significant change in dopamine turnover to HVA in the striatum,[40] thus slowing down noradrenaline synthesis in the same area. In this way, dopamine can be said to modulate noradrenaline stimulation.

III. ROLE OF THE STRIATUM IN MENTAL FUNCTION

The biochemistry of mental disorders has been the subject of a great many reports, symposia, and monographs. Lately, the field of cerebral monoamines has received increased attention, first with the "serotonin hypothesis" of

Brodie[42] and then with the "noradrenaline hypothesis" of Schildkraut.[43] Recently, I and others have added dopamine to the spectrum of involved neurotransmitters,[7,41] while GABA has been implicated in schizophrenia by Roberts.[72]

The striatum has been implicated in mental function for many years, particularly by Mettler *et al.*[44,45] The evidence reviewed in my recent paper[7] adds further weight to this intriguing possibility. If such is the case, dopamine is probably the main substance involved. Recent studies by Bliss and Ailion [46] indicate that the metabolism of dopamine is indeed accelerated by a variety of behaviors. These authors have shown that emotionally disturbing experiences, either when they are accompanied by augmented activity or when they are unassociated with it, will increase the metabolism of dopamine. In turn, an increase in activity in the absence of stress will again accelerate the metabolism of dopamine.

To date, we have put emphasis on the motor and external behavioral aspects of mental function. In a recent study, Botez and Barbeau[47] have reviewed the role of subcortical structures in the mechanism of speech, one of the highest intellectual functions of man. It was shown that speech, as the output side of information processing and the vehicle of language, requires constant modulation and control from subcortical mechanisms. After stereotaxic operations creating lesions in the so-called striatal loop output, the fluency of speech is often involved, but the conceptual system for language remains intact.

Studies on cognitive changes before and after L-dopa, such as those by Beardsley and Puletti[48] and others,[49] are still inconclusive. It is not yet possible to establish definitive patterns in IQ changes, and it is probable that it will be necessary to use much finer psychometric tests to form an opinion. However, we are certain that it will eventually be shown that many cortical cognitive parameters are directly under the modulating influence of cortico-subcortical loops[47] and that in those reverberating circuits the striatum, with its dominant dopamine metabolism, plays an important controlling role.

Akinetic patients appear bradypsychic, although most usual tests of intelligence are within normal limits for the age group. A closer analysis of this phenomenon, in addition to the previously described "puzzle test"[15] revealing a constructional defect (apraxia?), can be obtained from other simple tests, such as the Sorting-Test of Goldstein in which the patient is asked to separate a number of common objects in conceptual groups based, for example, on size, consistency, color, or other physical or utilitarian characteristics. Unless new possibilities are specifically pointed out to him, the akinetic Parkinsonism patient, like the frontal lobe patient,[50] is unable to *shift* to a new grouping. This is interpreted as a defect in the *strategy of learning* and not as a defect in memory, perception, intelligence, or capacity to learn, since, if prompted or guided, the patient can properly execute new groupings. For the akinetic Parkinsonism patient, once a repetitive pattern has been established, it is much easier to continue in the same direction than to change pattern.

L-Dopa therapy has been observed to produce intellectual "awakening" or alerting[51] and occasionally improved thinking capacity. Some authors have even reported actual improvement in learning ability, auditory perception, and intermediate memory functions as measured by verbal IQ and performance IQ scores.[52] This improvement appeared to be greatest in tests measuring perceptual organization such as block designs and object assembly,[53] although even within this test specialization only simple tasks were modified and the more complex procedures involving a sequence of actions, or making a decision, remained essentially unchanged. Riklan,[54] however, still maintains that an increase in behavioral activation or arousal is responsible for the observed increments in test scores of intellectual functions following short- and long-range L-dopa therapy. Finally, in the most recent studies on this subject, Garron et al.[55] confirm that late onset of Parkinson's disease and the symptom akinesia tend to be associated with intellectual deterioration and suggest that attention should be focused on such a group of patients to better evaluate changes in intellectual ability.

With more long-term use of L-dopa, many authors[56] are becoming aware of a subtle mental change occurring in many patients on chronic L-dopa therapy but apparently not correlated to the degree of physical impairment. This motor/psychological dissociation is potentially very important. Patients outwardly intellectually bright, mobile, well-oriented, and not depressed perform at a definitely low level on tasks involving constructive and perceptual organization. A recent study by our group utilizing the Kohs block design test[57] shows that there is an initial, highly significant improvement with L-dopa. It is of the same magnitude as that noted by others with other psychological tests. However, it is noteworthy that none of the patients with low pretreatment scores ever returned to a level of performance within the normal range. The general pattern for all Parkinsonism subgroups is one of marked impairment in general intellectual capacity, particularly within the sphere of perceptual organization. It thus appears that the striatum, and its dopamine system, plays a role in the normal elaboration of such functions.

Despite the initial "intellectual" improvement within the first few months of L-dopa therapy, there is an inexorable gradual decrease in performance in all groups of Parkinsonism patients, with a slow return to pretreatment levels. The slope of this progression is identical to that observed in patients not receiving L-dopa. Thus we can probably conclude that L-dopa does not stop the underlying progression of the disease, even while correcting the motor performance for long periods of time. Moreover, we can also conclude that L-dopa *per se* does not appear to be responsible for the progressive loss in intellectual performance, at least as measured with the Kohs block design test.

IV. ROLE OF THE STRIATUM IN AUTONOMIC FUNCTIONS

The role of the neostriatum in sensorimotor integration, and the part played by dopamine in this mechanism, could have been predicted from available knowledge. On the other hand, it has not been generally recognized that the striatum also plays an important role in regulating the homeostasis between ergotropic or trophotropic systems. As usually conceived, these systems are integrated at the level of their effector (output) centers in the brain stem and hypothalamus. A stimulatory input received in these centers will be reflected by changes either in the ergotropic or trophotropic systems. "Sensors" or receptor cells have been described in both areas (e.g., glucoreceptors). At the periphery, the ergotropic system is thought to be modulated by noradrenaline, while acetylcholine regulates the trophotropic system. However, recent studies indicate that, in the central nervous system, dopamine and serotonin play equally important roles in both these organizations. Long ascending and descending serotoninergic pathways in parallel with noradrenergic pathways have been described in the brain stem and spinal cord. The pattern now emerging within the confines of the nervous system is that of a chain of neurotransmitters. In the ergotropic system, dopamine appears to be the modulator for the short pathways of integrating areas, noradrenaline the transmitter for the long conduction pathways. In the trophotropic system, acetylcholine and serotonin play essentially similar alternating functions.

The brain stem and hypothalamus are classically considered the principal integrating areas for the autonomic system. There is no quarrel regarding the dual output of these centers, both through neural and through humoral pathways, and it is well known that a variety of inputs, from the cortex as well as from the periphery, will activate the output of these autonomic areas.

I shall now present evidence to support my contention[4] that the striatum should also be considered as one of the relay centers within the autonomic nervous system and that its function is to regulate the homeostasis of the effector centers in the brain stem and hypothalamus by negative feedback. There exists some old evidence from stimulation studies that the striatum can elicit autonomic responses,[58] but such responses are usually weak and difficult to obtain. As the "arrest reaction" was discovered only in moving animals, I propose that the striatum acts through negative feedbacks when the effector mechanisms elsewhere (brain stem and hypothalamus) are activated or in abnormal balance.

As seen in Fig. 3, I postulate that an input reaching the brain stem and hypothalamus and activating these centers will also inform the striatum of what is taking place at the periphery. In turn, by modulating activity within its own "black box," the striatum will attempt, by negative feedback, to correct the homeostatic imbalance between the ergotropic and trophotropic systems generated in lower centers. There are a number of feedback mechanisms which can play upon the striatum and regulate the turnover of dopamine within it.

A. Neuronal Regulation of Dopamine Metabolism

When the peripheral autonomic nervous system is interfered with through lesions caused by 6-hydroxydopamine,[59] the activity of the enzyme tyrosine hydroxylase is increased in the liver and adrenals,[60] in the brain,[61] and particularly in the caudate nucleus. This subject has recently been reviewed by Pletscher,[62] who concludes that hormonal influences probably predominate in the regulation of phenylethanolamine-N-methyltransferase (PNMT), whereas in the case of tyrosine hydroxylase, neuronal impulses are more important. Dopamine-β-hydroxylase seems to be controlled by both these factors to a similar extent.[63] Thus in the presence of a peripheral deficit, such as that induced by chemical or surgical sympathectomy, many centers, including the striatum, try to compensate by increasing the synthesis of catecholamines.

In the central nervous system, there is some evidence that the concentration of dopamine metabolites present at the receptor sites may play a role in the process of feedback regulation of the synthesis of dopamine. Although 3,4-dihydroxyphenylacetic acid (DOPAC) reflects intraneuronal metabolism and homovanillic acid (HVA) extraneuronal metabolism, we[64] have recently shown

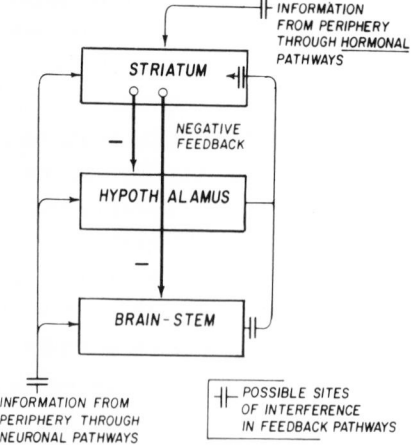

Fig. 3. Role of the striatum in autonomic regulation.

that large modifications in the concentration of these metabolites may result in significant changes in dopamine turnover.

B. Hormonal Regulation of Dopamine Metabolism

Tyrosine hydroxylase and dopamine-β-hydroxylase are diminished by hypophysectomy and can be restored, at least partly, by ACTH but not by gluco-corticoids.[65,66] There is now increasing evidence that a number of hormonal substances can modify the responsiveness of dopamine receptors. This has been well demonstrated for thyroid extracts, and recent results from our laboratory, obtained through the use of the apomorphine-aggression model in the rat (Minnich and Barbeau, unpublished results), indicate that many other homoral substances can modify the sensitivity of the dopamine receptors. This aspect of the problem will increase in importance as new knowledge is gathered from the hormonal effects of L-dopa in the hypothalamus.[67,68]

Recently, we have also focused attention on the presence within the brain of a renin-like substance which could act on angiotensin formation.[69] Further studies[70] have now shown that the renin activity is localized in the synaptoso-mal fraction of nerve endings within the brain and that the level of this activity in the caudate nucleus could be modified by aldosterone and progesterone. Angiotensin, when injected into the lateral ventricle of unanesthetized rats,[71] produced specific increases in the concentration of dopamine in the striatum.

Thus there is increasing evidence in favor of some involvement of the striatum in the regulation of autonomic function. Through mechanisms still unknown, the striatum is informed of events affecting the autonomic system at the periphery or even in the brain stem. At all times it reacts in the same way: by modifying (increasing or decreasing) the turnover of dopamine. A further refinement is the ability to direct this dopamine metabolism toward HVA or toward the formation of noradrenaline. This primary change can truly be called a modulation of feedback mechanisms of the autonomic nervous system.

Second, a change in striatal dopamine metabolism will affect the delicate balance in the striatum of all other neurotransmitters, but always toward re-establishing homeostatic balance between the ergotropic and trophotropic output systems. This influence of dopamine on other monoamines has recently been reviewed by me in detail[5] and is summarized in Fig. 4.

Brain stem and hypothalamic centers can react to *positive* stimulation with increased activity. Knowledge of this change in activity, and of the resulting state of peripheral autonomic imbalance, eventually also reaches the striatum. There, by first modifying dopamine turnover and thus affecting all other neuro-transmitters, this nucleus activates a *negative* feedback to the effector centers in the brain stem or hypothalamus with return toward autonomic balance. We

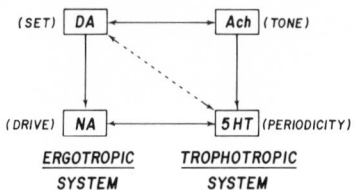

Fig. 4. Balance between neurotransmitters in the brain. DA: dopamine, Ach: acetylcholine, NA: noradrenaline, and 5HT: serotonin

are still ignorant of the mechanism through which the information about the state of autonomic imbalance reaches the striatum in this postulated feedback mechanism. Both neuronal and hormonal pathways could be involved, depending on the required speed of response.

The striatum is thus the regulator of autonomic homeostasis, and in this nucleus dopamine plays the role of modulator of the feedback system.

V. CONCLUSIONS

In this chapter, I have outlined the new data which permit a fresh look at the striatum and its connections. Many neurotransmitter systems are obviously involved in a complex interaction within this part of the brain. It is outdated to look at any one system, such as the nigrostriatal dopaminergic pathway, as acting alone. The resulting clinical manifestations of a defect in this pathway involve, in addition to akinesia, the interplay of GABA-ergic, cholinergic, and serotoninergic systems translated into rigidity, postural difficulties, and tremor.

From a number of clinical observations made since the introduction of L-dopa in the therapy of Parkinson's disease,[2,73] I have outlined what I consider are the main functions of the striatum: regulation of the set function in motility; filter mechanism for tone control; elaboration of the strategy of learning and feedback regulation of autonomic balance. The biology of these mechanisms involves dopamine at different levels. Our knowledge of each of these events is still embryonic, but the new interest in the metabolism of this amine will soon permit important progress.

There are already some clinical and pathophysiological conclusions drawn from the above observations. For example, in the treatment of Parkinson's disease, the long-term complications of concern are the abnormal involuntary movements and the oscillations in performance. We now know that both of these may be because of interference (overstimulation or competition) with dopamine receptors in the striatum.[6,36] New therapeutic approaches[74,77] to correct this interference are already being tested.

On the other hand, the demonstration[75] that in Huntington's chorea there is a specific dopamine deficit in the caudate nucleus, but not in the putamen, and that in this illness there is a decrease in concentration of GABA in the basal ganglia (Perry, unpublished observations) also permits new concepts of the pathophysiology of chorea and its treatment.[76]

Thus in the space of a scant 12 years our knowledge of the biology of basal ganglia disorders has progressed so rapidly that a completely new outlook on the management of these illnesses is now permitted. In the further comprehension of these developments, the principles enumerated herein should be helpful.

REFERENCES

1. A. Barbeau and F. H. McDowell, "L-DOPA and Parkinsonism," pp. 1–433, F. A. Davis, Philadelphia (1970).
2. A. Barbeau, L-DOPA therapy in Parkinson's disease: A critical review of nine years' experience, *Can. Med. Ass. J.* **101**:781–800 (1969).
3. N. E. Andén, A. Dahlström, K. Fuxe, and K. Larsson, Mapping out of catecholamine and 5-hydroxytryptamine neurons innervating the telencephalon and the diencephalon, *Life Sci.* **4**:1275–1280 (1965).
4. A. Barbeau, Functions of the striatum, *in* "Monoamines, noyaux gris centraux et syndrome de Parkinson," (J. de Ajuriaguerra and G. Gauthier, eds.) pp. 385–402, Masson et Cie, Paris (1971).
5. A. Barbeau, Role of dopamine in the nervous system, *in* "Monographs in Human Genetics" (J. François, ed.) Vol. 6, pp. 114–136, Karger, Basel (1972).
6. A. Barbeau, Contributions of levodopa therapy to the neuropharmacology of akinesia, *in* "Parkinson's Disease (Rigidity, Akinesia, Behavior)" (J. Siegfried, ed.) Vol. 1, pp. 152–174 Hans Huber, Publ., Berne (1972).
7. A. Barbeau, Dopamine and mental function, *in* "L-DOPA and Behavior" (S. Malitz, ed.) pp. 9–33, Raven Press, New York (1972).
8. N. A. Buchwald, C. D. Hull, L. M. Vernon, and G. A. Bernardi, Physiological and psychological aspects of basal ganglia functions, *in* "Psychotropic Drugs and Dysfunctions of the Basal Ganglia," pp. 82–92, P.H.S. Publication No. 1938, Government Printing Office, Washington, D.C. (1969).
9. D. Albe-Fessard, G. Guiot, Y. Lamarre, and G. Arfel, Activation of thalamocortical projections related to tremorogenic processes, *in* "The Thalamus" (D. Purpura and M. D. Yahr, eds.) pp. 237–253, Columbia University Press, New York (1966).
10. W. R. Hess, "Vegetative Funktionen und Zwischenhirn," Schwabe, Basel (1946).
11. D. Denny-Brown, Diseases of the basal ganglia. Their relation to disorders of movements, *Lancet* **2**:1099 (1960).
12. R. Jung and R. Hassler, The extrapyramidal motor system, *in* "Handbook of Physiology," Vol. 2, pp. 863–927, American Physiological Society, Washington, D.C. (1960).
13. A. Carlsson, The occurrence, distribution and physiological role of catecholamines in the nervous system, *Pharmacol. Rev.* **11**:490–493 (1959).
14. O. Hornykiewicz, Dopamine (3-hydroxytyramine) and brain function, *Pharmacol. Rev.* **18**:925–964 (1966).
15. M. Joubert and A. Barbeau, Akinesia in Parkinson's disease, *in* "Progress in Neuro-Genetics" (A. Barbeau and J. R. Brunette, eds.) pp. 366–376, Excerpta Medica Foundation, Amsterdam (1969).
16. J. de Ajuriaguerra, La notion d'akinésie, *in* "Monoamines, noyaux gris centraux et syndrome de Parkinson" (J. de Ajuriaguerra and G. Gauthier, eds.) pp. 565–579 Masson et Cie, Paris (1971).

17. J. Brumlik and B. Boshes, The mechanism of bradykinesia in Parkinsonism, *Neurology* **16**:337–344 (1966).
18. K. Kleist, "Gehirnpathologie," J. A. Barth, Leipzig (1934).
19. R. S. Schwab and I. Zieper, Effects of mood, motivation, stress and alertness on the performance of Parkinson's disease, *Psychiat. Neurol. (Basel)* **150**:345–357 (1965).
20. A. Barbeau, G. F. Murphy, and T. L. Sourkes, Excretion of dopamine in diseases of basal ganglia, *Science* **133**:1706–1707 (1961).
21. A. Barbeau, Dopamine and dopamine metabolites in Parkinson's disease. A review, *Proc. Aust. Ass. Neurologists* **5**:95–100 (1968).
22. A. Barbeau, La maladie de Parkinson et le métabolisme des amines cérébrales, *Rev. Praticien* **20**:5165–5173 (1970).
23. R. Papeschi, P. Molina-Negro, T. L. Sourkes, J. Hardy, and C. Bertrand, Concentration of homovanillic acid in the ventricular fluid of patients with Parkinson's disease and other dyskinesias, *Neurology* **20**:991–1001 (1970).
24. U. K. Rinne and V. Sonninen, Acid monoamine metabolites in the cerebrospinal fluid of patients with Parkinson's disease, *Neurology* **22**:62–69 (1972).
25. W. Birkmayer and O. Hornykiewicz, Der L-dioxyphenylalanin (L-DOPA)-Effekt beim Parkinson Syndrom des Menschen: Zur Pathogenese und Behandlung des Parkinson Akinese, *Arch. Psychiat. Nervenkrank.* **203**:560–572 (1962).
26. H. Corrodi, K. Fuxe, and U. Ungerstedt, Evidence for a new type of dopamine receptor stimulating agent, *J. Pharm. Pharmacol.* **23**:989–991 (1971).
27. A. M. Ernst, Mode of action of apomorphine and dexamphetamine on gnawing compulsion in rats, *Psychopharmacologia* **10**:316–323 (1967).
28. R. S. Schwab, L. V. Amador, and J. Y. Lettvin, Apomorphine in Parkinson disease, *Trans. Am. Neurol. Ass.* **76**:251–253 (1951).
29. J. Braham, I. Sarova-Pinhas, and Y. Goldhammer, Apromorphine in Parkinsonian tremor. *Brit Med. J.* **3**:768 (1970).
30. P. Castaigne, D. LaPlanne, and G. Dordain, Clinical experimentation with apomorphine in Parkinson's disease, *Res. Commun. Chem. Pathol. Pharmacol.* **2**:154–158 (1971).
31. G. C. Cotzias, P. S. Papavasiliou, C. Fehling, B. Kaufman, and I. Mena, Similarities between neurologic effects of L-DOPA and apomorphine, *New Engl. J. Med.* **232**:31–33 (1970).
32. S. E. Düby, L. K. Dahl, and G. C. Cotzias, Coupling of hypotensive and anti-Parkinson effects with two dopaminergic drugs, *Trans. Ass. Am. Physicians* **84**:289–296 (1971).
33. U. Ungerstedt, L. L. Butcher, S. G. Butcher, N. E. Anden, and K. Fuxe, Direct chemical stimulation of dopaminergic mechanisms in the neostriatum of the rat, *Brain Res.* **14**:461–470 (1969).
34. K. M. Taylor and S. H. Snyder, Amphetamine: Differentiation by D and L isomers of behavior involving brain norepinephrine or dopamine, *Science* **168**:1487–1488 (1970).
35. S. H. Snyder, K. M. Taylor, J. T. Coyle, and J. L. Meyerhoff, The role of brain dopamine in behavioral regulation and the actions of psychotropic drugs, *Am. J. Psychol.* **127**:199–207 (1970).
36. A. Barbeau, H. Mars, and L. Gillo-Joffroy, Adverse clinical side-effects of levodopa therapy, *in* "Recent Advances in Parkinson's Disease" (F. H. McDowell and C. H. Markham, eds.) Vol. 8, pp. 203–207, Contemporary Neurology, F. A. Davis, Philadelphia (1971).
37. A. Olivier, A. Parent, H. Simard, and L. J. Poirier, Cholinesterasic striatopallidal and striatonigral efferents in the cat and the monkey, *Brain Res.* **18**:273–282 (1970).
38. N. E. Andén, M. G. M. Jukes, and A. Lundberg, The effect of DOPA on the spinal cord. 2. A pharmacological analysis, *Acta Phys. Scand.* **67**:387–397 (1966).
39. N. E. Andén, A. Rubenson, K. Fuxe, and T. Hökfelt, Evidence for dopamine receptor stimulation by apomorphine, *J. Pharm. Pharmacol.* **19**:637–639 (1967).
40. A. Barbeau, Dopamine and disease, *Can. Med. Ass. J.* **103**:824–832 (1970).
41. G. Bryson, Biogenic amines in normal and abnormal behavioral states, *Clin. Chem.* **17**:5–26 (1971).

42. B. B. Brodie and W. D. Reid, Serotonin in brain: Functional considerations, *Advan. Pharmacol.* **6**:97 (1968).

43. J. J. Schildkraut, The catecholamine hypothesis of affective disorders: A review of supporting evidence, *Am. J. Psychol.* **122**:509–522. (1965).

44. F. A. Mettler and A. Crandell, Relation between parkinsonism and psychiatric disorder, *J. Nerv. Ment. Dis.* **129**:551–563 (1959).

45. N. S. Kline, and F. A. Mettler, The extrapyramidal system and schizophrenia, *in* "Extrapyramidal System and Neuroleptics" (J. M. Bordeleau, ed.) pp 487–491, Presses de l'Université de Montréal, Montréal (1961).

46. E. L. Bliss, and J. Ailion, Relationship of stress and activity to brain dopamine and homovanillic acid, *Life Sci.* **10**:1161–1169 (1971).

47. M. I. Botez and A. Barbeau, Role of subcortical structures and particularly of the thalamus in the mechanisms of speech and language. A review, *Intern. J. Neurol.* **8**:300–320 (1971).

48. J. V. Beardsley and F. Puletti, Personality (MMPI) and cognitive (WAIS) changes after levodopa treatment, *Arch. Neurol.* **25**:145–150 (1971).

49. R. Klaiber, J. Siegfried, W. H. Ziegler, and E. Perret, Psychomotor effects of L-DOPA combined with a decarboxylase inhibitor on parkinsonian patients, *Europ. J. Clin. Pharmacol.* **3**:172–175 (1971).

50. J. Barbizet, Rôle du lobe frontal dans les conduites mnésiques, *Presse Méd.* **79**:2033–2037 (1971).

51. G. C. Marsh, C. H. Markham, and R. Ansel, Levodopa's awakening effect on patients with parkinsonism, *J. Neurol. Neurosurg. Psychiat.* **34**:209–218 (1971).

52. T. C. Gutherie, H. S. Dunbar, and A. Weider, L-DOPA: Effect on highest integrative functions in parkinsonism, *Trans. Am. Neurol. Ass.* **95**:250–252 (1970).

53. A. W. Loranger, H. Goodell J. E. Lee, and F. H. McDowell, Levodopa treatment of Parkinson's syndrome. Improved intellectual functioning, *Arch. Gen. Psychiat.* **26**:163–168 (1972).

54. M. Riklan, Levodopa and behavior, *Neurology* **22**:(Part 2):43–55 (1972).

55. D. C. Garron, H. L. Klawans, and F. Narin, Intellectual functioning of persons with idiopathic parkinsonism, *J. Nerv. Ment. Dis.* **154**:445–452 (1972).

56. A. Barbeau, Long term appraisal of levodopa therapy, *Neurology* **22**:(Part 2):22–24 (1972).

57. M. I. Botez and A. Barbeau, The neuropsychology of akinesia. 1. Kohs block design test before and after levodopa therapy, *in* "Parkinson's Disease (Rigidity, Akinesia, Behavior)" (J. Siegfried, ed.) Vol. 2, Hans Huber Publ., Berne (1972).

58. E. A. Spiegel and E. G. Szekely, Prolonged sitmulation of the head of the caudate nucleus, *Arch. Neurol.* **4**:55–61 (1961).

59. U. Ungerstedt, 6-Hydroxydopamine-induced degeneration of central monoamine neurons, *Europ. J. Pharmacol.* **5**:107–110 (1968).

60. H. Thoenen, R. A. Mueller, and J. Axelrod, Trans-synaptic induction of tyrosine hydroxylase, *J. Pharmacol. Exptl.. Therap.* **169**:249–254 (1969).

61. B. Jacks, J. De Champlain, and J. P. Cordeau, Effect of 6-OH-dopamine on the central nervous system of rats, *Proc. Can. Fed. Biol. Soc.* **13**:136 (1970).

62. A. Pletscher, Regulation of catecholamine turnover by variations of enzyme levels, *Pharmacol Rev.* **24**:225–232 (1972).

63. P. B. Molinoff and J. Axelrod, Biochemistry of catecholamines, *Ann. Rev. Biochem.* **40**:465–500 (1971).

64. K. Yamada, J. L. Minnich, J. Donaldson, and A. Barbeau, Effect of 3,4-dihydroxyphenylacetic acid on catecholamines and serotonin in rat striatum, *J. Neurol. Sci.* **18**:311–315 (1973).

65. R. A. Mueller, H. Thoenen, and J. Axelrod, Effect of pituitary and ACTH on the maintenance of basal tyrosine hydroxylase activity in the rat adrenal gland, *Endocrinology* **86**:751–755 (1970).

66. R. Weinshilbaum and J. Axelrod, Dopamine-β-hydroxylase activity in the rat after hypophysectomy, *Endocrinology* **87**:894–899 (1970).

67. H. Friesen, H. Guyda, P. Hwang, J. E. Tyson, and A. Barbeau, Functional evaluation of prolactin secretion. A guide to therapy, *J. Clin. Invest.* **51**:706–709 (1972).

68. S. W. Spaulding, G. N. Burrow, R. Donabedian, and M. Van Woert, L-DOPA suppression of thyrotropin releasing hormone response in man, *J. Clin. Endocrinol. Metab.* **35**:182–185 (1972).

69. D. Ganten, J. L. Minnich, P. Granger, K. Hayduk, H. M. Brecht, A. Barbeau, R. Boucher, and J. Genest, Angiotensin-forming enzyme in brain tissue, *Science* **173**:64–65 (1971).

70. J. L. Minnich, D. Ganten, A. Barbeau, and J. Genest, Subcellular localization of cerebral renin-like activity, *in* "Hypertension '72" (J. Genest and E. Koiw, eds.) pp. 432–435, Springer-Verlag, Berlin (1972).

71. J. L. Minnich, J. Donaldson, and A. Barbeau, Modification des concentrations cérébrales en monoamines par l'angiotensine, *Union Med. Can.* **102**:903–906 (1973).

72. E. Roberts, An hypothesis suggesting that there is a defect in the GABA system in schizophrenia, *Neurosc. Res. Progr. Bull.* **10**:468–482 (1972).

73. A. Barbeau, Biochemistry of Parkinson's disease, *Excerpta Med. Intern. Congr. Ser.* **38**:152 (1961).

74. A. Barbeau, Treatment of Parkinson's disease with levodopa and Ro 4–4602. Review and present status, *in* "Advances in Neurology," vol. 2 (M. D. Yahr, ed.), Raven Press, New York, pp. 173–198 (1973).

75. A. Barbeau, Biochemistry of Huntington's chorea, *in* "Huntington's Chorea: 1872–1972" (A. Barbeau, T. N. Chase, and G. W. Paulson, eds.), Raven Press, New York (1973).

76. A. Barbeau, The biochemistry of Huntington's chorea. Recent developments, *Psychiat. Forum* (1973).

77. A. J. Kastin and A. Barbeau, Preliminary clinical studies with L-prolyl-L-leucyl-glycine amide in Parkinson's disease, *Can Med. Ass. J.* **107**:1079–1081 (1972).

78. O. Hornykiewicz, Biochemical and pharmacological aspects of akinesia, *in* "Parkinson's Disease (Rigidity, Akinesia, Behavior)" (J. Siegfried, ed.) Vol. 1, pp. 127–149, Hans Huber Publ., Berne (1972).

79. O. Hornykiewicz, Dopamine: Its physiology, pharmacology and pathological neurochemistry, *in* "Biogenic Amines and Physiological Membranes" (J. H. Biel and L. G. Abood, eds.) Part 2, pp. 173–258, Dokker, New York (1971).

80. O. Hornykiewicz, Neurochemistry of parkinsonism, *in* "Handbook of Neurochemistry" (A. Lajtha, ed.) Vol. 7, pp. 465–501, Plenum Press, New York. (1972).

Chapter 10

PATHOPHYSIOLOGY OF CENTRAL NERVOUS SYSTEM REGULATION OF ANTERIOR PITUITARY FUNCTION

Dorothy T. Krieger

Neuroendocrinology Laboratory
Division of Endocrinology, Department of Medicine
Mount Sinai School of Medicine of the City University of New York
New York

I. INTRODUCTION

Central nervous system (CNS) regulation of anterior pituitary function involves the interaction of two major bodily systems—the nervous and endocrine systems. A discussion of both the role of the CNS in hormonal regulation and of hormonal effects on CNS function is outside the scope of this chapter, which will focus mainly on selected aspects of the role of the CNS in hormonal regulation.

A. Experimental Approaches Utilized

The concept of CNS regulation of anterior pituitary function first received recognition through the work of Marshall,[1] who demonstrated the importance of environmental stimuli, acting through the nervous system and anterior pituitary gland, in the regulation of reproductive cycles and rhythms. It was subsequently shown by Harris,[2] DeGroot,[3] and others that electrical stimulation of the hypothalamus resulted in the discharge of various anterior pituitary hormones. The pathways mediating such control were not known.

The extensive experimentation designed to elucidate the mechanism(s) of such regulation has utilized diverse approaches; such as observation of endocrine changes following destruction and stimulation of specific hypothalamic and extrahypothalamic areas, observation of endocrine changes following pituitary stalk section or transplantation to isolate the pituitary from CNS influence, systemic or intrahypothalamic administration of various hormones or chemotransmitters, mapping of neuronal degeneration following CNS lesions, electron microscopy of the hypothalamus, autoradiography of CNS areas following steroid administration, histochemical fluorescence studies of CNS areas for amine localization, microscopic observation of the hypothalamic–pituitary vascular connections, and chemical fractionation of hypothalamic extracts.

These studies have led to new insights in diverse fields of investigation. Some examples include: a more detailed understanding of the substrate of environmental–brain interactions, isolation of polypeptides or polypeptide derivatives of CNS origin [hypothalamic releasing hormones (HRH)], which effect synthesis and release of pituitary hormones, demonstration of localized steroid uptake within specific areas of the CNS and the effect of such steroids on local electrical activity and HRH levels, demonstration of changes in concentration and/or turnover of specific neurotransmitter substances in specific CNS areas in altered hormonal states, demonstration of distinct neural pathways to the hypothalamus each utilizing specific neurotransmitters (alteration of whose concentrations leads to diverse endocrine changes affecting one or another target organ), demonstration of new cytoarchitectural pathways within the CNS, and demonstration of the role of neonatal hormone concentration in the regulation of various aspects of CNS maturation and function. As indicated above, such insights have developed from a combination of physiological and biochemical animal studies, the awareness of clinical endocrine disorders suggestive of a role of the CNS in endocrine regulation, and the demonstration of altered CNS function in "pure" endocrine disease as well as altered endocrine function in "pure" CNS disease.

In a field so inclusive and interdisciplinary, a review can merely indicate the broad outlines of such CNS–hormonal interrelations and indicate the pathophysiological mechanisms of clinical disorders. Of the numerous books and reviews on the subject, those cited[4-9] are extremely valuable compilations and interpretations of recent advances. Figure 1 presents an overview of the interrelationships of the CNS and the endocrine system. The further sections of this chapter will consider in greater detail the anatomical features and connections of the hypothalamus which translate neural stimuli from other parts of the CNS into the secretion of specific releasing factors. The neural transmitters involved in these pathways, the chemical nature and possible mode of action of the releasing factors, the periodic functioning of various components of the CNS–endocrine system, the importance of the steroid milieu at the time of pre- and neonatal development of the CNS with regard to its subsequent functioning, and clinical correlates illustrating CNS–endocrine relationships will also be considered.

B. Principles of Neuroendocrine Regulation

Before entering into this more detailed discussion, several points are worth noting: (1) Most feedback processes are considered to be inhibitory—i.e., the products of glandular secretion depress the concentration of or the release of the substance that normally stimulates that gland. Occasionally, feedback may be positive (e.g., a low concentration of estrogen can stimulate the release of FSH releasing factor, and there is recent evidence from Reichlin et al.[10] of a positive feedback of thyroid hormone on thyrotropin releasing factor). (2) Although the major part of recent interest in CNS regulation of endocrine function has centered on the hypothalamus, because of the isolation therefrom of releasing factors and also because of its anatomical localization as an end station of the autonomic nervous system, it should be recognized that the hypothalamus itself is under excitatory and inhibitory control of other portions of the CNS. (3) It should be noted that except for exteroceptive or interoceptive stimuli feeding into the nervous system, the hypothalamic–pituitary–adrenal system is essentially a "closed loop," allowing for self-regulation and avoiding overshoot in the production of any secretory product. (4) In a discussion of how a neurogenic or hormonal agent leads to alteration of secretion of a specific releasing factor, the concept of "set-point" has been invoked. This is similar to the concept involved in the setting of a thermostat. For example, an individual has a very discrete threshold at which he responds to an osmotic stimulus with the secretion of ADH (antidiuretic hormone). In a normal individual, plasma osmolality is maintained at approximately 291 mOsm/kg—i.e., the neurosecretory cells of the posterior pituitary are only activated to secrete above this blood osmolality. Similarly, the cell regulating ACTH secretion is so "programmed" that it will normally stop secreting ACTH above a certain level of steroid concentration and will initiate secretion when the steroid concentration falls below that level. (5) The diagram of Fig. 1 does not depict the time course of the pathways involved. It should be appreciated that certain reactions under hormonal regulation proceed rapidly—e.g., the reaction: stimulus $\longrightarrow$ ACTH $\longrightarrow$ corticoid

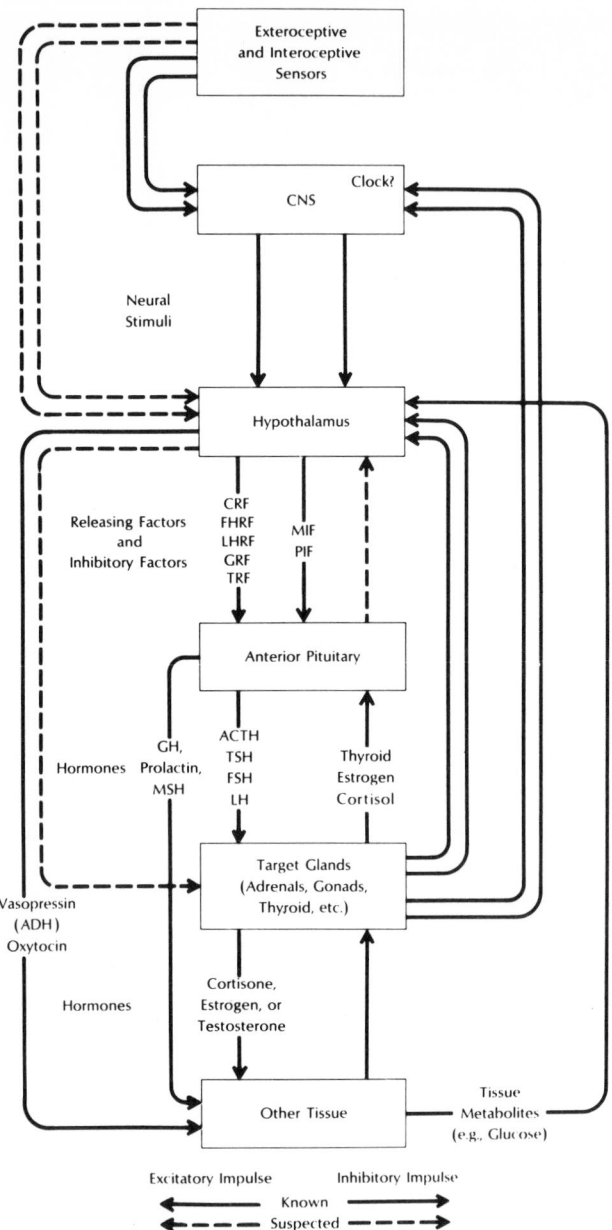

Fig. 1. Interrelationships involved in neuroendocrine regulation. [From Krieger,[212] p. 92. Reprinted by permission.] CRF: corticotropin-releasing factor, FHRF: follicular hormone-releasing factor, LHRF: luteinizing hormone-releasing factor, GRF: growth hormone-releasing factor, GH: growth hormone, FSH: follicular stimulating hormone, LH: luteinizing hormone, MSH: melanocyte stimulating hormone, ACTH: adrenocorticotropic hormone, PIF: prolactin inhibitory factor, and MIF: melanocyte stimulating hormone inhibitory factor. **Note added in proof:** Since this diagram was prepared a growth hormone inhibiting factor (somatostatin) has been described.[218]

secretion, can be effected within 5 min. In contrast, corticosteroid suppression of ACTH release or the increase in ACTH following adrenalectomy appears to require several hours. Whether or not this lag is related to a delay in CNS uptake or clearing of cortisol is unclear. Similarly, growth hormone stimulation of free fatty acid release requires approximately $1\frac{1}{2}$–2 hr, whereas epinephrine stimulation of such release occurs within minutes. (6) In Fig. 1, within the box labeled "CNS," a "clock?" is indicated. This refers to the well-known phenomenon of the circadian periodicity of the levels of many bodily constituents and functions. That is, these variables describe a sinusoidal curve, so that one peak and one nadir are reached in the course of a 24-hr period. Current investigation suggests that this "clock" is located within the CNS. (7) From the standpoint of bodily homeostasis, the question has been raised as to the advantage provided by the many links depicted in this "vertical organization of secretory activity." While it is obvious that no definite answer can be given to such a question, one suggestion is that such organization provides a means of amplification of a given signal, leading to a greater response than would be obtained if such a signal acted directly on the target organ. Furthermore, since the CNS integrates stimuli from a variety of sources and modulates responses from a number of target organs, it is obviously of value to have a given stimulus evaluated in terms of the remainder of the bodily economy. Thus a more appropriate response can be obtained than by direct action of the stimulus on a target gland alone.

II. CONCEPT OF ENDOCRINE REGULATION VIA RELEASING HORMONES OF NEURAL ORIGIN

A. Anatomical Factors

1. General Considerations

There has been ample demonstration from the wealth of animal experimental studies beginning in the 1930s that lesions or stimulation of the hypothalamus or afferent pathways which converge on the hypothalamus are associated with alterations in the structure and function of endocrine end organs previously believed to be solely under anterior pituitary control. It has further been demonstrated that these alterations in end-organ function are secondary to the effects of such CNS manipulation on the respective anterior pituitary tropic hormones. There was also evidence that the various tropic hormones could be affected differentially depending on the site of CNS manipulation.

The only known neural innervation of the anterior pituitary is via sym-

pathetic fibers entering in association with its blood supply. While such innervation may play a role in the regulation of anterior pituitary function by
altering the amount of blood flow to the pituitary in a unit time, there is no
evidence that there is direct innervation of pituitary secretory cells.* This would
remove known synaptic transmitters from consideration as the agents that could
act *directly* on the pituitary to mediate CNS regulation of anterior pituitary
function.

Several major questions then arise: (1) What is the nature of the hypothalamic or other neural substances which are active in effecting pituitary
tropic hormone synthesis and/or release? It is now generally accepted (see below) that there are specific chemical substances (hypophysiotropic hormones)
concerned with the secretion of each of the established anterior pituitary
hormones. Indeed, there may be stimulatory and inhibitory substances of neural
origin for each of these hormones. To date, extracts of hypothalamus and possibly of other CNS areas, of varying degrees of purity, have been shown to
have specific hypophysiotropic effects when administered *in vivo* or *in vitro*.
(2) Where in the CNS are such hypophysiotropic hormones produced? (3) What
types of cells secrete such hormones? (4) How do such hypophysiotropic hormones reach the anterior pituitary gland? The present section will deal with the
anatomical substrate relative to the last three of these questions.

*2. Gross Histological and Functional Anatomy of the Hypothalamus and
Its Major Connections*

The major gross anatomical relations of the hypothalamus are indicated
in Fig. 2. The anterior boundary is marked by the rostral edge of the optic
chiasm, the posterior boundary by the mammillary bodies, the lateral boundaries by the optic tracts and the sulci formed with the temporal lobes, and the
dorsal boundary by the overlying thalamus. The median eminence of the
hypothalamus is believed to be the major storage site for hypothalamic releasing factors and the area from which they are discharged into the hypothalamic–hypophyseal portal system, their conduit to the anterior pituitary
(see below). Grossly, the term "median eminence" refers to the external representation of the tuber cinereum, from which the pituitary stalk arises.

In coronal section (Fig. 3), it may be seen that the median eminence is the
portion of the hypothalamus occupying the area from the inferior extent of the
third ventricle to the base of the brain and extending laterally to the margins of
the infundibular recess of the third ventricle. The median eminence area is also

---

*Recently, however, MacLeod and Lehmeyer[219] have reported the presence of α-receptors
in the pituitary gland involved in the regulation of prolactin release. In addition, the presence
of dopamine-β-hydroxytase has been demonstrated in bovine pituitary glands.[220]

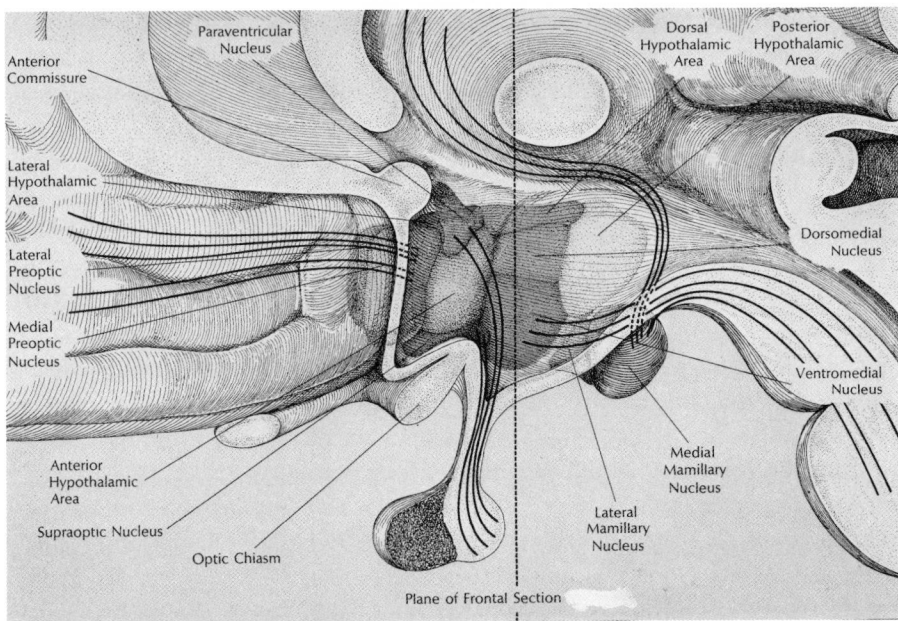

Fig. 2. Lateral view of hypothalamus indicating major nuclear centers and CNS interconnections. [From Krieger,[212] p. 90. Reprinted by permission.]

noteworthy in that it is one of the few areas in the CNS which does not appear to have a "blood–brain barrier," thereby implying a unique function with regard to possible hormonal and other feedback. The major fiber tracts are contained in the periventricular area (where they run in a dorsoventral direction) and in the lateral hypothalamic area, through which they pass longitudinally. The major nuclear groups are situated in the medial hypothalamus (Figs. 2 and 3). It should be noted that there are no nuclear groups in the median eminence area.

A tropic effect of the hypothalamus on the pituitary has been demonstrated in experiments[11] in which normal differentiation of cellular detail and granulation of transplanted pituitary fragments together with normal endocrine end-organ function occurred only when such pituitary fragments were transplanted into the hypothalamus, rather than into other CNS areas. Such experiments gave rise to the concept of hypothalamic hypophysiotropic hormones. To date, there has been no specific localization of a given hypophysiotropic hormone to a given hypothalamic nuclear group, and the exact location of the actual neurosecretory neurons (see below) that produce such hypophysiotropic hormones is not known. At present, it is felt that such secretory elements are located in a heterogeneous mass of cells lying in the region of the arcuate nucleus, tuberal nucleus, and the periventricular region extending as far forward as the preoptic area. Axons from these cells can be traced to their endings in the median eminence. Krulich *et al.*[12] have recently reported distribution of hypophysiotropic factors in specific areas of the rat hypothalamus, based on extraction of specific factors from 400-μ sections of freshly frozen hypothalami. Thyrotropin-releasing factor was found in the dorsal mediolateral part of the medial hypothalamus. Most of the luteinizing and follicle hormone-releasing factor and growth hormone-inhibitory factor activity was found in the median eminence area, with some activity reported in extracts of the preoptic area. Prolactin-inhibitory and releasing factor activity was also found in the preoptic area. This is in contrast to other evidence that releasing factors can be extracted from hypothalamic tissues with no correlation with specific hypothalamic nuclei.[13] Figure 4 depicts the attempted localization of releasing hormone activity and is a composite of findings derived from human and animal studies.

Although the hypothalamus is generally considered the major source of hypophysiotropic factors, recent experiments[14] have suggested that production of hypophysiotropic hormones may occur elsewhere within the CNS and be transported to the median eminence area of the hypothalamus by way of the cerebrospinal fluid. Such suggestions are based on (1) the specialized character of the ependymal cells (tanocytes) lining the infundibular recess of the rat, with villous protrusions facing the third ventricle and processes terminating in con-

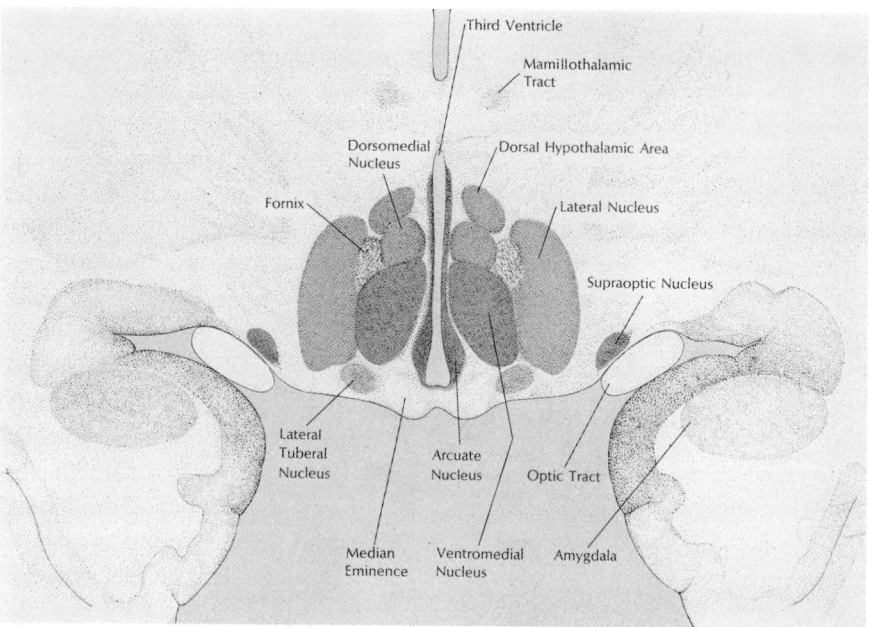

Fig. 3. Frontal section of hypothalamus indicating nuclear centers. [From Krieger,[212] p. 90. Reprinted by permission.]

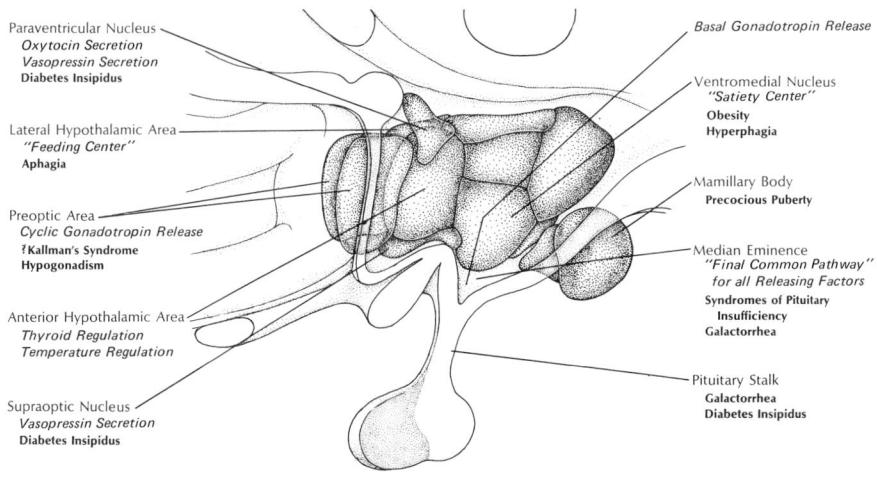

Paraventricular Nucleus
 Oxytocin Secretion
 Vasopressin Secretion
 Diabetes Insipidus

Lateral Hypothalamic Area
 "Feeding Center"
 Aphagia

Preoptic Area
 Cyclic Gonadotropin Release
 ?Kallman's Syndrome
 Hypogonadism

Anterior Hypothalamic Area
 Thyroid Regulation
 Temperature Regulation

Supraoptic Nucleus
 Vasopressin Secretion
 Diabetes Insipidus

Basal Gonadotropin Release

Ventromedial Nucleus
 "Satiety Center"
 Obesity
 Hyperphagia

Mamillary Body
 Precocious Puberty

Median Eminence
 "Final Common Pathway"
 for all Releasing Factors
 Syndromes of Pituitary
 Insufficiency
 Galactorrhea

Pituitary Stalk
 Galactorrhea
 Diabetes Insipidus

Effect of lesions indicated in **black**   *Normal function indicated in italic*

Fig. 4. Correlations of anatomical and physiological function within the hypothalamus. Although localization of specific neuroendocrine disorders to a specific anatomical area is not possible, this figure is a composite of results obtained from human and animal studies which suggest the major localizations observed to date. Lesions in afferent projections to these areas may also result in disturbances of endocrine function. [From Krieger,[213] p. 35. Reprinted by permission.]

tact with the capillaries of the primary plexus of the hypothalamic–hypophyseal portal system (Fig. 5) (see below), (2) the demonstration of neurosecretory material within the ependymal cells and their processes, and (3) the presence of "synaptoid" contacts between axons and ependymal processes within the median eminence[15] (Fig. 6).

3. Afferent Neural Connections of the Hypothalamus: The "Final Common Pathway"

The hypothalamus is intimately related both structurally and functionally to the midbrain, the limbic system, and the rhinencephalic nuclei of the forebrain. Pathways (both excitatory and inhibitory) from these areas (which have their own multitudinous afferent connections) converge on and traverse the hypothalamus, so that the hypothalamus may be considered to be a "nodal point" within these systems. Altered activity of these systems, therefore, could be expected to profoundly influence hypothalamic neurosecretion. Figure 7 indicates diagramatically and Fig. 8 schematically some of the interrelationships and projections of these systems. According to this concept, the hypothalamic neurosecretory cell may be considered to be a "final common pathway" (Figs. 9, 10, and 11) representing the terminus of a multisynaptic pathway involved in the regulation of neurohormone release.

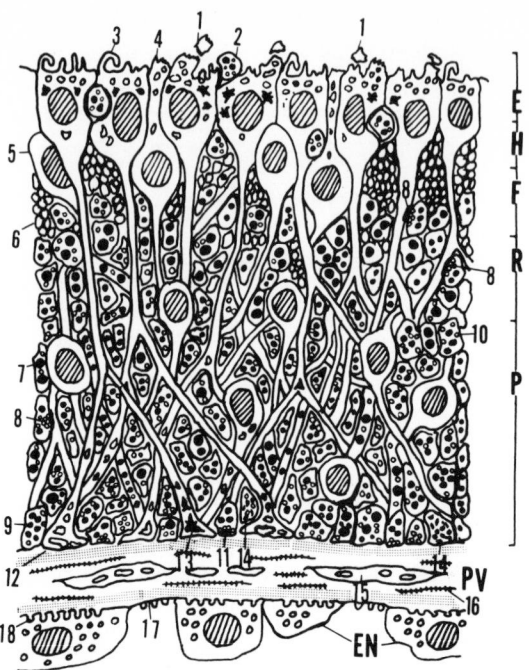

Fig. 5. Diagram of the fine structure of the rat median eminence. E, Ependymal layer; EN, endothelial cell; F, fiber
layer; H, hypendymal layer; P, palisade layer; PV, perivascular space; R, reticular layer. 1, Cytoplasmic masses
released from ependymal cells into the third ventricle; 2,
monoaminergic axons protruding into the third ventricle; 3,
marginal fold; 4, process of hypendymal cell; 5, hypendymal
cell; 6, mass of fine fibers; 7, large granules (these may be
grouped into two types); 8, synaptoid contacts; 9, small
granule; 10, small vesicle; 11, active point; 12, ependymal
endfoot containing vesicles; 13, ependymal or glial endfoot
containing dense bodies; 14, unidentified fiber ending containing peculiarly shaped granules; 15, fibroblast; 16, col-
lagen fiber; 17, fenestration; 18, pinocytosis. [From Kobayashi and Matsui,[15] p. 10. Reprinted by permission.]

4. "Aminergic" Pathways to the Hypothalamus

Serotoninergic, adrenergic, and dopaminergic pathways to the hypothalamus
have been identified by histochemical fluorescence techniques.[16] Noradrenergic
neurons are present in the olfactory bulb and brain stem, tegmentum of the
mesencephalon, pons, and medulla oblongata. Apparently, no adrenergic cell
bodies have been detected above the brain stem with this technique, although
axons in the median forebrain bundle within the hypothalamus have been found

to contain norepinephrine, and noradrenergic terminals within the median eminence have been identified. These terminals retain their fluorescence after sympathectomy, and it is suggested that they arise from cells in the arcuate and periventricular areas. Serotoninergic neurons are localized in the raphe nuclei of the mesencephalon. Although terminals within the median eminence have not yet been identified, large concentrations of serotonin in this area have been reported.[17] Dopaminergic neurons, in addition to being concentrated within the substantia nigra, have been demonstrated to arise from within the hypothalamus itself (arcuate and periventricular regions) and to send axons to the internal and external layers of the median eminence.

5. Neurosecretory Granules Within the Hypothalamus

Electron microscopic studies of the median eminence area have described three types of granules and one type of small vesicle within the axon terminals of this area,[15] in addition to four types of nerve endings. These endings differ as to whether they contain large, intermediate, small, or no granules in association with small vesicles. It has been suggested that the large and intermediate granules contain neurohypophyseal hormones and possibly releasing factors, that the small granules are carriers of monoamines, and that the small vesicles contain acetylcholine. Confirmation of these suggestions and study of changes with altered physiological states are needed.

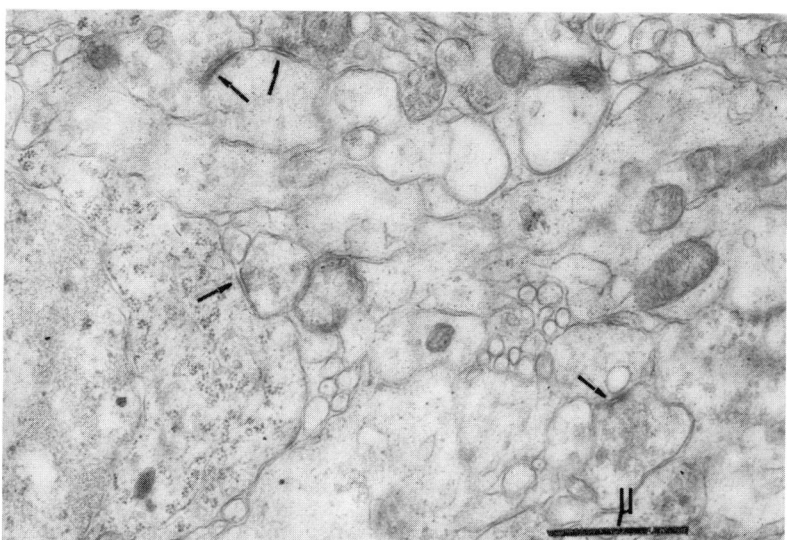

Fig. 6. Arrows indicate synaptoid contacts between hypendymal cells, ependymal process, and monaminergic fibers in the median eminence. [From H. Kobayashi, personal communication.]

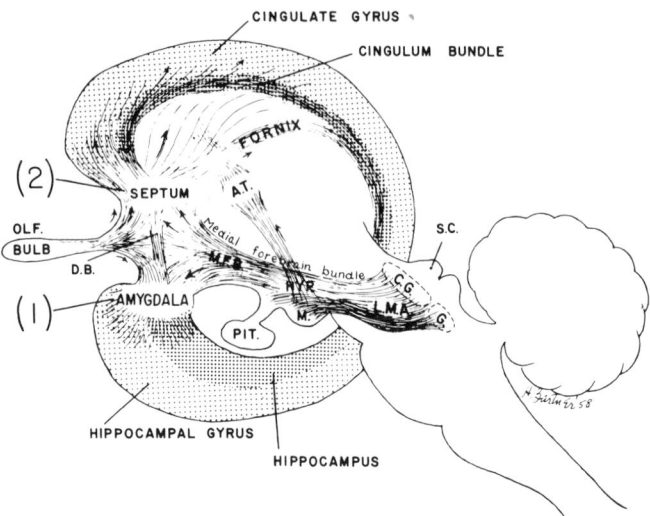

Fig. 7. Ascending connections from the midbrain and hypothalamus to the limbic lobe. In this diagram, only the ascending pathways are indicated and attention is focused on the divergence of two streams of fibers to the amygdala (1) and septum (2), where there are confluences, respectively, with fibers of the lateral and medial olfactory tracts. The concentric rings of archicortex and mesocortex of the limbic lobe are shown, respectively, in dark and light stipple. A.T., anterior thalamic nuclei; C.G., central gray of midbrain; D.B., diagonal band of Broca; G., ventral and dorsal tegmental nucleus of Gudden,; HYP., hypothalamus; L.M.A., limbic midbrain area of Nauta; M., mammillary body; PIT., pituitary; S.C., superior colliculus. [From Maclean.[214] Reprinted by permission.]

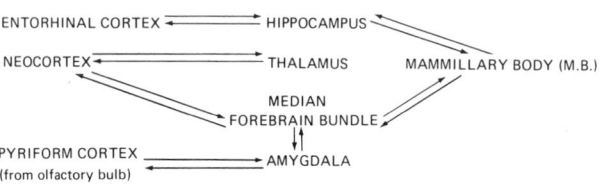

Fig. 8. Schematic representation of afferent connections of the hypothalamus.

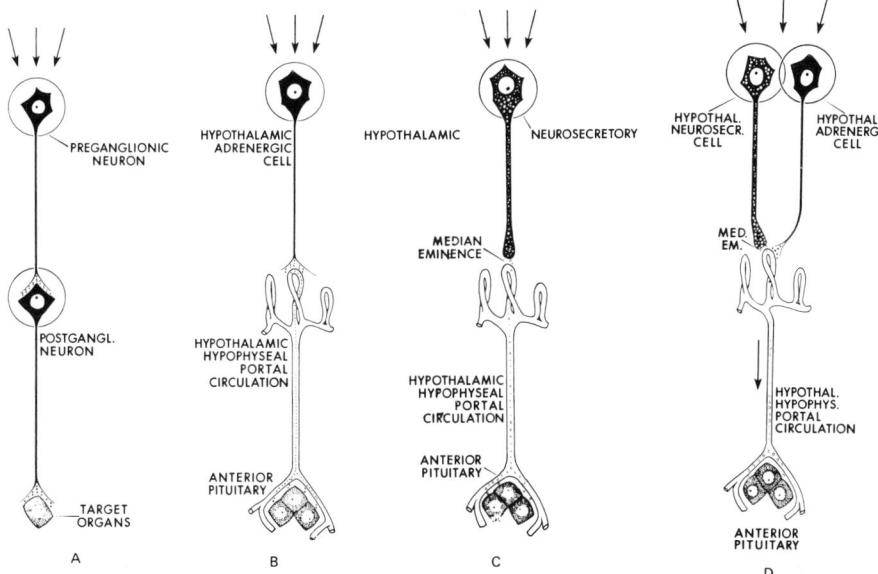

Fig. 9. Diagrammatic examples of final common pathways operating by means of neurotransmitter substances. Afferent neural and chemical stimuli, both extrinsic and intrinsic, that impinge on the final common pathway are indicated by arrows. A. In the sympathetic system, the transmitter substance at the synapse between pre- and postganglionic neurons is presumed to be acetylcholine, that at the postganglionic terminal adrenaline and/or noradrenaline. If the innervated organ is an endocrine tissue, the effect of the catecholamines is not secretomotor, but vasomotor; i.e., it may affect the secretory activity of the endocrine tissue via its blood supply rather than by direct action on the gland cells. B. Hypothalamic adrenergic and cholinergic cells may serve as the final common path by transmitting nervous stimuli to the adenohypophysis via the portal circulation as postulated by some investigators, but may actually play a subsidiary role in the release of neurosecretory material of polypeptide nature into the hypothalamic–hypophyseal portal vessels. C. In vertebrates other than fishes, a portal system of vessels occurs which carries neurohormones from axon terminals in the median eminence of the neurohypophysis to the anterior pituitary. D. Possible interrelations between "neurosecretory" and "neurotransmitter" substances on adenohypophysis. [Composite, from Scharrer.[215] Reprinted by permission.]

6. Hypothalamic–Hypophyseal Portal System

From the above discussion, it is evident that there is no direct neural connection between the hypothalamus and the anterior pituitary. The existence of vascular connections between the hypothalamus and pituitary had been known in the nineteenth century, but it was only in 1936 that Wislocki and King[18] made the important observation that the direction of blood flow in these vessels was from the hypothalamus to the pituitary (Fig. 12). It is now generally accepted that all blood reaching the pars distalis of the anterior pituitary has first been in contact with neural tissue. Branches of the hypophyseal arteries first break up into capillary plexi situated in the median eminence, the pituitary stalk, and the posterior pituitary. This is considered to be the primary plexus and is marked by the presence of numerous vascular coils which present a large surface area of contact with the median eminence. The vessels of the primary plexus then combine to form the portal veins. The long portal veins are derived from the capillary plexi of the stalk and median eminence, the short portal vessels from the capillary plexus within the posterior lobe. The long and short groups of portal vessels are believed to supply separate zones of the pars distalis, but it is not known whether or not substances released from specific nerve endings can be delivered to specific areas of the anterior pituitary by specific portal vessels. It has been estimated that in the rat[18] the blood flow through the combined median eminence and stalk is approximately 10 μl/min/ mg of tissue, which is among the highest of any tissue.

Ingenious experiments involving cannulation of pituitary portal veins in the intact animal (Fig. 13) or collection of stalk blood from such an animal have shown[19] that infusion of hypothalamic extracts into the portal vein will raise the levels of pituitary tropic hormones in the systemic circulation, and that the pituitary portal blood content of substances with releasing factor activity

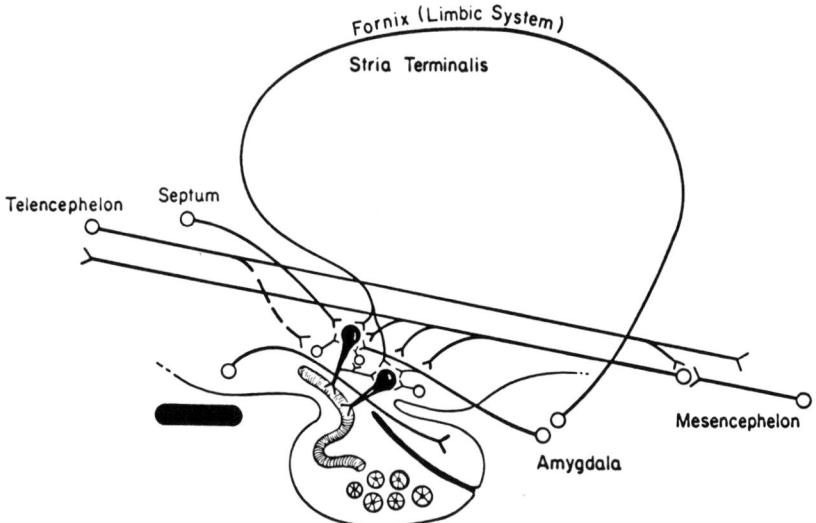

Fig. 10. Schematic representation of the hypothalamic neurosecretory cell as representing the final common pathway in translating neural impulses to endocrine secretion. Dashed line, inhibitory synapse; solid lines, pathways that may be either inhibitory or stimulatory to the neurosecretory cell.

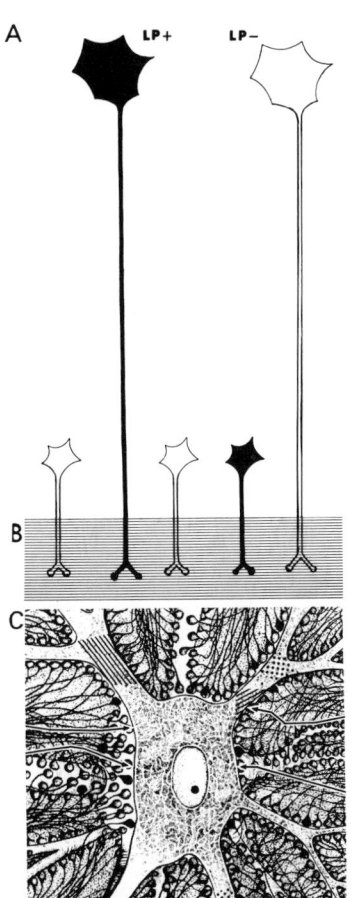

Fig. 11. Model of limbic and extralimbic influences on the CRF–neuron pool. A. Limbic pyramidal system; LP+ fibers (e.g., from amygdala), limbic pyramidal system which excites CRF–neuron pool; LP – fibers (e.g., dorsal hippocampus), limbic pyramidal system which inhibits CRF–neuron pool. B. Limbic extrapyramidal and other influences on the CRF–neuron pool. The hierarchy of A is higher than that of B. The striped area of B indicates interneuron pools. C. CRF–neuron pool. Excitatory synapses, black; inhibitory synapses, arrow. The stippled and striped areas indicate parts of the receptive surface of the neuron which are sensitive to circulating factors. (From Schade [2,16] pp. 4, 5, reprinted by permission.)

is increased following the placement of neurotransmitter substances into the third ventricle. This is the most direct proof to date of both the presence of hypophysiotropic hormones in portal blood and their stimulation by neurotransmitter agents.

B. Chemical Nature of Hypothalamic Releasing Hormones

Great progress has been made in recent years in the isolation and chemical characterization of different hypothalamic releasing hormones. Two of these (TRH, thyrotropin-releasing hormone, and LHRH, luteinizing hormone-

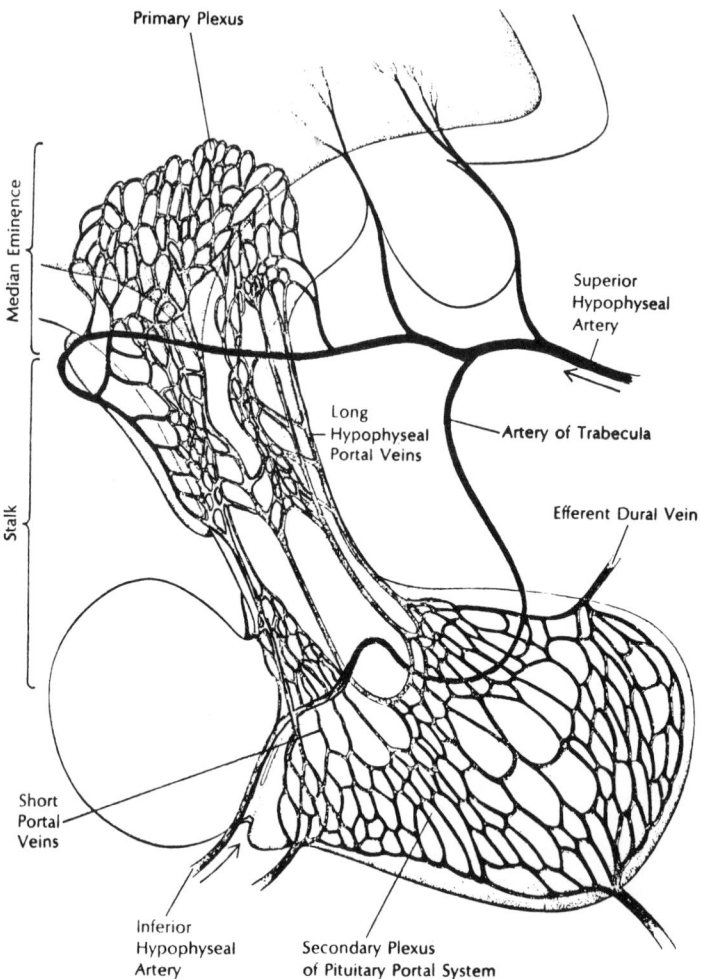

Fig. 12. Diagrammatic representation of the pituitary portal vasculature. Arterial blood first passes through a primary plexus in the median eminence area into long hypophyseal portal veins which then break up into a secondary plexus within the anterior pituitary gland before draining into the general venous circulation. The short portal veins receive blood from the stalk and posterior pituitary before breaking up into the secondary plexus within the anterior pituitary. [From Krieger,[212] p. 89. Reprinted by permission.]

releasing hormone) are currently available in synthetic form, which has resulted in significant advances in the understanding of their action and regulation. These releasing hormones have been administered in the presence of varying concentrations of end-organ (target and other) hormones, and the resultant concentration of various pituitary hormones has been determined. The availability of chemically pure hypothalamic releasing hormones should also soon lead to the development of radioimmunoassays for their determination in pituitary portal and peripheral blood and perhaps cerebrospinal fluid. Such determinations should be of enormous further aid in the understanding of CNS–endocrine interrelations.

Chemical identification of HRH was long hindered by the extremely low concentration of these substances in the hypothalamus. In 1965, Guillemin *et al.*[13] reported purification of ovine TRH from acetic acid extracts of an acetone powder prepared from 80,000 sheep hypothalami, this yielding 400 μg of material active at approximately a 0.1 μg dose *in vivo*. In 1966, Schally *et al.*[20] reported a yield of 900 μg of a TRH active at 10 ng *in vivo* from 20,000 porcine hypothalami. Over the past 5 years, heroic efforts have led to the accumulation of sufficient material to allow chemical characterization[21,22] and synthesis[23,24] of TRH. Both tritiated TRH and the C^{14}-labeled synthetic peptide have been shown to be preferentially concentrated in the pituitary, giving additional evidence of this site of releasing hormone action.

The availability of synthetic TRH has led to clinical studies which indicate its usefulness in distinguishing hypothyroidism of pituitary origin from that of hypothalamic origin.[25,26] With the use of TRH it has also been demonstrated that the negative feedback relationship between thyroid hormone and TSH secretion involves inhibition of TRH action at the pituitary level.[27,28] This appears to be mediated by the induction of new messenger RNA synthesis by thyroid hormone with subsequent production of a peptide or protein that inhibits some aspect of TRH stimulation of TSH secretion.[29,30] The inhibitory action of corticosteroids on TSH secretion has, however, been shown to take place at a higher level, namely, inhibition of TRH release.[31] It has been demonstrated that biosynthesis of TRH does not involve ribosomal protein synthetic systems.

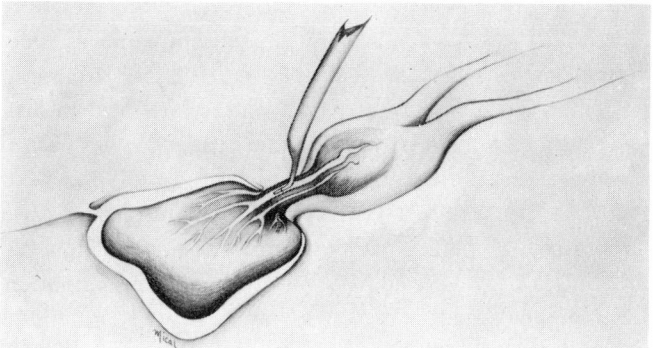

Fig. 13. Drawing indicating cannula inserted in a pituitary portal vessel of a rat with the fine tip of the cannula pointing toward the pituitary gland. [From Porter *et al.*,[217] p. 200. Reprinted by permission.]

Another unexpected finding with regard to the action of TRH has been demonstrated in clinical studies as well as in *in vitro* incubation studies—that TRH administration leads to a striking rise in prolactin levels.[32,33] It is known that prolactin secretion is normally inhibited by a prolactin-inhibitory hormone. The relationship of TRH to this factor or a postulated prolactin-releasing hormone (see below) still remains to be explored.

LHRH was the next hypothalamic releasing hormone to be characterized. At present, it appears that the naturally isolated and synthetic material has FSH (follicle stimulating hormone) as well as LH-releasing activity, although the latter by far predominates. Such releasing activity is dose dependent, smaller doses showing predominately LHRH activity. Whether there is just one releasing hormone for both of the gonadotropic substances or two such releasing hormones is still a matter for further investigation. It has been shown that the nature of the gonadotropin response to a releasing hormone is dependent on the hormonal state of the subject.[34-36] It may also be that there is some inherent FSHRH activity within the LHRH molecule but that another substance with major FSHRH activity still remains to be isolated.

The first isolation of LHRH was achieved by 12 successive purification steps.[37] The starting material consisted of ventral hypothalami of 165,000 pigs. LHRH has been shown[38] to be a decapeptide with the following sequence: pyroglutamyl-histidine-tryptophan-serine-tyrosine-glycine-leucine-arginine-proline-glycine amide. Clinical testing of LHRH has also been performed and indicates the usefulness of this substance in distinguishing hypogonadism of a pituitary source from that of a hypothalamic source.[39]

The existence of a MSH (melanocyte stimulating hormone) release-inhibiting hormone (MRIH or MIF) produced by the hypothalamus had been postulated on the basis of increased MSH secretion by pituitary glands removed from contact with the hypothalamus. Hypothalamic extracts have also been shown to increase pituitary MSH content. It has recently been demonstrated that MRIH is a tripeptide with the sequence prolyl-leucyl-glycine amide.[40] Of interest is that this tripeptide forms the side-chain of oxytocin. This tripeptide has now been synthesized and its biological activity found to be similar to that of MRIH as obtained by purification from hypothalamic extracts. It has been shown that microsomal preparations of rat hypothalami have an enzyme which may be concerned with the release of MRIH.[41] There are additional observations from the same laboratory which indicate that mitochondrial preparations of rat hypothalami contain enzymatic activity which leads to the formation of MSH-releasing hormone, the structure of which may be related to the remaining five peptides of the oxytocin molecule.

There is evidence that the hypothalamus normally inhibits pituitary prolactin secretion, like that of MSH. Such inhibition may be brought about by a prolactin-inhibiting factor (PIF) of hypothalamic origin. PIF activity has been shown to be present in dilute acetic acid extracts of ovine hypothalamic tissue.[42] The chemical nature of such a substance is unknown. Other extracts of rat hypothalamus may stimulate prolactin secretion, thereby suggesting the presence of prolactin-releasing hormone. The chemical nature of this postulated substance is unknown.

There is still considerable controversy about the chemical structure of growth hormone-releasing hormone(s) (GHRH). Schally has reported the purification and synthesis of a substance from porcine[43] and human hypothalamic tissue which causes depletion of pituitary growth hormone content (bioassay) in rats and causes increased plasma growth hormone–like activity (bioassay) in this species. GHRH has been identified in these studies as a decapeptide: valine-histidine-leucine-serine-alanine-glutamic acid-glutamic acid-lysine-glutamic acid-alanine. However, when administered systemically to the rhesus monkey, this purified preparation is not active in raising immunoassayable plasma growth hormone levels. Purified porcine, human, or synthetic GHRH also does not lead to any increase in plasma growth hormone levels in human subjects.[44] (There is one report[45] indicating a rise in immunoassayable plasma growth hormone when GHRH was administered by direct intrapituitary injection in the rat.) Therefore, although it is well established that crude or partially purified hypothalamic preparations can release growth hormone *in vivo* and *in vitro* as measured by radioimmunoassay, it is possible that the synthetic compound may require a cofactor for its action or that a different fraction will be identified in hypothalamic extracts which will stimulate release of immunoreactive growth hormone. Similarities between the synthetic decapeptide and the amino-terminal sequence of the β chain of porcine hemoglobin have been noted, raising the possibility that the latter may be a precursor of GHRH in a manner similar to α_2-globulin serving as a precursor for angiotensin.[46]

The existence of a growth hormone-inhibitory hormone has been suggested.[218]* On the basis of bioassay experiments, such a factor has been postulated by Dhariwal *et al.*[47] However, Schally *et al.*[48] have found no evidence for such inhibitory activity in extracts of porcine hypothalamus.

The characterization of corticotropin-releasing-factor (hormone) (CRF) has occupied much effort since the early studies demonstrating the existence of a substance of hypothalamic origin which regulates ACTH secretion. These were really the first studies which demonstrated the existence of substances of hypothalamic origin involved in endocrine control. However, no characterization of such a CRF has been reported.

*See note, p. 354.

Initial attempts to isolate and characterize CRF were performed on posterior pituitary extracts, since it was thought at the time that these studies were initiated that hypothalamic CRF, like vasopressin and oxytocin, was stored in the posterior pituitary. It has been shown that lysine-8-vasopressin and arginine-8-vasopressin can stimulate ACTH release *in vitro,* but there is some question whether the effective dose is within the range of physiological significance. Three other CRFs (α_1, α_2, and β) have been purified from posterior pituitary tissue. α_2-CRF has a proposed structure with an amino acid sequence identical to that of α-MSH, save for a blocking group on the *N*-terminal serine and possesses ACTH releasing activity without vasopressor activity. α_1-CRF appears to be closely related to α-MSH, but no conclusive amino acid sequencing is available. Pure α-MSH, however, has no CRF activity, although demonstration of some ACTH releasing activity *in vitro,* but not *in vivo,* by synthetic analogues of α-MSH has been reported.[49] It should also be noted that α-MSH and ACTH share the same amino acid sequence for their first 13 residues. β-CRF has the *C*-terminal sequence (lysine-glycine amide) of vasopressin in addition to having a ring structure reminiscent to that of vasopressin; it also embodies the *N*-terminal sequence of α-MSH. Perhaps, processes similar to those for oxytocin, which may be broken down to yield both MSH release-inhibiting and stimulating hormones, may occur for vasopressin.

Studies with hypothalamic extracts all indicate that material with potent CRF activity can be obtained with activity corresponding to α-CRF and β-CRF of neurohypophyseal origin. The relationship of such hypothalamic CRFs to either vasopressin, α-MSH, or the other neurohypophyseal CRFs still remains to be established. Although there are physiological indications of a possible ACTH inhibitory factor, there is as yet no report of isolation of such a factor from hypothalamic extracts.

C. Mode of Secretion of Hypothalamic Releasing Hormones

The concept of neurosecretion with regard to posterior pituitary function has been well established by the pioneer work of the Scharrers.[50] These investigators emphasized the special nature of neurosecretory cells within the CNS. They noted that these cells have the morphological and electrophysiological characteristics of nerve cells, but are also capable of liberating (secreting) an active substance into the bloodstream or tissue fluids which acts at a distance on other cell populations. These considerations also apply to the neuroregulation of anterior pituitary function. It has already been noted that various types of granules have been described in the axon terminals of the median eminence area and that some of these granules possibly contain releasing hormone(s). There is still some question whether the neurosecretory granules are broken down within the cell and their products subsequently released from the cytoplasm or whether the granules are released intact into the pituitary portal vasculature by a process of reverse pinocytosis.

D. Mode of Action of Hypothalamic Releasing Hormones

How these hypothalamic releasing hormones effect pituitary hormone release on contact with the appropriate pituitary cells must next be explained. Such effects occur within 5–10 min following injection of such releasing hormones into a peripheral vein. Figure 14 and Table 1 outline current theories of the mechanisms of action of releasing hormones. There is some evidence, especially with regard to TRH, that releasing hormones may act through the adenyl-cyclase–cyclic AMP system, and it has been shown that cyclic AMP can mimic the effects of hypothalamic releasing hormones on pituitary hormones.[51] It has been demonstrated that increased extracellular potassium is effective in stimulating release of all pituitary hormones, but there is no information about the metal-ion binding properties of the releasing factors. The presence of cal-

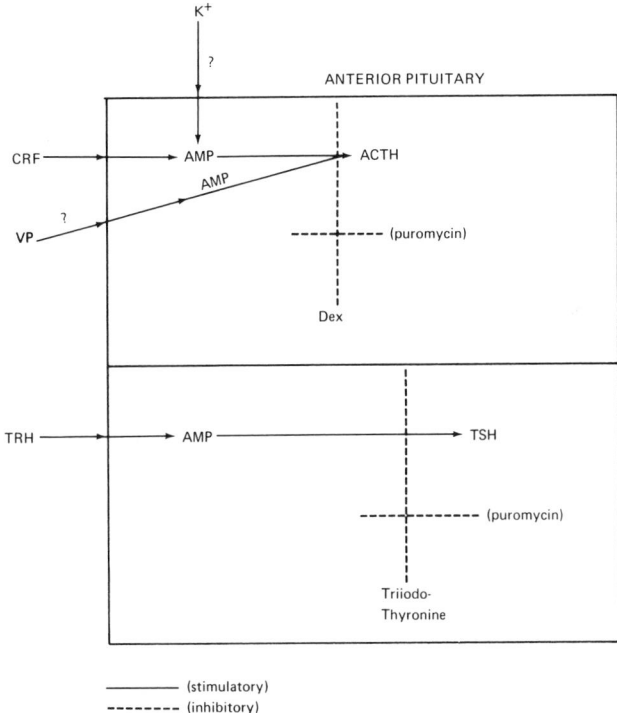

Fig. 14. Possible mechanism of action of releasing hormones. The presence of calcium ions is also required for the action of such releasing hormones. CRF, corticotropin releasing factor; TRH, thyrotropin releasing hormone; TSH, thyrotropin stimulating hormone; AMP, 3′5′-cyclic AMP; Dex, dexamethasone; VP, vasopressin.

TABLE 1

Pituitary hormone	Evidence that HRH effect is mediated via protein synthesis	Effect of cyclic AMP on level of pituitary hormone	Effect of HRH on pituitary cyclic AMP levels	Effect of High K+ (normal Ca²⁺) on pituitary hormone content	Effect of low Ca²⁺ on pituitary hormone content			
					Alone	On K⁺ effect	On HRH effect	On cyclic AMP effect
ACTH	?	↑		↑				
LH	No	↑		↑	↓	↓	Sl. ↓	
FSH	?	↑		↑	?	↓	Sl. ↓	
TSH	No	↑	↑	↑	↓			
GH	No	↑		↑	↓			

cium is also necessary for release of pituitary hormones by hypothalamic releasing hormones.

The action of releasing hormones is not prevented or inhibited by addition of actinomycin D, puromycin, or cycloheximide, thereby indicating that release of the pituitary hormone by the hypothalamic hormone is not dependent on the initiation of protein synthesis by the latter. However, it has been demonstrated both for thyroid hormone and for dexamethasone that the blocking of the hypothalamic releasing hormone action on the tropic hormone by the target organ hormone is dependent on protein synthesis—i.e., that such end-organ inhibition involves the synthesis of a protein which then blocks the action of the hypothalamic releasing hormone.

Although these hypothalamic hormones are called "releasing" hormones, because of the rapidity with which they increase plasma pituitary tropic hormones, there is evidence that they may also stimulate synthesis of the pituitary hormones. Electron microscopic studies have demonstrated an increase in the number of pituitary secretory granules following administration of releasing hormone. Releasing hormones incubated with pituitary tissue *in vitro* cause an increase in the total pituitary content of the appropriate hormone and an increased incorporation of radioactive precursors of the pituitary hormones. Finally, *in vivo* administration of releasing hormones to animals with transplanted pituitaries causes cellular differentiation of these previously inactive glands and pituitary hormone secretion.

E. Neurotransmitters Involved in the Release of Hypophysiotropic Releasing Hormones

Clearly, the hypothalamic neurosecretory cell (and perhaps other neurosecretory cells located in diverse areas of the CNS) has a mechanism for synthesis and release of hypophysiotropic hormones. What remains to be demonstrated is how the function of this neurosecretory cell is modulated, i.e., stimulated or inhibited. It is visualized that depolarization of the cell membrane would initiate processes resulting in HRH synthesis and release. One factor contributing to such depolarization would be the resultant of the effect of the excitatory and inhibitory neural transmitters acting on the cell synaptic membrane. The extent of this depolarization would also be governed by the local intracellular and extracellular concentrations of various ions. In the case of specialized receptors, local glucose concentration, pH, osmotic pressure, and steroid concentration would also be factors. Such specialized receptors may be localized to a very discrete area of the CNS involved in a specific aspect of hormonal regulation. For example, it has been shown by means of both autoradiography and determination of local steroid content that there are specialized areas of uptake for cortisone and gonadal hormones, the locations of which are anatomically different.[52–54] The final response of a given neurosecretory cell, as determined by initiation or cessation of secretion, therefore would be a resultant of all of these forces (synaptic transmitters and local environment) as they act on the cell. At present, although it is agreed that synaptic transmission occurs within the CNS as in the peripheral nervous system, substances serving as such central neurotransmitters are mostly identified by inference.

In view of the lack of anatomically or histologically demonstratable CNS areas involved in the regulation of each hypophysiotropic hormone (see above), it has become increasingly evident that there may instead be a neurotransmitter "coding" for each hormone. This would imply that pathways utilizing one type of neurotransmitter within a relatively broad anatomical area would lead to release of a specific hypophysiotropic hormone, as modified by local CNS concentrations of the target-organ hormone and perhaps of the pituitary hormone. In addition, since CNS-mediated pituitary hormone release occurs under different circumstances, such as periodic release, basal release, and stress-mediated release (see below), it may well be that different neurotransmitters are involved in each of these aspects of the regulation of a given pituitary hormone. As has been shown for the hypothalamic–pituitary–adrenal system,[55] a given neurotransmitter can activate ACTH release when implanted into one but not into another area of the hypothalamus or limbic system, and a given hypothalamic area can be stimulated (with resultant ACTH release) by one neurotransmitter but not by another. This is similar to the suggested concept of "chemical coding" within the CNS (rather than anatomical localization) for various aspects of eating, drinking, and social behavior. Thus stimulation of a number of areas of the rat brain by cholinergic substances causes water-satiated rats to drink and food-deprived rats to eat less, whearas stimulation of some of these sites by adrenergic substances has the effect of causing food-satiated rats to eat more and water-deprived rats to drink less.[56]

CNS regulation of growth hormone secretion appears to utilize adrenergic mechanisms. The rise in plasma growth hormone levels normally seen following induction of hypoglycemia is enhanced by β-adrenergic blocking agents and blocked by α-adrenergic blocking agents[57] as well as by reserpine, chlorpromazine, and monamine oxidase inhibitors. None of the last three agents has been effective in blocking the sleep-associated rise in plasma growth hormone levels.[58,59] Basal plasma growth hormone levels are increased following the administration of L-dopa[60]; this effect also appears to be mediated via α-adrenergic receptors.[61] Intrahypothalamic implantation of serotonin in baboons is reported to be ineffective in raising plasma GH levels,[62] and a moderate degree of cholinergic blockade did not alter growth hormone responsiveness to insulin induced hypoglycemia.[63]

Biogenic amines have been implicated in the regulation of both gonadotropin and prolactin release, levels of these hormones generally displaying a reciprocal relationship to each other. In the case of LH regulation, separate consideration has also to be given to transmitters that effect tonic hormone release in contradistinction to those that affect phasic (e.g., ovulatory) release. Pituitary LH release is increased[64,65] and prolactin release is decreased by dopamine,[66] whereas drugs that deplete hypothalamic norepinephrine block LH release[67] and facilitate prolactin release.[68] Administration of L-tryptophan, which presumably increases brain serotonin levels, also causes increased prolactin release, with only slight inhibition of LH release.[69] A cholinergic component in the neural mechanisms involved in ovulation is evidenced by the finding that atropine administered during a critical period on the day of proestrus can prevent ovulation.[70]

In the case of ACTH regulation, there is evidence for the involvement of several different neurotransmitters in the different mechanisms that can evoke ACTH release. Cholinergic and serotoninergic pathways have been demonstrated to be involved in the regulation of circadian periodicity of ACTH levels.[71,72] There is some controversy[71-73] as to the role of adrenergic pathways in this regard. Basal ACTH release can be stimulated in various species by either serotonin, carbachol (as an example of a cholinergic transmitter), or norepinephrine,[71,74,75] although, as pointed out above, not all of these transmitters are effective at the same hypothalamic location. Adrenergic inhibition of ACTH release has been reported, but these studies involved previously stressed animals,[76] and it may well be that different mechanisms are involved in this situation, although the same laboratory has also reported an increase in plasma corticosterone levels when brain catecholamine levels are depleted.[77] Stress-induced ACTH release was not affected in one study by systemically administered cholinergic blocking agents,[71] although intrahypothalamic implantation of such agents is reported to be effective in this regard.[78] In a different approach, stress and adrenalectomy both have been reported to be associated with increased CNS noradrenaline turnover,[79] but catecholamine depletion is not associated with a lessened adrenal secretory response to stress.[71,72,80] There is too little information to implicate either α-, or β-receptors as components of the adrenergic system.

There are no reported studies dealing with neurotransmitter-evoked TSH release.

F. Feedback Regulation of Pituitary Hormone Secretion

Prior to the recognition of the role of the CNS in the regulation of endocrine secretion, it was thought that most aspects of pituitary–target-organ interaction could be explained by a loop involving stimulation of the target organ by the pituitary tropic hormone and a resultant negative feedback action of the target-organ hormone on pituitary tropic hormone secretion. It is now recognized that the feedback action of target-organ hormones may be effected via neural mechanisms, presumably altering concentrations of releasing hormone and thereby affecting pituitary tropic hormone secretion. For some time, after experimental evidence for such neural feedback mechanism was presented, this was challenged by Bogdanove.[81] He rightly pointed out that these experiments ignored the possibility that when such hormones were implanted into the median eminence area they could more effectively reach all portions of the anterior pituitary (by being in contact with the pituitary portal system) than if they were actually implanted within the pituitary proper, where a more localized effect would be expected. Therefore, the possibility was raised that reported effects of intrahypothalamic implantation were actually manifestations of a pituitary locus of action. However, more recent studies showing feedback effects and regional uptake of hormonal implants far removed from either the ventricular or portal system support a neural locus of action of such implants. This is not to say that a direct feedback on the pituitary gland does not exist: such feedback has been demonstrated for adrenal,[82] thyroid,[83] gonadal,[84] and growth hormones.[85] This section will consider only the neural aspects of such feedback.

Such feedback may be positive or negative. Examples of negative feedback have been presented throughout this chapter. However, it should be pointed out that in certain situations (e.g., with stress in the case of CNS–pituitary–adrenal interrelations), the feedback action can be overriden by other neural drives, which perhaps evoke releasing hormone activity through mechanisms or pathways other than the feedback one. In the case of other hormones, such as the gonadal ones, and possibly thyroid hormone, positive feedback has been demonstrated: rising estrogen levels trigger LH release,[84] and increased levels of thyroid hormone may lead to TRH release,[10] perhaps to more finely modify regulation of TSH release than can be accomplished through mere direct negative feedback of T4 on TSH. There is also evidence that the pituitary hormones themselves can act directly on the hypothalamus to influence their own secretion.[86]

1. ACTH

Only the effects of high glucocorticoid levels on ACTH secretion will be considered, since little is known of the site or mode of action whereby low coritcosteroid levels stimulate ACTH secretion. There is evidence that feedback suppression of ACTH secretion may occur in a physiological setting.[82] and also may occur with the pharmacological amounts of steroids necessary to suppress stress-evoked ACTH release.[87] Cortisone implants into the amygdala, septal area, and hypothalamus (between optic chiasm and median eminence) suppressed compensatory adrenal hypertrophy and stress-induced ACTH secretion, whereas only implants in the median eminence area suppressed basal release.[88] A direct feedback action of ACTH on the hypothalamus has also been suggested.[86] This was based on studies which showed that following exogenous ACTH administration there is a reduction in the fall of pituitary ACTH stores which normally follows stress in adrenalectomized rats and a reduction in the extent of rise of CRH following adrenalectomy.[89]

Uptake of systemically administered labeled corticosteroid by hippocampal neurons has been demonstrated,[53] as well as concentration of such steroids within the septal area.[52] It is also known that corticosteroids increase CNS excitability[90] (i.e., decrease threshold for electrically induced seizures) as well as alter firing rates of specific neurons when administered microiontophoretically.[91] ACTH similarly has been reported to alter hypothalamic unit activity in the adrenalectomized animal.[92] Systemic administration of corticosteroids has been reported to alter CNS serotonin concentration.[93] Therefore, these studies provide some insight into the possible mode of action of steroid feedback on the CNS, although much remains to be done with regard to the actual cellular mechanisms involved.

2. Growth Hormone

Since growth hormone has no specific target organ, the question of feedback is a somewhat different one. Of the substrates on which growth hormone is known to act, it has been shown that intrahypothalamic implantation of glucose will block the normal growth hormone response to systemic hypoglycemia.[94] Implants of growth hormone into the median eminence area and other hypothalamic areas have been associated with a fall in pituitary weight and growth hormone concentration.[95]

3. Gonadotropic Hormone

Most of the investigations of feedback by gonadal hormones have dealt with the inhibitory and stimulatory effect of *estrogen* on gonadotropic hormone release. The findings have been interpreted as demonstrating that the medial basal hypothalamus is the main site for negative feedback of gonadotropin secretion by estrogen in females, with some evidence indicating a secondary feedback effect on the pituitary. This conclusion is based on findings demonstrating that anterior hypothalamic lesions prevent the gonadotropin inhibitory effects of estrogen administration,[96] that cytological changes occur in hypothalamic nuclei following changes in circulating estrogen levels,[97] that there is greater concentration of tritiated estradiol within the hypothalamus than in other cerebral tissues[98] and such radioactivity is localized to nuclei of specific neuronal groups,[54] and that there is alteration of the electrical excitability of specific neuronal groups following estrogen administration.[99]

The stimulatory effects of estrogen on gonadotropin secretion may take place within the same area, and, indeed, whether a stimulatory or inhibitory effect is obtained may well depend on the level of estrogen and the duration of stimulation. It has been shown[100] that implantation of estrogen into the median eminence in female rats increases plasma LH levels 5 days later, but subsequently leads to severe gonadotropic inhibition. Precocious vaginal opening in rats has been effected by estrogen implants in the anterior hypothalamic or preoptic region, although additional sensitive areas in the pituitary may be present.[101] There is also a suggestion that the stimulatory effect of decreased estrogen levels may operate at the median eminence level.[102]

a. Progesterone. Progesterone and norethindrone (a synthetic progestational compound) implanted into the median eminence area, but not elsewhere in the brain, inhibit ovarian and uterine development in the rat[101] and block ovulation in the rabbit.[103] Similar to the results cited for estrogen, uptake of tritiated progesterone by hypothalamus[104] and a biphasic effect of progesterone on the threshold for electrical activation of certain EEG patterns have been reported.[105] Very little information is available on the localization of progesterone stimulatory feedback.

b. Testosterone. Studies of implantation of both testosterone and cyproterone (an antiandrogen) into the basomedial hypothalamic area lead respectively, to decreased and increased LH secretion.[101] There is also evidence of neural uptake of tritiated testosterone,[106] although specific nuclear binding has not been demonstrated.

c. Gonadotropic Hormones. It has been reported that FSH and LH implantations into the hypothalamus depress pituitary content of endogenous gonadotropin.[86]

4. Thyrotropic Hormone

Although the bulk of data available indicates that the pituitary itself is capable of appropriate response to altered thyroid hormone level,[107] there is also evidence of a CNS feedback site. Preoptic area implantation of thyroxine[108] causes inhibition of TSH release, and intrahypothalamic injections of thyroxine inhibit thyroid function in hypophysectomized rats with heterotopic (transplanted) pituitary grafts.[109]

III. FACTORS INVOLVED IN THE PERIODIC RELEASE OF PITUITARY HORMONES

A. Introduction

It has become apparent that a great number of bodily functions and constituents undergo periodic fluctuations. The cycle of these functions may be ultradian (frequency less than 24 hr per cycle), circadian (frequency approximately 24 hr per cycle), infradian (frequency greater than 24 hr but less than 7 days), or it may have a frequency even longer than 7 days.[110] Such rhythms, in addition to being characterized by frequency, are characterized by amplitude and phase (the time of peaking relative to clock time). It is currently thought that such rhythms are endogenous and innate but are entrained (i.e., their phase determined) by a restricted class of environmental periodicities.[110] In the absence of such environmental periodicities (e.g., in the presence of constant light or constant dark), such rhythms may become free running.[1] Such circadian rhythms are usually remarkably independent of temperature. Although the existence of other geophysical controls has been postulated, it is generally accepted that light is the most important synchronizer. In this regard, it may be the light–dark transition, the duration of the light period, the specific time in the 24-hr cycle that light or dark is present, etc., that is the actual synchronizing influence. It is still unclear whether the sleep–wake cycle or the light–dark cycle is the important factor with regard to certain specific rhythms.

The mechanism by which these environmental influences actually regulate such periodicity is largely unknown. The participation of the CNS is thought to be of prime importance, perhaps through photoreceptors, not all of which are retinal in location.[112,113] The pathways from the retina involved are not necessarily those of vision: the physiological existence of retinohypothalamic path-

---

[1]In a light-entrained 24-hr cyclicity, under normal conditions all individuals peak at the same time of day. When the individual is placed in constant light or constant dark, the cycle length for a given individual may vary by plus or minus 2 hr from the 24-hr period, so that different individuals peak at different times of day. This change in cycle length is called a "free-running period."

ways, accessory optic tracts,[114] and pathways from the eye to the superior cervical ganglion of the sympathetic chain and thence to the pineal gland has been demonstrated.[115]

Although the activity of an individual cell can be demonstrated to fluctuate in a rhythmic manner, it is evident that there are hierarchial systems of controls making for rhythmic activity of whole organs and systems. However, there is as yet no evidence to permit choosing between the existence of one "master clock" in the organism (presumably located within the CNS) to which the rhythmic functioning of all body parameters is coupled, or of several "clocks," each with coupled oscillatory systems. It is also not known which parameters manifest primary rhythms (with those of others following secondarily). For example, it has been suggested that the rhythm of urinary potassium excretion is coupled with that of adrenal corticosteroids, since both rhythms manifest the same frequency and phase.[116] However, if an individual is exposed to an altered or inverted light–dark schedule,[117] the phase shift of these two parameters does not occur concomitantly, implying a lack of tight coupling and perhaps different regulatory mechanisms for each.

In the human, as well as in the experimental animal, there is evidence that rhythmic functioning of many body parameters is not present at birth but develops at a later time, perhaps associated with a particular level of CNS maturation. This is readily evident in the case of sleep–wake rhythms and also occurs with regard to periodicity of adrenal cortical function.[118] With regard to the ontogenetic development of such rhythms, it would be important to know if there is periodicity of central neurotransmitter concentration or turnover, and if such periodicity is age related. To date, such studies have been limited to determination of concentration rather than of turnover of neurotransmitter in the adult animal, and at only a few time points in the circadian cycle. A rhythmic variation in the concentration of norepinephrine has been observed in various anatomical portions of the cat brain, each region studied having a different time of peaking and in some instances a different frequency.[119] Similar findings have been reported for CNS concentration of serotonin in the rat.[120] While these studies merely represent an ability to correlate frequencies of CNS function with those of other parameters, they favor the primary role of the CNS in the maintenance of bodily rhythms.

B. Specific Examples of Periodic Hormone Release

The periodic release of a number of pituitary and target-organ hormones has been established. In some instances (e.g., ACTH), this appears to be a true circadian (approximately one cycle every 24 hr) rhythm, in others (e.g., growth hormone) a sleep-associated rhythm which shifts with the time of onset of sleep.[59]

1. ACTH

The circadian periodicity of ACTH release, usually as inferred from measurements of periodicity of adrenal corticoid levels, has been most extensively studied. This circadian pattern has been characterized as one in which, in the human at some time between 6 months and 4 years of age,[118] peak levels occur in the early morning hours, with a progressive decline over the remainder of the 24-hr period.[121] Reversal of the day–night living schedule results in phase reversal of the circadian pattern within a period of approximately 8 days.[122] By use of more frequent sampling,[123] it has become apparent that such a rise and fall of ACTH and corticosteroid levels do not occur in a smooth linear manner but that there are multiple peaks present throughout the day. However, the greatest number occur from midnight to 9:00 A.M., with an ensuing downward trend (Fig. 15).

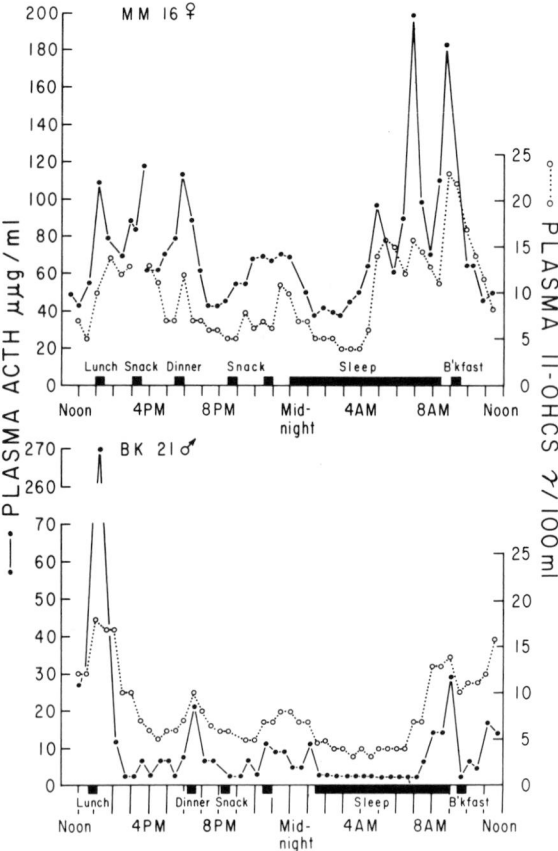

Fig. 15. Circadian periodicity of plasma 11-OHCS and plasma ACTH levels over a 24-hr period as determined by half-hourly sampling. Meal times and sleep as indicated.

There is evidence that ACTH periodicity persists in the absence of the adrenal gland[124] and that there is a periodicity in hypothalamic content of corticotropin releasing factor[125,126] which persists in hypophysectomized animals,[127] again indicating a prime role of the CNS in its regulation. The neural substrate for such periodicity is believed to be centered in the hypothalamic–limbic system areas. Lesions in these areas in the experimental animal[128,129] and disease of these areas in the human[130] have been reported to disrupt the circadian periodicity of corticosteroid levels. Alteration of CNS neurotransmitter is also reported to be associated with an absence of corticosteroid circadian periodicity—acetylcholine,[71] serotonin,[72] and norepinephrine[73] being implicated in this regard. (It is of interest that at least with regard to the effect of alteration of level of the first two transmitters cited there is no associated disruption of feedback responsiveness or stress-evoked pituitary–adrenal responses, implicating the existence of functionally different pathways with regard to these various parameters of ACTH regulation.)

Light–dark transition and sleep–wake transition both have been implicated as phase regulators of this periodicity. Enucleated animals[131] as well as blinded humans[132] manifest altered circadian patterns of both plasma ACTH and corticosteroid levels, as do animals reared either in constant light or in constant dark.[131,133] In humans, if the dark period is prolonged beyond the time of awakening, there appears to be a shift of the time when maximum levels are attained,[134] although in constant dark the normal rise with peak levels shortly after awakening still persists.

2. Growth Hormone

The occurrence of a sleep-associated rise in plasma growth hormone levels coincident with the first appearance of the electroencephalographic stage of slow-wave sleep is now well established.[59] As noted above, this is not a true circadian periodicity, since the onset of this rise may be delayed if sleep onset is delayed, and a second rise may be seen if the subject is awakened and then allowed to resume sleep. To date, with the possible exception of free fatty acids[135] this rise is not suppressible by any of the parameters (e.g., glucose,[58] α-adrenergic blocking agents,[58] corticosteroids[136] which suppress stress-provoked growth hormone release.

There are a few cited cases indicating absence of the sleep-associated rise in growth hormone levels in patients with hypothalamic disease.[137,138] Blindness has been reported to be associated with both abnormal[139] and normal[140] nocturnal growth hormone patterns.

3. Other Hormones

Although there is obviously *cyclic* periodicity of LH release in the female, there is still controversy about the circadian periodicity of gonadotropin release.

The reports concerning the presence of or absence of periodicity of FSH release[141-143] are conflicting, but there is general agreement concerning a lack of circadian periodicity of LH release,[143,144] although smaller cyclic variations may be present. The lack of reported periodicity of LH release is in keeping with evidence of no[145] or minimal[146] circadian variation in human plasma testosterone levels, although a recent report does describe such variation.[147] Very recently, a circadian periodicity of serum prolactin[148] and thyrotropin[149] levels has been reported in normal human subjects. It is not known whether these are true circadian or sleep-related periodicities, and nothing is known about the neural substrate for their control.

IV. EFFECT OF NEONATAL HORMONAL MILIEU ON NEUROENDOCRINE CONTROL MECHANISMS

A. Concept of a Critical Period

In many species, brain growth, characterized by growth of axons and dendrites and establishment of neuronal connections and myelin sheath deposition, continues after gestation. In addition, there are regional differences in the rate of development of different brain areas. It has been observed that the absence or supranormal presence of steroids in the fetus or early in neonatal life can have profound effects on neuroendocrine function, with alteration both of pituitary hormone secretion and of behavior. With regard to the effect of exogenous hormone administration to the neonatal animal, it has been found that alteration of neuroendocrine function is effected only if the hormone is given within a very finite period after birth, which is called the "critical period." No effect on neuroendocrine function is seen if such hormones are administered at a later stage of development.

B. Gonadal Steriods

Cyclicity of gonadotropin release, whether in an estrous or a menstrual rhythm, is characteristic of the female, whereas no such cyclic phenomenon occurs in the male. In view of what already has been presented concerning neural regulation of pituitary hormone release, one would expect that the locus controlling such periodic release would be within the CNS.

In the experimental animal, it is now clear[150] that sexual differentiation of the brain is completed shortly after birth. Male animals castrated late *in utero* or early in postnatal life (before day 10 in rats) as adults display a cyclic pattern of gonadotropin release, and if estrogen and progesterone are then administered they display female behavior on being mounted by a male. However, if female animals are treated with either androgen or estrogen during this neonatal period, loss of ovulation and cyclic ovarian function ensues, and as adults these animals manifest more "masculine" reactions in sexual behavior, which are not modified by female hormone administration. As adults such neonatally androgenized female animals may be made to ovulate if the median eminence area is electrically stimulated. Similar results are seen in adult animals bearing anterior hypothalamic lesions or undergoing anterior hypothalamic deafferentiation—i.e., such animals are acyclic, but ovulation will occur following electrical stimulation of the median eminence. These findings imply that the anterior hypothalamus is the sensitive area of the CNS with regard to cyclic gonadotropin release and that in the presence of male hormone in neonatal life such cyclic release is abolished. A nervous system locus for the effects of such androgenization is also suggested by experiments showing that reserpine and chlorpromazine when given together with neonatal androgen to female animals protect against the masculinizing effect of the androgen. There are no studies of alteration in neurotransmitter content in such neonatally androgenized female animals. Alteration in synaptic membrane sensitivity (see Section IIE) may also be a factor in the changes seen in such animals, since it has been demonstrated that the estradiol binding capacity of the anterior and middle hypothalamus, as well as that of the uterus, is depressed below that in control animals.*

C. Adrenal Steroids

Although pituitary–adrenal reactivity to stress is somewhat depressed after the first postnatal day in the rat, normal responsiveness returns by 7 days of age,[151] whereas circadian periodicity of adrenal corticosteroid levels does not appear until somewhere between days 21 and 25 of life.[152] (The time of onset of such periodicity can be advanced if animals are handled daily following birth.[153]) It has been reported[154] that injection of 1 mg of cortisol acetate on the day of birth severely slows the growth of such animals but that as adults, they manifest normal pituitary–adrenal responses to stress. However, the administration of hydrocortisone or dexamethasone on days 2–4 of neonatal life is associated with a virtual disappearance of circadian periodicity of plasma corticosteroid levels when such animals are studied at 30 days of age.[155] As in the case of neonatal gonadal steroid administration, there are no studies of alteration of central neurotransmitter content or turnover in such animals nor are there any studies on the effect of such treatment on CNS corticosteroid uptake and binding.

D. Thyroid Hormone

Neonatal systemic administration of L-thyroxine leads to impairment of thyroid and pituitary growth in the adult animal, reduced protein-bound iodine

---

*Recent findings, however, suggest that fetal and postnatal exposure of the hypothalamohypophyseal unit of primates to androgen does not prevent cyclic gonadotropin secretion.[221]

levels, diminished pituitary TSH content, and a subnormal rise in TSH following propylthiouracil challenge. Similar results were obtained with intrahypothalamic (medial and basal) implantation of systemically ineffective amounts of thyroxine.[156]

V. CLINICAL CORRELATES

The enormous amount of experimental data indicating the major role of the CNS in the control of anterior pituitary function has led to further studies designed to elucidate possible malfunctioning of this system (as well as other aspects of its role in normal physiological processes) in the human. Earlier studies[157] were concerned with the evaluation of basal indices of endocrine function in patients with gross lesions of the CNS, particularly the hypothalamus. With the knowledge that there are various levels of organizational control of a given pituitary hormone (i.e., with regard to basal level, stress-evoked release, feedback regulation, and circadian periodicity), it became evident that early abnormalities in endocrine function in such patients might not be apparent in studies in which only basal hormonal levels were measured. It also became apparent that functional alteration in neurotransmitter content or action, steroid–brain receptor interaction, etc., could result in altered endocrine function in the absence of any grossly discernible CNS lesion. These functional alterations may also be involved in the regulation of normal physiological processes such as puberty.

The development of clinical and physiological correlates is still in its early stages, but an impressive list of possible neuroendocrine disorders may be formulated (see Table II). Before considering these in somewhat greater detail, it would be best to delineate available methods of evaluating endocrine dysfunction in states of altered neural function. Although in many instances it is said that these tests evaluate hypothalamic function, in view of the widespread distribution of CNS loci involved in endocrine regulation it is more proper to consider these tests as evaluating "neuroendocrine" function.

A. Testing for Neuroendocrine (Hypothalamic) Dysfunction

In testing for neuroendocrine dysfunction, it is important to know the locus of action (in the chain, CNS–pituitary–target gland) of the test being employed. Since there is no currently available assay for any of the releasing hormones, one must also first demonstrate integrity of pituitary and target-organ function if the end point is to be a secretory product of either of these organs. It is also conceivable, although at present there is no supporting experimental evidence, that prolonged absence of hypothalamic releasing hormone stimulation can be associated with evidence of pituitary hypofunction on testing, which can only be reversed by prolonged administration of hypothalamic stimulatory substances. This is analogous to the encountering of thyroid or adrenal unresponsiveness to a single dose of tropic hormone in long-standing pituitary insufficiency. Such unresponsiveness can be reversed by more prolonged tropic hormone administration.

1. Testing for End-Organ Deficiency

Testing for end-organ deficiency would include measurement of urinary corticosteroid and 17-ketosteroid levels and their response to ACTH administration; assessment of vaginal smear or semen analysis and plasma testosterone levels, and response of plasma testosterone (in males) to gonadotropin administration; measurement of serum thyroxine levels, radioactive iodine uptake, and response of these parameters to TSH administration. Such tests would indicate whether or not a deficiency in basal adrenal, thyroid, or gonadal function is present, and if so whether it is primary in nature or secondary to lack of tropic hormone stimulation. Once end-organ function is shown to be normal, one can proceed to testing of pituitary function.

2. Testing for Pituitary Tropic Hormone Deficiency

Testing for pituitary tropic hormone deficiency is done both by measuring basal levels of these hormones and by assessing their (or target-organ hormone) responsiveness to HRH administration. HRH, as indicated above, is not available for testing with regard to all pituitary hormones. Therefore, when testing for growth hormone responsiveness (since there is no available HRH), one uses a stimulus which is presumed to act by first stimulating GHRH, in this case insulin hypoglycemia. In the presence of normal pituitary function, basal growth hormone levels would be low (the normal situation) but there would be a sharp rise in such levels following induction of hypoglycemia. An absent growth hormone response would indicate either pituitary or hypothalamic dysfunction. At present, there is no laboratory test which will distinguish between these two in regard to growth hormone release. When testing for TSH responsiveness in secondary (pituitary) hypothyroidism, basal TSH levels would be low and there would be no response to TRH administration. In hypothyroidism associated with hypothalamic disease, although basal TSH levels would be low a normal TSH response to TRH would be seen. Similar considerations apply to testing for LH with LHRH. In the case of ACTH, although there are no available purified preparations of CRF, vasopressin may be considered as a prototype of this substance (see Section IIB) so that its administration would be expected to lead to a rise of plasma ACTH and cortisol levels in normal patients and in those with hypothalamic disease, but not in patients with hypopituitarism.

It should be noted that in certain cases of hypothalamic disease, just as one would expect a decrease in levels of pituitary hormones normally stimulated by HRH, one would expect increases in the levels of pituitary hormones whose secretion is normally inhibited by hypothalamic factors, such as MSH and prolactin. Prolactin levels have been observed to be increased in hypothalamic disease.[158] Adequacy of hypothalamic–pituitary responsiveness can be assessed by administration of phenothiazine[159] (which normally causes an increase in prolactin levels). The absence of such a prolactin response would indicate either pituitary or hypothalamic disease. However, since TRH administration also increases prolactin levels in the normal individual, a normal prolactin response to TRH and an absent response to phenothiazine would suggest the presence of hypothalamic disease.

3. Testing for Releasing Factor Deficiency

Once normal pituitary and end-organ responsiveness is found to be present in cases of CNS disease associated with evidence of endocrine deficiency, although it would be logical to assume that such deficiency is secondary to hypothalamic disease it would be most desirable to actually measure levels of HRH. This is not feasible now: only recently has purified material become available (in the case of TSH and LHRH) to permit linkage of these peptides to protein carriers so that adequate antibodies could be produced to permit radioimmunoassay procedures. Under normal conditions levels of HRH in the peripheral blood (in contrast to those in pituitary portal blood) might be expected to be too low even for radioimmunoassay. Lacking such assays, HRH responsiveness can be tested through the use of stimuli known to affect pituitary function via a CNS intermediate. For example, lack of the normal pituitary–adrenal responsiveness to insulin hypoglycemia or to the injection of bacterial pyrogen in patients manifesting evidence of normal pituitary responsiveness to vasopressin might be taken as an index of CRF deficiency.

4. Altered Periodicity of Levels of Pituitary Tropic Hormones

As noted above and as will be noted subsequently, in many instances of CNS disease (especially hypothalamic and limbic system disease) basal endocrine function may be normal, and normal pituitary responsiveness may also be seen. In these instances, disruption of normal periodicity of pituitary hormone levels may be a clue to CNS pathology. The role of the CNS in the regulation of such periodicity has already been indicated. In view of the presumed greater complexity and diffuseness of the circuitry regulating such periodic release, as opposed to basal release, it would appear that a lesion affecting any part of this circuitry could result in altered periodicity of hormonal levels. This then might be an earlier endocrine manifestation of CNS dysfunction than alteration of basal hormone output or stress-evoked release, which would require dysfunction of a less complex pathway. The validity of this concept has already been demonstrated. In "endocrinologically normal" patients with hypothalamic–limbic system disease, alteration in the periodicity of plasma ACTH and corticosteroid levels is a more frequent finding than absence of the rise in growth hormone levels induced by hypoglycemia,[160] previously considered to be the most sensitive test of hypothalamic dysfunction. There is also suggestive evidence of altered growth hormone periodicity in CNS disease,[138] and as periodicity of other hormones is investigated in CNS disease other abnormalities may be anticipated.

5. Testing of Feedback Responsiveness

As indicated previously, there is evidence of both hypothalamic and pituitary participation in feedback mechanisms, so that these cannot be used for precise localization studies. With regard to adrenal function, administration of a potent glucocorticoid (dexamethasone) is associated with suppression of pituitary and adrenal secretion in the normal subject. No abnormality[161] in such suppressibility has been seen in patients with localized hypothalamic–limbic system disease. Administration of metyrapone (which blocks adrenocortical 11-hydroxylase activity and consequently lowers adrenal cortisone levels) leads to an increase in plasma ACTH levels in normal subjects.[162] Diminished responsiveness is more frequently seen in patients with pituitary rather than hypothalamic disease.[163]

Tests of feedback responsiveness of other pituitary end-organ systems have been less frequently utilized. With regard to gonadal function, clomiphene citrate (an antiestrogenic substance) is believed to compete with hypothalamic receptor sites for estrogen (and perhaps androgen) and thereby exert an antiestrogen or antiandrogen effect.[164] Whether by inhibiting the estrogen blockade of FSH or the androgen blockade of LH, or by having a direct stimulatory effect on gonadotropin releasing hormone, clomiphene administration leads to an increase in both FSH and LH plasma levels. Such increases would be considered evidence of normal hypothalamic LHRH–FSHRH function, although there have not been sufficient clinical studies in patients with presumed hypothalamic hypogonadism to validate this assumption. The effect of cyproterone acetate (an antiandrogen) on LH and FSH release in the human is still unclear, so that its administration cannot be used as a test of hypothalamic function.

Since a large part of thyroxine feedback action is exerted (see Section IIF) at the level of pituitary gland, such testing would not be of great value in uncovering possible hypothalamic dysfunction. There have likewise been no studies of the effect of administration of propylthiouracil (which lowers serum thyroxine and triiodothyronine levels, resulting in increased plasma TSH levels) to patients with hypothalamic disease.

B. Endocrine Disorders in CNS Disease

1. Localized Hypothalamic Disease

In Bauer's initial studies[157] of endocrine abnormalities associated with hypothalamic disease, the most common associated clinical finding was diabetes insipidus. Of endocrinopathies that could be associated with evidence of anterior pituitary dysfunction, the most common were those of gonadal dysfunction, either hypogonadism (usually associated with anterior hypothalamic lesions) or precocious puberty (usually seen with lesions involving the posterior hypothalamus or pineal gland). These findings were based on postmortem studies in cases in which the underlying CNS disease and/or its endocrine manifestations were sufficiently severe to result in death of the patient. The presence of associated normal pituitary histology and function in many instances was not ascertained, which would be necessary before ascribing the endocrine phenomena to hypothalamic dysfunction.

One of the first detailed case reports describing anterior pituitary dysfunction in hypothalamic disease was that of Selenkow et al.[165] Hypopituitarism (documented by the presence of absent urinary gonadotropin titers, low adrenal steroid levels, which were only slightly responsive to ACTH administration, and low normal indices of thyroid function unresponsive to TSH administration) was observed in a patient who at postmortem was found to have sarcoid infiltration of the meninges, hypothalamus, tuber cinereum, and sphenoid sinus. The anterior pituitary gland (save for a single minute granulomatous lesion in the region adjacent to the diaphragma sellae), the thyroid gland, and the adrenal glands were normal histologically. Other reports[166,167] have further documented the association of panhypopituitarism with discrete hypothalamic lesions and histologically normal, although somewhat small, pituitary glands.

Concomitant with these studies were others assessing the incidence of both clinically symptomatic and asymptomatic endocrine abnormalities in patients with hypothalamic disease. An initial review by Kahana et al.[168] was based on findings in small groups of patients having various categories of hypothalamic lesions. In this study, the more frequent association of hypothalamic lesions with hypogonadism and diabetes insipidus was again confirmed. Only suggestive evidence of thyroid and/or adrenal dysfunction was obtained in fewer than half of the patients. Another study[169] reported normal circadian periodicity of adrenal steroid levels in eight of nine patients with hypothalamic disease (however, sampling was limited to a 7-hr span of the 24-hr cycle). In this report, abnormal metyrapone responsiveness (three of nine patients tested) and abnormal dexamethasone suppressibility (two of three patients tested) were noted; however, two of the patients with both abnormalities were receiving diphenylhydantoin, which is known to alter responsiveness to both of these parameters.[170,171] In view of the common use of many medications which act on the CNS and which thereby might modify endocrine function, endocrine testing of patients with hypothalamic disease should be performed in the absence of any effective levels of these medications.

It was next suggested[172] that alteration in growth hormone response to insulin-induced hypoglycemia might be the earliest endocrine parameter to be affected in hypothalamic disease, similar to the finding that lack of growth hormone secretion is the earliest indicator of pituitary parenchymal disease. However, in studying 14 patients with localized hypothalamic disease without accompanying clinical or laboratory evidence of hypopituitarism, Krieger et al.[160] observed abnormal circadian patterns of plasma corticosteroid levels in 50% of these patients, whereas *normal* growth hormone responsiveness to insulin-induced hypoglycemia was seen in 77%. This indicates that abnormalities in the circadian periodicity of plasma corticosteroid levels (and perhaps in circadian periodicity of other hormonal variables) might be the earliest endocrine abnormality to be detected in patients with delimited hypothalamic disease. Similar conclusions were cited[173] in a study of nine patients with St. Louis encephalities, presenting mainly with hypothalamic dysfunction.

The above studies are all concerned with cases of obvious anatomically localized hypothalamic disease. In many instances, hypothalamic lesions may be too small to be detected neuroradiographically. In such instances, endocrinopathies may be diagnosed on testing as being secondary to hypothalamic disease, even though no organic lesion is identified. Although in some instances such endocrinopathy may be the result of a functional disturbance of hypothalamic regulation (see below), the possibility of a small lesion should be kept in mind. Sherwin et al.[174] reported 26 instances of hamartomatous nodules in the region of the mammillary body and tuber cinereum in 121 consecutive brain autopsies; in these 26 cases there was an increased frequency of thyroid, pituitary, and adrenal adenomata and adrenal hyperplasia. Although the data are not sufficient for valid statistical analysis, an interrelation of these pathological findings remains an intriguing possibility.

Diffuse cerebrovascular disease, which may or may not be apparent on angiographic studies, may also be associated with evidence of endocrine dysfunction. It has been speculated that this may be because areas involved in neuroendocrine regulation are affected.[175]

2. Diseases Which May Be Secondary to Neuroendocrine Dysfunction

Table 2 lists various clinical entities in the category of diseases possibly secondary to neuroendocrine dysfunction and is not meant to be all inclusive. In the previous section, emphasis was placed on endocrine abnormalities seen in the presence of demonstrable CNS (especially hypothalamic) lesions. The present section deals mainly with entities which, on theoretical and physiological grounds, might be secondary to either functional or relatively minute anatomical lesions of areas involved in neuroendocrine regulation. Therefore, they might be difficult to demonstrate neurologiocally or radiologically. In these instances, evidence of either hypo-, hyper-, or otherwise deranged regulation of pituitary hormone secretion might be expected. Not all of the clinical conditions listed in Table 2 will be discussed in detail. In those instances, the references cited may be consulted.

a. Diencephalic Syndrome of Infancy. Infantile diencephalic syndrome[176] combines the features of emaciation with initial growth acceleration, large hands and feet, and normal or advanced bone age, in association with either hypothalamic tumor or trauma. In one reported case, there is reference to a single "normal" growth hormone determination. Another patient manifested elevated growth hormone levels, with a normal rise in these levels following induction of insulin hypoglycemia but a paradoxical rise after glucose administration, which normally suppresses growth hormone levels.[177]

b. Cerebral Giantism. Cerebral giantism[178] is a disease of children characterized by initially rapid growth (although final height is usually within the normal range), acromegalic features, accelerated bone age when initially seen,

TABLE 2. SOME DISEASES OF POSSIBLE NEUROENDOCRINE ETIOLOGY

Diencephalic syndrome of infancy
Cerebral giantism
Acromegaly
Idiopathic growth retardation
Emotional deprivation and growth failure syndrome
Hypogonadotropic hypogonadism (with and without anosmia)
Precocious puberty
Functional (hypothalamic) amenorrhea
Stein–Leventhal syndrome
Nonpuerperal galactorrhea
Anorexia nervosa
Idiopathic hypopituitarism
Hypothalamic hypothyroidism
Hyperthyroidism
Cushing's syndrome
Congenital lipodystrophy

normal sella turcica, and a nonprogessive cerebral disorder with mental retardation. Pneumoencephalography has demonstrated ventricular enlargement, with no evidence of a focal CNS lesion. Tests of basal endocrine function have been normal. Growth hormone studies have not been performed, however, save for one instance[179] in which increased activity of sulfation factor (believed to be the metabolic intermediary in growth hormone action) was observed. It should be noted that cerebral giantism is probably different from the giantism reported in cases of neurofibromatosis (with associated hypothalamic involvement),[180] where paradoxical hypersecretion of growth hormone has also been reported.[181]

c. Acromegaly. In roughly one-third to one-fourth of patients with acromegaly there is no evidence of sellar enlargement, although this does not preclude the possibility that small pituitary adenomata (producing growth hormone) may be present. Another possibility, still to be proven, is that these instances may represent continued, inappropriate stimulation by growth hormone-releasing hormone or removal of inhibition by a postulated growth hormone-inhibiting factor. The validity of this concept must await characterization of such hormones and the development of assays for their measurement.

d. Idiopathic Growth Retardation (Isolated Growth Hormone Deficiency). There is a strong possibility that cases of isolated growth hormone deficiency which occur in the absence of any visible abnormality of the sella are secondary to absent or altered growth hormone-releasing factor (GRF). Proof of this will not be possible until GRF has been characterized and becomes available for clinical testing. To date, the peptide which has been characterized as GRF is ineffective in elevating plasma growth hormone levels in normal human subjects.

e. Emotional Deprivation and Growth Failure. The syndrome of emotional deprivation and growth failure is characterized by short stature and clinical findings suggestive of hypopituitarism (especially absent or diminished growth hormone levels, both basal and in response to insulin hypoglycemia, and diminished basal corticosteroid excretion and responsiveness to metyrapone) with X-ray evidence of normal sellar size in children with emotional disturbances and a history of abnormal home environments.[182] Such children grow at remarkable rates when removed from their home environments, concomitant with evidence of return of normal growth hormone release. It can only be assumed that, in these instances, emotional factors interfere with release of both CRF and GRF, perhaps by altering CNS neurotransmitter content or turnover.

f. Hypogonadotropic Hypogonadism, With and Without Anosmia. Hypogonadotropic hypogonadism has been reported in both male and female subjects, may be familial in nature, and may be associated with other defects in addition to anosmia, i.e., color blindness, synkinesias, and mental defect.[183]

In some cases,[184] hypothalamic hypoplasia as well as hypoplasia of the region of the anterior commissure and the olfactory tubercles has been reported. As noted previously, the anterior hypothalamus is believed to play a major role in the regulation of gonadotropic function, and the existence of neural connections between the olfactory lobes and the hypothalamus has also been demonstrated. The importance of olfactory stimuli in lower mammals in the regulation of sexual behavior in both sexes and gonadotropin cyclicity in females is well recognized,[185] and there is some evidence that olfactory stimuli may play a role in primate gonadotropin regulation.[186]

There are other diseases with evidence of CNS dysfunction associated with hypogonadism, in which the nature of the CNS pathology and etiology of the hypogonadism still remain to be elucidated. These include the Laurence–Moon–Biedl syndrome, which is characterized by retinitis pigmentosum, diabetes insipidus, polydactyly, mental retardation, obesity, and hypogonadism; and the Prader–Willi or HHHO syndrome, consisting of hypotonia, hypomentia, hypogonadism, and obesity.

g. *Precocious Puberty.* Although the mechanism(s) for the initiation of normal puberty is poorly understood, the timing of puberty is under CNS control. This has been shown in animal experiments in which immature ovaries and pituitaries function in a normal adult fashion when transplanted into adult recipients and in studies in which puberty is advanced in female animals following the production of hypothalamic, amygdalar, or hippocampal lesions. It has been hypothesized, but by no means proven, that puberty involves a change in the "set-point" at which the hypothalamus responds to circulating gonadal steroid hormone levels. By increasing this set-point at puberty, negative feedback would be decreased, making for increased levels of releasing hormones, gonadotropic hormones, and consequently gonadal hormones, with their ensuing physical and metabolic effects. There is no suggestion, however, of how this "reset" is accomplished.

Although most cases of true isosexual precocious puberty (where ovulation occurs in the female child and spermatozoa production in the male) are idiopathic (80–90% in female children, 70% in male children), the most frequent organic lesion associated with precocious puberty is a tumor in or near the pineal gland (in male children) or in the hypothalamus (in female children). Approximately one-third of the cases of precocity in males are found to be secondary to an organic lesion, whereas this is true in only 5% of female children with precocity. Of pineal tumors associated with precocity, 95% are in male children, 5% in female children. When hypothalamic tumors are seen in association with precocious puberty, 75% are in male children, 25% in female children. It should be stressed that not all pineal tumors in preadolescent children are associated with precocity, this occurring in only about 25% of cases, more often with tumors of nonparenchymal than parenchymal origin.[187] This might imply that these nonparenchymal tumors interfere, by overgrowth, with the production of pineal antigonadotropic substance. It has been suggested that melatonin (which is found uniquely in the pineal) may have such an effect[188] through interfering with either gonadotropin release (perhaps acting as an antimetabolite to serotonin) or gonadotropin action. More recently, antigonadotropic activity has been ascribed to 8-arginine-vasotocin, which has been isolated from beef pineal glands.[189] The presence of a gonadotropin-inhibiting substance in urine of normal children and adults has also been reported.[190] This does not appear to be melatonin, and such an inhibitor has been found to be absent only in cases of precocious puberty and in patients with the Stein–Leventhal syndrome (see below). These findings may therefore indicate that the etiology of the precocious puberty seen in pineal or hypothalamic disease may be related to the absence of a gonadotropin inhibitory substance normally produced or transmitted by these areas. This would allow enhanced gonadotropin release in advance of the time it would normally occur, with consequent pituitary and gonadal stimulation. There is also evidence that the incidence of precocious puberty seen with pineal tumors correlates with the extent of hypothalamic involvement of compression by the tumor.[191]

There is a form of precocious puberty, without evidence of a hypothalamic or pineal lesion, occurring more often in females in which there is coexistence of unilateral ostetis fibrosa and pigmented skin lesions on the same side as the bone lesions (polyostotic fibrous dysplasia or Albright's syndrome). It was originally suggested that the endocrine changes were secondary to pressure on the brain exerted by bony lesions at the base of the skull, but precocity has been seen in the absence of skull involvement. Discrete CNS lesions have not been associated thus far, although Albright *et al.*[192] commented on one case in which diminution in the size of one mammillary body and an "extra nucleus" in adjacent tissue were noted. Some of the patients with Albright's syndrome show evidence of hypothyroidism. Precocious puberty has also been reported in association with juvenile primary hypothyroidism.[193] Adequate thyroid replacement therapy results in restoration of prepubertal gonadotropic function. In these cases, it is not clear whether the hypothyroid state is causing specific changes in hypothalamic neuronal function or the increase in pituitary gonadotropin activity is a "cross-specific" type of response (similar to that of TSH) to the low levels of thyroid hormone.

Finally, many children classified as having idiopathic sexual precocity manifest diffuse, nonspecific EEG abnormalities.[194] If these studies are confirmed, they would indicate that some "functional" CNS disorder may be present in these instances now classified as idiopathic.

h. Functional (Hypothalamic) Amenorrhea. Secondary amenorrhea, usually of a temporary nature, is seen in women with otherwise normal pituitary function and has been ascribed to emotional disturbances. These women exhibit low urinary estrogen levels, normal urinary gonadotropin levels as measured by bioassay, low serum LH levels, and serum and FSH levels that are low in view of the low estradiol levels present.[39] The typically pulsatile fluctuation of serum LH levels seen in normal women is absent in these patients. Following administration of synthetic LHRH, a rise in both serum LH and FSH levels is observed, qualitatively similar but quantitatively greater than that in normal subjects. These data suggest a functional derangement in hypothalamic releasing mechanisms, perhaps secondary to a neurotransmitter defect, as postulated above for the growth hormone impairment seen in emotional deprivation.

i. Stein–Leventhal Syndrome (Polycystic Ovary Syndrome). Stein–Leventhal syndrome is characterized by sterility, amenorrhea or oligomenorrhea, and usually hirsutism, associated with the presence of polycystic ovaries. Bioassayable urinary gonadotropin levels are usually within the normal range. Specific studies of LH levels have shown that they tend to be consistently high, below that reached at the normal ovulatory peak (which is lacking in these subjects) but considerably higher than is normal during other phases of the menstrual cycle. The situation in such patients is somewhat analogous to that in the neonatally androgenized rat, which also develops polycystic ovaries. However, if these polycystic rat ovaries are transplanted into normal hosts, they revert to normal, indicating that the alterations in the neonatally androgenized rat are secondary to extraovarian mechanisms. These alterations probably occur within the CNS area responsible for the maintenance of LH cyclicity. In patients with the Stein–Leventhal syndrome, there may be a functional disturbance related to this area. There has been a report of association of a history of childhood CNS injury or "encephalitis" and the polycystic ovary syndrome in five instances.[195]

j. Nonpuerperal Galactorrhea. It has already been mentioned that prolactin secretion is normally kept in check by a hypothalamic prolactin-inhibiting factor (PIF). Lesions which destroy the site of production of this factor, or prevent its transmission to the pituitary, would be expected to be associated with elevated serum prolactin levels and possible ensuing galactorrhea. Such findings have been noted in association with hypothalamic disease, pituitary tumor, pituitary stalk section, and following the administration (to normal subjects) of a number of tranquilizers and related drugs, all of which seem to manifest in common an ability to lower CNS norepinephrine levels. This is presumably the mechanism by which these drugs lower hypothalamic output of PIF, since, as indicated previously, such release is believed to be mediated by adrenergic mechanisms. It may well be that in those cases of galactorrhea unassociated with a CNS lesion or drug administration a functional alteration in neurotransmitter content is also present. L-dopa, which increases hypothalamic concentrations of catecholamines, has been shown to stimulate hypothalamic secretion of PIF[196] and is effective in temporarily suppressing nonpuerperal galactorrhea and in lowering the elevated serum prolactin levels seen in these patients.[197]*

k. Anorexia Nervosa. Anorexia nervosa is characterized by emotional disturbance, severe weight loss, and amenorrhea, although without the evidence of breast atrophy and loss of axillary and pubic hair that is seen in other states of hypogonadism or malnutrition. Reports of endocrine assays in such patients have been variable. Urinary gonadotropin levels have been reported as being normal[198] or low or absent.[199] In the last instance, no release of gonadotropin following clomiphene administration was found, but pregnancy ensued in three subjects following combined treatment with human menopausal gonadotropin and human chorionic gonadotropin, indicating normal gonadal function in such patients. Plasma growth hormone levels have been variable, some patients showing normal[177] and others elevated levels[198]; similarly, unresponsiveness[198] and normal responsiveness[200] of plasma growth hormone levels to insulin hypoglycemia have both been reported. Other endocrine functions (thyroid and adrenal) may have low basal levels, but responsiveness of the hypothalamic–pituitary–end–organ axis is apparently normal. Although some of the noted hormonal abnormalities may be secondary to the diminished caloric intake, it is not clear that this explains all of the pathophysiological findings in this condition.

l. Idiopathic Hypopituitarism. The same considerations noted with regard to idiopathic growth hormone deficiency, gonadotropin deficiency, and thyroid deficiency are applicable to idiopathic hypopituitarism. That is, in the presence of a normal-appearing pituitary gland, one might encounter isolated or multiple deficiencies of pituitary tropic hormones secondary to abnormalities in the synthesis and/or release of one or more hypothalamic releasing hormones.

m. Hypothalamic Hypothyroidism. Isolated pituitary thyrotropin deficiency is a rare cause of hypothyroidism. Although it has been assumed that this condition could result from either pituitary or hypothalamic disease, it has only been recently, with the clinical availability of thyrotropin releasing hormone, that differentiation between these two etiologies has become possible. The cases of hypothalamic hypothyroidism that have been described[25,201] have demonstrated virtually undetectable serum thyrotropin levels, with normal rises following TRH administration. These patients also had evidence of deficiencies of other pituitary tropic hormones and, in some instances, of vasopressin as well. Some had histories of intracranial disease or trauma. Approximately 30 cases of isolated thyrotropin deficiency, including many patients with organic

---

*2-Br-α-Ergocryptine has recently been demonstrated to have greater effectiveness in suppressing prolactin secretion.[222]

brain disease, were reported before TRH was available. Some of these may also have been cases of hypothalamic hypothyroidism.

n. Hyperthyroidism. There has been a long-standing clinical belief that emotional trauma is a precipitating cause of Grave's disease. Experimental evidence, however, is conflicting as to whether or not stresses of different kinds increase, decrease, or produce no change in indices of thyroid function in rabbits, ewes, rats, or humans. Certainly in most cases of thyrotoxicosis in the human TSH levels are low or undetectable. The few reported patients having hyperthyroidism associated with chronic TSH overproduction[202,203] had enlarged pituitary fossae, implicating a pituitary etiology of the hyperthyroidism, although the possibility of pituitary adenoma formation as a result of chronic hyperstimulation by TRH cannot be excluded. There has been a recent case report of hyperthyroidism with high TSH levels in the absence of a pituitary tumor[204] but with no convincing proof that the elevated TSH levels were secondary to increased TRH secretion.

o. Cushing's Syndrome. The etiology of Cushing's syndrome associated with bilateral adrenocortical hyperplasia is still open to question. With the understanding of the role of the CNS in the regulation of various aspects of hypothalamic–pituitary–adrenal function, it would appear that the major endocrine abnormalities seen in this syndrome (loss of circadian periodicity of adrenal corticosteroid levels, lack of responsiveness of plasma corticosteroid levels to exogenous stress, and lack of normal suppressibility of corticosteroid levels following dexamethasone) could be explained by functional alterations in hypothalamic control mechanisms regulating periodicity, stress responsiveness, and the "set-point" at which steroid suppression occurs.[205] The occurrence of abnormalities in growth hormone responsiveness in such patients, whether with active disease or in remission,[206] lends support to the thesis of such a central etiology. There have been isolated case reports[207,208] of the association of Cushing's syndrome with intracranial disease, and we have encountered (unpublished observations) one such patient with very high plasma ACTH levels. A syndrome of periodic hypothalamic discharge, accompanied by evidence of periodic adrenocortical hypersecretion and unsuppressibility, together with elevated plasma ACTH levels has been reported.[209] The only positive neurological finding was that of slight ventricular dilatation and cortical atrophy.

p. Congenital Lipodystrophy. Congenital lipodystrophy is characterized by lack of subcutaneous fat, early accelerated skeletal development, acanthosis nigricans, hyperlipemia with hypertrygliceridemia, increased plasma insulin levels with evidence of insulin resistance and adolescent onset of diabetes, and variable changes in adrenal activity. It has been suggested[210] that the syndrome is secondary to a disturbance of the hypothalamic–hypophyseal system. In all of five reported cases, ventricular dilatation was identified by pneumoencephalography. One postmortem examination of a patient with acquired lipodystrophy (which is thought to be a variant of this condition) demonstrated lesions in the floor of the third ventricle and hyperplasia of the adrenal cortex and of the basophilic cells in the anterior lobe of the pituitary gland.

C. Clinical Use of Releasing Hormones

The synthesis thus far of two releasing hormones (TRH and LHRH) has opened many diagnostic and therapeutic possibilities. The use of TRH and LHRH in distinguishing hypothalamic from pituitary etiologies of hypothyroidism and hypogonadism, respectively, has already been mentioned. As other releasing and inhibiting hormones (factors) are synthesized, similar tests will become available. Should GHRH be synthesized, its use in the cases of growth failure secondary to its deficiency should make effective specific therapy available, without having to rely on the restricted supplies of human growth hormone now available (currently the only therapy for this condition). This will decrease the risk of antibody formation, which may obviate a therapeutic response in some instances of treatment with human growth hormone. Similar therapeutic considerations utilizing LHRH may pertain to some anovulatory states. It is also to be expected that as the structure of these releasing hormones becomes known analogues will be devised which will both be void of releasing hormone activity and block the biological action of the releasing hormone. Such effects have been reported *in vitro*[211] for LRF analogues. The implications of the use of LRF analogues for contraception and of other analogues in treatment of states of possible hypersecretion of releasing hormones (i.e., acromegaly and Cushing's syndrome) are most exciting.

REFERENCES

1. F. H. A. Marshall, Sexual periodicity and the causes which determine it, *Phil. Trans.* **B226**:423–456 (1936).
2. G. W. Harris, The induction of ovulation in the rabbit by electrical stimulation of the hypothalamo-hypophyseal mechanism, *Proc. Roy. Soc. Lond. Ser. B* **122**:374–394 (1937).
3. J. DeGroot and G. W. Harris, Hypothalamic control of the anterior pituitary gland and blood lymphocytes, *J. Physiol.* **111**:335–346 (1950).
4. R. Burgus and R. Guillemin, Hypothalamic releasing factors, *Ann. Rev. Biochem.* **39**:499–526 (1970).
5. J. Meites, "Hypothalamic Hormones of the Hypophysis," Williams Wilkins, Baltimore (1970).
6. B. Donovan, "Mammalian Neuroendocrinology," McGraw-Hill, New York, (1970).

7. S. M. McCann and J. C Porter, Hypothalamic–pituitary stimulating and inhibiting hormones, *Physiol. Rev.* **49**:240–284 (1969).

8. L. Martini and W. F. Ganong, "Neuroendocrinology," Vols. I and II, Academic Press, New York and London (1966).

9. G. W. Harris and B. T. Donovan, "The Pituitary Gland," University of California Press, Berkeley and Los Angeles (1966).

10. S. Reichlin, J. B. Martin, M. Mitnick, R. L. Boshans, V. Grimm, J. Bollinger, J. Gordon, and J. Malacars, The hypothalamus in pituitary–thyroid regulation, *Rec. Progr. Hormone Res.* **28**:229–286 (1972).

11. B. Halasz, L. Pupp, and S. Uhlarin, Hypophysiotropic area in the hypothalamus, *J. Endocrinol.* **25**:147–15(1962).

12. L. Krulich, P. Illner, C. P. Fawcett, M. Quijada, and S. M. McCann, *in* "Growth and Growth Hormone" (I. A. Pecile and E. Muller, eds.) pp. 306–316, Excerpta Medica, Amsterdam (1972).

13. R. Guillemin, E. Sakiz, and D. N. Ward, Further purification of TSH releasing factor (TRF) from sheep hypothalamic tissues, with observations on the amino acid composition, *Proc. Soc. Exptl. Biol. Med.* **118**:1132–1137 (1965).

14. K. M. Knigge and D. E. Scott, Structure and function of the median eminence, *Am. J. Anat.* **129**:223–244 (1970).

15. H. Kobayashi and T. Matsui, *in* "Frontiers in Neuroendocrinology" (W. F. Ganong and L. Martini, eds.) pp. 3–46, Oxford University Press, New York (1969).

16. B. Folck and C. Owman, A detailed methodological description of the fluorescence method for the cellular demonstration of biogenic amines, *Acta Univ. Lond. Sect. II* **7**:1–23 (1965).

17. R. S. Piezzi, F. Larin, and R. J. Wurtman, Serotonin, 5-hydroxyindoleacetic acid (5HIAA) and monoamine oxidase in the bovine median eminence and pituitary gland, *Endocrinology* **86**:1460–1462 (1970).

18. G. B. Wislocki and L. S. King, The permeability of the hypophysis and hypothalamus to vital dyes with a study of the hypophyseal vascular supply, *Am. J. Anat.* **58**:421–472 (1936).

19. J. C. Porter, I. A. Kamberi, and Y. A. Grazia, *in* "Frontiers in Neuroendocrinology" (W. F. Ganong and L. Martini, eds.) pp. 145–177, Oxford University Press, New York (1971).

20. A. V. Schally, T. W. Redding, J. F. Barrett, and C. Y. Bowers, Purification of porcine thyrotropin releasing factor (TRF), *Fed. Proc.* **25**:348 (1966).

21. R. Burgus, T. F. Dunn, D. Desiderio, and R. Guillemin, Derives polyptididiques de synthese doues d'activite hypophysiotrope TRF nouvelles observations, *Compt. Rend. Acad. Sci.* **269**:1870–1873 (1969).

22. K. Folkers, F. Enzmann, J. Boler, C. Y. Bowers, and A. V. Schally, Discovery of modification of the synthetic tripeptide-sequence of the thyrotropin releasing hormone having activity, *Biochem. Biophys. Res. Commun.* **37**:123–126 (1969).

23. R. Burgus, T. F. Dunn, D. Desiderio, D. N. Ward, W. Vale, R. Guillemin, A. M. Felix, D. Gillessen, and R. C. Studes, Biological activity of synthetic polypeptide derivatives related to the structure of hypothalamic TRF, *Endocrinology* **86**:573–589 (1970).

24. J. Boler, F. Enzmann, K. Folkers, C. Y. Bowers, and A. V. Schally, The identity of chemical and hormonal properties of the thyrotropin releasing hormone and pyroglutamyl-histidyl-proline-amide. *Biochem. Biophys. Res. Commun.* **37**:705–710 (1969).

25. J. A. Pittman, Jr., E. D. Haigler, Jr., J. M. Hershman, and C. S. Pittman, Hypothalamic hypothyroidism, *New Engl. J. Med.* **284**:844–845 (1971).

26. N. Fleischer, M. Lorente, J. Kirkland, R. Kirkland, G. Clayton, and M. Calderon, Increased secretion of prolactin after administration of synthetic thyrotropin releasing hormone (TRH) in man, *J. Clin. Endocrinol.* **34**:617–624 (1972).

27. W. Vale, R. Burgus, and R. Guillemin, Competition between thyroxine and TRF at the pituitary level in release of TSH, *Proc. Soc. Exptl. Biol. Med.* **125**:210–213 (1967).

28. C. Y. Bowers, A. V. Schally, G. A. Reynolds, and W. D. Hawley, Interactions of L-thyroxine or L-triiodothyronine and thyrotropin-releasing factor on the release and synthesis of thyrotropin from the anterior pituitary gland of mice, *Endocrinology* **81**:741–747 (1967).

29. C. Y. Bowers, K. L. Lee, and A. V. Schally, A study on the interaction of the thyrotropin-releasing factor and L-triiodothyronin: Effects of puromycin and cyclohexamide, *Endocrinology* **82**:75–82 (1968).

30. W. Vale, R. Burgus, and R. Guillemin, On the mechanism of action of TRF: Effects of cycloheximide and actinomycin on the release of TSH stimulated *in vitro* by TRF and its inhibition by thyroxine, *Neuroendocrinology* **3**:34–46 (1968).

31. J. P. Wilber and R D. Utiger, The effect of glucocorticoids on thyrotropin secretion, *J. Clin. Invest.* **48**:2096–2103 (1969).

32. L. S. Jacobs, P. J. Snyder, J. F. Wilber, R. D. Utiger, and W. H. Daughaday, Increased serum prolactin after administration of synthetic thyrotropin releasing hormone (TRH) in man, *J. Clin. Endocrinol.* **33**:996–998 (1971).

33. C. Y. Bowers, H. G. Friesen, P. Hwang, H. J. Guyda, and K. Folkers, Prolactin and thyrotropin release in man by synthetic pyroglytamyl-histidyl-prolinamide, *Biochem. Biophys. Res. Commun.* **45**:1033–1041 (1971).

34. A. Arimura and A. V. Schally, Progesterone suppression of LH releasing hormone-induced stimulation of LH release in rats, *Endocrinology* **87**:653–657 (1970).

35. A. Arimura and A. V. Schally, Augmentation of pituitary responsiveness to LH-releasing hormone (LH-RH) by estrogen, *Proc. Soc. Exptl. Biol. Med.* **136**:290–293 (1971).

36. V. Hillard, A. V. Schally, and C. H. Sawyer, Progesterone blockade of the ovulatory response to intrapituitary infusion of LH-RH in rats, *Endocrinology* **88**:730–736 (1971).

37. A. V. Schally, A. Arimura, Y. Baba, R. M. G. Nair, H. Matsuo, T. W. Redding, and L. Debeljuk, Isolation and properties of the FSH and LH-releasing hormone, *Biochem. Biophys. Res. Commun.* **43**:393–399 (1971).

38. H. Matsuo, Y. Baba, R. M. G. Nair, A. Arimura, and A. V. Schally, Structure of the porcine LH- and FSH-releasing hormone in the proposed amino acid sequence, *Biochem. Biophys. Res. Commun.* **43**:1334–1339 (1971).

39. S. S. C. Yen, R. Rebar, G. Vandenberg, and H. Judd, Hypothalamic amenorrhea and hypogonadotropism: Responses to synthetic LRH, *New Engl. J. Med.* (in press).

40. R. M. G. Nair, A. J. Kastin, and A. V. Schally, Isolation and structure of hypothalamic MSH release-inhibiting hormone, *Biochem. Biophys. Res. Commun.* **43**:1376–1381 (1971).

41. M. E. Celis, S. Taleisnik, I. L. Schwartz, and R. Walter, Proposed structure of melanocyte stimulating hormone release inhibitory factor, *Biophys. Soc. Abst.* **11**:98a (1971).

42. A. P. Dhariwal, C. E. Grosvenor, J. Antunes-Rodrigues, and S. M. McCann, Studies on the purification of ovine prolactin-inhibing factor, *Endocrinology* **82**:1236–1241 (1968).

43. A. V. Schally, Y. Baba, and R. M. G. Nair, The amino acid sequence of a peptide with growth hormone–releasing activity isolated from porcine hypothalamus, *J. Biol. Chem.* **246**:6647–6650 (1971).

44. A. J. Kastin, A. V. Schally, C. Gual, S. Glick, and A. Arimura, Clinical evaluation in man of a substance with growth hormone releasing activity in rats, *J. Clin. Endocrinol.* **35**:326–329 (1972).

45. L. A. Frohman, J. W. Maran, and A. P. S. Dhariwal, Plasma growth hormone responses to intrapituitary injections of growth hormone releasing factor (GRF) in the rat, *Endocrinology* **88**:1483–1488 (1971).

46. D. F. Veber, R. D. Bennett, J. D. Milkowski, G. Gal, R. G. Denkewalter, and R. Hirschmann, Synthesis of a proposed growth hormone releasing factor, *Biochem. Biophys. Res. Commun.* **45**:235–239 (1971).

47. A. P. S. Dhariwal, L. Krulich, and S. M. McCann, Purification of a growth hormone-inhibitory factor (GIF) from sheep hypothalamus, *Neuroendocrinology* **4**:282–288 (1969).

48. A. V. Schally, S. Sawano, A. Arimua, J. F. Barrett, I. Wakabayashi, and C. Y. Bowers, Isolation of growth hormone–releasing hormone (GRH) from porcine hypothalami, *Endocrinology* **84**:1493–1506 (1969).
49. R. Guillemin, Hypothalamic factors releasing pituitary hormones, *Rec. Progr. Hormone Res.* **20**:89–121 (1964).
50. E. Scharrer and B. Scharrer, "Neuroendocrinology," Columbia University Press, New York (1963).
51. I. I. Geschwind, *in* "Hypophysiotropic Hormones of the Hypothalamus" (J. Meites, ed.) pp. 298–319, Williams and Wilkins, Baltimore (1970).
52. B. S. McEwen, J. M. Weiss, and L. Schwartz, Uptake of corticosterone by rat brain and its concentration by certain limbic structures, *Brain Res.* **6**:227–241 (1969).
53. J. L. Gerlach and B. S. McEwen, Rat brain binds adrenal steroid hormone—Radio-autography of hippocampus with corticosterone, *Science* **175**:1133–1136 (1972).
54. W. E. Stumpf, Estradiol-concentrating neurons: Topography in the hypothalamus by dry-mount autoradiography, *Science* **159**:1001–1003 (1968).
55. H. P. Krieger and D. T. Krieger, Chemical stimulation of the brain: Effect on adrenal corticoid release, *Am. J. Physiol.* **218**:1632–1641 (1970).
56. N. E. Miller, Chemical coding of behavior in the brain, *Science* **148**:328–338 (1965).
57. W. G. Blackard and S. A. Heidingsfelder, Adrenergic receptor control mechanism for growth hormone secretion, *J. Clin. Invest.* **47**:1407–1414 (1968).
58. C. Lucke and S. M. Glick, Experimental modification of the sleep induced peak of growth hormone secretion, *J. Clin. Endocrinol.* **32**:729–736 (1971).
59. Y. Takahashi, D. M. Kipnis, and W. H. Daughaday, Growth hormone secretion during sleep, *J. Clin. Invest.* **47**:2079–2090 (1968).
60. A. E. Boyd, III, H. E. Levobitz, and J. B. Pfeiffer, Stimulation of human growth hormone secretion by L-DOPA, *New Engl. J. Med.* **283**:1425–1429 (1970).
61. P. C. Kansal, J. Buse, O. R. Talbert, and M. G. Buse, The effect of L-DOPA on plasma growth hormone, insulin and thyroxine, *J. Clin. Endocrinol.* **34**:99–105 (1972).
62. P. T. K. Toivolia and C. C. Gale, Stimulation of growth hormone release by microinjection of norepinephrine into hypothalamus of baboons, *Endocrinology* **90**:895–902 (1972).
63. W. G. Blackard and C. C. Waddell, Cholinergic blockade and growth hormone responsiveness to insulin hypoglycemia, *Proc. Soc. Exptl. Biol. Med.* **131**:192–196 (1969).
64. H. P. G. Schneider and S. M. McCann, Mono- and indolamines and control of LH secretion, *Endocrinology* **86**:1127–1133 (1970).
65. I. A. Kamberi, R. S. Mical, and J. C. Porter, Effect of anterior pituitary perfusion and intraventricular injection of catecholamines and indoleamines on LH release, *Endocrinology* **87**:1–12 (1970).
66. I. A. Kamberi, R. S. Mical, and J. C. Porter, Effect of anterior pituitary perfusion and intraventricular injection of catecholamines on prolactin release, *Endocrinology* **88**:1012–1020 (1971).
67. C. Kordon and J. Glowinski, Selective inhibition of superovulation by blockade of dopamine synthesis during the "critical period" in the immature rat, *Endocrinology* **85**:924–931 (1969).
68. K. H. Lu, Y. Amenomori, C. L. Chen, and J. Meites, Effects of central acting drugs on serum and pituitary prolactin levels in rats, *Endocrinology* **87**:667–672 (1970).
69. R. W. Turkington and J. H. McIndoe, Stimulation of human prolactin secretion by intravenous infusion of L-tryptophan, *Am. Soc. Clin. Invest.*, p. 98a (1972) (abst.).
70. J. W. Everett, C. H. Sawyer, and J. E. Markee, A neurogenic timing factor in control of the ovulatory discharge of luteinizing hormone in the cyclic rat, *Endocrinology* **44**:234–250 (1949).
71. D. T. Krieger, A. I. Silverberg, F. Rizzo, and H. P. Krieger, Abolition of circadian periodicity of plasma 17-OHCS levels in the cat, *Am. J. Physiol.* **215**:959–967 (1968).
72. D. T. Krieger and F. Rizzo, Circadian periodicity of plasma 17-OHCS: Mediation by serotonin dependent pathways, *Am. J. Physiol.* **217**:1703–1707 (1969).

73. G. R. VanLoon, W. Nicholson, and R. Brown, Drug-induced hypersecretion of ACTH in rats and its treatment with L-DOPA, *in* "Fifty-third Meeting of the Endocrine Society," San Francisco, p. A-129 (1971). (abst.).

74. E. V. Naumenko, Hypothalamic chemoreactive structures and the regulation of pituitary–adrenal function, *Brain Res.* **11**:1–10 (1968).

75. E. Endroczi, G. Schreiberg, and K. Lissak, The role of central nervous activating and inhibitory structures in the control of pituitary–adrenocortical function. Effects of intracerebral cholinergic and adrenergic stimulation, *Acta. Physiol. Acad. Sci. Hung.* **24**:211–221 (1963).

76. G. R. VanLoon, L. Hilger, A. B. King, A. T. Boryczka, and W. F. Ganong, Inhibitory effect of L-dihydroxyphenylalanine on the adrenal venous 17-hydroxycorticosteroid response to surgical stress in dogs, *Endocrinology* **88**:1404–1414 (1971).

77. G. R. VanLoon, U. Scapagnini, G. P. Moberg, and W. F. Ganong, Evidence for central adrenergic neural inhibition of ACTH secretion in the rat, *Endocrinology* **89**:1464–1469 (1971).

78. G. A. Hedge and P. G. Smelik, Corticotropin release: Inhibition by intrahypothalamic implantation of atropine, *Science* **159**:891–892 (1968).

79. K. Fuxe, H. Corrodi, T. Hokfelt, and G. Jonsson, Central monoamine neurons and pituitary–adrenal activity, *Progr. Brain Res.* **32**:42–56 (1970).

80. P. G. Smelik, ACTH secretion after depletion of hypothalamic monoamines by reserpine implants, *Neuroendocrinology* **2**:247–254 (1967).

81. E. M. Bogdanove, Direct gonad–pituitary feedback: An analysis of effects of intracranial estrogenic depots on gonadotropin secretion, *Endocrinology* **73**:696–712 (1963).

82. J. W. Kendall, *in* "Frontiers in Neuroendocrinology" (W. F. Ganong and L. Martini, eds.) pp. 177–207, Oxford University Press, New York 1969).

83. S. Reichlin, *in* "Neuroendocrinology" (W. F. Ganong and L. Martini, eds.) Vol. 1, pp. 445–536, Academic Press, New York (1967).

84. R. L. Van deWiele, J. Bogumil, I. Dyrenfurth, M. Ferin, R. Jewelewicz, M. Warren, and G. M. Khall, Mechanisms regulating the menstrual cycle in women, *Rec. Progr. Hormone Res.* **26**:63–94 (1970).

85. E. E. Muller, S. Sawano. A. Arimura, and A. V. Schally, Mechanism of action of growth hormone altering its own secretion rate: Comparison with the action of dexamethasone, *Acta Endocrinol.* **56**:499–509 (1967).

86. M. Motta, F. Fraschini, F. Piva, and L. Martini, *in* "Frontiers in Neuroendocrinology" (W. F. Ganong and L. Martini, eds.) pp. 211–253, Oxford University Press, New York (1969).

87. F. E. Yates, *in* "The Adrenal Cortex." (A. B. Eisenstein, ed.) pp. 133–183, Little Brown, Boston (1967).

88. B. Bohus, C. Nyakas, and K. Lissak, Involvement of suprahypothalamic structures in the hormonal feedback of corticosteroids, *Acta Physiol. Hung.* **34**:1–8 (1968).

89. B. Donovan, *in* "Mammalian Endocrinology," p. 71, McGraw-Hill, New York (1970).

90. D. M. Woodbury and A. Vernadakis, *in* "Methods in Hormone Research" (R. I. Dorfman ed.) Vol. 5, pp. 1–57, Academic Press, New York (1966).

91. K. Ruf and F. A. Steiner, Steroid sensitive single neurons in rat hypothalamus and midbrain; identification by microelectrophoresis, *Science* **156**:667–669 (1967).

92. C. H. Sawyer, M. Kawakami, B. Myerson, D. J. Whitmayer, and J. J. Lilley, Effects of ACTH, dexamethasone and asphyxia on electrical activity of rat hypothalamus, *Brain Res.* **10**:213–226 (1968).

93. A. R. Green and G. Curzon, Decrease of 5-hydroxytryptamine in the brain provoked by hydrocortisone and its prevention by allopurinol, *Nature* **22**:1095–1097 (1968).

94. S. Blanco, D. S. Schalch, and S. Reichlin, Control of growth hormone secretion by glucoreceptors in the hypothalamic–pituitary unit. *Fed. Proc.* **25**:191 (1966).

95. L. Krulich and S. M. McCann, Influence of stress on the growth hormone content of the pituitary of the rat, *Proc. Soc. Exptl. Biol. Med.* **121**:1114–1119 (1966).

96. B. Flerko and J. Szentagothai, Oestrogen sensitive nervous structures in the hypothalamus, *Acta Endocrinol.* **26**:121–127 (1957).

97. J. Szentagothai, B. Flerko, B. Mess, and B. Halasz, "Hypothalamic Control of the Anterior Pituitary," pp. 192–265, Publ. House Hung. Acad. Sci, Budapest (1962).

98. D. W. Pfaff, Uptake of ³H-estradiol by the female rat brain. An autoradiographic study, *Endocrinology* **82**:1149–1155 (1968).

99. V. D. Ramirez, B. R. Komisaruk, D. I. Whitmayer, and C. H. Sawyer, Effects of hormones and vaginal stimulation on the EEG and hypothalamic units in rats, *Am. J. Physiol.* **212**: 1376–1384 (1967).

100. Y. Palka, V. D. Ramirez, and C. H. Sawyer, Distribution and biological effects of tritiated estradiol implanted in the hypothalamo-hypophyseal region of female rats, *Endocrinology* **78**:487–499 (1966).

101. J. Davidson, *in* "Frontiers in Neuroendocrinology" (W. F. Ganong and L. Martini, eds.) pp. 343–388, Oxford University Press, New York (1969).

102. M. Igarashi, Y. Ibuki, H. Kubu, J. Kumioka, N. Yokota, Y. Ebara, and S. Matsumoto, Mode and site of action of clomiphene, *Am. J. Obstet. Gynecol.* **97**:120–123 (1967).

103. S. Kannematsu and C. H. Sawyer, Blockade of ovulation in rabbits by hypothalamic implants of norethindrone, *Endocrinology* **76**:691–699 (1965).

104. K. Seiki, M. Miyamota, A. Yamashita, and M. Kotani, Further studies on the uptake of labelled progesterone by the hypothalamus and pituitary of rats, *J. Endocrinol.* **43**:129–130 (1969).

105. C. H. Sawyer, M. Kawakami, and S. Kannematsu, Neuroendocrine aspects of reproduction, *Proc. Assc. Res. Nerv. Ment. Dis.* **43**:59–84 (1966).

106. B. S. McEwen, D. W. Pfaff, and R. E. Zigmond, Factors influencing sex hormone uptake by rat brain regions. III. Effects of competing steroids on testosterone uptake, *Brain Res.* **21**:29–38 (1970).

107. S. Reichlin, *in* "Neuroendocrinology" (L. Martini and W. Ganong, eds.) Vol. I, pp. 445–536, Academic Press, New York (1966).

108. K. M. Knigge, *in* "Major Problems in Neuroendocrinology" (E. Bajusz and G. Jasmin, eds.) pp. 261–285, Karger, Basel (1964).

109. J. W. Kendall, S. I. Shimoda, and M. Greer, Brain dependent TSH secretion from heterotopic pituitaries, *Neuroendocrinology* **2**:76–87 (1967).

110. H. Halberg, Chronobiology, *Ann. Rev. Physiol.* **31**:675–725 (1969).

111. J. Aschoff, Exogenous and endogenous components in circadian rhythms, *Cold Spring Harbor Symp. Quant. Biol.* **25**:11–28 (1960).

112. J. Benoit, The role of the eye and of the hypothalamus in the photostimulation of gonads in the dark, *Ann. N.Y. Acad. Sci.* **117**:204–216 (1964).

113. W. F. Ganong, M. D. Shepherd, J. R. Wall, E. T. Van Brunt, and M. T. Clegg, Penetration of light into the brain of mammals, *Endocrinology* **72**:962–963 (1963).

114. V. Critchlow, *in* "Advances in Neuroendocrinology" (A. Nalbandov, ed.) pp. 377–402, University of Illinois Press, Urbana, Ill. (1963).

115. R. Y. Moore, A. Heller, R. K. Bhatnager, R. J. Wurtman, and J. Axelrod, Central control of the pineal gland—Visual pathways, *Arch. Neurol.* **18**:208–218 (1968).

116. L. G. Wesson, Jr., Electrolyte excretion in relation to diurnal cycles of renal function, *Medicine* **43**:547–592 (1964).

117. D. T. Krieger, J. Kreuzer, and F. Rizzo, Constant light: Effect on circadian pattern and phase reversal of steroid and electrolyte levels in man, *J. Clin. Endocrinol.* **29**:1634–1638 (1969).

118. R. Franks, Diurnal variation of plasma 17-OHCS in children, *J. Clin. Endocrinol.* **27**:75–78 (1967).

119. D. J. Reis, A. Corvelli, and J. Conners, Circadian and ultradian rhythms of serotonin regionally in cat brain, *J. Pharmacol. Exptl. Therap.* **167**:328–333 (1969).

120. F. Hery, E. Rouer, and J. Glowinski, Daily variations of serotonin metabolism in the rat brain, *Brain Res.* **43**:445–465 (1972).

121. C. J. Migeon, F. H. Tyler, J. P. Mahoney, A. A. Florentin, H. Castle, E. K. Bliss, and L. T. Samuels, The diurnal variation of plasma levels and urinary excretion of 17-hydroxycorticosteroids in normal subjects, night workers and blind subjects in man, *J. Clin. Endocrinol.* **16**:622–633 (1956).
122. G. T. Perkoff, K. Eik-Nes, C. A. Nugent, H. L. Fred, R. A. Nimer, L. Rush, L. T. Samuels, and F. H. Tyler, Studies of the diurnal variation of plasma 17-hydroxycorticosteroids, *J. Clin. Endocrinol.* **16**:432–443 (1959).
123. D. T. Krieger, W. Allen, F. Rizzo, and H. P. Krieger, Characterization of the normal pattern of plasma corticosteroid levels, *J. Clin. Endocrinol.* **32**:266–284 (1971).
124. C. T. Nichols and F. H. Tyler, Diurnal variation in adrenal cortical function, *Ann. Rev. Med.* **18**:313–324 (1967).
125. K. Takebe, M. Sakakura, and K Mashimo, Continuance of diurnal rhythmicity of CRF activity in hypophysectomized rats, *Endocrinology* **90**:1515–1520 (1972).
126. T. Hiroshige and M. Sakakura, Circadian rhythm of corticotropin releasing activity in the hypothalamus of normal and adrenalectomized rats, *Neuroendocrinology* **7**:25–36 (1971).
127. G. Seiden and A. Brodish, Presence of a diurnal rhythm in hypothalamic corticotrophin releasing factor (CRF) in the absence of hormone feedback *Endocrinology* **90**:1401–1403 (1972).
128. M. A. Slusher, Effect of chronic hypothalamic lesions on diurnal and stress corticosteroid levels, *Am. J. Physiol.* **206**:1161–1164 (1964).
129. G. P. Moberg, U. Scapagnini, J. DeGroot, and W. F. Ganong, Effect of sectioning the fornix on diurnal fluctuation in plasma corticosterone levels in the rat, *Neuroendocrinology* **7**:11–15 (1971).
130. D. T. Krieger and H. P. Krieger, Circadian variation of the plasma 17-hydroxycorticosteroids in central nervous disease, *J. Clin. Endocrinol.* **26**:929–940 (1966).
131. P. Chiefetz, N. Garrud, and J. F. Dingman, Effect of bilateral adrenalectomy and continuous light on the circadian rhythm of corticotropin in female rats, *Endocrinology* **82**:1117–1124 (1967).
132. D. T. Krieger and F. Rizzo, Circadian periodicity of plasma 11-hydroxycorticosteroid levels in subjects with partial and absent light perception, *Neuroendocrinology* **8**:165–179 (1971).
133. D. T. Krieger and H. P. Krieger, Effect of neo- and postnatal alteration of light environment on circadian periodicity of plasma cortiocosteroid levels (11-OHCS) in the rat, *in* "Proc. Internat. Union Physiol. Sci." (1971), (Abst.).
134. D. N. Orth and D. P. Island, Light synchronization of the circadian rhythm in plasma cortisol (17-OHCS) concentration in man, *J. Clin. Endocrinol.* **29**:479–486 (1969).
135. C. Lucke, N. Adelman, and S. M. Glick, The effect of elevated free fatty acids (FFA) on the sleep-induced human growth hormone peak, *J. Clin. Endocrionol.* **35**:407–412 (1972).
136. D. T. Krieger, J. Albin, S. Paget, and S. Glick, Failure of suppression of nocturnal growth hormone rise by acute corticosteroid administration, *Horm. Metab. Dis.* (1972) (in press).
137. D. T. Krieger, *in* "Biorhythms and Human Reproduction," Wiley, New York (1972).
138. J. W. Finkelstein, J. Kream, A. Ludan, and L. Hellman, Sulfation factor (somatomedin): An explanation for continued growth in the absence of immunoassayable growth hormone in patients with hypothalamic tumors, *J. Clin. Endocrinol.* **35**:13–17 (1972).
139. D. T. Krieger and S. Glick, Absent sleep peak of growth hormone release in blind subjects, *J. Clin. Endocrinol.* **33**:847–850 (1971).
140. E. D. Weitzman, M. Perlow, J. F. Sassin, D. Fukushima, B. Burack, and L. Hellman, Persistence of the twenty-four hour pattern of episodic cortisol secretion and growth hormone release in blind subjects, *in* "Proc. 97th Meeting Am. Neurol. Ass.," p. 52 (1972).
141. C. Faiman and J. S. Winter, Diurnal cycles in plasma FSH, testosterone and cortisol in men, *J. Clin. Endocrinol.* **33**:186–192 (1971).

142. B. B. Saxena, H. Demura, H. M. Gandy, and R. E. Peterson, Radioimmunoassay of human follicle stimulating and luteinizing hormones in plasma, *J. Clin. Endocrinol.* **28**:519–534 (1968).

143. D. T. Krieger, R. Ossowski, M. Fogel, and W. Allen, Lack of circadian periodicity of human serum FSH and LH levels, *J. Clin. Endocrinol.* **35**:619–632 (1972).

144. R. T. Rubin, A. Kales, R. Adler, T. Fagan, and W. Odell, Gonadotropin secretion during sleep in normal adult men, *Science* **175**:196–198 (1972).

145. M. A. Kirshner, M. B. Lipsett, and D. R. Collins, Plasma ketosteroids and testosterone in man: A study of the pituitary–testicular axis, *J. Clin. Invest.* **44**:657–665 (1965).

146. J. A. Resko and K. B. Eik-Nes, Diurnal testosterone levels in peripheral plasma of human male subjects, *J. Clin. Endocrinol.* **26**:573–576 (1966).

147. J. L. Evans, A. M. MacLean, A. A. Ismail, and D. Love, Circulating levels of plasma testosterone during sleep, *Proc. Royal Soc. Med.* **64**:841–842 (1971).

148. J. F. Sassin, A. G. Frantz, S. Kapen, and E. D. Weitzman, Human prolactin: 24 hour pattern with increased release during sleep, *Science* **75**:1205–1207 (1972).

149. L. Vanhaelst, E. VanGauter, J. P. Degaute, and J. Goldstein, Circadian variations of serum thyrotropin levels in man, *J. Clin. Endocrinol.* **35**:479–482 (1972).

150. R. A. Gorski, *in* "Frontiers in Neuroendocrinology" (L. Martini and W. F. Ganong, eds.) pp. 237–290, Oxford University Press, New York (1971).

151. K. Milkovic and S. Milkovic, Functioning of the pituitary–adrenocortical axis in rats at and after birth, *Endocrinology* **73**:535–539 (1963).

152. C. Allen and J. W. Kendall, Maturation of the circadian rhythm of plasma corticosterone in the rat, *Endocrinology* **80**:926–930 (1967).

153. R. Ader, Early experiences accelerate maturation of the 24-hour adrenocortical rhythm, *Science* **163**:1225–1226 (1969).

154. S. Schapiro, Neonatal cortisol administration: Effect on growth, the adrenal gland and pituitary–adrenal response to stress, *Proc. Soc. Exptl. Biol. Med.* **120**:771–774 (1965).

155. D. T. Krieger, Circadian corticosteroid periodicity: "Critical period" for abolition by neonatal corticosteroids, *Science.* **178**:1205–1207 (1972).

156. J. L. Bakke, N. Lawrence, and S. Robinson, Late effects of thyroxine injected into the hypothalamus of the neonatal rat, *Neuroendocrinology* **10**:183–195 (1972).

157. H. G. Bauer, Endocrine and metabolic conditions related to pathology in the hypothalamus: A review, *J. Neur. Ment. Dis.* **128**:323–338 (1959).

158. R. W. Turkington, Secretion of prolactin by patients with hypothalamic tumors, *J. Clin. Endocrinol.* **34**:159–164 (1972).

159. M. Apostolakis, S. Kapetanakis, G. Lazos, and A. Madena-Pyrgaki, *in* "Lactogenic Hormones" (G. E. W. Wolstenhome and J. Knight, eds.) pp. 349–354, Churchill Livingston, Edinburgh (1972).

160. D. T. Krieger, S. Glick, A. Silverberg, and H. P. Krieger, A comparative study of endocrine tests in hypothalamic disease, *J. Clin. Endocrinol.* **28**:1589–1598 (1967).

161. D. T. Krieger, F. R. Ross, and H. P. Krieger, Response to dexamethasone suppression in central nervous system disease, *J. Clin. Endocrinol.* **26**:227–230 (1966).

162. S. A. Berson and R. Yalow, Immunoassay of ACTH in plasma, *J. Clin. Invest.* **47**:2725–2751 (1968).

163. D. T. Krieger, H. P. Kolodnỳ, and H. P. Krieger, Metopirone studies in hypothalamic–pituitary disease, *J. Clin. Endocrinol.* **24**:1169–1177 (1964).

164. P. Franchimont, *in* "Frontiers in Neuroendocrinology" (L. Martini and W. F. Ganong, eds.) pp. 331–359, Oxford University Press, New York (1971).

165. H. A. Selenkow, H. R. Tyler, D. D. Matson, and D. H. Nelson, Hypopituitarism due to hypothalamic sarcoidosis, *Am. J. Med. Sci.* **238**:456–463 (1959).

166. S. D. Gailani, A. L. Rogue, P. Band, and C. Ross, Hypopituitarism due to localized hypothalamic lesions, *Arch. Int. Med.* **126**:284–286 (1970).

167. F. A. Killeffer and W. E. Stern, Chronic effects of hypothalamic injury, *Arch. Neurol.* **22**:419–428 (1970).

168. L. Kahana, M. Lebovitz, W. Lusk, H. T. McPherson, E. T. Davidson, J. H. Oppenheimer, F. L. Engel, B. Woodhall, and G. Odon, Endocrine manifestations of intracranial extrasellar lesions, *J. Clin. Endocrinol.* **21**:304–324 (1962).

169. J. H. Oppenheimer, L. V. Fisher, and J. W. Jailer, Disturbance of the pituitary–adrenal interrelationship in diseases of the central nervous system, *J. Clin. Endocrinol.* **21**:1023–1036 (1961).

170. D. T. Krieger, The effect of diphenylhydantoin on pituitary–adrenal interrelations, *J. Clin. Endocrinol.* **22**:490–493 (1962).

171. V. H. Asfeldt and J. Buhl, Inhibitory effect of diphenylhydantoin on the feedback control of corticotrophin release, *Acta Endocrinol.* **61**:551–560 (1969).

172. M. T. Rabkin and A. G. Frantz, Hypopituitarism: A study of growth hormone and other endocrine functions, *Ann. Int. Med.* **64**:1197–1207 (1966).

173. J. Drewry, R. Unger, N. Kaplan, and J. Sanford, The pituitary–adrenal axis in encephalitis, *in* "Proc. 49th Meeting Endocrine Soc." (1967) (abst. 126).

174. R. P. Sherwin, J. E. Grassi, and S. C. Sommers, Hamartomatous malformation of the posterolateral hypothalamus, *Lab. Invest.* **11**:89–97 (1962).

175. J. Gilroy and J. S. Meyer, *in* "An Introduction to Clinical Neuroendocrinology" (E. Bajusz, ed.) pp. 340–355, Karger, Basel (1967).

176. I. Gamstorp, B. Kjellman, and B. Palmgren, Diencephalic syndromes of infancy, *J. Pediat.* **70**:383–390 (1967).

177. G. T. Peake and W. H. Daughaday, Disturbance of pituitary function in central nervous system disease, *Med. Clin. N. Am.* **52**:357–369 (1968).

178. J. F. Sotos, P. R. Dodge, D. Muirhead, J. D. Crawford, and N. B. Talbot, Cerebral gigantism in childhood, *New Engl. J. Med.* **271**:109–116 (1964).

179. B. Kjellman, Cerebral gigantism, *Acta Paediat. Scand.* **54**:603–609 (1965).

180. F. R. Ford, *in* "Diseases of the Nervous System," p. 156, Thomas, Springfield, Ill. (1966).

181. R. A. Fefferman, G. Costin, and M. D. Kogut, Hypothalamic gigantism, *Clin. Res.* **20**:253 (1972).

182. G. E. Powell, J. A. Brasel, S. Raiti, and R. M. Blizzard, Emotional deprivation and growth retardation simulating idiopathic hypopituitarism. II. Endocrinologic evaluation of the syndrome, *New Engl. J. Med.* **276**:1279–1283 (1967).

183. F. Kallmann, W. A. Schonfeld, and S. E. Barrera, The genetic aspects of primary eunuchoidism, *Am. J. Ment. Defic.* **48**:203–236 (1944).

184. S. Gauthier, La dyplasie olfacto-genetale (agenesis des lobes olfactifs avec absence de developpment gonadique a la puberte), *Neuroveg.* **21**:345–394 (1961).

185. H. M. Bruce, Pheromones, *Brit. Med. Bull.* **26**:10–13 (1970).

186. R. P. Michael and E. B. Keverne, Pheromones in the communication of sexual status in primates, *Nature (London)* **218**:746–749 (1968).

187. J. I. Kitay and M. D. Altschule, "The Pineal Gland," Harvard University Press, Cambridge, Mass. (1954).

188. R. J. Wurtman, J. Axelrod, and D. E. Kelly, "The Pineal," Academic Press, New York (1968).

189. D. W. Cheesman and B. L. Fariss, Isolation and structure elucidation of a gonadotropin inhibiting peptide from the beef pineal gland, *Clin. Res.* **17**:167 (1970).

190. B. Landau, H. S. Schwartz, and L. J. Soffer, Presence of a gonadotropin-inhibiting factor in urine of young children, *Metabolism* **9**:85–87 (1960).

191. J. F. Bing, J. M. Globus, and H. Simon, Pubertas praecox: A survey of the reported cases and verified anatomical findings, *J. Mt. Sinai Hosp.* **4**:935–965 (1964).

192. F. Albright, A. M. Butler, A. Hampton, and P. Smith, Syndrome characterized by osteitis fibrosa disseminata, areas of pigmentation and endocrine dysfunction with precocious puberty in females, *New Engl. J. Med.* **216**:724–726 (1937).

193. J. J. Van Wyck and M. Grumbach, Syndrome of precocious menstruation and galactorrhea in juvenile hypothyroidism: An example of hormonal overlap in pituitary feedback. *J. Pediat.* **57**:416–435 (1960).

194. N. Liu, M. M. Grumbach, R. A. DeNapoli, and A. Morishima, Prevalence of electroencephalographic abnormalities in idiopathic precocious puberty and premature pubarche: Bearing on pathogenesis and neuroendocrine regulation of puberty, *J. Clin. Endocrinol.* **25**:1296–1308 (1965).
195. D. G. Barturka, B. A. Eskin, E. M. Smith C. Dacou, and M. B. Dratman, Brain damage, hypertrichosis and polycystic ovaries, *Am. J. Obstet. Gynecol.* **99**:387–389 (1967).
196. I. A. Kamberi, R. S. Mical, and J. C. Porter, Prolactin inhibiting activity in hypophyseal stalk blood and elevation by dopamine, *Experientia* **26**:1150–1151 (1970).
197. W. B. Malarky, L. S. Jacobs, and W. H. Daughaday, Levo-Dopa suppression of prolactin in nonpuerperal galactorrhea, *New Engl. J. Med.* **285**:1160–1163 (1971).
198. V. Marks, *in* "An Introduction to Clinical Neuroendocrinology" (E. Bajusz, ed.) pp. 328–339, Karger, Basel (1967).
199. T. Hart, Jr., N. Kase, and C. P. Kimball, Induction of ovulation and pregnancy in patients with anorexia nervosa, *Am. J. Obstet. Gynecol.* **108**:580–584 (1970).
200. J. Landon, F. C. Greenwood, T. C. B. Stamp, and V. Wynn, The plasma sugars, free fatty acid, cortisol, and growth hormone response to insulin and the comparison of this procedure with other tests of pituitary and adrenal function. II. In patients with hypothalamic or pituitary dysfunction or anorexia nervosa, *J. Clin. Invest.* **45**:437–449 (1966).
201. L. Shenkman, T. Mitsuma, A. Suphavai, and C. S. Hollander, Hypothalamic hypothyroidism, *J. Am. Med. Ass.* **222**:480–481 (1972).
202. C. R. Hamilton, L. C. Adams, and F. Maloof, Hyperthyroidism due to thyrotropin-producing pituitary chromophobe adenoma, *New Engl. J. Med.* **283**:1077–1080 (1970).
203. G. Faglia, C. Ferrari, V. Neri, P. Beck-Peccoz, B. Ambrosi, and F. Valentini, High plasma thyrotropin levels in two patients with pituitary tumor, *Acta Endocrinol. Copen.* **69**:649–658 (1972).
204. C. E. Emerson and R. D. Utiger, Hyperthyroidism and excessive thyrotropin secretion, *New Engl. J. Med.* **287**:328–333 (1972).
205. D. T. Krieger, The central nervous system and Cushing's syndrome, Mount *Sinai J. Med.* **39**:416–428 (1972).
206. D. T. Krieger and S. Glick, Growth hormone and cortisol responsiveness in Cushing's syndrome: Relation to a possible central nervous system etiology, *Am. J. Med.* **52**:25–40 (1972).
207. P. Heinbecker, Pathogenesis of Cushing's syndrome, *Medicine* **23**:225–247 (1944).
208. K. R. Crispell and W. Parson, Coexistence of Cushing's syndrome and internal hydrocephalus produced by a cerebellar tumor, *Am. J. Med.* **13**:247–250 (1952).
209. S. M. Wolf, R. C. Adler, E. R. Buskirk, and R. H. Thompson, A syndrome of periodic hypothalamic discharge, *Am. J. Med.* **36**:956–967 (1964).
210. M. Seip, *in* "An Introduction to Clinical Neuroendocrinology" (E. Bajusz, ed.) pp. 414–427, Karger, Basel (1967).
211. W. Vale, G. Grant, J. Rivier, M. Monahan, M. Amors, R. Blackwell, R. Burgus, and R. Guillemin, Synthetic polypeptide antagonists of the hypothalamus luteinizing releasing factor, *Science* **176**:933–934 (1972).
212. D. T. Krieger, The hypothalamus and *Hosp. Practice* neuroendoerinology. **6**:87–99 (1971).
213. D. T. Krieger, *Hosp. Practice* (**6**:127–138 1971).
214. P. D. Maclean, *Am. J. Med.* **25**:611–626 (1958).
215. E. Scharrer, *Arch. Anat. Microscop. Morphol. Exptl.* **54**:364–365 (1965).
216. J. P. Schadé, in "Pituitary, Adrenal, and The Brain." Progress in Brain Research (D. de Wied and J. A. W. M. Weijaen, eds.) pp. 2–10, Elsevier, Amsterdam (1970).
217. J. C. Porter, R. C. Mical, I. Kamberi, and Y. R. Grazia, *Endocrinology* **87**:197–200 (1970).

218. W. Vale, P. Brazeau, C. Rivier. J. Rivier, G. Grant, R. Burgus, and R. Guillemin, Biological activities of somatostatin, Program 55th Annual Meeting Endocrine Society, A–118 (1973).
219. R. M. MacLeod and J. E. Lehmeyer, Pituitary gland alpha-adrenergic receptors and their function in prolactin secretion, Program 55th Annual Meeting Endocrine Society, A–50 (1973).
220. E. A. Zimmerman, K. C. Hsu, L. Cote, M. Tannenbaum, L. S. Freedman, M. Roffman, and M. Goldstein, Studies of dopamine-B-hydroxylase in bovine pituitary and adrenal glands using an immunoperoxidase technique, *Fed. Proc.* **32**:296 (1973).
221. F. J. Karsch, D. J. Dierschke, and E. Knobil, Sexual differentiation of pituitary function: apparent difference between primates and rodents, *Science* **179**:484–486 (1973).
222. G. Tolis, E. del Pozo, and M. S. Goldstein, Dopamine and ergocryptine effects on hyperprolactinemic states, Program 55th Annual Meeting Endocrine Society, A–50 (1973).

INDEX

CONTENTS OF VOLUME 1

ARTICLES PLANNED FOR FUTURE VOLUMES

Pathogenesis of Intrauterine Infections of the Brain
John L. Sever and Jerome Kurent

Wilson's Disease
I. Herbert Schienberg and Irving Sternlieb

Effects of Perinatal Anoxia on Brain Development
Fred Plum and Thomas E. Duffy

Disorders of Organic Acid Metabolism Associated with Brain Dysfunction
Kay Tanaka

Effects of Radiation on Brain Function
Ernest Furchtgott

Pathogenesis of Brain Dysfunction in Inborn Errors of Amino Acid Metabolism
Gerald Gaull, Harris Tallan, David Rassin, and Abel Lajtha

Pathogenesis of Seizure Discharge
Joseph Wilder

Effects of Thyroid Hormones on Brain Development
Louis Sokoloff and Charles Kennedy

Effects of Heavy Metals on the Nervous System
Martin Krigman and Paul Mushak

Pathogenesis of Slow Infections of the Central Nervous System
Richard I. Carp and Halldor Thormar

Teratological Aspects of Brain Dysfunction
Anatole Dekaban